T0320419

Interest Rate Modeling

Containing many results that are new, or which exist only in recent research articles, this thoroughly revised third edition of *Interest Rate Modeling: Theory and Practice, Third Edition* portrays the theory of interest rate modeling as a three-dimensional object of finance, mathematics, and computation. It introduces all models with financial-economical justifications, develops options along the martingale approach, and handles option evaluations with precise numerical methods.

Features
- Presents a complete cycle of model construction and applications, showing readers how to build and use models
- Provides a systematic treatment of intriguing industrial issues, such as volatility smiles and correlation adjustments
- Contains exercise sets and a number of examples, with many based on real market data
- Includes comments on cutting-edge research, such as volatility-smile, positive interest-rate models, and convexity adjustment

New to the Third edition
- Introduction of Fed fund market and Fed fund futures
- Replacement of the forward-looking USD LIBOR by the backward-looking SOFR term rates in the market model, and the deletion of dual-curve market model developed especially for the post-crisis derivatives markets
- New chapters on LIBOR Transition and SOFR Derivatives Markets

Chapman & Hall/CRC Financial Mathematics Series
Series Editors

M.A.H. Dempster
Centre for Financial Research
Department of Pure Mathematics and Statistics
University of Cambridge, UK

Dilip B. Madan
Robert H. Smith School of Business
University of Maryland, USA

Rama Cont
Mathematical Institute
University of Oxford, UK

Robert A. Jarrow
Ronald P. & Susan E. Lynch Professor of
Investment Management aSamuel Curtis
Johnson Graduate School of Management
Cornell University

Recently Published Titles

Financial Mathematics: A Comprehensive Treatment in Discrete Time
Giuseppe Campolieti and Roman N. Makarov

Introduction to Financial Derivatives with Python
Elisa Alòs and Raúl Merino

The Handbook of Price Impact Modeling
Dr. Kevin Thomas Webster

Sustainable Life Insurance: Managing Risk Appetite for Insurance Savings & Retirement Products
Aymeric Kalife with Saad Mouti, Ludovic Goudenege, Xiaolu Tan, and Mounir Bellmane

Geometry of Derivation with Applications
Norman L. Johnson

Foundations of Quantitative Finance
Book I: Measure Spaces and Measurable Functions
Robert R. Reitano

Foundations of Quantitative Finance
Book II: Probability Spaces and Random Variables
Robert R. Reitano

Foundations of Quantitative Finance
Book III: The Integrais of Riemann, Lebesgue and (Riemann-)Stieltjes
Robert R. Reitano

Foundations of Quantitative Finance
Book IV: Distribution Functions and Expectations
Robert R. Reitano

Foundations of Quantitative Finance
Book V: General Measure and Integration Theory
Robert R. Reitano

Computational Methods in Finance, Second Edition
Ali Hirsa

Interest Rate Modeling
Theory and Practice, Third Edition
Lixin Wu

For more information about this series please visit: Chapman and Hall/CRC Financial Mathematics Series - Book Series - Routledge & CRC Press

Interest Rate Modeling
Theory and Practice
Third Edition

Lixin Wu

Hong Kong University of Science and Technology

CRC Press
Taylor & Francis Group
Boca Raton London New York

CRC Press is an imprint of the
Taylor & Francis Group, an **informa** business

A CHAPMAN & HALL BOOK

Designed cover image: Lixin Wu

Third edition published 2025
by CRC Press
2385 NW Executive Center Drive, Suite 320, Boca Raton FL 33431

and by CRC Press
4 Park Square, Milton Park, Abingdon, Oxon, OX14 4RN

CRC Press is an imprint of Taylor & Francis Group, LLC

© 2025 Lixin Wu

Second Edition published by CRC Press 2019
First Edition published by CRC Press 2009

Reasonable efforts have been made to publish reliable data and information, but the author and publisher cannot assume responsibility for the validity of all materials or the consequences of their use. The authors and publishers have attempted to trace the copyright holders of all material reproduced in this publication and apologize to copyright holders if permission to publish in this form has not been obtained. If any copyright material has not been acknowledged please write and let us know so we may rectify in any future reprint.

Except as permitted under U.S. Copyright Law, no part of this book may be reprinted, reproduced, transmitted, or utilized in any form by any electronic, mechanical, or other means, now known or hereafter invented, including photocopying, microfilming, and recording, or in any information storage or retrieval system, without written permission from the publishers.

For permission to photocopy or use material electronically from this work, access www.copyright.com or contact the Copyright Clearance Center, Inc. (CCC), 222 Rosewood Drive, Danvers, MA 01923, 978-750-8400. For works that are not available on CCC please contact mpkbookspermissions@tandf.co.uk

Trademark notice: Product or corporate names may be trademarks or registered trademarks and are used only for identification and explanation without intent to infringe.

Library of Congress Cataloging-in-Publication Data
Names: Wu, Lixin, 1961- author.
Title: Interest rate modeling : theory and practice / Lixin Wu, Hong Kong
University of Science and Technology.
Description: Third edition. | Boca Raton : C&H/CRC Press, 2025. | Series:
Chapman and Hall/CRC financial mathematics series | Includes
bibliographical references and index.
Identifiers: LCCN 2024008660 (print) | LCCN 2024008661 (ebook) | ISBN
9781032483559 (hardback) | ISBN 9781032484440 (paperback) | ISBN
9781003389101 (ebook)
Subjects: LCSH: Interest rates--Mathematical models. | Interest rate
futures--Mathematical models.
Classification: LCC HG6024.5 .W82 2025 (print) | LCC HG6024.5 (ebook) |
DDC 332.801/5195--dc23/eng/20240223
LC record available at https://lccn.loc.gov/2024008660
LC ebook record available at https://lccn.loc.gov/2024008661

ISBN: 978-1-032-48355-9 (hbk)
ISBN: 978-1-032-48444-0 (pbk)
ISBN: 978-1-003-38910-1 (ebk)

DOI: 10.1201/ 9781003389101

Typeset in Nimbus font
by KnowledgeWorks Global Ltd.

Publisher's note: This book has been prepared from camera-ready copy provided by the authors.

To my parents,
To Molly,
Dorothy and Derek

Contents

Preface to the First Edition

Motivations

This book was motivated by my teaching of the subject to graduate students at the Hong Kong University of Science and Technology (HKUST). My interest-rate class usually consists of students working toward both research degrees and professional degrees; their interests and levels of mathematical sophistication vary quite a bit. To meet the needs of the students with diverse backgrounds, I must choose materials that are interesting to the majority of them, and strike a balance between theory and application when delivering the course materials. These considerations, together with my own preferences, have shaped a coherent course curriculum that seems to work well. Given this success, I decided to write a book based on that curriculum.

Interest-rate modeling has long been at the core of financial derivatives theory. There are already quite a number of monographs and textbooks on interest-rate models. It is a good idea to write another book on the subject only if it will contribute significant added value to the literature. This is why I thought about this book. This book portrays the theory of interest-rate modeling as a three-dimensional object of finance, mathematics, and computation. In this book, all models are introduced with financial and economical justifications; options are modeled along the so-called martingale approach; and option evaluations are handled with fine numerical methods. With this book, the reader may capture the interdisciplinary nature of the field of interest-rate (or fixed-income) modeling, and understand what it takes to be a competent quantitative analyst in today's market.

The book takes the top-down approach to introducing interest-rate models. The framework for no-arbitrage models is first established, and then the story evolves around three representative types of models, namely, the Hull–White model, the market model, and affine models. Relating individual models to the arbitrage framework helps to achieve better appreciation of the motivations behind each model, as well as better understanding of the interconnections among different models. Note that these three types of models coexist in the market. The adoption of any of these models or their variants may often be determined by products or sectors rather than by the subjective will. Hence, a quant must have flexibility in adopting models. The premise, of course, is a thorough understanding of the models. It is hoped that, through the top-down approach, readers will get a clear picture of the status of this important subject, without being overwhelmed by too many specific models.

This book can serve as a textbook. Inherited from my lecture notes for a diverse pool of students, the book is not written in a strict mathematical style. Were that the case, there would be a lot more lemmas and theorems in the text. But efforts were indeed made to make the book self-contained in mathematics, and rigorous justifications are given for almost all results. There are quite a number of examples in the text; many of them are based on real market data. Exercise sets are provided for all but one chapter. These exercises often require computer implementation. Students not only learn the martingale approach for interest-rate modeling, but they also learn how to implement various models on computers. The adoption of materials is influenced by my experiences as a consultant and a lecturer for industrial courses. My early students often noted that materials in the course were directly relevant to their work in institutions. Those materials are included here.

To a large extent, this book can also serve as a research monograph. It contains many results that are either new or exist only in recent research articles, including, as only a few examples, adaptive Hull–White lattice trees, market model calibration by quadratic programming, correlation adjustment, and swaption pricing under affine term structure models. Many of the numerical methods or schemes are very efficient or even optimized, owing to my original background as a numerical analyst. In addition, notes are given at the end of most chapters to comment on cutting-edge research, which includes volatility-smile modeling, convexity adjustment, and so on.

Study Guide

When used as a textbook, this book can be covered in two 14-week semesters. For a one-semester course, I recommend the coverage of Chapters 1–4, half of Chapter 5 on lattice trees, and Chapter 6. As a reference book for self-learning, readers should study short-rate models, market models, and affine models in Chapters 5, 6, and 9, respectively. For applications, Chapter 8 is also very useful. Chapter 7, meanwhile, is special in this book, as it is particularly prepared for those readers who are interested in market model calibration.

Chapters 1 and 2 contain the mathematical foundations for interest-rate modeling, where we introduce Ito's calculus and the martingale representation theorem. The presentation of the theories is largely self-contained, except for some omissions in the proof of the martingale representation theorem, which I think is technically too demanding for this type of book.

In Chapter 3, I introduce bonds and bond yields, which constitute the underlying securities or quantities of the interest-rate derivatives markets. I also discuss the composition of bond markets and how they function. For completeness, I include the classical theory of risk management that is based on parallel yield changes.

Chapter 4 is a cornerstone of the book, as it introduces the Heath–Jarrow–Morton (HJM) model, the framework for no-arbitrage pricing models. With market data, we demonstrate the estimation of the HJM model. Forward measures, which are important devices for interest-rate options pricing, are introduced. Change of measures, a very useful technique for option pricing, is discussed in general.

Chapter 5 consists of two parts: a theoretical part and a numerical part. The theoretical part focuses on the issue of when the HJM model implies a Markovian short-rate model, and the numerical part is about the construction and calibration of short-rate lattice models. A very efficient methodology to construct and calibrate a truncated and adaptive lattice is presented with the Hull–White model, which, after slight modifications, is applicable to general Markovian models with the feature of mean reversion.

Chapter 6 is another cornerstone of the book, where I introduce the LIBOR market and the LIBOR market model. After the derivation of the market model, I draw the connection between the model and the no-arbitrage framework of HJM. This chapter contains perhaps the simplest yet most robust formula for swaption pricing in the literature. Moreover, with the pricing of Bermudan swaptions, I give an enlightening introduction to the popular Longstaff–Schwartz method for pricing American options in the context of Monte Carlo simulations.

Chapter 7 discusses an important aspect of model applications in the markets—model calibration. Model calibration is a procedure to fix the parameters of a model based on observed information from the derivatives market. This issue is rarely dealt with in academic or theoretical literature, but in the real world it cannot be ignored. With the LIBOR market model, I show how a problem of calibration can be set up and solved.

In Chapter 8, I address two intriguing industrial issues, namely, volatility and correlation adjustments. Mathematically, these issues are about computing the expectation of a financial quantity under a non-martingale measure. With unprecedented generality and clarity, I offer analytical formulae for these evaluation problems. The adjustment formulae have widespread applications in pricing futures, non-vanilla swaps, and swaptions.

Finally, in Chapter 9, I introduce the class of affine term structure models for interest rates. Rooted in general equilibrium theory for asset pricing, the affine term structure models are favored by many people, particularly those in academic finance. These models are parsimonious in parameterization, and they have a high degree of analytical tractability. The construction of the models is demonstrated, followed by their applications to pricing options on bonds and interest rates.

Preface to the Second Edition

It has been almost ten years since the publication of the first edition of this book. In responses to the 2008 financial crisis, major changes have taken place over the past ten years in financial markets, from changes in regulations to the practice of derivatives pricing and risk management. Regulators have been pushing OTC trades to go through central counterparty clearing houses, which are subject to initial margin (IM) and variable margin (VM). For the remaining OTC trades, collaterals have become market standard, in addition to risk capital requirements. The funding costs for IM, VM, collaterals, and risk capital have become a burden to many firms. How to take into account the funding costs in trade prices has been a central issue to the practitioners, regulators, and researchers. The current solution is to make various valuation adjustments, so-called xVA, to either the trade prices or accounting books, which has been controversial and is still debated today.

Major changes also occurred to the modeling of the interest rate derivatives. Pre-crisis term structure models, which were based on a single forward rate curve – so-called single curve modeling – were replaced by multiple curve models, which simultaneously model multiple forward rate curves. Yet, most multi-curve models are at odds with the basis swap curves, which suggest that the forward rate curves cannot evolve separately in any usual ways. There is an affine solution of multi-curve modeling which is compatible with the basis curves, but such a model is very different from models quants are used to, such as the SABR-LIBOR market model (SABR-LMM) that is popular owing to its capacity to manage volatility smile risks.

In the second edition, we will offer our solutions to xVA and post-crisis interest rate modeling. Specifically, we want to achieve three objectives. First, we will introduce the theories of major smile models for interest rate derivatives, and then adapt the most important one, the SABR-LMM model, to the post-crisis markets. Second, we will introduce models for inflation rate derivatives and credit derivatives. Third, we will introduce our solution to the issue of xVA. Altogether, six new chapters will be added. With exception of the last chapter on xVA, all new chapters will be developed around the central theme: the LIBOR market model (this is a distinguished feature of the second edition).

In Chapter 10, we will introduce the SABR model and the Heston's type LMM model that feature the role of stochastic volatility in the formation of volatility smiles. Through these two models, we try to demonstrate the methodologies and techniques of smile models that are based on stochastic volatilities.

In Chapter 11, we will introduce the Lévy market model, a framework of models that captures volatility smiles based on the dynamics of jumps and diffusions. Although this topic has theoretically been complex, we will offer a simple exposition of model construction and pricing.

In Chapter 12, we take readers to inflation derivatives modeling and pricing, a two-decade-old theoretical subject not yet fully understood. Our aim is to first build a solid foundation, and then on top of that develop the inflation market model to justify the current market practice in pricing and hedging inflation derivatives.

In Chapter 13, we deal with single-name credit derivatives, with the intention of pricing credit instruments, bonds, credit default swaps (CDS), CDS options (or credit swaption), and even collateralized debt obligation (CDO) using an LMM type model. We will redefine risky zero-coupon bonds using tradable securities, and then risky forward rates, and eventually the credit market model. This model allows us to price all instruments except CDOs, for which we need additional tools like copulas to model correlated defaults.

In Chapter 14, we will rebuild the foundation, namely, the risky zero-coupon bonds, for the post-crisis interest rate derivative markets. Based on the new foundation, we redefine LIBOR in the presence of credit risk of LIBOR panel banks, and demonstrate that such a risk is responsible for the emergence of the basis curves. We then define a dual-curve LMM and, more notably, the dual-curve SABR-LMM. A large portion of the chapter is then devoted to the pricing of caplets and swaptions under the dual-curve SABR-LMM, along the approach of the heat kernel expansion method of Henry-Labordère for the SABR-LMM model.

Finally, in Chapter 15, we present an xVA theory, which is applicable to general derivatives pricing, including interest rate derivatives. We will prove that the bilateral credit valuation adjustment is part of the fair price, and demonstrate how funding costs enter the P&L of trades. We show that only the market funding liquidity risk premium can enter into pricing, otherwise price asymmetry will occur.

Preface to the Third Edition

Over the last ten years, global fixed-income markets have undergone a transition, as LIBOR, the indexes underlying the interbank lending and derivatives markets, were being phased out and replaced by risk-free overnight rates, accompanied by the fading of the LIBOR derivatives markets and the rise of derivatives based on alternative indexes. Nowadays, the interbank derivatives market on USD, both onshore and offshore, is indexed to SOFR, the secured overnight financing rate, and similar changes have taken place simultaneously in derivatives markets on other major currencies. Given these changes, a large part of the second edition of this book becomes obsolete and requires substantial updates to catch up with the reality of today's markets.

The major additions to this edition are the description of the LIBOR scandal and the subsequent replacement of USD LIBOR, a set of USD interest rate indexes, by SOFR, a single interest rate index. The major changes I have made are the replacement of the forward-looking USD LIBOR by the backward-looking SOFR term rates in the market model, and the deletion of the dual-curve market model developed especially for the post-crisis derivatives markets. In this new edition, Chapters 4 and 9 are new, and Chapters 7 and 14 of the second edition are deleted, thus keeping the total number of chapters unchanged. Enhancements or updates are also made to other chapters, particularly in Chapters 6, 8 and 11.

In Chapter 3, we have added the Federal fund market and Federal fund futures to the list of U.S. fixed-income market. We highlight the role of Federal Open Market Committee in the making of monetary policies, as well as the usage of Federal fund futures to predict the possible changes of Federal fund rate.

Chapter 4 is new, where we explain the LIBOR scandal, the interactive process between regulators and market participants that has resulted in the replacement of USD LIBOR by SOFR, and the introduction of SOFR derivatives and derivatives market. Additional details on the U.S. repo market are supplemented, in order to explain the construction of the SOFR as an index.

In Chapter 5, we derive the Black formula for derivatives pricing and hedging under stochastic interest rates. Through the derivation we highlight the advantages brought forward by adopting a proper numeraire asset and its martingale measure for derivatives pricing. Forward rates are introduced alongside with futures rates, as part of the preparation for interest-rate derivatives modeling in subsequent chapters.

In Chapter 6, we address the pricing of major SOFR derivatives, except swaptions, under the Heath-Jarrow-Morton model.

In Chapter 7, we illustrate the construction of calibrated lattice trees for Ho-Lee and Hull-White model. The difference from the previous editions is the nodal values of the term rates are backward-looking.

In Chapter 8, we have added the pricing of SOFR futures under the affine term structure models, which now plays a role more important than in the LIBOR era.

Chapter 9 is also new, where we introduce the SOFR market model, which is expected to play the role of the LIBOR market model in the past. On Monte Carlo simulations with the SOFR market model, we have added the upper bound estimate. A recent method to construct the initial term structure of SOFR term rates for production use is supplemented.

In Chapter 11, we have added the technical derivation of the HKLW formula, for pricing call options under the SABR model.

The rest of the chapters are essentially carried over from the second edition, with adaptations to the SOFR derivatives markets, wherever necessary.

Acknowledgments to the Third Edition

First I want to thank Mr. Callum Fraser, the editor at Taylor & Francis who took the initiative to contact me in September 2022 for the third edition, when I was thinking about the same thing in view of the substantial changes that had occurred in the interest rate derivatives markets over the years.

Special thank goes to my daughter, Dorothy, for proof-reading six chapters during her precious winter break at home. Her corrections and suggestions have definitely made this edition better.

Lixin Wu
Hong Kong

Author

Lixin Wu earned his PhD in applied mathematics from UCLA in 1991. Originally a specialist in numerical analysis, he switched his area of focus to financial mathematics in 1996. Since then, he has made notable contributions to the area. He co-developed the PDE model for soft barrier options and the finite-state Markov chain model for credit contagion. He is, perhaps, best known in the financial engineering community for a series of works on market models, including an optimal calibration methodology for the standard market model, a market model with square-root volatility, a market model for credit derivatives, a market model for inflation derivatives, and a dual-curve SABR market model for post-crisis derivatives markets. He also has made valuable contributions to the topic of xVA. Over the years, Dr. Wu has been a consultant for financial institutions and a lecturer for Risk Euromoney and Marco Evans, two professional education agencies. He is currently a full professor at the Hong Kong University of Science and Technology.

Chapter 1

The Basics of Stochastic Calculus

The seemingly random fluctuation in stock prices is the most distinctive feature of financial markets, and it creates both risks and opportunities for investors. How to model random fluctuations in stock prices as well as in other financial time series data (indexes, interest rates, exchange rates, etc.) has long been a central issue in the discipline of quantitative finance. In mathematical terms, a financial time series is a stochastic process. In quantitative modeling, a financial time series is treated as a function of other standardized stochastic processes, and these stochastic processes serve as the engines for the random evolution of financial time series. With these models, we can assess risk, value risky securities, and design hedging strategies, and we can make decisions on asset allocations in a scientific way. Among the many standardized stochastic time series that are available, the so-called Brownian motion is no doubt the most basic yet the most important one. This chapter is devoted to describing Brownian motions and its calculus.

1.1 Brownian Motions

Financial time series data actually exist in discrete form, for example, tick data, daily data, weekly data, and so on. It may be intuitive to describe these data with discrete time series models. Alternatively, we can also describe them with continuous time series models, and then take discrete steps in time. It turns out that working with continuous-time financial time series is much more efficient than working with discrete time series models, due to the existence of an arsenal of stochastic analysis tools in continuous time. The theory of Brownian motions is the single most important building block of continuous-time financial time series. We proceed by introducing Brownian motions through a limiting process, starting with simple random walks.

DOI: 10.1201/9781003389101-1

1.1.1   Simple Random Walks

Simple random walks are discrete time series, $\{X_i\}$, defined as

$$X_0 = 0,$$

$$X_{n+1} = \begin{cases} X_n - \sqrt{\Delta t}, & p = \dfrac{1}{2} \\ X_n + \sqrt{\Delta t}, & 1 - p = \dfrac{1}{2} \end{cases}, \quad n = 0, 1, 2, \ldots, \qquad (1.1)$$

where $\Delta t > 0$ stands for the interval of time for stepping forward. One can verify that $\{X_i\}$ has the following properties:

1. The increment of $X_{n+1} - X_n$ is independent of $\{X_i\} \, \forall i \leq n$.

2. $E[X_n \mid X_m] = X_m, m \leq n$.

An interesting feature of the simple random walk is the linearity of X_i's variance in time: given X_0, the variance of X_i is equal to $i\Delta t$, the time it takes the time series to evolve from X_0 to X_i.

Out of the simple Brownian random walk, we can construct a continuous-time process through linear interpolation:

$$\bar{X}(t) = X_i + \frac{t - i\Delta t}{\Delta t}(X_{i+1} - X_i), \quad t \in [i\Delta t, (i+1)\Delta t]. \qquad (1.2)$$

We are interested in the limiting process of $\bar{X}(t)$ as $\Delta t \to 0$, in the hope that the limit remains a meaningful stochastic process. The next theorem confirms just that.

Theorem 1.1.1 (The Lindeberg–Lévy Central Limit Theorem). *For the continuous process, $\bar{X}(t)$, there is*

$$\lim_{\Delta t \to 0} P\left\{\bar{X}(s+t) - \bar{X}(s) \leq x\right\} = \frac{1}{\sqrt{2\pi t}} \int_{-\infty}^{x} \exp\left(-\frac{u^2}{2t}\right) du. \qquad (1.3)$$

Proof: The proof is a matter of applying the central limit theorem. Without loss of generality, we let $s = 0$ and take $\Delta t = t/n$. Apparently, there are $\bar{X}(0) = X_0, \bar{X}(t) = X_n$, and

$$P\{X_n - X_0 \leq x\} = P\left\{\sum_{i=1}^{n}(X_i - X_{i-1}) \leq x\right\}$$

$$= P\left\{\frac{1/n \sum_{i=1}^{n}(X_i - X_{i-1}) - 0}{\sqrt{\Delta t/n}} \leq \frac{x}{\sqrt{n\Delta t}}\right\}$$

$$= P\left\{\frac{1/n \sum_{i=1}^{n}(X_i - X_{i-1}) - 0}{\sqrt{\Delta t/n}} \leq \frac{x}{\sqrt{t}}\right\}. \qquad (1.4)$$

According to the central limit theorem, we have

$$\lim_{\Delta t \to 0} P\left\{ \frac{1/n \sum_{i=1}^{n} (X_i - X_{i-1}) - 0}{\sqrt{\Delta t/n}} \leq \frac{x}{\sqrt{t}} \right\} = P\left\{ \varepsilon \leq \frac{x}{\sqrt{t}} \right\}$$

$$= \int_{-\infty}^{x/\sqrt{t}} \frac{1}{\sqrt{2\pi}} \exp\left(-\frac{v^2}{2} \right) dv$$

$$\left(\text{let } u = \sqrt{t}v \right) = \frac{1}{\sqrt{2\pi t}} \int_{-\infty}^{x} \exp\left(-\frac{u^2}{2t} \right) du, \tag{1.5}$$

where ε is a standard normal random variable. This completes the proof. $\square$

We call the limited process,

$$W(t) = \lim_{\Delta t \to 0} \bar{X}(t), \quad 0 \leq t \leq \infty, \tag{1.6}$$

a Wiener process in honor of Norbert Wiener, a pioneer in stochastic control theory. It is also called a Brownian motion, after Robert Brown, a Scottish botanist who in 1827 observed that pollen grains suspended in liquid moved irregularly in motions that resembled simple random walks. The formal definition of Brownian motion is given in the next section.

1.1.2 Brownian Motions

A continuous stochastic process is a collection of real-valued random variables, $\{X(t, \omega), 0 \leq t \leq T\}$ or $\{X_t(\omega), 0 \leq t \leq T\}$, that are defined on a probability space $(\Omega, \mathcal{F}, \mathbb{P})$. Here Ω is the collection of all ω's, which are so-called sample points, $\mathcal{F}$ is the smallest σ-algebra that contains Ω, and $\mathbb{P}$ is a probability measure on Ω. Each random outcome, $\omega \in \Omega$, corresponds to an entire time series

$$t \to X_t(\omega), \quad t \in T, \tag{1.7}$$

which is called a path of X_t. In view of Equation 1.7, we can regard $X_t(\omega)$ as a function of two variables, ω and t. For notational simplicity, however, we often suppress the ω variable when its explicit appearance is not necessary.

In the context of financial modeling, we are particularly interested in the Brownian motion introduced earlier. Its formal definition is given below.

Definition 1.1.1. *A Brownian motion or a Wiener process is a real-value stochastic process, W_t or $W(t), 0 \leq t \leq \infty$, that has the following properties:*

1. *$W(0) = 0$.*

2. *$W(t + s) - W(t)$ is independent of $\{W(u), 0 \leq u \leq t\}$.*

3. *For $t \geq 0$ and $s > 0$, the increment $W(t + s) - W(t) \sim N(0, s)$.*

4. *$W(t)$ is continuous almost surely (a.s.).*

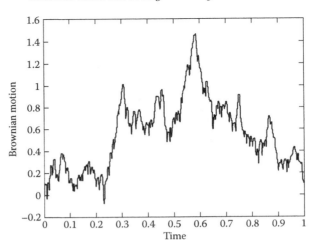

FIGURE 1.1: A sample path of a Brownian motion.

Here $N(0, s)$ stands for a normal distribution with mean zero and variance s. Note that in some literature, property 4 is not part of the definition, as it can be proved to be implied by the first three properties (Varadhan, 1980 or Ikeda and Watanabe, 1989). A sample path of $W(t)$ is shown in Figure 1.1, which is generated with a step size of $\Delta t = 2^{-10}$.

Brownian motion plays a major role in continuous-time stochastic modeling in physics, engineering and finance. In finance, it has been used to model the random behavior of asset returns. Several major properties of the Brownian motion are listed below.

Lemma 1.1.1. *A Brownian motion, $W(t)$, has the following properties:*

1. *Self-similarity:* $\forall \lambda > 0$ $W(t) \mapsto (1/\sqrt{\lambda})W(\lambda t) = \tilde{W}(t)$ *is also a Brownian motion.*

2. *Unbounded variation: for any $T \geq 0$, $\lim_{\Delta t \to 0} \sum_{j, t_j \leq T} |\Delta W_{t_j}| = \infty$, where $\Delta W_{t_j} = W(t_{j+1}) - W(t_j)$.*

3. *Non-differentiability: $W(t)$ is not differentiable at any $t \geq 0$.*

The self-similarity property implies that $W(t)$ is a fractal object. This can be straightforwardly proven and the proof is left to readers. We will see that unbounded variation implies non-differentiability. To prove unbounded variation, we will need the following lemma.

Lemma 1.1.2. *Let $0 = t_0 < t_1 < \cdots < t_n = T$ represent an arbitrary partition of time interval $[0, T]$ and $\Delta t = \max_j (t_{j+1} - t_j)$. Then $\lim_{\Delta t \to 0} \sum_{j=0}^{n-1} \left(\Delta W_{t_j}\right)^2 = T$ almost surely.*

Proof: Let $\Delta t_j = t_{j+1} - t_j, \forall j$. Then we can write

$$\Delta W_{t_j} = \varepsilon_j \sqrt{\Delta t_j}, \quad \varepsilon_j \sim N(0,1) \text{ iid.} \tag{1.8}$$

Here, iid stands for "independent with identical distribution." By Kolmogorov's large number theorem,

$$\sum_{j=0}^{n-1} \left(\Delta W_{t_j}\right)^2 = \sum_{j=0}^{n-1} \Delta t_j \varepsilon_j^2 \xrightarrow{\text{a.s.}} \sum_{j=0}^{n-1} \Delta t_j E\left[\varepsilon_j^2\right] = T, \tag{1.9}$$

since $E[\varepsilon_j^2] = 1$, $\forall j$. $\qquad\square$

[**Proof** of unbounded variation and non-differentiability]
We will do the proof by the method of contradiction. Suppose that $W(t)$ has bounded variation over a finite interval $[0, T]$ such that

$$\lim_{\Delta t \to 0} \sum_{j, t_j \leq T} \left|\Delta W_{t_j}\right| = C < \infty \tag{1.10}$$

for some finite constant C. Since a continuous function is uniformly continuous over a finite interval, we have

$$\left|\Delta W_{t_j}\right| = o(1) \quad \text{as} \quad \Delta t \to 0. \tag{1.11}$$

Using Lemma 1.1.2 and Equation 1.11, we then arrive at

$$T \approx \sum_{j, t_j \leq T} \left|\Delta W_{t_j}\right|^2 \leq \left(\max\left|\Delta W_{t_j}\right|\right) \sum_{j, t_j \leq T} \left|\Delta W_{t_j}\right| \leq C \cdot o(1) \xrightarrow{\Delta t \to 0} 0, \tag{1.12}$$

which is a contradiction, so the variation must be unbounded.
The property of non-differentiability follows from the unbounded variation property. In fact, if $W'(t)$ existed and was finite over any interval, say $[0, T]$, then there would be

$$\lim_{\Delta t \to 0} \sum_{j, t_j \leq T} \left|\Delta W_{t_j}\right| \triangleq \int_0^T |W'(t)|\, dt < \infty, \tag{1.13}$$

where "$\triangleq$" means "is defined as." Equation 1.13 contradicts the property of unbounded variation we have just proved. $\qquad\square$

1.1.3 Adaptive and Non-Adaptive Functions

We now define the class of functions of stochastic processes such that their values at time t can be determined based on information available up to time t. Formally, we introduce the notion of filtration.

Definition 1.1.2. *Let $\mathcal{F}_t$ denote the smallest σ-algebra containing all sets of the form*

$$\{\omega; W_{t_1}(\omega) \in B_1, \ldots, W_{t_k}(\omega) \in B_k\} \subset \Omega, \qquad (1.14)$$

where $t_j \leq t$ and $B_j \subset \mathbf{R}$ are Borel sets, with $j \leq k$ and $\mathbf{R}$ standing for the set of real numbers, for $k = 1, 2, \ldots$. Denote the σ-algebra by $\mathcal{F}_t = \sigma(W(s), 0 \leq s \leq t)$; we call the collection of $(\mathcal{F}_t)_{t \geq 0}$ a Brownian filtration.

For applications in mathematical finance, it suffices to think of $\mathcal{F}_t$ as "information up to time t" or "history of W_s up to time t." According to the definition, $\mathcal{F}_s \subset \mathcal{F}_t$ for $s \leq t$, meaning that a filtration is an increasing stream of information. Readers can find thorough discussions of Brownian filtration in many previous works, for example, Øksendal (1992).

Definition 1.1.3. *A function, $f(t)$, is said to be $\mathcal{F}_t$-adaptive if*

$$f(t) = \tilde{f}(\{W(s), 0 \leq s \leq t\}, t), \quad \forall t, \qquad (1.15)$$

that is, the value of the function at time t depends only on the path history up to time t.

Adaptive functions[1] are natural candidates to work with in finance. Suppose that the value of a function represents a decision in investment. Then, such a decision has to be made based on the available information up to the moment of making decision. The next example helps to illustrate of what kind of function is or is not an $\mathcal{F}_t$-adaptive function.

Example 1.1. *Function*

$$f(t) = \begin{cases} 0 & \min_{0 \leq s \leq t} W(s) < 2 \\ 1 & \min_{0 \leq s \leq t} W(s) \geq 2 \end{cases} \qquad (1.16)$$

is $\mathcal{F}_t$-adaptive, whereas

$$f(t) = \begin{cases} 0 & \min_{0 \leq s \leq 1} W(s) < 2 \\ 1 & \min_{0 \leq s \leq 1} W(s) \geq 2 \end{cases} \qquad (1.17)$$

is not $\mathcal{F}_t$-adaptive, because $f(t)$ cannot be determined at any time $t < 1$.

1.2 Stochastic Integrals

Stochastic calculus considers the integration and differentiation of general $\mathcal{F}_t$-adaptive functions. The purpose of developing such a stochastic calculus is to

---

[1] They are also called non-predictable or non-anticipative functions.

model financial time series (with random dynamics) using either integral or differential equations. According to Lemma 1.1.1, a Brownian motion, $W(t)$, is nowhere differentiable in the usual sense of differentiation for deterministic functions. To define differentials of stochastic processes in a proper sense, we must first study the notion of stochastic integrals.

Stochastic integrals can be defined for functions in the square-integrable space, $H^2[0,T] = L^2(\Omega \times [0,T], d\mathbb{P} \times dt)$, which is defined to be the collection of functions satisfying

$$E\left[\int_0^T |f(t,\omega)|^2 \, dt\right] < \infty. \tag{1.18}$$

Note that, unless indicated otherwise, $E[\cdot]$ means $E^{\mathbb{P}}[\cdot]$, the unconditional expectation under $\mathbb{P}$. The definition consists of a three-step procedure. First, we define elementary or piecewise constant functions in an intuitive way. Second, we define the integrals of a bounded continuous function as a limit of integrals of elementary functions. Finally, we define the integral of a general square-integrable function as a limit of integrals of bounded continuous functions. The key in this three-step procedure is to ensure the convergence of the limits in $L^2(\Omega, \mathcal{F}, \mathbb{P})$, the Hilbert space of random variables satisfying

$$E\left[X^2(\omega)\right] < \infty.$$

This definition approach is taken by Øksendal (1992). Alternative treatments of course also exist; see, for example, Mikosch (1998).

An elementary function has the form

$$\varphi(t,\omega) = \sum_j c_j(\omega)\chi_{(t_j,t_{j+1}]}(t), \tag{1.19}$$

where $\chi_A(t)$ is the indicator function such that $\chi_A(t) = 1$ if $t \in A$, or otherwise $\chi_A(t) = 0$, and $c_j(\omega)$ is adapted to $\mathcal{F}_{t_j}$. For the elementary function, the integral is defined in a rather natural way:

$$\int_0^T \varphi(t,\omega) \, dW(t,\omega) = \sum_j c_j(\omega)\Delta W(t_j,\omega), \tag{1.20}$$

where $\Delta W(t_j,\omega) = W(t_{j+1},\omega) - W(t_j,\omega)$ is the increment of the over time interval $(t_j, t_{j+1}]$. The next result plays a crucial role in defining the stochastic integral for general functions.

Lemma 1.2.1 (Ito isometry). *If $\varphi(t,\omega)$ is a bounded elementary function, then*

$$E\left[\left(\int_0^T \varphi(t,\omega) \, dW(t,\omega)\right)^2\right] = \int_0^T E\left[\varphi^2(t,\omega)\right] dt. \tag{1.21}$$

Proof: We proceed straightforwardly:

$$
E\left[\left(\int_0^T \varphi(t,\omega)\,\mathrm{d}W(t,\omega)\right)^2\right] = E\left[\sum_{i,j=0}^{n-1} c_i c_j \Delta W(t_i,\omega)\Delta W(t_j,\omega)\right]
$$

$$
= E\left[\sum_{j=0}^{n-1} c_j^2 (\Delta W(t_j,\omega))^2\right]
$$

$$
= \sum_{j=0}^{n-1} E\left[c_j^2\right]\Delta t_j
$$

$$
= \int_0^T E[\varphi^2(t,\omega)]\,\mathrm{d}t.
$$

This completes the proof. □

For bounded continuous functions, we can define the stochastic integral intuitively through a limiting process:

$$
\int_0^T f(t)\,\mathrm{d}W(t) \triangleq \lim_{\Delta t \to 0} \sum_{j=0}^{n-1} f(t_j)\Delta W(t_j). \tag{1.22}
$$

The summation in Equation 1.22 is called a Riemann–Stieltjes sum. The existence and uniqueness of the limit is assured by the next lemma.

Lemma 1.2.1. *Let $f(t)$ be a bounded continuous and $\mathcal{F}_t$-adaptive function. Then, given any partition of $[0,T]$ with $\Delta t = \max_j(t_{j+1} - t_j)$, the limit of the Riemann–Stieltjes sum in Equation 1.22 exists and is unique in $L^2(\Omega, \mathcal{F}, \mathbb{P})$. Moreover, Ito's isometry holds for bounded continuous functions:*

$$
E\left[\left(\int_0^T f(t)\,dW(t,\omega)\right)^2\right] = \int_0^T E[\,f^2(t)]\,\mathrm{d}t. \tag{1.23}
$$

Proof: Construct an elementary function:

$$
\varphi_n(t,\omega) = \sum_{j=0}^{n-1} f(t_j,\omega)\chi_{(t_j,t_{j+1}]}(t). \tag{1.24}
$$

Note that $\varphi_n(t)$ satisfies Equation 1.21, the Ito's isometry. Due to the continuity of $f(t)$, there is

$$
\int_0^T E[(\,f(t) - \varphi_n(t))^2]\,\mathrm{d}t \xrightarrow{\Delta t \to 0} 0, \tag{1.25}
$$

which implies that $\varphi_n(t)$ is a Cauchy sequence in $L^2(\Omega, \mathcal{F}, \mathbb{P})$. By making use again of Ito's isometry for $\varphi_n - \varphi_m$, we can see that

$$
\int_0^T \varphi_n(t,\omega)\,\mathrm{d}W_t(\omega) \tag{1.26}
$$

is a Cauchy sequence in $L^2(\Omega, \mathcal{F}, \mathbb{P})$. Hence its limit exists as an element in the space, which, also by Ito isometry, is independent of the partition $\{t_i\}$. We denote the limit as

$$\int_0^T f(t)\, \mathrm{d}W(t).$$

Finally, by taking limit $n \to \infty$ for the equality

$$E\left[\left(\int_0^T \varphi_n(t, \omega)\, \mathrm{d}W(t, \omega)\right)^2\right] = \int_0^T E\left[\varphi_n^2(t, \omega)\right]\mathrm{d}t,$$

we will arrive at Equation 1.23. Hence, Ito isometry holds for continuous functions as well. □

For a general function in $L^2(\Omega, \mathcal{F}, \mathbb{P})$, the definition (Equation 1.22) for stochastic integrals is no longer valid. Nonetheless, we can approximate a general function of $L^2(\Omega, \mathcal{F}, \mathbb{P})$ by a sequence of bounded continuous functions in the sense that

$$\int_0^T E\left[(\,f(t) - f_n(t))^2\right]\mathrm{d}t \to 0 \quad \text{as } n \to \infty, \tag{1.27}$$

and thus we define the stochastic integral or Ito's integral for $f(t)$ as the limit

$$\int_0^T f(t)\, \mathrm{d}W(t) \stackrel{\Delta}{=} \lim_{n\to\infty} \int_0^T f_n(t)\, \mathrm{d}W(t). \tag{1.28}$$

Furthermore, by Ito's isometry for continuous functions, Equation 1.23, we assert that $\int_0^T f_n(t)\, \mathrm{d}W(t)$ is a Cauchy sequence in $L^2(\Omega, \mathcal{F}, \mathbb{P})$, and thus its limit exists. The details of the justification are found in Øksendal (1992).

We finish this section by presenting additional properties of Ito's integrals. The proofs are straightforward and thus omitted for brevity.

Properties of Stochastic Integrals:
For an $\mathcal{F}_t$-adaptive function, $f \in L^2(\Omega, \mathcal{F}, \mathbb{P})$,

1. $E\left[\int_t^T f(s)\, \mathrm{d}W(s)\,\middle|\, \mathcal{F}_t\right] = 0$, where $E[\cdot|\mathcal{F}_t]$ stands for the expectation conditional on $\mathcal{F}_t$.

2. Continuity: $\int_0^t f(s)\, \mathrm{d}W(s)$ is also an $\mathcal{F}_t$-adaptive function and is continuous almost surely.

1.2.1 Evaluation of Stochastic Integrals

We now consider the evaluation of stochastic integrals. Suppose that we know the anti-derivative of a function, $f(t)$, such that

$$\frac{\mathrm{d}F(t)}{\mathrm{d}t} = f(t). \tag{1.29}$$

Could there be

$$\int_0^t f(W(s))\,\mathrm{d}W(s) = F(W(t)) - F(W(0))? \tag{1.30}$$

The answer to this question is no. Consider, for example, $f(t) = W(t)$. If Equation 1.30 was correct, then there would be

$$\int_0^t W(s)\,\mathrm{d}W(s) = \frac{1}{2}\left[W^2(t) - W^2(0)\right] = \frac{1}{2}W^2(t). \tag{1.31}$$

Taking expectations on both sides and applying the first property of the stochastic integrals, we would obtain

$$0 = E\left[\int_0^t W(s)\,\mathrm{d}W(s)\right] = E\left[\frac{1}{2}W^2(t)\right] = \frac{1}{2}t, \tag{1.32}$$

which is a contradiction. This result suggests that general rules in deterministic calculus is not applicable to stochastic integrals.

As a showcase of integral evaluation, we try to work out the integral of $f(t) = W(t)$ according to its definition. Let $t_j = jt/n$ and denote W_j for $W(t_j)$, $j = 0, \ldots, n$. Start from the partial sum as follows:

$$
\begin{aligned}
S_n &= \sum_{j=0}^{n-1} W(t_j)\Delta W(t_j) = \sum_{j=0}^{n-1} W_j(W_{j+1} - W_j) \\
&= \sum_{j=0}^{n-1} W_j W_{j+1} - W_j^2 \\
&= \sum_{j=0}^{n-1} -W_{j+1}^2 + 2W_{j+1}W_j - W_j^2 + W_{j+1}^2 - W_{j+1}W_j \\
&= \sum_{j=0}^{n-1} -(W_{j+1} - W_j)^2 + W_{j+1}^2 - W_j^2 - W_j\Delta W_j \\
&= -\sum_{j=0}^{n-1} (\Delta W_j)^2 + W_n^2 - W_0^2 - S_n.
\end{aligned} \tag{1.33}
$$

We then have, by Kolmogorov's large number theorem,

$$
\begin{aligned}
S_n &= \frac{1}{2}\left(W_n^2 - W_0^2\right) - \frac{1}{2}\sum_{j=0}^{n-1}(\Delta W_j)^2 \\
&\xrightarrow{\Delta t \to 0} \frac{1}{2}\left[W^2(t) - W^2(0)\right] - \frac{1}{2}t;
\end{aligned} \tag{1.34}
$$

that is,

$$\int_0^t W(s)\,\mathrm{d}W(s) = \frac{1}{2}\left[W^2(t) - W^2(0)\right] - \frac{1}{2}t. \tag{1.35}$$

Now, both sides of Equation 1.35 vanish under expectations. Compared with deterministic calculus, there is an additional term, $-t/2$, in Equation 1.35. The above procedure of integral valuation, however, is inefficient and cumbersome for general functions. In the next section, we introduce the theory for efficient evaluation of stochastic integrals.

Similar to deterministic calculus, an integral equation implies a corresponding differential equation. To see that, we differentiate Equation 1.35 with respect to t, obtaining

$$W(t)\,\mathrm{d}W(t) = \frac{1}{2}\mathrm{d}\left[W^2(t) - t\right],\qquad(1.36)$$

or

$$\mathrm{d}W^2(t) = 2W(t)\,\mathrm{d}W(t) + \mathrm{d}t.\qquad(1.37)$$

Equation 1.37 is the first *stochastic differential equation* (SDE) to appear in this book; it relates the differential of $f(t) = W^2(t)$ to the differential of $W(t)$. Knowing Equation 1.37, we can calculate the stochastic integral $\int_0^t W(s)\,\mathrm{d}W(s)$ easily, without going through the procedure from Equations 1.33 through 1.35. In the next section, we study the dynamics of general functions in a broader context.

1.3 Stochastic Differentials and Ito's Lemma

In this section, we study the differentials of functions of other stochastic processes. In stochastic calculus, the so-called Ito's process is most often used as the basic stochastic process.

Definition 1.3.1. *Ito's process is a continuous stochastic process of the form:*

$$X(t) = X_0 + \int_0^t \sigma(s)\,dW(s) + \int_0^t \mu(s)\,ds,\qquad(1.38)$$

where $\sigma(s)$ and $\mu(s)$ are adaptive functions satisfying

$$E\left[\int_0^t \left(\sigma^2(s) + |\mu(s)|\right)ds\right] < \infty,\quad \forall t.\qquad(1.39)$$

The corresponding differential of Ito's process is

$$dX(t) = \sigma(t)\,dW(t) + \mu(t)\,dt.\qquad(1.40)$$

We call $\sigma(t)$ and $\mu(t)$ the volatility and drift of the SDE, respectively.

We now consider a function of $X(t)$, $Y(t) = F(X(t), t)$. The next lemma describes the SDE satisfied by $Y(t)$.

Lemma 1.3.1 (Ito's Lemma). *Let $X(t)$ be Ito's process with drift $\mu(t)$ and volatility $\sigma(t)$, and let $F(x,t)$ be a smooth function with bounded second-order derivatives. Then $Y(t) = F(X(t),t)$ is also an Ito's process with drift*

$$N(t) = \frac{\partial F}{\partial t} + \frac{1}{2}\sigma^2(t)\frac{\partial^2 F}{\partial x^2} + \mu(t)\frac{\partial F}{\partial x}, \tag{1.41}$$

and volatility

$$\Sigma(t) = \sigma(t)\frac{\partial F}{\partial x}. \tag{1.42}$$

Proof: By Taylor's expansion,

$$\Delta Y(t_i) = F(X(t_i + \Delta t), t_i + \Delta t) - F(X(t_i), t_i)$$
$$= F_x \Delta X + F_t \Delta t + \frac{1}{2}F_{xx}(\Delta X)^2 + F_{xt}\Delta X \Delta t + \frac{1}{2}F_{tt}(\Delta t)^2$$
$$+ \text{ higher order terms.} \tag{1.43}$$

Because

$$\Delta W_t = \sqrt{\Delta t} \cdot \varepsilon, \quad \varepsilon \sim N(0,1), \tag{1.44}$$

we generally have

$$E\left[|\Delta W_t|^p \Delta t^q\right] \propto \Delta t^{(p/2)+q}. \tag{1.45}$$

Here "$\propto$" means "of the order of." Based on Equation 1.45, we know that the order of magnitude of both the cross term and the higher-order terms in Equation 1.43 is $O(\Delta t^{3/2})$, and thus we can rewrite Equation 1.43 as

$$\Delta Y(t_i) = F_x \Delta X + F_t \Delta t + \frac{1}{2}F_{xx}\sigma^2(t_i)(\Delta W_i)^2 + O(\Delta t^{3/2})$$

$$= F_x \Delta X + F_t \Delta t + \frac{1}{2}F_{xx}\sigma^2(t_i)\Delta t$$

$$+ \frac{1}{2}F_{xx}\sigma^2(t_i)\left(\Delta W_i^2 - \Delta t\right) + O(\Delta t^{3/2}), \tag{1.46}$$

where $\Delta W_i = \Delta W_{t_i}$. Sum up the increment for $i = 0, 1, \ldots, n-1$, we obtain

$$Y(t) - Y(0) = \sum_{i=0}^{n-1} F_x(X_i, t_i)\Delta X_i + F_t(X_i, t_i)\Delta t_i + \frac{1}{2}F_{xx}(X_i, t_i)\sigma^2(t_i)\Delta t_i$$

$$+ \frac{1}{2}\sum_{i=0}^{n-1} F_{xx}(X_i, t_i)\sigma^2(t_i)[(\Delta W_i^2 - \Delta t_i] + O(\Delta t^{\frac{1}{2}}), \tag{1.47}$$

where $X_i = X(t_i)$ and $\Delta X_i = X(t_{i+1}) - X(t_i)$. Next, we will show that the fourth term on the right-hand side converges to zero in $L^2(\Omega, \mathcal{F}, \mathbb{P})$ as the partition of time shrinks to zero. Clearly, the mean of the fourth term is

$$E\left[\sum_{i=0}^{n-1} F_{xx}(X_i, t_i)\sigma^2(t_i)(\Delta W_i^2 - \Delta t_i)\right]$$

$$= \sum_{i=0}^{n-1} E\left[F_{xx}(X_i, t_i)\sigma^2(t_i)\right]\left[E\left(\Delta W_i^2\right) - \Delta t_i\right] = 0. \tag{1.48}$$

Here, we have used the fact that $F(X_t, t)$ and $\sigma(t)$ are both $\mathcal{F}_t$-adaptive functions. What is left to show is that the variance of the term converges to zero. For notational simplicity, we denote

$$a_i = F_{xx}(X_i, t_i)\sigma^2(t_i). \tag{1.49}$$

We then have

$$\text{Var}\left[\sum_{i=0}^{n-1} a_i\left((\Delta W_i)^2 - \Delta t_i\right)\right]$$

$$= E\left[\left(\sum_{i=0}^{n-1} a_i\left((\Delta W_i)^2 - \Delta t_i\right)\right)^2\right]$$

$$= E\left[\sum_{i,j=0}^{n-1} a_i a_j \left((\Delta W_i)^2 - \Delta t_i\right)\left((\Delta W_j)^2 - \Delta t_j\right)\right]$$

$$= E\left[\sum_{i=0}^{n-1} a_i^2 \left((\Delta W_i)^2 - \Delta t_i\right)^2\right]$$

$$= E\left[\sum_{i=0}^{n-1} a_i^2 \left((\Delta W_i)^4 - 2\Delta t_i(\Delta W_i)^2 + (\Delta t_i)^2\right)\right]$$

$$= \sum_{i=0}^{n} E\left[a_i^2\right] \cdot \left\{E\left[(\Delta W_i)^4\right] - \Delta t_i^2\right\}$$

$$= \sum_{i=0}^{n} E\left[a_i^2\right] \cdot \left(3(\Delta t_i)^2 - \Delta t_i^2\right)$$

$$\leq 2\left(\max_j \Delta t_j\right) \cdot \sum_{i=0}^{n-1} E\left[a_i^2\right] \Delta t_i$$

$$\approx 2\left(\max_j \Delta t_j\right) \cdot \int_0^T E[a^2]\,\mathrm{d}t \to 0 \quad \text{as } \max_j \Delta t_j \to 0. \tag{1.50}$$

Based on Equations 1.48 and 1.50, we conclude that the fourth term of Equation 1.47 vanishes in $L^2(\Omega, \mathcal{F}, \mathbb{P})$. Hence, as $\Delta t = \max_j \Delta t_j \to 0$, Equation 1.47 becomes

$$Y(t) - Y(0) = \int_0^t F_x \mathrm{d}X(s) + F_t \mathrm{d}s + \frac{1}{2}F_{xx}\sigma^2(s)\mathrm{d}s. \tag{1.51}$$

In differential form, the above equation becomes

$$\mathrm{d}Y(t) = F_x \mathrm{d}X(t) + F_t \mathrm{d}t + \frac{1}{2}F_{xx}\sigma^2(t)\mathrm{d}t$$

$$= \left(F_t + \mu(t)F_x + \frac{1}{2}\sigma^2(t)F_{xx}\right)\mathrm{d}t + \sigma(t)F_x\mathrm{d}W(t). \tag{1.52}$$

This finishes the proof. □

Next, we study the application of Ito's Lemma with two examples.

Example 1.2. *Consider again the differential of the function* $f(t) = W_t^2$. *We have*

$$\frac{\partial f}{\partial t} = 0, \qquad \frac{\partial f}{\partial W} = 2W_t, \quad and \quad \frac{\partial^2 f}{\partial W^2} = 2. \tag{1.53}$$

According to Ito's Lemma, we have

$$df = 2W_t \, dW_t + dt, \tag{1.54}$$

which reproduces Equation 1.37.

Example 1.3. *(A lognormal process). A lognormal model is perhaps the most popular model for asset prices in financial studies. This model is based on the assumption that the return on an asset over a fixed horizon obeys a normal distribution. Let S_t denote the price of an asset at time t. Then, the return over a horizon, $(t, t + \Delta t)$, is defined as*

$$\ln \frac{S_{t+\Delta t}}{S_t} = \mu_t \Delta t + \sigma_t \varepsilon \sqrt{\Delta t} \sim N(\mu_t \Delta t, \sigma_t^2 \Delta t). \tag{1.55}$$

Here, $\mu_t = \mu(t)$ and $\sigma(t) = \sigma_t$, and similar variants are also used interchangeably for other functions, whenever no confusion is caused. Note that the random term is proportional to $\sqrt{\Delta t}$, whereas the drift term is proportional to Δt. Hence, for a small Δt, the random term dominates the deterministic term. This is consistent with the widely held belief that, in an efficient market, asset price movements over a short horizon are pretty much random. Taking the limit $\Delta t \to dt$, we arrive at the model with a differential equation:

$$d(\ln S_t) = \mu_t \, dt + \sigma_t \, dW_t. \tag{1.56}$$

By integrating the above equation over the interval $(0, t)$, we then obtain

$$\ln S_t = \ln S_0 + \int_0^t \mu_s \, ds + \sigma_s \, dW_s, \tag{1.57}$$

or

$$S_t = S_0 e^{\int_0^t \mu_s ds + \sigma_s dW_s}. \tag{1.58}$$

To derive the differential equation for S_t, we let $X_t = \ln S_t / S_0$ and write $S_t = S_0 e^{X_t}$. Obviously, there are

$$\frac{\partial S_t}{\partial t} = 0, \qquad \frac{\partial S_t}{\partial X} = \frac{\partial^2 S_t}{\partial X^2} = S_0 e^{X_t} = S_t. \tag{1.59}$$

According to Ito's Lemma,

$$dS_t = \left(\mu_t + \frac{1}{2}\sigma_t^2 \right) S_t \, dt + \sigma S_t \, dW_t. \tag{1.60}$$

This is the very popular lognormal process or geometric Brownian motion for asset price modeling.

Remark 1.3.1. *In formalism, Equation 1.50 may imply*

$$\int_0^t a(s) \, (dW_s)^2 = \int_0^t a(s) \, ds \qquad (1.61)$$

for any $\mathcal{F}_t$-adaptive square-integrable function, $a(s)$. Based on Equation 1.61, we define the following operational rules:

$$(dW_t)^2 = dt \qquad (1.62)$$

and

$$dW_t dt = 0 \quad and \quad dt dt = 0 \qquad (1.63)$$

for the Brownian motion. Note that Equation 1.62 is obviously at odds with $dW_t = \varepsilon\sqrt{\Delta t}$, $\varepsilon \sim N(0,1)$, so the operational rule 1.62 has to be interpreted according to Equation 1.61.

The above operation rules offer a great deal of convenience to stochastic differentiations. Taking the derivation of Ito's Lemma, for example, we may now proceed as

$$dY(t) = \frac{\partial F}{\partial t}dt + \frac{\partial F}{\partial x}dX(t) + \frac{1}{2}\frac{\partial^2 F}{\partial x^2}\left(dX(t)\right)^2$$

$$= \frac{\partial F}{\partial t}\,dt + \frac{\partial F}{\partial x}\left(\sigma(t)\,dW_t + \mu(t)\,dt\right) + \frac{1}{2}\frac{\partial^2 F}{\partial x^2}\sigma^2(t)dt \qquad (1.64)$$

$$= \left(\frac{\partial F}{\partial t} + \mu(t)\frac{\partial F}{\partial x} + \frac{1}{2}\sigma^2(t)\frac{\partial^2 F}{\partial x^2}\right)dt + \sigma(t)\frac{\partial F}{\partial x}dW_t.$$

Equation 1.64 carries an important insight for stochastic differentiation: when calculating the differential of a general function, we must retain the second-order terms in stochastic variables.

1.4   Multi-Factor Extensions

The need to extend Ito process and Ito's Lemma to multiple dimensions stems from the fact that the values of both a single asset and a portfolio of assets can be affected by multiple sources of risks. This is obvious to see for a portfolio. For an asset, we sometimes need to distinguish *idiosyncratic* risk from *systematic* risk. Idiosyncratic risk is asset-specific, whereas systematic risk is a market-wide risk, or risk of a macroeconomic nature.

1.4.1 Multi-Factor Ito's Process

A multiple-factor Ito's process takes the form

$$\mathrm{d}X_i(t) = \mu_i(t)\,\mathrm{d}t + \sum_{j=1}^{n} \sigma_{ij}(t)\,\mathrm{d}W_j(t)$$

$$= \mu_i(t)\,\mathrm{d}t + \boldsymbol{\sigma}_i^{\mathrm{T}}(t)\,\mathrm{d}\mathbf{W}_t,$$

where

$$\mathbf{W}(t) = \begin{pmatrix} W_1(t) \\ W_2(t) \\ \vdots \\ W_n(t) \end{pmatrix}$$

is a vector of independent Brownian motion, and

$$\boldsymbol{\sigma}_i(t) = \begin{pmatrix} \sigma_{i,1}(t) \\ \sigma_{i,2}(t) \\ \vdots \\ \sigma_{i,n}(t) \end{pmatrix}$$

is called the volatility vector. Let $\mathbf{X}(t) = (X_1(t), X_2(t), \ldots, X_n(t))^{\mathrm{T}}$. In integral form, the multi-factor Ito's process is

$$\mathbf{X}_t = \mathbf{X}_0 + \int_0^t \boldsymbol{\mu}_s\,\mathrm{d}s + \int_0^t \boldsymbol{\Sigma}(s)\,\mathrm{d}\mathbf{W}_s,$$

where

$$\boldsymbol{\mu}_t = \begin{pmatrix} \mu_1(t) \\ \mu_2(t) \\ \vdots \\ \mu_n(t) \end{pmatrix}$$

is the vector of drifts, and

$$\boldsymbol{\Sigma}(t) = \begin{pmatrix} \boldsymbol{\sigma}_1^{\mathrm{T}}(t) \\ \boldsymbol{\sigma}_2^{\mathrm{T}}(t) \\ \vdots \\ \boldsymbol{\sigma}_n^{\mathrm{T}}(t) \end{pmatrix}$$

is the volatility matrix. Note that both $\boldsymbol{\mu}(t)$ and $\boldsymbol{\sigma}_i(t)$, $i = 1, \ldots, n$ are $\mathcal{F}_t$-adaptive processes, and they satisfy

$$E\left[\int_0^t \left(\sum_{j=1}^{n} \| \boldsymbol{\sigma}_j(s) \|^2 + |\boldsymbol{\mu}_s|_1 \right) \mathrm{d}s \right] < \infty, \quad \forall t.$$

Namely, $\boldsymbol{\sigma}_j(s), j = 1, \ldots, n$ are square integrable and $\boldsymbol{\mu}_s$ has bounded variation.

1.4.2 Ito's Lemma

The one-factor Ito's Lemma can be generalized directly to a multi-factor situation. A proof parallel to that of Lemma 1.2 can be assembled by using vector notations.

Lemma 1.4.1 (Ito's Lemma). *Let f be a deterministic twice continuous differentiable function, and let $\mathbf{X}_t$ be an n-factor Ito's process. Then $Y_t = f(\mathbf{X}_t, t)$ is also an n-factor Ito's process, and it satisfies*

$$dY_t = \left(\frac{\partial f}{\partial t} + \sum_{j=1}^{n} \mu_j(t)\frac{\partial f}{\partial X_j} + \frac{1}{2}\sum_{i,j=1}^{n} \boldsymbol{\sigma}_i^{\mathrm{T}}(t)\boldsymbol{\sigma}_j(t)\frac{\partial^2 f}{\partial X_i \partial X_j} \right) dt$$
$$+ \sum_{j=1}^{n} \frac{\partial f}{\partial X_j}\boldsymbol{\sigma}_j^{\mathrm{T}}(t)\,d\mathbf{W}_t.$$

1.4.3 Correlated Brownian Motions

Let $W(t)$ and $\tilde{W}(t)$ be two Brownian motions under the probability space $(\Omega, \mathcal{F}, \mathbb{P})$. We say that $W(t)$ and $\tilde{W}(t)$ are correlated if

$$\mathrm{Cov}\left[\Delta W(t), \Delta \tilde{W}(t)\right] = \rho \Delta t \quad \text{for} \quad \rho \neq 0. \tag{1.65}$$

Equivalently, we can write

$$\begin{aligned} W(t) &= W_1(t), \\ \tilde{W}(t) &= \rho W_1(t) + \sqrt{1-\rho^2}W_2(t), \end{aligned} \tag{1.66}$$

where $W_1(t)$ and $W_2(t)$ are independent Brownian motions. We have the following additional operation rule for correlated Brownian motions:

$$dW(t)\,d\tilde{W}(t) = \rho\,dt. \tag{1.67}$$

With the help of the above operation rule, we can derive the processes of the product and the quotient of two Ito's processes. The following results are very useful for financial modeling, and for this reason we call them the *product rule* and the *quotient rule*, respectively.

Product rule: Let $X(t)$ and $Y(t)$ be two Ito's processes such that

$$\begin{aligned} dX(t) &= \sigma_X(t)\,dW(t) + u_X(t)\,dt, \\ dY(t) &= \sigma_Y(t)\,d\tilde{W}(t) + u_Y(t)\,dt, \end{aligned} \tag{1.68}$$

where $dW(t)\,d\tilde{W}(t) = \rho\,dt$. Then,

$$\begin{aligned} d\left(X(t)Y(t)\right) &= X(t)\,dY(t) + Y(t)\,dX(t) + dX(t)\,dY(t) \\ &= X(t)\,dY(t) + Y(t)\,dX(t) + \sigma_X(t)\sigma_Y(t)\rho\,dt. \end{aligned} \tag{1.69}$$

Quotient rule: Let $X(t)$ and $Y(t)$ be two Ito's processes. Then,

$$d\left(\frac{X(t)}{Y(t)}\right) = \frac{dX(t)}{Y(t)} - \frac{X(t)\,dY(t)}{Y^2(t)} - \frac{dX(t)\,dY(t)}{Y^2(t)} + \frac{X(t)\,(dY(t))^2}{Y^3(t)}. \quad (1.70)$$

The proofs for both rules are left as exercises.

1.4.4 The Multi-Factor Lognormal Model

We now introduce the classic model of a financial market with multiple assets, an important area of application for the multi-factor Ito's Lemma. This financial market consists of a money market account (also called a savings account), B_t, and n risky assets, $\{S_t^i\}_{i=1}^n$. The price evolutions of these $n + 1$ assets are governed by the following equations:

$$dB_t = r_t B_t dt,$$
$$dS_t^i = S_t^i \left(\mu_t^i dt + \boldsymbol{\sigma}_i^{\mathrm{T}}(s)\,d\mathbf{W}_t\right), \quad i = 1, 2, \ldots, n.$$

Here r_t is the risk-free interest rate, μ_t^i and $\boldsymbol{\sigma}_i(s)$ the rate of return and volatility of the ith asset. Driving the market is the n-dimensional Brownian motion. We therefore call the above model an n-factor model. Note that the savings account is considered a riskless asset so it is not driven by any Brownian motion.

By the multi-factor Ito's Lemma, we can derive the equations for the log of asset prices:

$$d \ln S_t^i = \left(\mu_t^i - \frac{1}{2}\|\boldsymbol{\sigma}_i(t)\|^2\right)dt + \boldsymbol{\sigma}_i^{\mathrm{T}}(t)\,d\mathbf{W}_t.$$

The above equation readily allows us to solve for the asset price:

$$S_t^i = S_0^i \exp\left(\int_0^t \boldsymbol{\sigma}_i^{\mathrm{T}}(s)\,d\mathbf{W}_s + \left(\mu_s^i - \frac{1}{2}\|\boldsymbol{\sigma}_i(s)\|^2\right)ds\right),$$

for $i = 1, 2, \ldots, n$. The value of the money market account, meanwhile, is simply

$$B_t = \exp\left(\int_0^t r_s\,ds\right).$$

1.5 Martingales

The notion of martingales is key to derivatives modeling. The definition is given below.

Definition 1.5.1. *A stochastic process, M_t, is called a $\mathbb{P}$-martingale if and only if it has the following properties:*

1. *$E^{\mathbb{P}}[|M_t|] < \infty, \quad \forall t$.*

2. *$E^{\mathbb{P}}[M_t | \mathcal{F}_s] = M_s, \quad \forall s \leq t$.*

The martingale properties are associated with fair games in investments or speculations. Let us think of $M_t - M_s$ as the profit or loss (P&L) of a gamble between two parties over the time period (s, t). Then the game is considered fair if the expected P&L is zero. Examples of fair games in everyday life include the coin tossing game and futures investments in financial markets. In mathematics, there are plenty of examples as well. In fact, we have already seen several of them so far, of which we remind readers below.

Example 1.4. *1. The simple random walk, X_n, is a martingale because $E[|X_n|] < n\sqrt{\Delta t}$ and $E[X_n | \mathcal{F}_m] = X_m, \ m \leq n$.*

2. A $\mathbb{P}$-Brownian motion, W_t, is a martingale by definition.

3. The stochastic integral $X_t = \int_0^t f(u)\, dW_u$ is a martingale, since

$$E^{\mathbb{P}}[X_t | \mathcal{F}_s] = E^{\mathbb{P}}\left[\int_0^s + \int_s^t f(u)\, dW_u \,\middle|\, \mathcal{F}_s\right]$$
$$= \int_0^s f(u)\, dW_u = X_s, \quad \forall s \leq t. \tag{1.71}$$

Here, we have applied the first property of stochastic integrals (see page 9).

4. The process $M_t = \exp\left(\int_0^t \sigma_s\, dW_s - \frac{1}{2}\sigma_s^2\, ds\right)$ is an exponential martingale. In fact, using the Ito's Lemma, we can show that

$$dM_t = \sigma_t M_t\, dW_t, \tag{1.72}$$

which is an Ito's process without drift. It follows that

$$M_t = M_s + \int_s^t M_u \sigma_u\, dW_u. \tag{1.73}$$

Based on the conclusion of the last example, we know that M_t is a martingale.

We emphasize here that an Ito's process is a martingale process if and only if its drift term is zero. Finally, we present two additional examples.

5. $M_t = W_t^2 - t$ *is a martingale. Here is the justification: for* $s \leq t$,

$$
\begin{aligned}
E^{\mathbb{P}}\left[W_t^2 - t \,\middle|\, \mathcal{F}_s\right] &= E^{\mathbb{P}}\left[(W_t - W_s + W_s)^2 - t \,\middle|\, \mathcal{F}_s\right] \\
&= E^{\mathbb{P}}\left[(W_t - W_s)^2 + 2W_s(W_t - W_s)\right. \\
&\quad \left. + W_s^2 - t \,\middle|\, \mathcal{F}_s\right] \\
&= (t - s) + 0 + W_s^2 - t = W_s^2 - s. \qquad (1.74)
\end{aligned}
$$

6. *Let* X_T *be a contingent claim depending on information up to time* T. *Define*

$$
N_t = E^{\mathbb{P}}[X_T \mid F_t].
$$

Then N_t *is a* $\mathbb{P}$-*martingale. In fact, for any* $s \leq t$, *we have*

$$
\begin{aligned}
E^{\mathbb{P}}[N_t \mid \mathcal{F}_s] &= E^{\mathbb{P}}\left[E^{\mathbb{P}}[X_T \mid \mathcal{F}_t] \,\middle|\, \mathcal{F}_s\right] \\
&= E^{\mathbb{P}}[X_T \mid \mathcal{F}_s] \\
&= N_s. \qquad (1.75)
\end{aligned}
$$

Equation 1.75 demonstrates the so-called tower law, namely, an expectation conditioned first on history up to time t and then on history up to an earlier time, s, is the same as that conditioned only on history up to the earlier time, s.

1.6 Feynman-Kac Theorem

To prepare for option pricing in the subsequent chapters, we introduce the Feynman-Kac Theorem (Kac, 1949).

Theorem 1.6.1 (Feynman-Kac). *Let* X_t *be an Ito's process such that*

$$
dX_t = \mu(t, X_t)dt + \sigma(t, X_t)dW_t.
$$

Let $H(x)$ *and* $r(t, x)$ *be deterministic functions. Define a function*

$$
F(t, x) = E\left[e^{-\int_t^T r(s, X_s)ds} H(X_T) \mid X_t = x\right].
$$

Then $F(t, x)$ *satisfies the following partial differential equation:*

$$
\frac{\partial F}{\partial t} + \mu(t, x)\frac{\partial F}{\partial x} + \frac{1}{2}\sigma^2(t, x)\frac{\partial^2 F}{\partial x^2} - r(t, x)F(t, x) = 0,
$$

and the terminal condition

$$
F(T, x) = H(x)
$$

for all x.

Proof: Consider the auxiliary function

$$\tilde{F}(t,x) = e^{-\int_0^t r(s,X_s)ds} F(t,x) = E\left[e^{-\int_0^T r(s,X_s)ds} H(X_T)|X_t = x\right],$$

By Ito's Lemma, the dynamics of $\tilde{F}(t,X_t)$ is

$$d\tilde{F}(t,X_t) = \left(\frac{\partial \tilde{F}}{\partial t} + \mu(t,x)\frac{\partial \tilde{F}}{\partial x} + \frac{1}{2}\sigma^2(t,x)\frac{\partial^2 \tilde{F}}{\partial x^2}\right)dt + \sigma(t,x)\frac{\partial \tilde{F}}{\partial x}dW_t.$$

$$(1.76)$$

By definition, $\tilde{F}(t,X_t)$ is an martingale, so the drift term of its dynamics must be zero, thus yielding a partial differential equation (PDE):

$$\frac{\partial \tilde{F}}{\partial t} + \mu(t,x)\frac{\partial \tilde{F}}{\partial x} + \frac{1}{2}\sigma^2(t,x)\frac{\partial^2 \tilde{F}}{\partial x^2} = 0.$$

In terms of $F(t,x)$, the equation becomes

$$\frac{\partial F}{\partial t} + \mu(t,x)\frac{\partial F}{\partial x} + \frac{1}{2}\sigma^2(t,x)\frac{\partial^2 F}{\partial x^2} - r(t,x)F(t,x) = 0,$$

subject to terminal condition

$$F(T,x) = H(x).$$

This completes the proof. □

The Feynman-Kac Theorem enables mathematical modeling in terms of partial differential equation (PDE) to many applied areas, including financial derivatives. The PDE approach has been particularly useful for exotic derivatives pricing.

Exercises

1. Solve subproblem (a) with the help of Excel® or MATLAB®.

   (a) Draw N standard normal random numbers, $X_i, i = 1,\ldots,N$ and verify that

   $$\frac{1}{N}\sum_{i=1}^{N} X_i^2 \to 1 \quad \text{as } N \to \infty,$$

   by increasing N from 10 to 1000. Plot the above average value versus N.

   (b) Let $W(t)$ be a Brownian motion. For fixed t, prove that

   $$\sum_{i=1}^{N} (\Delta W(i\Delta t))^2 \to t \quad \text{a.s.}$$

   Here $\Delta t = t/N$.

2. Prove directly from the definition of Ito's integrals that

$$\int_0^t s\,dW(s) = tW(t) - \int_0^t W(s)\,ds.$$

$\left(\text{Hint: } \sum_j \Delta(s_j W_j) = \sum_j s_j \Delta W_j + \sum_j W_j \Delta s_j. \right)$

3. Justify the following equality with and without using Ito's Lemma.

$$\int_0^t W^2(s)\,dW(s) = \frac{1}{3}W^3(t) - \int_0^t W(s)\,ds.$$

4. Let $f(t)$ be a continuous and adaptive function. Prove that

$$E\left[\int_0^T f(t)\,dW(t) \right] = 0,$$

$$E\left[\left(\int_0^T f(t)\,dW(t) \right)^2 \right] = E\left[\int_0^T |f(t)|^2\,dt \right].$$

5. Let X_t and Y_t be Ito's processes and $f(x,y,t)$ be a twice continuous differentiable function of x, y, and t. Use a two-dimensional Ito's Lemma of the form

$$df = \frac{\partial f}{\partial t}dt + \frac{\partial f}{\partial x}dX_t + \frac{\partial f}{\partial y}dY_t$$
$$+ \frac{1}{2}\left(\frac{\partial^2 f}{\partial x^2}(dX_t)^2 + 2\frac{\partial^2 f}{\partial x \partial y}(dX_t dY_t) + \frac{\partial^2 f}{\partial y^2}(dY_t)^2 \right)$$

to prove the *Quotient rule*:

$$d\left(\frac{X_t}{Y_t} \right) = \frac{dX_t}{Y_t} - \frac{X_t dY_t}{Y_t^2} - \frac{dX_t dY_t}{Y_t^2} + \frac{X_t(dY_t)^2}{Y_t^3}.$$

Chapter 2

The Martingale Representation Theorem

In this chapter, we introduce the martingale approach to derivatives pricing. This approach consists of two major steps: the derivation of the martingale probability measure and the construction of the replication strategy. The mathematical foundation of this approach is formed by two theorems, namely, the Cameron–Martin–Girsanov (CMG) theorem and the martingale representation theorem. The derivation of the martingale probability measure is achieved by using the CMG theorem, while the construction of the replication strategy is based on the martingale representation theorem. A significant portion of this chapter is devoted to establishing these two theorems. We motivate our discussions with a simple binomial model for tradable assets and then proceed to establish the two theorems in continuous time for a complete market with multiple underlying securities. Once the pricing formula for general options has been established, we proceed to price call options and derive the famous Black–Scholes formulae. Until up to Chapter 10, we limit ourselves to models in complete markets where every source of risk can be traded or hedged. The proof of the martingale representation theorem is provided in the appendix of this chapter. At the end of the chapter, we make comments on derivative pricing in incomplete markets.

2.1 Changing Measures with Binomial Models

2.1.1 A Motivating Example

Consider the simplest option-pricing model with an underlying asset following a one-period binomial process, as depicted in Figure 2.1. In Figure 2.1, $0 \leq p \leq 1$ and $\bar{p} = 1 - p$. The option's payoffs at time 1, $f(S_u)$ and $f(S_d)$, are given explicitly, and we want to determine $f(S)$, the value of the option at time 0. Without loss of generality, we assume that there is a zero interest rate in the model. To avoid arbitrage, we must impose the order $S_d \leq S \leq S_u$. We call $\mathbb{P} = \{p, \bar{p}\}$ the objective measure of the underlying process.

DOI: 10.1201/9781003389101-2

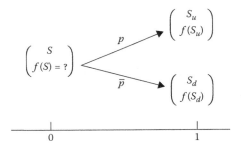

FIGURE 2.1: A binomial model for option pricing.

It may be tempting to price the option by expectation under $\mathbb{P}$:

$$
\begin{aligned}
f(S) &= E^{\mathbb{P}}[f(S_1)] \\
&= p\,f(S_u) + \bar{p}\,f(S_d).
\end{aligned}
\tag{2.1}
$$

However, except for a special p, the above price generates arbitrage and thus is wrong. To see that, we replicate the payoff of the option at time 1 using a portfolio of the underlying asset and a cash bond, with respective numbers of units, α and β, such that, at time 1,

$$
\begin{aligned}
\alpha S_u + \beta &= f(S_u), \\
\alpha S_d + \beta &= f(S_d).
\end{aligned}
\tag{2.2}
$$

Solving for α and β, we obtain

$$
\begin{aligned}
\alpha &= \frac{f(S_u) - f(S_d)}{S_u - S_d}, \\
\beta &= \frac{S_u f(S_d) - S_d f(S_u)}{S_u - S_d}.
\end{aligned}
\tag{2.3}
$$

Equation 2.2 implies that the time-1 values of the portfolio and option are identical. To avoid arbitrage, their values at time 0 must be identical as well[1] which yields the arbitrage price of the option at time 0:

$$
\begin{aligned}
f(S) &= \alpha S + \beta \\
&= q\,f(S_u) + \bar{q}\,f(S_d) \\
&= E^{\mathbb{Q}}[f(S_1)],
\end{aligned}
\tag{2.4}
$$

where $\mathbb{Q} = \{q, \bar{q}\}$, and

$$
q = \frac{S - S_d}{S_u - S_d}, \quad \bar{q} = 1 - q
\tag{2.5}
$$

---

[1]Such an argument is also called the dominance principle.

is a different set of probabilities. Note that Equation 2.4 gives the no-arbitrage price of the option. Any other price will induce arbitrage to the market. Hence, the expectation price, in Equation 2.1, is correct only if $p = q$. In fact, $\{q, \bar{q}\}$ is the only set of probabilities that satisfies

$$S = qS_u + \bar{q}S_d = E^{\mathbb{Q}}(S_1). \tag{2.6}$$

The price formulae, Equations 2.4 and 2.6, have rather general implications. First, the price of the option can be expressed as an expectation of the payoff under a special probability distribution. Second, this special probability distribution is nothing else but the "martingale measure" for the underlying asset. As a result, the original objective measure plays no role in derivatives pricing and it needs to be changed into the "martingale measure" for such a purpose. If we introduce a stochastic variable, ζ, such that

$$\zeta = \begin{cases} \zeta_u = \dfrac{q}{p}, & \text{if } S_1 = S_u, \\[2mm] \zeta_d = \dfrac{\bar{q}}{\bar{p}}, & \text{if } S_1 = S_d, \end{cases} \tag{2.7}$$

we then can rewrite the price formula as

$$
\begin{aligned}
E^{\mathbb{Q}}[f(S_1)] &= p\zeta_u f(S_u) + \bar{p}\zeta_d f(S_d) \\
&= E^{\mathbb{P}}[\zeta f(S_1)].
\end{aligned} \tag{2.8}
$$

In the finance literature, ζ is called a pricing kernel. Hence, finding the pricing measure is equivalent to finding the pricing kernel. In the next section, we elaborate the pricing kernel with a multi-period binomial tree.

2.1.2 Binomial Trees and Path Probabilities

Let us move one step forward and consider the binomial tree model up to two time steps, as shown in Figure 2.2, where each pair of numbers represents a state (which can be associated with the price of an asset if necessary). Out of each state at time j, two possible states are generated at time $j + 1$. Hence, there are 2^j states at time j, starting with a single state at time 0. The branching probabilities for reaching the next two states from one state, (i, j), are $p_{i,j} \in [0, 1]$ and $\bar{p}_{i,j} = 1 - p_{i,j}$, respectively. The collection of branching probabilities, $\mathbb{P} = \{p_{i,j}, \bar{p}_{i,j}\}$, is again called a measure. As is shown in Figure 2.2, there are two paths over the time horizon from 0 to 1, whereas there are four paths over the time horizon from 0 to 2. The corresponding path probabilities for the horizon from 0 to 1 are

$$\pi_{0,1} = \bar{p}_{0,0} \quad \text{and} \quad \pi_{1,1} = p_{0,0}, \tag{2.9}$$

whereas for the horizon from 0 to 2, they are

$$\pi_{0,2} = \bar{p}_{0,0}\bar{p}_{0,1},\, \pi_{1,2} = \bar{p}_{0,0}p_{0,1},\, \pi_{2,2} = p_{0,0}\bar{p}_{1,1},\, \text{and } \pi_{3,2} = p_{0,0}p_{1,1}. \tag{2.10}$$

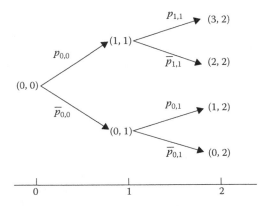

FIGURE 2.2: A two-period binomial tree.

The path probabilities can also be marked in a binomial tree as is shown in Figure 2.3.

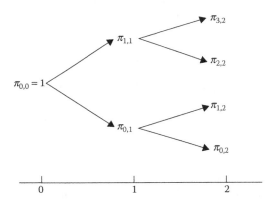

FIGURE 2.3: A path probability tree under $\mathbb{P}$.

Consider now another set of branching probabilities, $\mathbb{Q} = \{q_{i,j}, \bar{q}_{i,j} = 1 - q_{i,j}\}$, for the same tree. The corresponding path probabilities are

$$\pi'_{0,1} = \bar{q}_{0,0} \quad \text{and} \quad \pi'_{1,1} = q_{0,0} \tag{2.11}$$

up to time 1, and

$$\pi'_{0,2} = \bar{q}_{0,0}\bar{q}_{0,1}, \pi'_{1,2} = \bar{q}_{0,0}q_{0,1}, \pi'_{2,2} = q_{0,0}\bar{q}_{1,1}, \text{ and } \pi'_{3,2} = q_{0,0}q_{1,1} \tag{2.12}$$

up to time 2. Suppose that the $\mathbb{P}$-probability of paths $\pi_{i,j} \neq 0$ for all i, j. We then can define the ratio of path probabilities as follows:

$$\zeta_{i,j} = \frac{\pi'_{i,j}}{\pi_{i,j}}. \tag{2.13}$$

These ratios can then be equipped with the original measure, $\mathbb{P}$, and thus be treated as possible values of a random process, ζ, as shown in Figure 2.4.

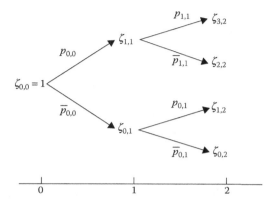

FIGURE 2.4: A Radon–Nikodym random process for the two-period tree.

The purpose of introducing ζ is to allow a change of measures. Consider pricing a contingent claim with maturity j that pays $x_{i,j}$ if the (i,j) state is realized. For $j = 1$, the expected price of the claim under $\mathbb{Q}$ is

$$E^{\mathbb{Q}}[X_1 \mid \mathcal{F}_0] = \sum_{i=0}^{1} \pi'_{i,1} x_{i,1} = \sum_{i=0}^{1} \pi_{i,1} \left(\frac{\pi'_{i,1}}{\pi_{i,1}} \right) x_{i,1} = E^{\mathbb{P}}[\zeta_1 X_1 \mid \mathcal{F}_0]. \quad (2.14)$$

For $j = 2$, the expectation price is

$$E^{\mathbb{Q}}[X_2 \mid \mathcal{F}_0] = \sum_{i=0}^{3} \pi'_{i,2} x_{i,2} = \sum_{i=0}^{3} \pi_{i,2} \left(\frac{\pi'_{i,2}}{\pi_{i,2}} \right) x_{i,2} = E^{\mathbb{P}}[\zeta_2 X_2 \mid \mathcal{F}_0]. \quad (2.15)$$

The role of ζ is self-explanatory in Equations 2.14 and 2.15: if $\mathbb{Q}$ is the pricing measure, then ζ is the pricing kernel. In the mathematical literature, the kernel is often called a Radon–Nikodym derivative between the two measures, and it is denoted as

$$\zeta \triangleq \frac{d\mathbb{Q}}{d\mathbb{P}}. \quad (2.16)$$

An alternative viewpoint is to treat the derivative ζ as a mapping of the filtration, $\mathcal{F}_t$, to a number:

$$\left. \frac{d\mathbb{Q}}{d\mathbb{P}} \right|_{\mathcal{F}_t} = \zeta_t. \quad (2.17)$$

Intuitively, Equations 2.14 and 2.15 can be generalized to binomial processes with multiple time steps.

Next, we consider the change of measures under filtrations larger than $\mathcal{F}_0$.

According to the definition of $\zeta_t, t \geq 0$, the ratio of path probabilities over (t, T) is

$$\frac{\zeta_T}{\zeta_t}. \tag{2.18}$$

Conditional on the filtration, $\mathcal{F}_t$, ζ_t is known for certainty and thus we obtain the formula

$$E^{\mathbb{Q}}[X_T \,|\, \mathcal{F}_t] = E^{\mathbb{P}}\left[\frac{\zeta_T}{\zeta_t} X_T \,\middle|\, \mathcal{F}_t\right] = \zeta_t^{-1} E^{\mathbb{P}}[\zeta_T X_T \,|\, \mathcal{F}_t]. \tag{2.19}$$

Let us take the binomial model in Figure 2.2 as an example of Equation 2.19. Suppose that, at $t = 1$, we have reached the lower state, $(0, 1)$, and we thus have filtration $\mathcal{F}_1 = \{(0, 0), (0, 1)\}$. Then,

$$\begin{aligned}
E^{\mathbb{Q}}[X_2 \,|\, \mathcal{F}_1] &= \bar{q}_{0,1} x_{0,2} + q_{0,1} x_{1,2} \\
&= \left(\frac{\bar{q}_{0,0}}{\bar{p}_{0,0}}\right)^{-1} \left(\bar{p}_{0,1} \frac{\bar{q}_{0,0}\bar{q}_{0,1}}{\bar{p}_{0,0}\bar{p}_{0,1}} x_{0,2} + p_{0,1} \frac{\bar{q}_{0,0}q_{0,1}}{\bar{p}_{0,0}p_{0,1}} x_{1,2}\right) \\
&= \left(\frac{\pi'_{0,1}}{\pi_{0,1}}\right)^{-1} \left(\bar{p}_{0,1} \frac{\pi'_{0,2}}{\pi_{0,2}} x_{0,2} + p_{0,1} \frac{\pi'_{1,2}}{\pi_{1,2}} x_{1,2}\right) \\
&= \zeta_1^{-1} E^{\mathbb{P}}[\zeta_2 X_2 \,|\, \mathcal{F}_1]. \tag{2.20}
\end{aligned}$$

We next make a remark on the martingale property of ζ_t. By substituting $X_T = 1$ in Equation 2.19, we obtain

$$1 = \zeta_t^{-1} E^{\mathbb{P}}[\zeta_T \,|\, \mathcal{F}_t], \quad \forall t \leq T, \tag{2.21}$$

or

$$\zeta_t = E^{\mathbb{P}}[\zeta_T \,|\, \mathcal{F}_t], \tag{2.22}$$

which indicates that the Radon–Nikodym derivative, ζ_t, is a martingale process under the measure $\mathbb{P}$.

Finally, we discuss the existence of the Radon–Nikodym derivative, ζ_t. In our binomial tree model in Figure 2.2, the Radon–Nikodym derivatives can obviously be defined if $\pi_{i,j} \neq 0$ for all i, j. Even if $\pi_{i,j} = 0$, the derivative can still be defined if we simultaneously have $\pi'_{i,j} = 0$. In fact, with the following definition for the Radon–Nikodym derivative:

$$\zeta_j = \begin{cases} \dfrac{\pi'_{i,j}}{\pi_{i,j}}, & \text{if the } i\text{th path is taken and } \pi_{i,j} \neq 0, \\ 0, & \text{if the } i\text{th path is taken and } \pi_{i,j} = \pi'_{i,j} = 0, \end{cases} \tag{2.23}$$

previous arguments on the change of measures remain valid. In general, a Radon–Nikodym derivative, $d\mathbb{Q}/d\mathbb{P}$, exists as long as $\mathbb{Q}$ is said to be absolutely continuous with respect to $\mathbb{P}$.

Definition 2.1.1. *Measures* $\mathbb{P}$ *and* $\mathbb{Q}$ *operate on the same sample space,* S. *We say that* $\mathbb{Q}$ *is absolutely continuous with respect to* $\mathbb{P}$ *if, for any subset* $A \subseteq S$,

$$\mathbb{P}(A) = 0 \Longrightarrow \mathbb{Q}(A) = 0, \tag{2.24}$$

and it is denoted as $\mathbb{Q} \ll \mathbb{P}$.

For later use, we also introduce the concept of equivalence between two measures.

Definition 2.1.2. *Measures* $\mathbb{P}$ *and* $\mathbb{Q}$ *operate in the same sample space,* S. *If* $\mathbb{Q} \ll \mathbb{P}$ *and* $\mathbb{P} \ll \mathbb{Q}$ *hold simultaneously, we say that the two measures are equivalent and we write* $\mathbb{Q} \approx \mathbb{P}$.

An identical definition for equivalence is

$$\mathbb{P}(A) = 0 \Longleftrightarrow \mathbb{Q}(A) = 0. \tag{2.25}$$

That is, a $\mathbb{P}$-null set is also a $\mathbb{Q}$-null set, and vice versa.

2.2 Change of Measures under Brownian Filtration

2.2.1 The Radon–Nikodym Derivative of a Brownian Path

Consider a path of $\mathbb{P}$-Brownian motion over $(0, t)$ with discrete time stepping,

$$\{W(0) = 0, W(\Delta t), W(2\Delta t), \ldots, W(n\Delta t)\}, \tag{2.26}$$

where $\Delta t = t/n$. With the probability ratio in mind, our immediate question is what the path probability is. The answer, unfortunately, is zero. The implication that we cannot define the notion of the probability ratio given that the same path is realized under two different probability measures. To circumvent this problem, we first seek to calculate the probability for the Brownian motion to travel in a corridor (the so-called corridor probability), as is shown in Figure 2.5, and then we define the ratio of the corridor probabilities. The ratio of the path probabilities is finally defined through a limiting procedure. The corridor can be represented by the intervals $A_i = (x_i - \Delta x/2, x_i + \Delta x/2), i = 1, 2, \ldots, n$, where $x_i = W(i\Delta t)$ and $\Delta x > 0$ is a small number.

For a Brownian motion, the marginal distribution at $t_i = i\Delta t$ is known to be

$$f_{\mathbb{P}}(x) = \frac{1}{\sqrt{2\pi\Delta t}} e^{-(1/2)[(x-x_i)^2/\Delta t]} \sim N(x_i, \Delta t).$$

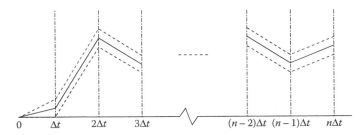

FIGURE 2.5: A corridor for Brownian motion with step size Δt.

Hence, the probability for the next step to fall in A_{i+1} is

$$\text{Prob}_{\mathbb{P}}(A_{i+1}) = \int_{x_{i+1}-\Delta x/2}^{x_{i+1}+\Delta x/2} f_{\mathbb{P}}(x)\mathrm{d}x$$

$$\approx f_{\mathbb{P}}(x_{i+1})\Delta x = \frac{\Delta x}{\sqrt{2\pi\Delta t}}e^{-(1/2)[(x_{i+1}-x_i)^2/\Delta t]}. \qquad (2.27)$$

Approximately, we can define the corridor probability to be

$$\prod_{i=1}^{n}\text{Prob}_{\mathbb{P}}(A_i) = \left(\frac{\Delta x}{\sqrt{2\pi\Delta t}}\right)^n e^{-(1/2\Delta t)\sum_{i=0}^{n-1}(x_{i+1}-x_i)^2}. \qquad (2.28)$$

Next, suppose that the same path is realized under a different marginal probability,

$$f_{\mathbb{Q}}(x) = \frac{1}{\sqrt{2\pi\Delta t}}e^{-(1/2)[(x-x_i+\gamma\Delta t)^2/\Delta t]} \sim N(x_i - \gamma\Delta t, \Delta t), \quad \forall i, \qquad (2.29)$$

where γ is taken to be constant for simplicity. Then the corresponding corridor probability can be similarly obtained to be

$$\prod_{i=1}^{n}\text{Prob}_{\mathbb{Q}}(A_i) = \left(\frac{\Delta x}{\sqrt{2\pi\Delta t}}\right)^n e^{-(1/2\Delta t)\sum_{i=0}^{n-1}(x_{i+1}-x_i+\gamma\Delta t)^2}. \qquad (2.30)$$

It follows that the ratio of the two corridor probabilities is

$$\zeta_t = \exp\left(-\frac{1}{2\Delta t}\sum_{i=0}^{n-1}\left[(x_{i+1}-x_i+\gamma\Delta t)^2 - (x_{i+1}-x_i)^2\right]\right)$$

$$= \exp\left(-\frac{1}{2\Delta t}\sum_{i=0}^{n-1}\left[2(x_{i+1}-x_i)\cdot\gamma\Delta t + \gamma^2\Delta t^2\right]\right)$$

$$= \exp\left(-\gamma\sum_{i=0}^{n-1}(x_{i+1}-x_i) - \frac{1}{2}\gamma^2\Delta t\cdot n\right)$$

$$(2.31)$$

$$= \exp\left(-\gamma(x_n - x_0) - \frac{1}{2}\gamma^2 t\right)$$

$$= e^{-\gamma W_t - (1/2)\gamma^2 t},$$

which depends neither on the path history nor on Δt, but only on the terminal point, W_t, of the Brownian motion. Hence, if we take the limit, $\Delta t \to 0$, for the corridor probabilities, the result will remain unchanged. Note that ζ_t is an exponential martingale, and can be used to define a new measure, $\mathbb{Q}$, such that

$$\left.\frac{d\mathbb{Q}}{d\mathbb{P}}\right|_{\mathcal{F}_t} = \zeta_t. \tag{2.32}$$

Now let us consider the following question: what is the distribution of the $\mathbb{P}$-Brownian motion, W_t, under the new measure, $\mathbb{Q}$? To answer this question, we evaluate the moment-generating function of W_t under $\mathbb{Q}$:

$$\begin{aligned}
E^{\mathbb{Q}}\left[e^{\lambda W_t}\right] &= E^{\mathbb{P}}\left[\zeta_t e^{\lambda W_t}\right] \\
&= E^{\mathbb{P}}\left[e^{-\gamma W_t - (1/2)\gamma^2 t + \lambda W_t}\right] \\
&= E^{\mathbb{P}}\left[e^{(\lambda - \gamma)W_t}\right]e^{-(1/2)\gamma^2 t} \\
&= e^{(1/2)(\lambda - \gamma)^2 t - (1/2)\gamma^2 t} \\
&= e^{(1/2)\lambda^2 t - \lambda\gamma t}.
\end{aligned} \tag{2.33}$$

The above result indicates that, under $\mathbb{Q}$,

$$W_t \sim N(-\gamma t, t), \tag{2.34}$$

or, equivalently, we can say that $\tilde{W}_t \triangleq W_t + \gamma t$ is a $\mathbb{Q}$-Brownian motion. A formal statement on the relationship between the measures and Brownian motions is given in the following section.

2.2.2 The CMG Theorem

The following theorem states that a Brownian motion with a drift is in fact a standard Brownian motion under a different measure. The Radon–Nikodym derivative can be expressed in terms of the drift, which can be a time-dependent adaptive function.

Theorem 2.2.1 (The CMG Theorem). *Let W_t be a $\mathbb{P}$-Brownian motion and γ_t be an $\mathcal{F}_t$-adaptive process satisfying the Novikov condition,*

$$E^{\mathbb{P}}\left[\exp\left(\frac{1}{2}\int_0^T \gamma_t^2\, dt\right)\right] < \infty. \tag{2.35}$$

Define a new measure, $\mathbb{Q}$, as

$$\left.\frac{d\mathbb{Q}}{d\mathbb{P}}\right|_{\mathcal{F}_t} = \exp\left(\int_0^t -\gamma_s\, dW_s - \frac{1}{2}\gamma_s^2\, ds\right). \tag{2.36}$$

Then $\mathbb{Q}$ is equivalent to $\mathbb{P}$ and

$$\tilde{W}_t = W_t + \int_0^t \gamma_s \, ds \tag{2.37}$$

is a $\mathbb{Q}$-Brownian motion.

Proof: The equivalence is because the Radon–Nikodym derivative is positive. To prove the remaining parts of the theorem, we need to show that $\tilde{W}_t$ satisfies the following three properties under $\mathbb{Q}$:

1. $\tilde{W}_0 = 0$;

2. $\tilde{W}_{t+s} - \tilde{W}_t$ is independent of $\{\tilde{W}_u, 0 \le u \le t\}$; and

3. $\tilde{W}_{t+s} - \tilde{W}_t \sim N(0, s)$.

Properties 1 and 2 are easy to show and thus their proofs are omitted. Let us proceed to establish the third property by checking the moment-generating function of the increment. For any real constant, λ, we have

$$\begin{aligned}
&E^{\mathbb{Q}}\left[\left. e^{\lambda\left(\tilde{W}_{t+s} - \tilde{W}_t\right)} \right| \mathcal{F}_t\right] \\
&= E_t^{\mathbb{P}}\left[\frac{d\mathbb{Q}}{d\mathbb{P}} e^{\lambda\left(\tilde{W}_{t+s} - \tilde{W}_t\right)}\right] \\
&= E_t^{\mathbb{P}}\left[\exp\left(\int_t^{t+s} -\gamma_u dW_u - \frac{1}{2}\gamma_u^2 du + \lambda\left(\tilde{W}_{t+s} - \tilde{W}_t\right)\right)\right] \\
&= E_t^{\mathbb{P}}\left[\exp\left(\int_t^{t+s} -\gamma_u dW_u - \frac{1}{2}\gamma_u^2 du \right.\right. \\
&\qquad\qquad \left.\left. + \lambda\left(W_{t+s} - W_t\right) + \lambda \int_t^{t+s} \gamma_u du\right)\right] \\
&= e^{(1/2)\lambda^2 s} E_t^{\mathbb{P}}\left[\exp\left(\int_t^{t+s} -(\gamma_u - \lambda)\, dW_u - \frac{1}{2}(\gamma_u - \lambda)^2\, du\right)\right] \\
&= e^{(1/2)\lambda^2 s}. \tag{2.38}
\end{aligned}$$

Here $E_t^{\mathbb{P}}[\cdot]$ is the short form of $E^{\mathbb{P}}[\cdot|\mathcal{F}_t]$. Based on the property of independent increments of Brownian motion, we can also show that Equation 2.38 holds for any filtration, $\mathcal{F}_u, u \le t$. The theorem is thus proved. □

We remark here that, conventionally, $E_0^{\mathbb{P}}[\cdot]$ or $E^{\mathbb{P}}[\cdot|\mathcal{F}_0]$ is considered an unconditional expectation, and is simply written as $E^{\mathbb{P}}[\cdot]$.

2.3   The Martingale Representation Theorem

The martingale representation theorem plays a critical role in the so-called martingale approach to derivatives pricing. This theorem has two important consequences. First, it leads to a general principle for derivatives pricing. Second, it implies a replication or hedging strategy of a derivative using its underlying security. We first present a simple version of the theorem based on a single Brownian filtration, $\mathcal{F}_t = \sigma(W_s, 0 \leq s \leq t)$. We begin with a martingale process, M_t, such that

$$dM_t = \sigma_t dW_t, \tag{2.39}$$

and we call σ_t the volatility of M_t.

Theorem 2.3.1 (The Martingale Representation Theorem). *Suppose that N_t is a $\mathbb{Q}$-martingale process that is adaptive to $\mathcal{F}_t$ and satisfies $E^{\mathbb{Q}}[N_T^2] < \infty$ for some T. If the volatility of M_t is non-zero almost surely, then there exists a unique $\mathcal{F}_t$-adaptive process, φ_t, such that $E^{\mathbb{Q}}[\int_0^T \varphi_t^2 \sigma_t^2 \, dt] < \infty$ almost surely, and*

$$N_t = N_0 + \int_0^t \varphi_s \, dM_s, \quad t \leq T, \tag{2.40}$$

or, in differential form,

$$dN_t = \varphi_t \, dM_t. \tag{2.41}$$

A proof combining the techniques of Steele (2000) and Øksendal (2003) is provided in the appendix of this chapter. A different proof can be found in Korn and Korn (2000).

The following lemma characterizes function ϕ_t for the martingale representation theorem. The proof is straightforward and thus is left to readers.

Lemma 2.3.1. *If a martingale N_t is represented by another martingale such that*

$$dN_t = \varphi_t \, dM_t.$$

then φ_t is given by

$$\phi_t = \frac{\langle dN_t, dM_t \rangle}{\langle dM_t, dM_t \rangle}, \tag{2.42}$$

where $\langle \cdot, \cdot \rangle$ stands for covariance.

2.4   A Complete Market with Two Securities

We consider the first "complete market" in continuous time, which consists of a money market account and a risky security. The price processes for the two securities, B_t and S_t, are assumed to be

$$dB_t = r_t B_t \, dt, \qquad\qquad B_0 = 1,$$
$$dS_t = S_t(\mu_t \, dt + \sigma_t \, dW_t), \quad S_0 = S_0.$$

Here, the volatility of the risky asset is $\sigma_t \neq 0$ almost surely, and the short rate, r_t, can be stochastic. Denote the discounted price of the risky asset as $Z_t = B_t^{-1} S_t$, which can be shown to follow the process

$$dZ_t = Z_t \left((\mu_t - r_t)dt + \sigma_t dW_t \right)$$

$$= Z_t \sigma_t d \left(W_t + \int_0^t \frac{(\mu_s - r_s)}{\sigma_s} ds \right). \qquad (2.43)$$

By introducing

$$\gamma_t = \frac{\mu_t - r_t}{\sigma_t}, \qquad (2.44)$$

which is $\mathcal{F}_t$-adaptive, and by defining a new measure, $\mathbb{Q}$, according to Equation 2.36, we have

$$\tilde{W}_t = W_t + \int_0^t \gamma_s \, ds,$$

which is a $\mathbb{Q}$-Brownian motion. In terms of $\tilde{W}_t$, Z_t satisfies

$$dZ_t = \sigma_t Z_t \, d\tilde{W}_t,$$

which is a lognormal $\mathbb{Q}$-martingale. Recall that in the binomial model for option pricing, we also derived the martingale measure for the underlying security.

2.5 Replicating and Pricing of Contingent Claims

Let $H(S_T)$ be a contingent claim (or option) with payoff day or maturity T. The claim is an $\mathcal{F}_T$-adaptive function whose value depends on $\{S_t, 0 \leq t \leq T\}$. Define first a $\mathbb{Q}$-martingale with the discounted payoff:

$$N_t = E^{\mathbb{Q}}(B_T^{-1} H(S_T) \,|\, \mathcal{F}_t).$$

Without loss of generality, we assume that $E^{\mathbb{Q}}[N_t^2] < \infty$. According to the martingale representation theorem, there exists an $\mathcal{F}_t$-adaptive function, φ_t, such that

$$dN_t = \varphi_t \, dZ_t, \qquad (2.45)$$

where Z_t, defined in the last section, is the discounted price of S_t. Next, we define

$$\psi_t = N_t - \varphi_t Z_t. \qquad (2.46)$$

Consider now the portfolio with φ_t units of the stock and ψ_t units of the money market account, denoted as (φ_t, ψ_t). According to the definition of ψ_t, the discount value of the replication portfolio is

$$\tilde{V}_t = \varphi_t Z_t + \psi_t = N_t. \qquad (2.47)$$

This portfolio has two important properties. First, at time T, when the option matures,

$$\tilde{V}_T = N_T = B_T^{-1}H(S_T),$$ (2.48)

which implies that the spot value of the portfolio equals that of the option. In other words, the portfolio replicates the payoff of the contingent claim. Second, the replicating portfolio is a *self-financing* one, meaning that it can track the asset allocation, (φ_t, ψ_t), without the need for either capital infusion or capital withdrawal. In fact, based on Equations 2.45 and 2.47, we have

$$d\tilde{V}_t = dN_t = \varphi_t dZ_t.$$ (2.49)

In terms of the spot value, B_t, S_t and V_t, Equation 2.49 becomes

$$
\begin{aligned}
dV_t &= d(\tilde{V}_t B_t) \\
&= B_t d\tilde{V}_t + \tilde{V}_t dB_t \\
&= B_t \varphi_t dZ_t + (\varphi_t Z_t + \psi_t)dB_t \\
&= \varphi_t(B_t dZ_t + Z_t dB_t) + \psi_t dB_t \\
&= \varphi_t d(B_t Z_t) + \psi_t dB_t \\
&= \varphi_t dS_t + \psi_t dB_t.
\end{aligned}
$$ (2.50)

A direct consequence of the above equation is the equality

$$
\begin{aligned}
\varphi_{t+dt}S_{t+dt} + \psi_{t+dt}B_{t+dt} &= \varphi_t S_t + \psi_t B_t + \varphi_t dS_t + \psi_t dB_t \\
&= \varphi_t S_{t+dt} + \psi_t B_{t+dt},
\end{aligned}
$$ (2.51)

which says that the values of the portfolio before and after rebalancing at time $t + dt$ are equal, and no cash flow is generated when we update the asset allocation from (φ_t, ψ_t) to $(\varphi_{t+dt}, \psi_{t+dt})$. This is what we mean by a self-financing portfolio. By the dominance principle, the value of the contingent claim is nothing but the value of the replication portfolio:

$$V_t = B_t E^{\mathbb{Q}}\left[B_T^{-1}H(S_T)\,|\,\mathcal{F}_t\right] = E^{\mathbb{Q}}\left[e^{-\int_t^T r_s ds}H(S_T)\,|\,\mathcal{F}_t\right].$$ (2.52)

By already knowing the pricing measure, $\mathbb{Q}$, we have thus established a general pricing principle or pricing formula for contingent claims.

By Ito's Lemma, there is also

$$d\tilde{V}_t = \left(\frac{\partial \tilde{V}_t}{\partial t} + \frac{1}{2}\sigma^2(t)Z_t^2\frac{\partial^2 \tilde{V}_t}{\partial Z^2}\right)dt + \frac{\partial \tilde{V}_t}{\partial Z}dZ_t.$$ (2.53)

This equation has two implications. First, the function φ_t for representing $\tilde{V}_t$ using Z_t is

$$\varphi = \frac{\partial \tilde{V}_t}{\partial Z_t} = \frac{\partial V_t}{\partial S_t},$$

according to Lemma 2.3.1. Second, owing to its martingale property of $\tilde{V}_t$ under the risk-neutral measure, the discounted value function $\tilde{V}_t$ satisfies the partial differential equation:

$$\frac{\partial \tilde{V}_t}{\partial t} + \frac{1}{2}\sigma^2(t)Z^2\frac{\partial^2 \tilde{V}_t}{\partial Z^2} = 0. \tag{2.54}$$

When the interest rate r_t is a deterministic function, Equation 2.54 leads to the following PDE for the spot value function V_t:

$$\frac{\partial V_t}{\partial t} + \frac{1}{2}\sigma_t^2 S^2\frac{\partial^2 V_t}{\partial S_t^2} + r_t S\frac{\partial V_t}{\partial S} - r_t V_t = 0, \tag{2.55}$$

which is the celebrated Black–Scholes–Merton equation (Black and Scholes, 1973; Merton, 1973). To evaluate V_t, we may solve Equation 2.55 with the terminal condition

$$V_T(S) = H(S). \tag{2.56}$$

We emphasize here that Equation 2.55 is valid only for the short rate to be a deterministic function of time, so it is not that useful for fixed-income derivatives. Even for deterministic spot rate r_t, it is often more convenient to obtain the value of conventional European options by directly evaluating the expectation (Equation 2.52).

2.6   Multi-Factor Extensions

In derivatives pricing, we often need to simultaneously model the dynamics of multiple risky securities, using multiple risk factors. Because of that, we must extend several major results established so far to the setting of multiple risk sources or assets. These results include the CMG theorem, the martingale representation theorem, and the option pricing formula, as in Equation 2.52. The proofs are parallel to those for the one-dimensional case and thus are omitted for brevity. Hereafter, we use a superscript "T" to denote the transposition of a matrix.

Theorem 2.6.1 (The CMG Theorem). *Let* $\mathbf{W}_t = (W_1(t), W_2(t), \ldots, W_n(t))^{\mathrm{T}}$ *be an n-dimensional* $\mathbb{P}$-*Brownian motion, and let* $\boldsymbol{\gamma}_t = (\gamma_1(t), \gamma_2(t), \ldots, \gamma_n(t))^{\mathrm{T}}$ *be an n-dimensional* $\mathcal{F}_t$-*adaptive process, such that*

$$E^{\mathbb{P}}\left[\exp\left(\frac{1}{2}\int_0^T \|\boldsymbol{\gamma}_t\|_2^2\, dt\right)\right] < \infty.$$

Define a new measure, $\mathbb{Q}$*, with a Radon–Nikodym derivative*

$$\left.\frac{d\mathbb{Q}}{d\mathbb{P}}\right|_{\mathcal{F}_t} = \exp\left(\int_0^t -\boldsymbol{\gamma}_s^{\mathrm{T}}\, d\mathbf{W}_s - \frac{1}{2}\int_0^t \|\boldsymbol{\gamma}_s\|_2^2\, ds\right). \tag{2.57}$$

Then $\mathbb{Q}$ *is equivalent to* $\mathbb{P}$, *and*

$$\tilde{\mathbf{W}}_t = \mathbf{W}_t + \int_0^t \boldsymbol{\gamma}_s \, ds$$

is an n-*dimensional* $\mathbb{Q}$-*Brownian motion.*

Theorem 2.6.2 (The Martingale Representation Theorem). *Let* $\mathbf{W}_t$ *be an* n-*dimensional Brownian motion and suppose that* $\mathbf{M}_t$ *is an* n-*dimensional* $\mathbb{Q}$-*martingale process,* $\mathbf{M}_t = (M_1(t), M_2(t), \dots, M_n(t))^{\mathrm{T}}$, *such that*

$$dM_i(t) = \sum_{j=1}^n a_{ij}(t) \, dW_j(t).$$

Let $\mathbf{A} = (a_{ij})$ *be a non-singular matrix. If* N_t *is any one-dimensional* $\mathbb{Q}$-*martingale with* $E^{\mathbb{Q}}[N_t^2] < \infty$, *there exists an* n-*dimensional* $\mathcal{F}_t$-*adaptive process,* $\Phi_t = (\varphi_1(t), \varphi_2(t), \dots, \varphi_n(t))^{\mathrm{T}}$, *such that*

$$E^{\mathbb{Q}}\left[\int_0^t \left(\sum_j a_{ij}^2(s) \varphi_j^2(s) \, ds \right) \right] < \infty, \quad \forall i, \tag{2.58}$$

and

$$N_t = N_0 + \sum_{j=1}^n \int_0^t \varphi_j(s) \, dM_j(s)$$

$$\stackrel{\Delta}{=} N_0 + \int_0^t \Phi^{\mathrm{T}}(s) \, d\mathbf{M}(s).$$

2.7 A Complete Market with Multiple Securities

Based on the content introduced in the previous sections, we are ready to address the pricing of a contingent claim whose value depends on the values of n risky securities. Similar to the pricing of options on a single asset, the pricing of options on multiple assets consists of two steps: the construction of a martingale measure for the assets and the construction of the replication strategy.

2.7.1 Existence of a Martingale Measure

We consider a standard model of a complete financial market with a money market account and n risky securities. Let the time t prices be B_t and S_t^i,

$1 \le i \le n$, respectively. We assume lognormal price processes for all assets:

$$dB_t = r_t B_t dt,$$

$$dS_t^i = S_t^i \left(\mu_t^i dt + \sum_{j=1}^{n} \sigma_{ij} dW_j(t) \right) \tag{2.59}$$

$$= S_t^i \left(\mu_t^i dt + \boldsymbol{\sigma}_i^{\mathrm{T}}(t) d\mathbf{W}_t \right), \quad i = 1, 2, \ldots, n.$$

Here,

$$\boldsymbol{\sigma}_i^{\mathrm{T}}(t) = (\sigma_{i,1}, \sigma_{i,2}, \ldots, \sigma_{i,n}).$$

Let $Z_t^i = B_t^{-1} S_t^i$ denote the discounted asset price of the ith asset. It then follows that

$$dZ_t^i = Z_t^i \left[\boldsymbol{\sigma}_i^{\mathrm{T}}(t) d\mathbf{W}_t + \left(\mu_t^i - r_t \right) dt \right], \quad i = 1, 2, \ldots, n. \tag{2.60}$$

To construct a martingale measure for $Z_t^i, \forall i$, we must "absorb" the drift terms in Equation 2.60 into the Brownian motion. For that reason, we define an $\mathcal{F}_t$-adaptive function, $\boldsymbol{\gamma}_t$, via the following equations:

$$\boldsymbol{\sigma}_i^{\mathrm{T}}(t) \boldsymbol{\gamma}_t = \mu_t^i - r_t, \quad i = 1, 2, \ldots, n. \tag{2.61}$$

Suppose that $\boldsymbol{\gamma}_t$, the solution to Equation 2.61, exists and satisfies

$$E^{\mathbb{P}} \left[\exp \left(\int_0^T \|\boldsymbol{\gamma}_t\|^2 dt \right) \right] < \infty \tag{2.62}$$

for some $T > 0$. We then can define a new measure, $\mathbb{Q}$, according to Equation 2.57. Under this newly defined $\mathbb{Q}$,

$$\tilde{\mathbf{W}}_t = \mathbf{W}_t + \int_0^t \boldsymbol{\gamma}_s \, ds$$

is a multi-dimensional Brownian motion, with which we can rewrite the price processes for the discounted assets into

$$dZ_t^i = Z_t^i \boldsymbol{\sigma}_i^{\mathrm{T}}(t) d\tilde{\mathbf{W}}_t, \quad i = 1, 2, \ldots, n,$$

and $Z_t^i, i = 1, 2, \ldots, n$ are lognormal $\mathbb{Q}$-martingales.

We now study the existence of $\boldsymbol{\gamma}_t$ and condition 2.62. In matrix form, Equation 2.61 can be recast into

$$\boldsymbol{\Sigma} \boldsymbol{\gamma}_t = \boldsymbol{\mu}_t - r_t \mathbf{I}, \tag{2.63}$$

where

$$\boldsymbol{\Sigma} = \begin{pmatrix} \boldsymbol{\sigma}_1^{\mathrm{T}} \\ \boldsymbol{\sigma}_2^{\mathrm{T}} \\ \vdots \\ \boldsymbol{\sigma}_n^{\mathrm{T}} \end{pmatrix}, \quad \boldsymbol{\mu}_t = \begin{pmatrix} \mu_t^1 \\ \mu_t^2 \\ \vdots \\ \mu_t^n \end{pmatrix}, \quad \text{and} \quad \mathbf{I} = \begin{pmatrix} 1 \\ 1 \\ \vdots \\ 1 \end{pmatrix}.$$

Consider first the case when the inverse of the volatility matrix, $\mathbf{\Sigma}^{-1}$, exists and is bounded. Then we can determine $\boldsymbol{\gamma}_t$ uniquely as

$$\boldsymbol{\gamma}_t = \mathbf{\Sigma}^{-1}(\boldsymbol{\mu}_t - r_t\mathbf{I}),$$

and the boundedness condition 2.62 is obviously satisfied as well. The condition of non-singularity of $\mathbf{\Sigma}$ is, however, sufficient but not necessary for the existence of $\boldsymbol{\gamma}_t$.

Next, we show that the absence of arbitrage is a sufficient condition for the existence of $\boldsymbol{\gamma}_t$ as a solution to Equation 2.61. We may interpret the notion of the absence of arbitrage as follows: any riskless portfolios will earn a return equal to the risk-free rate. Let $\{\theta_i\}$ be a portfolio, where θ_i is the number of units in the ith asset. If the portfolio is a riskless one, then, by the absence of arbitrage, the return of the portfolio is zero; that is,

$$\mathrm{d}\left(\sum_i \theta_i Z_t^i\right) = \left(\sum_i \theta_i Z_t^i \left(\mu_t^i - r_t\right)\right)\mathrm{d}t + \left(\sum_i \theta_i Z_t^i \boldsymbol{\sigma}_i^{\mathrm{T}}\right)\mathrm{d}\mathbf{W}_t = 0.$$

The last equation implies that, as long as

$$\left(\theta_1 Z_t^1, \ldots, \theta_n Z_t^n\right)\begin{pmatrix}\boldsymbol{\sigma}_1^{\mathrm{T}} \\ \vdots \\ \boldsymbol{\sigma}_n^{\mathrm{T}}\end{pmatrix} = \left(\theta_1 Z_t^1, \ldots, \theta_n Z_t^n\right)\mathbf{\Sigma} = (0, \ldots, 0),$$

there will be

$$\left(\theta_1 Z_t^1, \ldots, \theta_n Z_t^n\right)\begin{pmatrix}\mu_t^1 - r_t \\ \vdots \\ \mu_t^n - r_t\end{pmatrix} = 0.$$

The statements above imply that $(\mu_t^1 - r_t, \ldots, \mu_t^n - r_t)^{\mathrm{T}}$ must lie in the linear space spanned by the columns of $\mathbf{\Sigma}$, meaning that there exists a coefficient vector, $\boldsymbol{\gamma}_t = (\gamma_t^1, \ldots, \gamma_t^n)$, such that

$$\boldsymbol{\mu}_t - r_t\mathbf{I} = \mathbf{\Sigma}\boldsymbol{\gamma}_t. \tag{2.64}$$

In terms of components, Equation 2.64 reads as follows:

$$\mu_t^i - r_t = \sum_{j=1}^n \sigma_{i,j}\gamma_t^j, \quad 1 \leq i \leq n. \tag{2.65}$$

The above arguments justify the existence of $\boldsymbol{\gamma}_t$ in the absence of arbitrage.

The solution, $\boldsymbol{\gamma}_t$, to Equation 2.64 is not unique unless $\mathbf{\Sigma}$ is nonsingular almost surely. Regardless of if $\mathbf{\Sigma}$ is singular, we can always find a bounded solution of $\boldsymbol{\gamma}_t$ provided that all non-zero singular values of $\mathbf{\Sigma}$ stay away from zero, thus satisfying the Novikov condition, Equation 2.62. When $\mathbf{\Sigma}$ is non-singular, we say that the market is non-degenerate. In a financial market, non-degeneracy means that none of the securities can be dynamically replicated by other securities, and thus none of them is redundant. Note that when there is only one risky security, $n = 1$, Equation 2.65 reduces to Equation 2.44.

In the finance literature, the components of $\boldsymbol{\gamma}_t$ are called the market prices of risks and each of these components can be interpreted as the excess of returns per unit of risk. Under the objective measure, $\mathbb{P}$, the discounted price process of an asset is not necessarily a martingale process. This means that trading an asset at the market price may not be a fair game in the sense that the expected return does not equal to the risk-free rate. In fact, empirical studies often suggest $\mu_t^i > r_t, \forall i$, which reflects an important reality of our financial markets where typical investors are risk averse and demand a premium for taking risks.

We finish this section with the remark that the market prices of risks of any two risky assets depending on a single risk source are equal. To see this, we let the prices of two tradable assets, S_t^1 and S_t^2, be driven by the same Brownian motion:

$$dS_t^i = S_t^i(\mu_t^i dt + \sigma_t^i dW_t), \quad i = 1, 2.$$

According to the no-arbitrage principle, there is an $\mathcal{F}_t$-adaptive process, γ_t, such that

$$\mu_t^i - r_t = \sigma_t^i \gamma_t, \quad i = 1, 2. \tag{2.66}$$

It follows that

$$\frac{\mu_t^1 - r_t}{\sigma_t^1} = \frac{\mu_t^2 - r_t}{\sigma_t^2} = \gamma_t. \tag{2.67}$$

In much of the literature, Equation 2.67 is used as a starting point for the derivation of the Black–Scholes–Merton equation.

2.7.2   Pricing Contingent Claims

Now we are ready to address the pricing of a contingent claim depending on the prices of multiple underlying securities. Having found the martingale measure, $\mathbb{Q}$, for the underlying securities, we define a $\mathbb{Q}$-martingale as

$$N_t = E^{\mathbb{Q}}(B_T^{-1} H(\boldsymbol{S}_T) \mid \mathcal{F}_t),$$

using the discounted value of $H(\boldsymbol{S}_T)$, the payoff function of the claim at time T. Without loss of generality, we assume that the volatility matrix of the underlying risky securities, $\boldsymbol{\Sigma}$, is non-singular.[2] According to the martingale representation theorem, there exists an $\mathcal{F}_t$-adaptive function, $\boldsymbol{\Phi}_t = (\varphi_1(t), \ldots, \varphi_n(t))^{\mathrm{T}}$, such that

$$dN_t = \boldsymbol{\Phi}_t^{\mathrm{T}} d\mathbf{Z}_t,$$

where $\mathbf{Z}_t$ is the vector of the discounted prices. We now define another process,

$$\psi_t = N_t - \boldsymbol{\Phi}_t^{\mathrm{T}} \mathbf{Z}_t,$$

[2]Otherwise, the market is degenerate and the replication portfolio is not unique.

and form a portfolio with ψ_t units of the money market account and $\phi_i(t)$ units of the ith risky security, $i = 1, \ldots, n$. The discounted value of the portfolio is

$$\tilde{V}_t = \mathbf{\Phi}_t^{\mathrm{T}} \mathbf{Z}_t + \psi_t = N_t. \tag{2.68}$$

The last equation implies replication of the payoff of the contingent portfolio. Furthermore, from Equation 2.68, we can derive

$$dV_t = \mathbf{\Phi}_t^{\mathrm{T}} d\mathbf{S}_t + \psi_t dB_t,$$

which implies that the replication strategy is a self-financing one. So, we conclude that the value of the contingent claim equals that of the portfolio and thus is given by

$$V_t = B_t E^{\mathbb{Q}} \big[B_T^{-1} H(\boldsymbol{S}_T) \,|\, \mathcal{F}_t \big] = E^{\mathbb{Q}} \Big[\mathrm{e}^{- \int_t^T r_s \mathrm{d}s} H(\boldsymbol{S}_T) \,|\, \mathcal{F}_t \Big]. \tag{2.69}$$

Formally, Equation 2.69 is identical to Equation 2.52, the formula for options on a single underlying security.

2.8 Notes

This chapter provides a rather comprehensive introduction to arbitrage pricing theory in a complete market, a market where the prices of risky assets are driven by lognormal processes. We have shown that the existence of a unique martingale measure that is equivalent to the physical measure is both necessary and sufficient for the absence of arbitrage in the complete market. We emphasize that the mathematical underpinning fundamental to arbitrage pricing is the martingale representation theory. For a discussion of arbitrage pricing theory in incomplete markets, we refer readers to Harrison and Kreps (1979) and Harrison and Pliska (1981).

Exercises

1. Check if the following stochastic processes are $\mathbb{P}$-martingales.

$$X_t = W_t^2 - t,$$
$$N(t) = W_t^3 - 3tW_t,$$
$$S_t = S_0 \mathrm{e}^{\gamma W_t - (1/2)\gamma^2 t},$$
$$S_t = S_0 \mathrm{e}^{\int_0^t \sigma(s) \mathrm{d}W_s - (1/2)\sigma^2(s) \mathrm{d}s}.$$

Here W_t is a $\mathbb{P}$-Brownian motion.

2. Numerically verify the CMG theorem. First, simulate 1000 terminal values of W_t for $t = 1$, denote the simulated values as $W^{(i)}, i = 1, \ldots, 1000$, and then do the following:

    (a) Compute the mean and variance of $\{W^{(i)}\}_{i=1}^{1000}$ using the uniform probability (or weight), $1/1000$.

    (b) For $\gamma = 0.2$, compute the mean and variance of the Brownian motion with a drift, $\{W^{(i)} + \gamma\}_{i=1}^{1000}$, with non-uniform weights $1/1000 \times \exp\{-\gamma W^{(i)} - \frac{1}{2}\gamma^2\}$.

   Discuss your results.

3. The moment-generating function of a normal random variable, $X \sim N(0, \sigma^2)$, is
$$E\left[e^{\lambda X}\right] = e^{(1/2)\lambda^2 \sigma^2},$$
   where λ is a constant. Let W_t be a $\mathbb{P}$-Brownian motion. Prove that
$$X_t = \int_0^t \sigma(s)\, dW_s \sim N\left(0, \int_0^t \sigma^2(s)\, ds\right)$$
   by checking its moment-generating function.

4. Let W_t be a $\mathbb{P}$-Brownian motion and γ_t be a $\mathcal{F}_t$-adaptive process, such that
$$E^{\mathbb{P}}\left[\exp\left(\int_0^t \gamma_s^2\, ds\right)\right] < \infty, \quad \forall t > 0.$$
   Define a new measure, $\mathbb{Q}$, such that
$$\left.\frac{d\mathbb{Q}}{d\mathbb{P}}\right|_{\mathcal{F}_t} = \exp\left(\int_0^t -\gamma_s\, dW_s - \frac{1}{2}\gamma_s^2\, ds\right).$$
   Let
$$\tilde{W}_t = W_t + \int_0^t \gamma_s\, ds.$$
   Prove that, under $\mathbb{Q}$,

    (a) for any $t \geq 0$ and $a \geq 0$, $\tilde{W}_{t+a} - \tilde{W}_t$ is independent of $\left\{\tilde{W}_s, s \leq t\right\}$;

    (b) $\tilde{W}_{t+a} - \tilde{W}_t \sim N(0, a)$ by checking the moment-generating function.

5. Let $\tilde{X}_t = X_t/B_t$ and $\tilde{S}_t = S_t/B_t$ be the discounted price of two assets. If
$$d\tilde{X}_t = \phi_t\, d\tilde{S}_t,$$
   prove that
$$dX_t = \phi_t\, dS_t + \psi_t\, dB_t$$
   for $\psi_t = X_t - \phi_t S_t$.

6. Let $\mathbf{Z}(t) = (Z_1(t), Z_2(t), \ldots, Z_n(t))^{\mathrm{T}} \in \mathbf{R}^n$ be an independent $\mathbb{P}$-Brownian motion and $\mathbf{f}(t) = (f_1(t), f_2(t), \ldots, f_n(t))^{\mathrm{T}} \in \mathbf{R}^n$ be a $\mathcal{F}_t$-adaptive function with respect to $\mathbf{Z}(t)$. Define the stochastic integral,

$$\int_0^t \mathbf{f}^{\mathrm{T}}(s)\,\mathrm{d}\mathbf{Z}(s) = \sum_{i=1}^n \int_0^t f_i(s)\,\mathrm{d}Z_i(s).$$

Prove that

$$E^{\mathbb{P}}\left[\int_0^t \mathbf{f}^{\mathrm{T}}(s)\,\mathrm{d}\mathbf{Z}(s)\right] = 0,$$

$$E^{\mathbb{P}}\left[\left(\int_0^t \mathbf{f}^{\mathrm{T}}(s)\,\mathrm{d}\mathbf{Z}(s)\right)^2\right] = E^{\mathbb{P}}\left[\int_0^t \|\mathbf{f}(s)\|_2^2\,\mathrm{d}s\right].$$

7. For $\mathcal{F}_t$-adaptive functions, $c(t), \boldsymbol{\alpha}(t) = (\alpha_1(t), \ldots, \alpha_n(t))^{\mathrm{T}}$, and independent $\mathbb{P}$-Brownian motion, $\mathbf{Z}(t) = (Z_1(t), Z_2(t), \ldots, Z_n(t))^{\mathrm{T}} \in \mathbf{R}^n$, define

$$X_t = \exp\left(\int_0^t c(s)\,\mathrm{d}s + \boldsymbol{\alpha}^{\mathrm{T}}(s)\,\mathrm{d}\mathbf{Z}(s)\right).$$

Prove that

$$\mathrm{d}X_t = \left(c(t) + \frac{1}{2}\|\boldsymbol{\alpha}(t)\|_2^2\right)X_t\mathrm{d}t + X_t\boldsymbol{\alpha}^{\mathrm{T}}(t)\,\mathrm{d}\mathbf{Z}(t).$$

For what function of $c(t)$ is the random process X_t an exponential martingale? Why?

8. (CMG theorem in n-dimensional space) Consider a probability measure, $\mathbb{P}$, on the space of paths $\mathbf{Z}(t) = (Z_1(t), Z_2(t), \ldots, Z_n(t))^{\mathrm{T}} \in \mathbf{R}^n, t \leq T$, such that $\mathbf{Z}(t)$ is a vector of an independent Brownian motion. Assume that $\boldsymbol{\lambda}(t)$ is a vector of $\mathcal{F}_t$-adaptive functions and set

$$M(t) = \exp\left(\int_0^t \boldsymbol{\lambda}^{\mathrm{T}}(s)\,\mathrm{d}\mathbf{Z}(s) - \frac{1}{2}\|\boldsymbol{\lambda}(s)\|_2^2\,\mathrm{d}s\right), \quad \forall t \leq T.$$

Define a new measure, $\mathbb{Q}$, such that

$$\left.\frac{\mathrm{d}\mathbb{Q}}{\mathrm{d}\mathbb{P}}\right|_{\mathcal{F}_t} = M(t), \quad \forall t \geq 0.$$

Prove that the random processes

$$W_j(t) = Z_j(t) - \int_0^t \lambda_j(s)\,\mathrm{d}s, \quad 1 \leq j \leq n$$

are independent Brownian motions under the measure $\mathbb{Q}$.

Appendix: The Martingale Representation Theorem

We first prove a slightly more restricted version of the martingale representation theorem, where one of the martingale processes is a Brownian motion.

Theorem 2.8.1. *Suppose that N_t is a $\mathbb{P}$-martingale such that $E^{\mathbb{P}}[N_t^2] < \infty, t \leq T$ and $N_0 = 0$. Then, there is a unique $\mathcal{F}_t$-adaptive function $\varphi(\omega, t) \in H^2[0, T]$, such that*

$$N_t = \int_0^t \varphi(\omega, s)\, dW_s, \quad \forall 0 \leq t \leq T. \tag{2.70}$$

To prove this theorem, we need a series of lemmas. But, first of all, we prove the uniqueness ahead of the existence.

Suppose that N_t can be expressed in terms of another function, $\psi(\omega, t)$. Then, there will be

$$0 = \int_0^t (\varphi(\omega, s) - \psi(\omega, s))\, dW_s, \quad \forall 0 \leq t \leq T. \tag{2.71}$$

By Ito's isometry, there is

$$0 = E^{\mathbb{P}}\left[\left(\int_0^t (\varphi(\omega, s) - \psi(\omega, s))\, dW_s\right)^2\right]$$

$$= \int_0^t E^{\mathbb{P}}\left[(\varphi(\omega, s) - \psi(\omega, s))^2\right] dt. \tag{2.72}$$

Hence,

$$\varphi(\omega, s) = \psi(\omega, s) \quad \text{in } H^2[0, T]. \tag{2.73}$$

The uniqueness is thus proved.

Lemma 2.8.1. *Let S be the set of all linear combinations of random variables of the form*

$$e^{\int_0^T h\, dW_t - \frac{1}{2}\int_0^T h^2\, dt}, \quad h \in L^2[0, T], \tag{2.74}$$

then S is dense in $L^2(\emptyset, \mathcal{F}_T, \mathbb{P})$.

To prove this lemma we need

Lemma 2.8.2. *If $\mathcal{D}$ is a closed linear subspace of $L^2(d\mathbb{P})$, an indefinite dimensional vector space, and $S \subset \mathcal{D}$, then*

$$\mathcal{D} \cap S^{\perp} = \emptyset, \quad \Rightarrow \bar{S} = \mathcal{D}. \tag{2.75}$$

Proof: For an arbitrary vector $f \in \mathcal{D}$ we have, uniquely, $f = g + h$, where $g \in \mathcal{S}$ and $h \in \mathcal{S}^\perp$. Since $\mathcal{D}$ is closed and $\mathcal{S} \subset \mathcal{D}$, there are $f - g = h \in \mathcal{D}$ and $f - g = h \in \mathcal{S}^\perp$. The hypothesis $\mathcal{D} \cap \mathcal{S}^\perp = 0$ implies $h = 0$ and $f = g \in \bar{\mathcal{S}}$, and it follows that $\bar{\mathcal{S}} = \mathcal{D}$.

[**Proof of Lemma 2.8.1**]: Let $g \in L^2(\emptyset, \mathcal{F}_T, \mathbb{P}) \cap \mathcal{S}^\perp$, we want to show that $g = 0$. We take a set $\mathcal{T} = \{t_i\}_{i=0}^n$ such that $0 = t_0 < t_1 < \ldots < t_n = T$ and consider a simple choice of simple function $h = \sum_{j=1}^n u_j 1_{(t_{j-1}, t_j]}$ for any $\mathbf{u} = (u_1, u_2, \ldots, u_n) \in \mathbb{R}^n$, then there is

$$f(\mathbf{u}) = E^P \left[g e^{u_1 W_{t_1} + u_2(W_{t_2} - W_{t_1}) + \ldots + u_n(W_{t_n} - W_{t_{n-1}})} \right] = 0.$$

The left-hand side is analytical in $\mathbf{u}$ and hence has analytic extension to $\mathbf{u} \in \mathbb{C}^n$. Since $f(\mathbf{u})$ is analytic and vanishes on the real axis, it is zero for $\mathbf{u} \in \mathbb{C}^n$, in particular, there is

$$E^P \left[g e^{i(y_1 W_{t_1} + y_2(W_{t_2} - W_{t_1}) + \ldots + y_n(W_{t_n} - W_{t_{n-1}}))} \right] = 0.$$

Let $\phi \in C_0^\infty(\mathbb{R}^n)$, with Fourier transform

$$\hat{\phi}(\mathbf{y}) = (2\pi)^{-\frac{n}{2}} \int_{\mathbb{R}^n} \phi(\mathbf{x}) e^{-i\mathbf{x} \cdot \mathbf{y}} d\mathbf{x}.$$

Then,

$$E^P \left[g\phi(W_{t_1}, W_{t_2} - W_{t_1}, \ldots, W_{t_n} - W_{t_{n-1}}) \right]$$
$$= (2\pi)^{-\frac{n}{2}} \int_{\mathbb{R}^n} \hat{\phi}(\mathbf{y}) E^P \left[g e^{i(y_1 W_{t_1} + y_2(W_{t_2} - W_{t_1}) + \ldots + y_n(W_{t_n} - W_{t_{n-1}}))} \right] dy = 0.$$

The above equality implies that $g = 0$. According to Lemma 2.8.2, we conclude that $\mathcal{S}$ is dense in $L^2(\emptyset, \mathcal{F}_T, \mathbb{P})$ $\qquad \square$

Lemma 2.8.3. *Let* $F \in L^2(\emptyset, \mathcal{F}_T, \mathbb{P})$, *then there is* $f(t, \omega) \in H^2[0, T]$ *such that*

$$F_T = E[F_T] + \int_0^T f(t, \omega) dW_t. \tag{2.76}$$

Proof: First we prove that result for F of the form in Equation 2.74. Define $F_t = e^{\int_0^T h dW_t - \frac{1}{2} \int_0^T h^2 dt}$, then

$$dF_t = hF_t dW_t, \quad F_0 = 1,$$

so that

$$F_T = 1 + \int_0^T hF_t dW_t.$$

For functions F which are linear combinations of functions of the form in Equation 2.74, a similar result follows by linearity. Because $\mathcal{S}$ in dense in

$L^2(\emptyset, \mathcal{F}_T, \mathbb{P})$, for any $F \in L^2(\emptyset, \mathcal{F}_T, \mathbb{P})$, there exists a sequence of $F_n \in \mathcal{S}$ such that $F_n \to F$ while

$$F_n = E[F_n] + \int_0^T f_n dW_t, \qquad (2.77)$$

for some $f_n \in H^2[0,T]$. Since $\{F_n\}$ are a Cauchy sequence in $L^2(\emptyset, \mathcal{F}_T, \mathbb{P})$, there is

$$E\left[(F_n - F_m)^2\right] = (E[F_n - F_m])^2 + \int_0^T (f_n - f_m)^2 dt \to 0, \quad \text{as } n, m \to \infty,$$

which implies $E[F_n] \to E[F]$ and $\{f_n\}$ is a Cauchy sequence in $H^2[0,T]$. Since $H^2[0,T]$ is a Hilbert space, there is an $f \in H^2[0,T]$ such that $f_n \to f$. Taking the limit for $n \to \infty$ in Equation 2.77 we complete the proof. □

When F takes the martingale process N_T in Theorem 2.3.1 with $N_0 = E[N_T] = 0$, Equation 2.76 becomes

$$N_T = \int_0^T \phi(\omega, s) dW_s.$$

By taking expectation conditional on $\mathcal{F}_t$, we obtain the desired result. Finally, based on Theorem 2.3.1, we can proceed to the proof of Lemma 2.3.1.

[**Proof** for Lemma 2.3.1] According to the statements of the theorem, there are $\mathcal{F}_t$-adaptive functions, σ and σ_N, in $H^2[0,T]$, such that

$$\begin{aligned} dM_t &= \sigma \, dW_t \\ dN_t &= \sigma_N \, dW_t. \end{aligned} \qquad (2.78)$$

If $\sigma \neq 0$ almost surely, we let $\varphi = \sigma_N / \sigma$ and then write

$$dN_t = \varphi\sigma \, dW_t = \varphi \, dM_t. \qquad (2.79)$$

By definition, $\varphi\sigma = \sigma_N \in H^2[0,T]$. The proof is completed. □

Chapter 3

U.S. Fixed-Income Markets

The U.S. fixed-income market is undoubtfully the dominant and most sophisticated piece of the global financial market. There are a number of sectors that constitute the U.S. fixed-income market: the Federal fund market, the bond market, the interbank lending market and the interbank derivatives market. The bond market is further divided into three sub-sectors: the sovereign bond market, the agency bond market and the corporate bond market. Owing to the dominant role of the U.S. economy as well as of the U.S. dollar for world trades, these U.S. sectors are the largest in size. Next, we will introduce these markets one by one.

3.1   The U.S. Federal Fund Market

Federal funds, or Fed funds for short, refer to reserves that commercial banks and other financial institutions deposit at the regional Federal Reserve banks; these funds can be lent to other financial institutions that have insufficient cash on hand to meet their lending and reserve needs. In 2023, the volume of Fed fund trades each day ranges between \$60 to 140 billion. The loans are unsecured and are made at a relatively low-interest rate, called the Federal funds rate, which is an overnight rate for a lending period of one day, the period with which most such loans are made.

The Fed fund rate is essentially determined by the Federal Open Market Committee (FOMC), the policy making body of the Federal Reserve System. Specifically, FOMC sets a target range for the Fed funds rate in routine FOMC meetings over each year. The setting of the target range is based on economic and monetary conditions. Consequently, the changes of target range affect broad economic conditions in the country, including inflation, growth and employment, so that the power to set the Fed fund rate is regarded as the major policy tool of the FOMC. The increase or decrease of the Fed fund rate is an indication of tightening or easing of money supply by the FOMC. Its impacts are immediately felt by the fixed-income market, Federal fund market and beyond. It is fair to state that the Fed funds rate is one of the most important interest rates for the U.S. economy.

The FOMC, however, cannot force banks to charge any specific Fed fund rate within the target range. The actual interest rate a lending bank will charge

DOI: 10.1201/9781003389101-3

is determined through supply/demand or negotiations between the two banks. The FOMC, however, can affect the Fed fund rate for transactions through direct open market operations by acting as a major player of lending or borrowing. At the end of the trading day, the weighted average of interest rates across all transactions of this type is calculated and such an average is called the effective Federal funds rate (EFFR). Figure 3.1 displays the EFFR from 1995 - 2023.

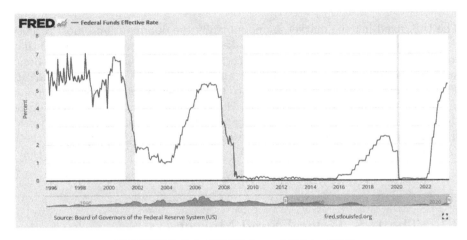

FIGURE 3.1: EFFR from 1995 to 2023.

The EFFR serves as an index for Fed fund derivatives, including Fed fund futures and overnight-index swaps. The Fed Fund futures are futures on one-month EFFR over the nearest sixty calendar months, i.e., the Fed fund futures have maturities up to five years. At the maturity of a Fed Fund futures, the settlement rate is calculated by taking the arithmetic average of the daily EFFR over the past calendar month, which is the reference period of the futures. On the EFFR, there are also overnight index swaps (OIS). The trading of Fed fund futures started around 1990. In an OIS, the fixed leg makes periodic interest payments for a certain notional value based on a fixed rate, while the floating leg makes interest payments accrued through daily compounding using certain overnight rate. For a long time, since late 1990s and until 2018, the EFFR had been the only choice of the overnight rate for OIS trades. The alternative overnight rate is secured overnight financing rate (SOFR), which will be introduced in the next chapter.

3.1.1 Fed Fund Futures

The one-month or 30-Day Fed fund futures and futures options are the most widely used tools for hedging exposure to Federal fund rate, and they are the direct reflection of collective market insight on the future course of FOMC's monetary policy. Next, let us take a close look at the Federal fund futures.

First is the contract specification of the Federal fund futures.

Contract Unit: $4,167 \times$ Contract IMM Index
Price Quotation: Contract IMM Index $=100(1 - R)$

Here, IMM is for International Monetary Market and R is the arithmetic average of daily effective Federal fund rate during the contract month:

$$R = \frac{1}{N_c} \sum_{i=1}^{N_c} f_i,$$

where N_c denotes the number of calendar days in the contract month, and f_i is the annualized EFFR of the ith day. The P&L from investing in a Fed fund futures is marked to market on the daily basis. From one day to the next, the mark to market P&L is given by

$$\begin{aligned} \text{P\&L} &= 4,167 \times 100(1 - R_1) - 4,167 \times 100(1 - R_0) \\ &= 416,700 \times (R_0 - R_1) \\ &= 5,000,000 \times \frac{1}{12} \times (R_0 - R_1). \end{aligned}$$

The expression above can be interpreted as follows. The notional value of the Fed fund futures is $5 millions, being long a Fed fund futures means being short the Fed fund rate, and one basis increase (decrease) of Fed fund rate results in a loss (gain) of $41.67.

Example 3.1. *An investor is long a Fed fund futures for the quote price of 92.75, signifying an average daily Fed fund rate of 7.25% per annum. If in the next day, the price changes to 92.55, then the investor will have a*

$$\text{P\&L} = \$4,167 \times (92.55 - 92.75) = -\$833.4,$$

which is a loss to the investor.

Fed fund futures are naturally a tool to hedge against exposures to Fed fund rate, and it also serves other purposes. Here we highlight perhaps the most important applications of the Fed fund futures: to predict the course forward of the Fed fund rate. Note that there are nine regular FOMC meetings within a year, when the target range of the Fed fund rate forward is reset. To learn how Fed fund futures predict the outcome of the reset, we start with, for example, the quotation of Fed fund futures as of 13 Sep 2023 for the first three nearest months, shown in Table 3.1.

It is a convention that the FOMC changes the Fed fund rate in units of 25 basis points. Suppose that in the Tth month there is a FOMC meeting for rate resetting. The market convention is to estimate the potential rate

TABLE 3.1: Quotation of Fed fund futures for maturities of Sep., Oct. and Nov., 2023

🔒 ⭕ AUTO-REFRESH IS OFF　　*Last Updated 13 Sep 2023 08:41:48 PM CT.　Market data is delayed by at least 10 minutes.*

MONTH	OPTIONS	CHART	LAST	CHANGE	PRIOR SETTLE	OPEN	HIGH	LOW	VOLUME	UPDATED
SEP 2023 ZQU3	OPT	ıl	94.665	UNCH (UNCH)	94.665	94.665	94.665	94.665	3	20:22:29 CT 13 Sep 2023
OCT 2023 ZQV3	OPT	ıl	94.66	UNCH (UNCH)	94.66	94.66	94.66	94.66	204	20:22:29 CT 13 Sep 2023
NOV 2023 ZQX3	OPT	ıl	94.57	+0.005 (+0.01%)	94.565	94.57	94.575	94.57	387	20:22:29 CT 13 Sep 2023

(Source: https://www.cmegroup.com/markets/interest-rates/stirs/30-day-federal-fund.quotes.html#venue=globex).

change by taking the difference between the beginning-of-the-month and the end-of-the-month EFFR rates for the Tth month:

$$\text{Change} = \text{EFFR(End)}_T - \text{EFFR(Start)}_T.$$

While a positive Change points to a tendency of rate hike, a negative Change points to a tendency of reduction. And the absolute value of $n_h = \lfloor \text{Change}/0.25\% \rfloor$ gives the amount of possible rate hike/reduction in units of 25 basis points, where $\lfloor x \rfloor$ is the integer floor of x. The excess value of $|\text{Change}/0.25\%|$ over $|n_h|$ defines the probability for a rate hike/reduction to be more than $|n_h|$ unit of 25 basis point:

$$P(\text{Change more than } |n_h| \text{ unis of 25 bps}) = \left| \frac{\text{Change}}{0.25\%} - n_h \right|,$$

while the probability of a rate change of $|n_h|$ units of 25 basis point is defined by

$$P(\text{Change of } |n_h| \text{ unis of 25 bps})$$
$$= 1 - P(\text{Change more than } |n_h| \text{ unis of 25 bps}).$$

To calculate EFFR(End)_T and EFFR(Start)_T, we make use of the Fed fund futures rate of the month, EFFR(Avg)_T, as well as the Fed fund futures rate for the previous month, EFFR(End)_{T-1}, or the following month, EFFR(Start)_{T+1}, depending on whether there is a rate-setting meeting in the previous and following months. Let M be the number of days before the rate setting meeting in the Tth month, and let N be the number of days remaining for the month, then the relationship between the Fed fund futures rate and the average EFFR before and after the meeting is

$$\text{EFFR(Avg)}_T = \frac{M}{M+N}\text{EFFR(Start)}_T + \frac{N}{M+N}\text{EFFR(End)}_T. \quad (3.1)$$

If there is no meeting in the $T-1^{st}$ month, we have

$$\text{EFFR(Start)}_T = \text{EFFR(Avg)}_{T-1},$$

so we can figure out EFFR(End)_T from Equation 3.1. If there is no meeting in the $T+1^{st}$ month, we have

$$\text{EFFR(End)}_T = \text{EFFR(Avg)}_{T+1},$$

so we can figure out EFFR(Start)_T from the same equation. The ways to figure out both EFFR(Start)_T and EFFR(End)_T are summarized in Table 3.2.

TABLE 3.2: Calculating EFFR(Start)_T and EFFR(End)_T.

	No Meeting in Following Mo.	No Meeting in Previous Mo.
N	Days in mo. before the Meeting	
M	Days in mo. after the Meeting (including meeting date)	
EFFR(Avg)_T	100-FF Contract Price	
EFFR(Start)_T	$\dfrac{\text{EFFR(Avg)}_T - \frac{M}{M+N}\text{EFFR(End)}_T}{\frac{N}{M+N}}$	EFFR(Avg)_{T-1}
EFFR(End)_T	EFFR(Avg)_{T+1}	$\dfrac{\text{EFFR(Avg)}_T - \frac{N}{M+N}\text{EFFR(Start)}_T}{\frac{M}{M+N}}$

Using the Fed fund futures prices of 13 Sep 2023, we can confirm that the probability for the Fed fund rate to remain in the range of $[5\%, 5.25\%]$ is 97%, as shown in Figure 3.2.

With the same method stated above, we can also calculate the probabilities of rate hike or rate reduction for the next rate setting month. Then, the cumulative probabilities can be calculated for each possible level of Federal fund rate. As time horizon increases, the number of possible levels of Federal fund rate increases and the possibilities for these levels become less concentrated.

3.2 The U.S. Bond Markets

The bulk of fixed-income investment has to do with bonds. A bond is a financial contract that promises to pay a stream of cash flows over a certain time horizon. The cash flows consist of payments of interest and a payment of the principal. The interest payments are made periodically until the maturity of the bond, and the principal is, based on which interest payments are

FIGURE 3.2: Quotation of Fed fund futures for maturities of Sep., Oct. and Nov., 2023 (Source: https://www.cmegroup.com/markets/interest-rates/cme-fedwatch-tool.html).

calculated, paid back at the maturity date. The interest payments are also called *coupons*. Figure 3.3 illustrates the cash flows of a typical coupon bond, where c and Pr stand for the coupon rate and the principal value for the bond, respectively, and ΔT is the gap of time, measured in units of a year, between two consecutive coupon payments. The principal value is also called the face value or par value.

The above cash flow structure is considered standard for bonds, and we call bonds with such a cash flow pattern bullet bonds or straight bonds. For non-bullet bonds, there are a variety of cash flow patterns or structures. One important example of a non-bullet bond is the class of floating-rate bonds, where the coupon rates are not fixed, but are indexed to three- or six-month rates for Certificate of Deposits (CD). Comprehensive discussions of CDs and the floating-rate bonds are provided later in this chapter.

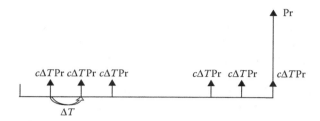

FIGURE 3.3: Cash flows of a coupon bond.

Bonds are tools to raise capital for their issuers, who can be central governments, local governments, or industrial/commercial corporations. In most developed countries, bond markets are mature and have notional values larger than those of their corresponding equity markets, except at times of equity bubbles. The major reasons for the attractiveness of the bond markets are the stable returns and lower risk (i.e., smaller price fluctuation) of bonds. With the globalization of capital markets, investors now have many types of bonds to choose from, such as between domestic bonds and international bonds, or between government bonds or corporate bonds. The profiles of the risks and returns of the investment alternatives vary. For an extensive discussion, please see other works, for example, Fabozzi (2007).

Let us take a closer look at the U.S. domestic bond markets. The U.S. domestic bond markets consist mainly of three sub-sectors:

1. *Government bonds:* Bonds issued by the Treasury Department of the U.S. Federal government and local governments.

2. *Agency bonds:* Bonds issued by certain agencies of the U.S. government or guaranteed or sponsored by the U.S. government.

3. *Corporate bonds:* Bonds issued by U.S. corporations.

3.2.1 Treasury Securities

The U.S. Treasury Department is the largest single issuer of debt in the world. Bonds in this sector are deemed riskless as they are guaranteed by the full faith of the Federal government of U.S.A. This sector is also distinguished by the large volume of total debt and the large size of any single issue, two factors that have contributed to making the Treasuries market the most active as well as most liquid bond market in the world.

Treasury securities include Treasury bills, Treasury notes, Treasury bonds, and Treasury Inflation-Protected Securities (TIPS). The Treasury bills are bonds without coupons (the so-called zero-coupon bonds). Upon issuance, the maturities of these zero-coupon bonds vary from 4 weeks (one month), 13 weeks (three months), 26 weeks (six months) to 52 weeks (12 months). The Treasury notes are coupon bonds with maturities between 2 and 10 years upon issuance, while the Treasury bonds have original maturities greater than 10 years. TIPS are issued with maturities of 5, 10, and 20 years. Thirty-year TIPS have also been issued occasionally. TIPS are issued with fixed coupon rates. The principal of a TIPS is adjusted semiannually based on the realized inflation rate over the proceeding six months. There is a floor to the principal, set equal to the initial notional value of the TIPS, so that the repayment of notional values or principals is always guaranteed.

The Treasury securities are issued through an auction process that takes place in a regular cycle of time. Four-week bills are auctioned every Thursday, whereas 13- and 26-week bills are auctioned every Monday. Fifty-two-week bills are auctioned every 4 weeks on Tuesday. Treasury notes are issued much less frequently. Two- and five-year notes are offered at the end of each month, and 3- and 10-year notes on the 15th of February, May, August, and November. The 30-year bonds are auctioned twice, in February and August. TIPS are auctioned in different cycles. Five-year TIPS are generally auctioned in the last week of April; 10-year TIPS are generally auctioned in the second week of January and July; and 20-year TIPS are generally auctioned in the last week of January. The 10-year TIPS dominate the liquidity of the TIPS market.

In the U.S, local governments, including state, city and county governments, also issue bonds, for financing their capital expenditures. These bonds form the so-called Municipal bond market. Municipal bonds are often exempt from most taxes yet, unlike the Treasury securities, they are not guaranteed and carry default risk. As such, municipal bonds are not in the same asset class of Treasury securities.

In terms of notional value, the Treasury bond sector is however not the largest sector of bonds, and it trails behind the agency bond sector. The agency bond sector includes fixed-income securities backed by student loans, public power systems and, in particular, mortgages. Bonds issued by The Federal National Mortgage Association (Fannie Mae) and The Federal Home Loan Mortgage Corporation (Freddie Mac) and bonds backed by student loans, at least, are guaranteed by the Federal government. Other agency bonds, however, are likely subject to default risk. The corporate bond market is also called a credit market, where the uncertainty of credit worthiness is the major source of risk as well as returns. The modeling of credit risk is a very big subject at its own right and will be dealt with in Chapter 14.

3.2.2   Quotation and Interest Accrual

The price of Treasury bills and Treasury coupon bonds are quoted using different conventions. The former is quoted using a discount yield, with which we can conveniently calculate their dollar prices, whereas the latter is quoted as a percentage of the principal value, using a price tick of 1/32nd of a percentage point. As examples, the quotes of newly issued Treasury securities are given in Table 3.3.

Some explanations are needed on how to figure out the dollar values of Treasury bills and notes/bonds from the quotes. We begin with Treasury bills.

TABLE 3.3: Quotes for U.S. Treasuries as of July 3, 2008

U.S. Treasuries Bills		Maturity Date	Discount/Yield	Discount/ΔY
3-month		6/5/2008	1.42/1.44	$-0.01/0.087$
6-month		9/4/2008	1.51/1.54	$0.03/-0.031$
Notes/Bonds	Coupon	Maturity Date	Price/Yield	Price/ΔY
2-year	2	2/28/2010	100-29$\frac{3}{4}$/1.52	$-0\text{-}00\frac{3}{4}/0.012$
3-year	4.75	3/31/2011	103-21$\frac{1}{4}$/1.42	$-0\text{-}02/0.018$
5-year	2.75	2/28/2013	101-16/2.43	$0\text{-}06/-0.040$
10-year	3.5	2/15/2018	99-23+/3.53	$0\text{-}14/-0.053$
30-year	4.375	2/15/2038	97-08$\frac{1}{2}$/4.54	$0\text{-}08+/-0.017$

Available at http://www.bloomberg.com/markets/rates/index.html.

The dollar value of a Treasury bill is calculated using the discount yield, Y_d, according to the formula

$$V = \Pr \cdot \left(1 - \frac{\tau}{360} Y_{\mathrm{d}}\right), \tag{3.2}$$

where τ is the number of days remaining to maturity. Suppose, for instance, that the six-month Treasury bill has a time to maturity of $\tau = 100$ days left and the discount yield remains unchanged. Then its price is

$$P = 100 \times \left(1 - \frac{100}{360} \times 1.51\%\right) = \$99.5806.$$

Note that the discount yield is merely a quoting device rather than a proper measure of investment return on the Treasury bills. In the next section, we discuss the measure of investment returns for Treasury bills, notes and bonds.

Next, we discuss the calculation of the dollar price of Treasury notes/bonds. One may notice that a "+" appears in the quotation of 10-year notes. Such a "+" stands for 0.5, or half a tick. The dollar values of the five Treasury notes and bonds are calculated as follows:

$$\begin{aligned}
100\text{-}29\tfrac{3}{4} &= 100 + \frac{29.75}{32} = 100.9297, \\
103\text{-}21\tfrac{1}{4} &= 103 + \frac{21.25}{32} = 103.6641, \\
101\text{-}16 &= 101 + \frac{16}{32} = 101.5, \\
99\text{-}23+ &= 99 + \frac{23.5}{32} = 99.7344, \\
97\text{-}08\tfrac{1}{2} &= 97 + \frac{8.5}{32} = 97.2656.
\end{aligned} \tag{3.3}$$

The above prices are not the dollar prices for transactions, however. For transactions, we must add the interest values accrued since the last coupon payment, to which the bond holders are entitled. The calculation of the accrued interest is demonstrated as follows. Consider a coupon bond with coupon rate c and coupon dates $\{T_i\}$. Suppose that we are in a day, t, between two consecutive coupon dates, T_j and T_{j+1} (i.e., $T_j < t \le T_{j+1}$), then the accrued interest is defined as

$$\text{AI}(t, T_j) = \Delta T \cdot c \cdot \text{Pr} \cdot q,$$

where q is the fraction of days elapsed since the last coupon date over the days of the current coupon period:

$$q = \frac{t - T_j}{T_{j+1} - T_j}. \tag{3.4}$$

The transaction price of the bond is simply the sum of (the dollar value of) the quote price and the accrued interest:

$$\text{Transaction price} = \text{Quote price} + \text{AI}(t, T_j).$$

Note that the industrial jargon for quote prices is clean prices, whereas the transaction prices are also called dirty prices, full prices or invoice prices.

Let us look at an example of accrued interest calculations.

Example 3.2. *Consider a 10-year Treasury note maturing on February 15, 2018. On March 7, 2008, the bond quote is 99-23+ (= 99.7344). We need to compute the accrued interest and then the dirty price.*

The coupon dates are February 15 and August 15. There are 182 days between the coupon dates, and on March 7, 21 days have elapsed since February 15, the last coupon date. Hence,

$$q = \frac{21}{182} = 0.115385,$$
$$\text{AI} = 0.5 \times 3.5\% \times 100 \times 0.115385 = 0.2019.$$

Then the full price is then

$$B^c = 99.7344 + 0.2019 = 99.9363. \tag{3.5}$$

It should be indicated that there exists an alternative quotation convention for Treasury notes and bonds, under which Table 3.3 is replace by 3.4.

For the quote price of notes and bond, there are four digits after the "-" sign, which remain in the unit of 1/32nd. While the last digit is "0", the third digit is peculiar, as it takes the number of 2, 4 or 6 only, constituting $2/8 = 1/4$, $4/8 = 1/2$ or $6/8 = 3/4$. The alternative quotation convention are seen in the web site of CME and the *The Wall Street Journal*.

TABLE 3.4: Quotes for U.S. Treasuries as of July 3, 2008

U.S. Treasuries Bills		Maturity Date	Discount/Yield	Discount/ΔY
3-month		6/5/2008	1.42/1.44	$-0.01/0.087$
6-month		9/4/2008	1.51/1.54	$0.03/-0.031$

Notes/Bonds	Coupon	Maturity Date	Price/Yield	Price/ΔY
2-year	2	2/28/2010	100-2960/1.52	$-0\text{-}00\frac{3}{4}/0.012$
3-year	4.75	3/31/2011	103-2120/1.42	$-0\text{-}02/0.018$
5-year	2.75	2/28/2013	101-1600/2.43	$0\text{-}06/-0.040$
10-year	3.5	2/15/2018	99-2340/3.53	$0\text{-}14/-0.053$
30-year	4.375	2/15/2038	97-0860/4.54	$0\text{-}08+/-0.017$

3.3 Bond Mathematics

3.3.1 Yield to Maturity

In a free market, the price of a bond is determined by supply and demand. Due to discounting, the full price is normally smaller than the aggregated value of the coupons and the principal. Hereafter we denote the full price of a bullet bond by B^c. Suppose that all cash flows are discounted by a uniform rate, y, of compounding frequency ω. Let the bond maturity be $T = n\Delta T$, and then y should satisfy the following equation:

$$B^c = \Pr \cdot \left(\sum_{i=1}^{n} \frac{c\Delta T}{(1+y\Delta t)^{i\Delta T/\Delta t}} + \frac{1}{(1+y\Delta t)^{n\Delta T/\Delta t}} \right), \quad (3.6)$$

where n is the number of coupons and $\Delta t = 1/\omega$. In bond mathematics, the compounding frequency is taken to be $\omega = 1/\Delta T$ by default, when $\Delta t = \Delta T$. This discount rate, which can be easily solved by a root-finding algorithm or even a trial-and-error procedure using Equation 3.5, is called the *yield to maturity* (YTM), as well as the *internal rate of return* (IRR) of the bond, and it is often simply called the *yield*.

As the function of the yield (for $\omega = 1/\Delta T$), the formula for a general time, $t \leq T$, is

$$B_t^c = \Pr \cdot \left(\sum_{i;i\Delta T>t}^{n} \frac{c\Delta T}{(1+y\Delta T)^{(i\Delta T-t)/\Delta T}} + \frac{1}{(1+y\Delta T)^{(n\Delta T-t)/\Delta T}} \right). \quad (3.7)$$

Assuming that $t \in (T_j, T_{j+1}]$, then the fraction of time elapsed since the last coupon payment is

$$q = \frac{t-T_j}{T_{j+1}-T_j} = \frac{t-T_j}{\Delta T},$$

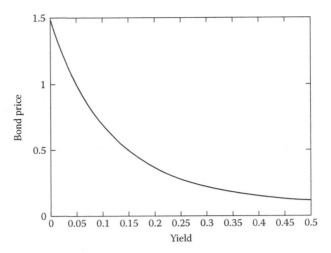

FIGURE 3.4: The price–yield relationship for a 10-year bond with $c = 0.05$, $\Delta T = 0.5$.

we then can write

$$t = T_j + \Delta Tq = (j + q)\Delta T \text{ and } i\Delta T - t = (i - j - q)\Delta T, \quad \forall i.$$

After some tidying up, we have the following concise price formula for the bond:

$$
\begin{aligned}
B_t^c &= \text{Pr} \cdot \left(\sum_{i=j+1}^{n} \frac{c\Delta T}{(1 + y\Delta T)^{i-j-q}} + \frac{1}{(1 + y\Delta T)^{n-j-q}} \right) \\
&= \text{Pr} \cdot (1 + y\Delta T)^q \left(\sum_{i=1}^{n-j} \frac{c\Delta T}{(1 + y\Delta T)^i} + \frac{1}{(1 + y\Delta T)^{n-j}} \right).
\end{aligned}
\tag{3.8}
$$

Given B_t^c, the bond price at any time t, the bond yield is implied by Equation 3.7. A rough way to compare the relative cheapness/richness of two bonds with the same coupon frequency is to compare their yields. Intuitively, a bond with a higher yield is cheaper and thus may be more attractive.

There is a one-to-one price–yield relationship owing to functional monotonicity, as shown in Figure 3.4. Because of this relationship, a bond price is also quoted using its yield in the industry. Furthermore, a bond price is a convex function of the yield, as we can see in Figure 3.4. Such a feature will be utilized later when we try to determine the price of interest rate futures and non-vanilla swaps.

3.3.2   Par Bonds, Par Yields, and the Par Yield Curve

The summation in Equation 3.6 can be worked out so that

$$B^c = \Delta T \cdot c \cdot \mathrm{Pr} \sum_{i=1}^{n} (1 + y\Delta T)^{-i} + \mathrm{Pr}(1 + y\Delta T)^{-n}$$
$$= \mathrm{Pr} \left[1 - \left(1 - \frac{c}{y} \right) \left(1 - \frac{1}{(1 + y\Delta T)^n} \right) \right]. \tag{3.9}$$

From the above expression, we can tell when the price is smaller, equal to, or larger than the principal value.

1. When $c < y$, $B^c < \mathrm{Pr}$. In such a case, we say that the bond is sold at discount (of the par value).

2. When the coupon rate is $c = y$, then $B^c = \mathrm{Pr}$, that is, the bond price equals the par value of the bond. In such a case, we call the bond a *par bond*, and the corresponding coupon rate a *par yield*.

3. When $c > y$, $B^c > \mathrm{Pr}$. In such a case, we call the bond a premium bond, which is traded at a premium to par.

Par yields play an important role in today's interest-rate derivatives market. As we shall see later, there are many derivatives based on the par yields.

3.4   Discount Curve and Zero-Coupon Yields

3.4.1   Discount Curve

Let us refer to the cash flow of the coupon bond shown in Figure 3.1 again. When there is no coupon, $c = 0$, the principal is the only cash flow and the coupon bond is reduced to a zero-coupon bond. The corresponding yield is called a zero-coupon yield. Conventionally, the time-t price of a zero-coupon bond maturing at time T into a par value of one dollar is denoted as $P(t, T)$ or P_t^T. In terms of its yield, the price of the zero-coupon bond is

$$P(t, T) = \frac{1}{(1 + y\Delta T)^{(T-t)/\Delta T}}, \tag{3.10}$$

where the time to maturity, $T - t$, does not have to be a multiple of ΔT. The collection of P_t^T for $T \geq t$ is called a discount curve.

With the discount curve, one can price any bond portfolio with deterministic cash flows. This is because any such portfolio can be treated as a portfolio of zero-coupon bonds. For example, we can express the price of a coupon bond in terms of those of zero-coupon bonds:

$$B^c(0) = \sum_{i=1}^{n} c \cdot \Delta T \operatorname{Pr} \cdot P_0^{i\Delta T} + \operatorname{Pr} \cdot P_0^{n\Delta T}. \qquad (3.11)$$

The discount curve is used for pricing off-the-run Treasury securities, and marking to market Treasury portfolios. Moreover, the discount curve is also essential for pricing future cash flows of any security, either deterministic or stochastic. To price a portfolio of interest-rate derivatives, we may model the dynamics of the entire yield curve, in contrast to modeling the dynamics of a stock price for stock options.

In most fixed-income markets, zero-coupon bonds (ZCB) are rare, with the exception of the U.S. Treasury markets, where they are actually quoted and traded, yet they occupy only a small fraction of the daily turnover of the Treasury bond market. Other than Treasury bills, the Treasury Department does not directly issue ZCBs. ZCBs other than Treasury bills first appeared in 1982, when both Merrill Lynch and Salomon Brothers created synthetic zero-coupon Treasury receipts backed by coupons and principals of selected Treasury coupon bonds. In 1985, the U.S. Treasury launched its Separate Trading of Registered Interest and Principal of Securities (STRIPS) program. This program facilitated the stripping of designated Treasury securities. Today, all Treasury notes and bonds, both fixed-principal and inflation-indexed principal, are eligible for stripping. In principle, one can replicate Treasury coupon bonds using STRIPS, but the relatively low liquidity of STRIPS makes such replication not always doable. Because of this, prices of STRIPS for $1 notional are not taken as discount factors. The market convention is to extract the discount curve from the liquid Treasury securities through a bootstrapping procedure. The prices of the STRIPS are actually subordinated to the discount curve so obtained.

In continuous-time finance, it is often favorable to work with continuous compounding, that is, by letting the term $\Delta T \to 0$ in Equation 3.10. At this limit, we have

$$P(t, T) = e^{-y \times (T-t)}.$$

Given $P(t, T)$, the corresponding zero-coupon yield can be calculated from the last equation:

$$y_{T-t} = -\frac{1}{T-t} \ln P(t, T).$$

3.4.2   On-the-Run Treasury Securities

A bond issuer may routinely issue bonds of various maturities and, in a market, there can be many bonds from the same issuer being traded. For various reasons, some bonds are more liquid than others. The most liquid ones are often called benchmark bonds for the issuer. Their yields reflect the level of

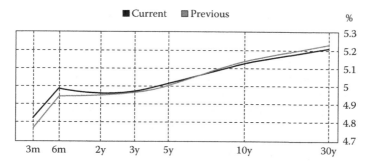

FIGURE 3.5: The U.S. Treasury yield curve for April 28, 2006 (gray) and May 1, 2006 (black) (data from http://www.bloomberg.com/markets/rates/index.html).

borrowing costs the market demands from the issuer. Moreover, the prices of the benchmark bonds imply a discount curve for cash flows from the issuer, and the discount curve can be used to gauge the relative cheapness/expensiveness of the issuer's other bonds. If a relatively cheaper or more expensive bond is found, one may trade against this bond using the benchmark bonds and is likely to earn an arbitrage profit. Hence, the prices or yields of the benchmark bonds carry essential information for the arbitrage pricing of the issuer's other bonds, and they are treated as a summary of the status quo of all bonds offered by the same issuer.

In the U.S. Treasury market, newly issued bills and notes/bonds are called on-the-run Treasury securities. Traditionally, the on-the-run issues enjoy higher liquidity and are thus treated as benchmarks. Table 3.1 provides the closing price quotes of the on-the-run Treasury issues for July 3, 2008. As can be seen in the table, the on-the-run issues have maturities of 3 months, 6 months, 2 years, 3 years, 5 years, 10 years, and 30 years. When we connect the yields of the benchmark bonds through interpolation, we obtain a so-called yield curve. Since bond yields vary from day to day, so does the yield curve. Figure 3.5 shows the yield curves for the U.S. Treasuries constructed by linear interpolation for April 28 and May 1, 2006, two consecutive trading days.

The yield curve provides a rough idea of the level of yields for various maturities. More importantly, based on the Treasury yield curve, we can extract the discount curve. In the next section, we describe the technique for extracting the discount curve from the yield curve.

3.4.3 Bootstrapping the Zero-Coupon Yields

The determination of the zero-coupon yield curve (or discount curve) based on the yields of the on-the-run issues is an under-determined problem: we need to solve for infinitely many unknowns based on a few inputs. To obtain a meaningful solution, one must parameterize the zero-coupon yield curve. The simplest parameterization that is financially acceptable is to assume piecewise linear and continuous functional forms for the zero-coupon yield curve. Under such a parameterization, the zero-coupon yield curve can be derived sequentially. Such a procedure is often called bootstrapping in finance. Next, we describe the bootstrapping procedure with the construction of the zero-coupon yield curve for U.S. Treasuries.

Let $\{B_j^c, T_j\}_{j=1}^7$ be the prices and maturities of the seven on-the-run issues. Let $T_0 = 0$ and $\Delta T = 0.5$. We assume that the zero-coupon yield for maturities between $[T_0, T_7]$ is a piecewise linear function. The determination of the YTMs is done sequentially. Because the first two on-the-run issues are zero-coupon bonds, we first back out $y(0.25)$ and $y(0.5)$, the zero yields for $(0, T_1]$ and $(T_1, T_2]$, using formula 3.10. This will require a root-finding procedure. Once $y(0.5)$ is found, we proceed to determining $y(i/2), i = 2, 3, 4$ from the following equation:

$$B_3^c = \frac{c_3/2}{(1 + y(1/2)/2)^i} + \sum_{i=2}^4 \frac{c_3/2}{(1 + y(i/2)/2)^i} + \frac{1}{(1 + y(2)/2)^4}, \qquad (3.12)$$

where

$$y(i/2) = y(0.5) + \alpha_i \times (i/2 - 0.5), \quad i = 2, 3, 4.$$

So, our zero-coupon yield is a linear function over $T \in [T_2, T_3]$. Equation 3.12 is an equation for α_i, which can be determined through a root-finding procedure. This procedure can continue all the way to $j = 7$. The entire zero-coupon yield curve for maturity $T \le 30$ so-determined is displayed in Figure 3.6.

A zero-coupon yield curve implies a discount curve. Suppose that the y_T is the zero-coupon yield for maturity T. Then the corresponding zero-coupon bond price is calculated according to Equation 3.10. With discount bond prices, we can value any coupon bond using Equation 3.11.

Associated with the discount curve are the notions of present value and future value. Owing to the (mostly) positive interest rates, the notional value of I_t today will become $I_{t+T}(>I_t)$ T years later. In terms of the discount curve, we can define the present value of a future cash flow, I_{t+T}, to be

$$I_t = I_{t+T} P(t, t+T). \qquad (3.13)$$

Conversely, we can rewrite Equation 3.13 as

$$I_{t+T} = \frac{I_t}{P(t, t+T)}, \qquad (3.14)$$

and refer I_{t+T} as the future value of I_t.

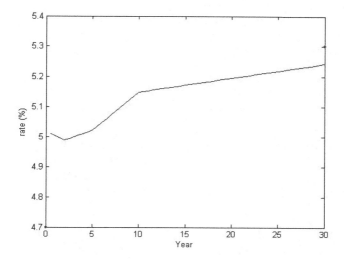

FIGURE 3.6: The zero-coupon yield curve of U.S. Treasuries on May 1, 2006.

3.5 Yield-Based Bond Risk Management

In previous sections, we have completely characterized the price–yield relationship of bonds. Bond price fluctuations can be attributed to yield fluctuations, and price risks can be treated as yield risks. It has become a common sense that yields to maturities of all bonds from the same issuer are highly correlated: yields (of different maturities) often move in the same direction such that the percentage bond price changes are comparable. This reality makes it possible to hedge against the price change of one bond using another bond. In this section, we introduce a yield-based theory of bond risk management, which is about hedging the bond price risk against the parallel shift of the yield curve. The theory was initially developed for bonds, yet it can also be applied to managing the yield risk of other interest-rate-sensitive instruments.

3.5.1 Duration and Convexity

In the bond market, bond prices change unpredictably on a daily basis. The changes in bond prices can be interpreted as the consequence of unpredictable changes in yields. It has been observed that the prices of long-maturity bonds are more sensitive to change in yields than are the prices of short-maturity bonds, and the impact of yield changes on bond prices seems proportional to the cash flow dates of the bonds. Intuitively, Macaulay (1938) introduced the weighted average of the cash flow dates as a measure of price sensitivity with respect to the bond yield:

$$D_{\mathrm{mac}} = \frac{\mathrm{Pr}}{B_t^c} \left[\sum_{i,T_i>t}^n \frac{\Delta T \cdot c}{(1+y\Delta T)^{(T_i-t)/\Delta T}} (T_i - t) \right. \tag{3.15}$$

$$\left. + \frac{1}{(1+y\Delta T)^{(T_n-t)/\Delta T}} (T_n - t) \right].$$

This measure is called the *Macaulay duration* in the bond market. Note that, for a zero-coupon bond, the duration is simply its maturity. It was later understood that the Macaulay duration is closely related to the derivative of the bond price with respect to its yield. In fact, differentiating Equation 3.7 with respect to y yields

$$\frac{\mathrm{d}B_t^c}{\mathrm{d}y} = -\frac{\mathrm{Pr}}{1+y\Delta T} \left[\sum_{i;T_i>t}^n \frac{\Delta T \cdot c}{(1+y\Delta T)^{(T_i-t)/\Delta T}} (T_i - t) \right.$$

$$\left. + \frac{1}{(1+y\Delta T)^{(T_n-t)/\Delta T}} (T_n - t) \right]. \tag{3.16}$$

In terms of D_{mac}, the Macaulay duration just defined, we have

$$\frac{\mathrm{d}B_t^c}{B_t^c} = -\frac{D_{\mathrm{mac}}}{1+y\Delta T} \mathrm{d}y \quad \text{or} \quad \frac{1}{B_t^c} \frac{\mathrm{d}B_t^c}{\mathrm{d}y} = -\frac{D_{\mathrm{mac}}}{1+y\Delta T}. \tag{3.17}$$

According to Equation 3.16, the Macaulay duration is essentially the rate of change with respect to the yield for each dollar of the market value of the bond. After multiplying by the change in the yield, the Macaulay duration gives the percentage change in the value of the bond. For convenience, we define

$$D_{\mathrm{mod}} = \frac{D_{\mathrm{mac}}}{1+y\Delta T},$$

and call it the *modified duration*. Then the first equation of Equation 3.17 can be written in the following simple form:

$$\frac{\mathrm{d}B_t^c}{B_t^c} = -D_{\mathrm{mod}} \, \mathrm{d}y. \tag{3.18}$$

Both D_{mac} and D_{mod} are called durations of bonds, which serve as measures of risk exposure with respect to a possible change in the bond yield. Modified durations are more often used in today's markets.

By using Equation 3.9, the succinct bond formula, we can obtain the following formula for the modified duration:

$$D_{\mathrm{mod}} = \frac{\mathrm{Pr}}{B^c} \left[\frac{c}{y^2} \left(1 - \frac{1}{(1+y\Delta T)^n} \right) + \left(1 - \frac{c}{y} \right) \frac{n\Delta T}{(1+y\Delta T)^{n+1}} \right]. \tag{3.19}$$

The above expression is simplified for par bonds, when $c = y$ and $B^c = \text{Pr}$, then the expression of modified duration becomes

$$D_{\text{mod}} = \frac{1}{y}[1 - (1 + y\Delta T)^{-n}]. \tag{3.20}$$

Note that Treasury bonds are quoted in yields and that recent issues are usually traded close to par, so Equation 3.19 gives us an approximate value of the durations for bonds being traded close to par.

The next example shows how much the dollar value of a bond changes given its duration.

Example 3.3. *Given a 30-year Treasury yielding 5% and trading at par, we can calculate the modified duration using Equation 3.20 and obtain $D_{\text{mod}} = 15.45$ years. This means that a one basis point variation in the yield will cause a change of 15.45 cents in the bond price for 100-dollar face value.*

From a mathematical viewpoint, the estimation of price changes using duration is equivalent to estimating functional value using a linear approximation. The accuracy of such an approximation becomes poorer when the change in the yield becomes larger. To see that, we consider the second-order expansion of the bond price change with respect to the yield:

$$\Delta B_t^c = \frac{dB_t^c}{dy}\Delta y + \frac{1}{2}\frac{d^2 B_t^c}{dy^2}(\Delta y)^2. \tag{3.21}$$

In view of Equation 3.21, we introduce the *convexity measure*:

$$C = \frac{1}{B_t^c}\frac{d^2 B_t^c}{dy^2} = \frac{\text{Pr}}{B_t^c}\left[\sum_{i;T_i > t}^{N} \Delta T \cdot c(T_i - t)(T_{i+1} - t) \right.$$
$$\times (1 + y\Delta T)^{-((T_{i+1}-t)/\Delta T)-1}$$
$$\left. + (T_N - t)(T_{N+1} - t)(1 + y\Delta T)^{-((T_{N+1}-t)/\Delta T)-1} \right]. \tag{3.22}$$

Equation 3.21 can then be recast to

$$\frac{\Delta B_t^c}{B_t^c} = -D_{\text{mod}}\Delta y + \frac{1}{2}C\Delta y^2. \tag{3.23}$$

Note that C captures the convexity or the curvature of the bond–price curve. Given the duration measure, D_{mod}, and the convexity measure, C, we can readily calculate the percentage change in the price with Equation 3.23. When $\Delta y^2 \ll \Delta y$, we can neglect the second-order term in Equation 3.23 and thus return to Equation 3.18. However, for relatively large Δy, the inclusion of the convexity term in Equation 3.23 will be necessary to produce a good approximation of the percentage change. The curves of the exact price, the linear approximation, and the quadratic approximation are shown in Figure 3.7.

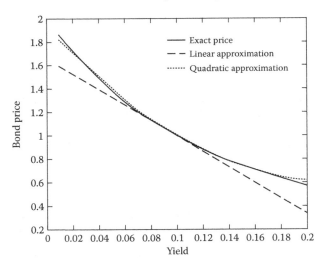

FIGURE 3.7: Linear and quadratic approximations of bond prices.

We remark here that although the duration measure and the convexity measure are introduced here for bonds, their applications are not restricted to bonds. We can apply them to any interest-rate-sensitive instruments.

3.5.2   Portfolio Risk Management

We can also calculate the duration and convexity of a portfolio of fixed-income instruments. Consider a portfolio of N instruments, with n_i units and price B_i^c for the ith instrument. Then, the absolute change in the portfolio value upon a parallel yield shift is given by

$$dV = \sum_i n_i dB_i^c = \sum_i n_i B_i^c \cdot \left(-D_{\mathrm{mod}}^i dy + \frac{1}{2} C^i dy^2 \right),$$

where D_{mod}^i and C^i are the duration and convexity of the ith instrument, respectively. The percentage change is then

$$\frac{dV}{V} = -\left(\sum_i x_i D_{\mathrm{mod}}^i \right) dy + \frac{1}{2} \left(\sum_i x_i C^i \right) dy^2, \tag{3.24}$$

where $x_i = n_i B_i^c / V$ is the percentage of the value in the ith instrument. Equation 3.24 indicates that the duration and convexity of a portfolio are the weighted average of the duration and convexity of its components, respectively.

In classical risk management, a portfolio manager can limit his or her exposure to interest-rate risk by reducing the duration while increasing the convexity of the portfolio. To avoid possible losses in cases of large yield moves, the manager usually will not tolerate negative net convexity. A portfolio with very small duration is called a duration-neutral portfolio. Practically, interest-rate futures and swaps are often used as hedging instruments for duration management.

The basic premise of the duration and convexity technology is that the yield curves shift in parallel, either upward or downward by the same amount. This is, however, a very crude assumption about the yield curve movement as, in reality, points in a yield curve do not often shift by the same amount and sometimes they do not even move in the same direction. For a more elaborate model of yield curve dynamics, we will have to resort to stochastic calculus in a multi-factor setting.

3.6 The Interbank Lending Market

To retail investors, the term "interest rate" likely means the overnight rate that is used for interest accrual in a savings account, as well as the rates for certificate of deposits of various terms. However, to professional investors of fixed-income markets, this term has a much broader meaning, and it encompasses an entire class of *rates of return* that are associated with various fixed-income instruments. In this section, we introduce instruments in the interbank market and then define various interest rates as the rates of returns for investing in these instruments, starting with the savings accounts.

3.6.1 Overnight Rate and Saving Account

The overnight rate or short rate is associated with a savings account in a bank. The short rate at day t is conventionally denoted as r_t. Interest on a savings account is accrued daily, using the actual/365 convention. Let B_t denote the account balance at the end of day t, and let $\Delta t = 1$ day $= 1/365$ year. Then the new balance at the end of the next day, $t + \Delta t$, is

$$B_{t+\Delta t} = B_t(1 + r_t\Delta t). \qquad (3.25)$$

Because $\Delta t \ll 1$, daily compounding is very well approximated by *continuous compounding*: in the limit of $\Delta t \to 0$, Equation 3.25 becomes

$$\mathrm{d}B_t = r_t B_t \mathrm{d}t. \qquad (3.26)$$

As r_t is applied to $(t, t + dt)$, an infinitesimal interval of time, it is also called the instantaneous interest rate. As a mathematical approximation and idealization, continuous compounding is necessary for continuous-time finance.

Suppose that a sum of money is deposited at $t = 0$ into a savings account and that there has not been a deposit or withdrawal since, then the balance at a later time, t, is

$$B_t = B_0 e^{\int_0^t r_s \, \mathrm{d}s}. \tag{3.27}$$

In the real world, the balance, B_t, is not known in advance due to the stochastic nature of the short rate. Nonetheless, the deposit in the savings account is considered a risk-free security, and its return is used as a benchmark to measure the profits and losses of other investments.

In reality, savings accounts for institutions and for retail customers offer different interest rates, which reflect different overhead management costs for institutional and individual clients. To distinguish it from an individual's account, we call the savings account for an institution a *money market account*. Note that this is somewhat an abuse of terminology. In the United States, a money market account is also a type of savings account for retail customers, which offers higher interest rates under some restrictions, including minimum balance requirements and limited numbers of monthly withdrawals, and its compounding rule is also different from continuous compounding, which will be introduced later under repurchasing agreement, an instrument equivalent to collateralized lending/borrowing. Hence, we need to emphasize here that, in fixed-income modeling, a money market account means a savings account for institutions that compounds continuously. Such a money market account plays an important role in continuous-time modeling of finance.

3.6.2   Term Rates and Certificates of Deposit

Term rates are associated with certificates of deposit (CD). A CD is a deposit that is committed to a fixed period of time, and the interest rate applied to the CD is called a term rate. For retail customers, the available terms are typically one month, three months, six months, and one year. Usually, the longer the term, the higher the term rate, as investors are awarded a higher premium for committing their money for a longer period of time. The interest payments of CDs use simple compounding. Let $r_{t,\Delta t}$ be the interest rate for the term Δt and I_t be the value of the deposit at time t. Then the balance at the maturity of the CD is

$$I_{t+\Delta t} = I_t(1 + r_{t,\Delta t}\Delta t). \tag{3.28}$$

Investors of CDs often roll over their CDs, meaning that after a CD matures, the entire amount (principal plus interest) is deposited into another CD with the same terms but with the prevailing term rate at the time when the rolling over takes place. Suppose that a CD is rolled over n times. Then the terminal balance at time $t + n\Delta t$ is

$$I_{t+n\Delta t} = I_t \cdot \prod_{i=1}^{n} (1 + r_{t+(i-1)\Delta t, \Delta t}\Delta t). \tag{3.29}$$

If the Δt term rate remains unchanged over the investment horizon, that is, $r_{t+(i-1)\Delta t,\Delta t} = r_{t,\Delta t}$, $i = 1, \ldots, n$, then there is

$$I_{t+n\Delta T} = I_t(1 + r_{t,\Delta t}\Delta t)^n, \qquad (3.30)$$

and we say that the deposit is compounded n times with interest rate $r_{t,\Delta t}$. We call $\omega = 1/\Delta t$ the compounding frequency, which is the number of compoundings per year. For example, when $\Delta t = 3$ months or 0.25 year, we have $\omega = 1/\Delta t = 4$, corresponding to the so-called quarterly compounding. By the way, a savings account is compounded daily, corresponding to $\omega = 365$.

Different term rates mean different rates of return. One way to compare CDs of different terms is to check their *effective annual yields* (EAY), defined as the dollar-value return over a year for a \$1 initial investment:

$$\text{EAY} = (1 + r_{t,\Delta t}\Delta t)^{1/\Delta t} - 1. \qquad (3.31)$$

Should interest rates stay constant over the investment horizon, then a higher EAY gives a higher return in value. In reality, term rates change in a correlated yet random way. Hence, for any fixed investment horizon when rolling over is needed, it is difficult to judge in advance which term is optimal for an investor. In fact, investors often choose terms based on cash flow considerations.

3.7 The LIBOR Market

LIBOR is an acronym for London Interbank Offer Rates. It was a set of interest rate indexes at which banks lent unsecured loans to other banks in the London wholesale money market. The so-called LIBOR market was largely an over-the-counter (OTC) market for loans and interest-rate derivatives of various currencies based on LIBOR; it crossed the geographical boundaries of countries and was outside the jurisdiction of the banking authority of any single government. Because of that, the LIBOR market had enjoyed a high degree of flexibility in financial innovations and had evolved rapidly since 1986, when the indexes were formally launched. Until 2012, for almost all major currencies, the turnover of interest-rate derivatives in the LIBOR market had surpassed that of each domestic market. In this chapter, we describe the major instruments of the LIBOR market.

3.7.1 The LIBOR Indexes

LIBOR was a set of reference interest rates at which banks lent unsecured loans to other banks in the London wholesale money market. The LIBOR rates were benchmark rates for overnight lending/borrowing and CDs. The terms for CD could vary from 1 month to 12 months. The most popular and important terms had been 1 month, 3 months, 6 months, and 12 months (1 year). For example, the LIBOR rates of December 31, 2007 for USD are listed in Table 3.5.

TABLE 3.5: USD LIBOR Rates of December 31, 2007

Term	Rate (Annualized)
overnight	4.823%
1 week	4.488 %
2 week	4.500 %
1 month	4.600 %
2 month	4.653 %
3 month	4.702 %
4 month	4.668 %
5 month	4.631 %
6 month	4.596 %
7 month	4.526 %
8 month	4.454 %
9 month	4.378 %
10 month	4.324 %
11 month	4.274 %
12 month	4.224 %

Taken from www.global-rate.com.

The day-count convention for USD is actual/360.

3.7.2   Derivative Instruments in LIBOR Market

The LIBOR market began when USD was deposited into non-U.S. banks in Europe. After the Second World War, the amount of USD in Europe increased enormously, as a result of both trading with the United States and the Marshall Plan. During the Cold War period (1950–1989), the Eastern Bloc countries deposited most of their USD assets into British or other European banks, for fear of the possibility that the United States would freeze these assets if they were held in U.S. banks. The USD owned by non-U.S. entities and circulated in Europe formed the basis of the so-called Eurodollar market. Over the years, as a result of the United States's successive commercial deficits, dollar-denominated assets also came to be held in many countries around the world. Such a situation effectively turned the Eurodollar market into a global market.

Since the Eurodollar market was not subject to the banking regulations of the U.S. government, banks in the Eurodollar market could operate on narrower margins than banks in the United States. Thus, the Eurodollar market has expanded largely as a means of avoiding the regulatory costs involved in dollar-denominated financial intermediation. Such financial intermediation has fostered various financial innovations, particularly derivatives. Many currently popular interest-rate derivative instruments were originated in the Eurodollar market.

Gradually, derivatives on interest rates of other major currencies also became tradable in the LIBOR market, including those on the German Mark, the Japanese Yen, the Swiss Franc, and of course, the Pound Sterling. In 1999, the Euro became the official currency of the European Union. Nowadays, 20 European countries have adopted the Euro, making it another major currency parallel to the USD. To some extent, the LIBOR market acted like an offshore market for these currencies. Thanks to globalization, the LIBOR market crossed the geographical boundaries of most industrialized nations, and LIBOR instruments of various currencies could be traded in any major financial centers of the world. In the next section, we introduce LIBOR instruments, from standardized ones to exotic ones, that were once very popular. We begin with standardized instruments.

3.7.2.1 Forward Rates and Forward-Rate Agreements

Nominally, a forward rate can be regarded as the interest rate for a forward loan, a contract that requires no upfront payment. In the absence of counterparty risk, the forward rate can be determined through arbitrage arguments and expressed in terms of discount factors. For simplicity, we let the term of the loan be $(T, T + \Delta T)$, the notional value be \$1, and the time of entering the forward loan be $t = 0$. Then at $t = 0$, the party who makes the forward loan is supposed to secure the loan capital by

- short $P(0, T)/P(0, T + \Delta T)$ unit of $T + \Delta T$-maturity ZCB, and

- long 1 unit of T-maturity ZCB.

Note that the forward loan plus the above transactions constitute a set of zero-net transactions. At time T, a loan of \$1 is made for the forward rate, denoted as $f(0; T, T + \Delta T)$, typically for simple compounding. Then at time $T + \Delta T$, the party who makes the loan will wind up with a

$$\text{P\&L} = 1 + \Delta T f(0; T, T + \Delta T) - \frac{P(0, T)}{P(0, T + \Delta T)},$$

which is certain at time $t = 0$ when the forward loan was initiated. To avoid arbitrage, the forward loan rate must be taken as

$$f(0; T, T + \Delta T) = \frac{1}{\Delta T} \left(\frac{P(0, T)}{P(0, T + \Delta T)} - 1 \right).$$

With the same arguments, we establish the forward rate for the loan at any time $t \leq T$ to be

$$f(t; T, T + \Delta T) = \frac{1}{\Delta T} \left(\frac{P(t, T)}{P(t, T + \Delta T)} - 1 \right).$$

A forward-rate agreement (FRA) is a contract with maturity T between two parties to bet on the LIBOR rate for the tenor $(T, T + \Delta T)$ to be above or below $f(0; T, T + \Delta T)$. At the maturity of the FRA, the P&L to the long position is defined to be

$$\text{P\&L} = P(T, T + \Delta T) \times \text{Notional} \times \Delta T \times (f(T; T, T + \Delta T) - f(0; T, T + \Delta T)), \tag{3.32}$$

where $f(T; T, T + \Delta T)$ is the LIBOR rate for the term ΔT realized at the maturity T. It is not hard to derive the mark-to-market (MtM) value of a long FRA at any time $t \leq T$ to be

$$\text{P\&L} = P(t, T + \Delta T) \times \text{Notional} \times \Delta T \times (f(t; T, T + \Delta T) - f(0; T, T + \Delta T)), \tag{3.33}$$

which is the net PV by shorting an additional FRA at time t. Apparently, the MtM value of the FRA at $t = 0$ is zero, reflecting the fact that the FRA is a zero-sum game to the counterparties.

In the marketplace, the typical maturities (T) for FRAs are 1 month, 3 months, 6 months, 9 months, and 12 months, and the typical tenors (ΔT) are 1 month, 2 months, 3 months, and 6 months. The 3m-tenor FRAs had the highest liquidity.

3.7.2.2 Eurodollar Futures

Eurodollar futures contracts were perhaps the most popular futures contracts in global capital markets. These contracts were traded on the International Monetary Market (IMM) of the Chicago Mercantile Exchange (CME) and the London International Financial Futures Exchanges (LIFFE). Underlying these futures contracts were the interest payments of three-month Eurodollar CDs of one million dollars notional. Let t be the current time, T be the maturity of a Eurodollar futures, and $\Delta T = 0.25$, then the simple interest for the CD (to be initiated at time T) would be

$$\$\,1,000,000 \times \Delta T \times \tilde{f}(t; T, T + \Delta T), \tag{3.34}$$

where $\tilde{f}(t; T, T + \Delta T)$ is the annualized interest rate for the CD seen at time t. At the maturity of the futures contract, $t = T$, the futures rate will settle into the three-month LIBOR rate: $\tilde{f}(T; T, T + \Delta T) = f(T; T, T + \Delta T)$.

Like most futures contracts, Eurodollar futures contracts are cash-settled and marked to market on a daily basis. There is no delivery of a cash instrument upon expiration because cash Eurodollar time deposits are not transferable. For a long futures position, the marking-to-market P&L is

$$\$1,000,000 \times \Delta T \times \left(\tilde{f}(t; T, T + \Delta T) - \tilde{f}(t + \Delta t; T, T + \Delta T) \right), \tag{3.35}$$

where Δt represents one trading day. According to Equation 3.35, one basis point decrement in the futures rate will generate a profit of $25 to the holder, which is the usual price tick for the futures contract. Trading can also occur in the minimum ticks of 0.0025%, or 1/4 ticks, representing $6.25 per contract and in 0.005%, or 1/2 ticks, representing $12.50 per contract.

Let us see an example of a Eurodollar futures investment for profit and loss. Assume that, upon initiation in October 2018, the rate for six-month maturity Eurodollar futures (on three-month Eurodollar CDs) is 5.5%; and, at maturity in April 2019, the three-month LIBOR is set or fixed at 6%. Then the final P&L is

$$\$1{,}000{,}000 \times 0.25 \times (5.5\% - 6\%) = -\$1250 \qquad (3.36)$$

plus a small amount of accrued interest in the margin account due to marking to market. We emphasize here that, because of the daily marking to market, the implied futures rate is not equal to but bigger than the corresponding LIBOR rate, as otherwise an arbitrage opportunity can occur. The issue regarding the difference of the two rates is model-based and, as we shall see later, very interesting.

In CME, the contracts had maturities in March, June, September, and December for up to 10 years into the future.

3.7.2.3 Floating-Rate Notes

A floating-rate note (FRN) is a coupon bond that pays floating-rate coupons indexed to some specific interest rate. The most basic FRN is indexed to LIBOR,

$$f\left(T_j; T_j, T_{j+1}\right) = \frac{1}{\Delta T_j}\left(\frac{P(T_j, T_j)}{P(T_j, T_{j+1})} - 1\right), \qquad (3.37)$$

such that, at times T_{j+1}, $j = 0, 1, \ldots, n-1$, a bond holder receives a coupon payment in the dollar amount of

$$\mathrm{Pr} \times \Delta T_j \times f\left(T_j; T_j, T_{j+1}\right). \qquad (3.38)$$

Here, $\Delta T_j = T_{j+1} - T_j$ and Pr stands for the principal of the bond. At the maturity of the bond, T_n, the bondholder is also paid back the principal.

The value or price of the FRN can be easily obtained by payment replications. Consider a FRN with tenor (T_0, T_N). Then, the cash flows for the FRN can be generated by rolling forward a CD with principal Pr at times $T_j, j = 1, \ldots, N$. At any moment t between T_k and T_{k+1}, the value of the CD is

$$V_t = \frac{P(t, T_{k+1})}{P(T_k, T_{k+1})}, \qquad (3.39)$$

which is also the price of the FRN.

3.7.2.4 Swaps

A swap is a contract to exchange interest payments out of a notional principal. Most interest-rate swap contracts exchange floating-rate payments for

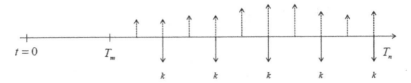

FIGURE 3.8: The cash flow pattern for a payer's swap.

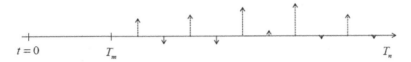

FIGURE 3.9: Net cash flows of a payer's swap.

fixed-rate payments, and only the net payment is made. The party who swaps a fixed-rate payment for a floating-rate payment (to pay the fixed rate and to receive the floating rate) is said to hold a payer's swap, whereas the counterparty is said to hold a receiver's swap. While an FRA or Eurodollar futures contract allows its holder to lock in a short-term rate in the future, a payer's swap allows its holder to lock in a long-term yield.

Let us take a look at the cash flow of a payer's swap as depicted in Figure 3.8. For generality, we let the period over which interest payments are to be exchanged be (T_m, T_n), which is called the term of the swap. The cash flows occur at the end of each accrual period. In Figure 3.8, the solid-line arrows represent fixed cash flows, whereas the dotted-line arrows represent floating and uncertain cash flows.

As was indicated earlier, only the net payment is made. The net cash flow is depicted in Figure 3.9, which is uncertain at the initiation of the swap contract (except, when $m = 0$, for the first piece of the cash flow due at T_1).

Next, let us consider the pricing of the payer's swaps, which take LIBOR, given in Equation 3.37, as the reference rate for the floating leg. The floating-leg cash flow can be generated by being long the T_m-maturity ZCB and being short the T_n-maturity ZCB, of the swap's notional value. At time T_m when the first bond matures, the notional is deposited into a CD to generate the first interest payment, with the notional rolled over to another CD until T_{n-1}. At time T_n, the notional is canceled by the T_n-maturity ZCB. Let the fixed rate be K and the number of interest-rate payments be n. Then the value of the swap at $t \leq T_m$ is the price difference between the floating leg and the fixed leg:

$$\text{Swap}(t; T_m, T_n, K) = P(t, T_m) - P(t, T_n) - \sum_{j=m}^{n-1} \Delta T_j K P(t, T_{j+1}). \quad (3.40)$$

As yet another market convention, when a swap is initiated, it takes zero value, meaning that there is no upfront payment by either party. The corresponding fixed rate of the contract is called the prevailing swap rate of the market. Denote the prevailing swap rate for the term (T_m, T_n) as $R_{m,n}(t)$. According to Equation 3.41, the prevailing swap rate is given by the formula

$$R_{m,n}(t) = \frac{P(t, T_m) - P(t, T_n)}{\sum_{j=m}^{n-1} \Delta T_j P(t, T_{j+1})}. \tag{3.41}$$

When $t = T_m$, we say that the swap contract is "spot-starting." Otherwise, if $t < T_m$, we say that it is "forward-starting."

A swap rate, in fact, is a synonym of a par yield. According to the swap rate formula, Equation 3.41, there is

$$P(t, T_m) = \sum_{j=m}^{n-1} \Delta T_j R_{m,n}(t) P(t, T_{j+1}) + P(t, T_n), \tag{3.42}$$

which implies that $R_{m,n}(t)$ is the coupon rate for a par bond (or a forward-starting par bond if $t < T_m$). In the LIBOR era, par bonds had certain liquidity.

3.7.2.5 Caps and Floors

A cap is a contract that consists of a series of call options on LIBOR and, sequentially, these options matures at time T_{j+1}, with payoff

$$\Delta T_j \left(f\left(T_j; T_j, T_{j+1}\right) - K \right)^+, \quad j = 0, 1, \ldots, n-1. \tag{3.43}$$

In the derivatives business, each option was called a caplet. When $t < T_0$, we call this cap forward-starting. Most caps traded in the market, however, were spot-starting with $t = T_0$. For a spot-starting cap, the payoff of the first caplet was known with certainty.

A floor is another contract that consists a series of put options on LIBOR with payoff

$$\Delta T_j \left(K - f\left(T_j; T_j, T_{j+1}\right) \right)^+ \tag{3.44}$$

at time $T_{j+1}, j = 0, 1, \ldots, n-1$. Each of these put options was called a floorlet. A pair of caplet and floorlet with the same maturity and strike rate satisfy put-call parity: a long caplet and a short floorlet constitute an FRA, so that given the price of a caplet one can easily determine the price of the floorlet.

The pricing of caplets is model-based. Since the LIBOR caps are no longer traded, we omit discussion of their pricing.

3.7.2.6 Swaptions

A swaption entitles its holder to enter into a swap contract at a specific fixed rate in the future. Let the specific fixed rate be K and the value of the corresponding payer's swap with tenor (T_m, T_n) be swap$(t; T_m, T_n, K)$. Then the payoff of the swaption at its maturity, $T \leq T_m$, is

$$\text{swap}(T; T_m, T_n, K)^+ = \max(\text{swap}(T; T_m, T_n, K), 0). \tag{3.45}$$

Swaptions allow investors to hedge interest-rate risk for longer terms. An investor who is concerned about higher interest rates over a period from T_m to T_n could opt to buy a swaption on the payer's swap with the prevailing swap rate for the period. If interest rates do go up, the holder will be compensated by the appreciation of the swaption.

3.7.2.7 Bermudan Swaptions

A Bermudan swaption allows its holder to exercise the option at a set of pre-specified dates prior to the maturity of the contract. There are two possibilities for the term of the underlying swap, namely, variable term and fixed term. Note that the former is much more popular. To describe the term, we let T_i be a date when early exercise is allowed. The variable term corresponds to the swap term of $(T_{i \vee m}, T_n)$, whereas the fixed term corresponds to the swap term of (T_i, T_{i+n}). Here T_m is the earliest starting date of the swap. Let T_b and T_e be the first and the last dates for exercise, respectively, and τ be the stochastic time of optimal early exercise. Then the value of the variable-term Bermudan swaption can be expressed as

$$\sup_{T_b \leq \tau \leq T_e} \mathbf{E}^{\mathbb{Q}} \left[B_{\tau}^{-1} \left(\sum_{j=i \vee m}^{n-1} \Delta T P(\tau, T_{j+1}) \right) (R_{i \vee m, n}(\tau) - K)^+ \right]. \tag{3.46}$$

while the value of the fixed-term Bermudan swaption is

$$\sup_{T_b \leq \tau \leq T_e} \mathbf{E}^{\mathbb{Q}} \left[B_{\tau}^{-1} \left(\sum_{j=i}^{i+n-1} \Delta T P(\tau, T_{j+1}) \right) (R_{i,i+n}(\tau) - K)^+ \right], \tag{3.47}$$

Here, $\mathbb{Q}$ stands for the risk-neutral measure.

Note that Bermudan swaptions are not considered to be vanilla derivative instruments like the ones we have described so far. However, their popularity or liquidity warrants them a place in this list of standard LIBOR products.

3.7.3 LIBOR Exotics

The LIBOR market was known for its vivid financial innovations due to relatively few regulations and the high level of investor participation. Many interesting exotic derivative products were first invented in this market, and these products had gained reputations as "LIBOR exotics." In this section, we introduce some of them; our list is by no means exhaustive.

Flexible Cap

A flexible cap is an interest-rate cap with a predetermined number of possible exercises or uses. The cap is automatically used if the reference interest rate is above the strike level. If the reference interest rate is fixed above the strike more times than the agreed number of exercises, then the flexible cap is terminated. If the predetermined number of possible exercises equals the total number of caplets, then the flexible cap reduces to a standard cap.

Chooser Cap

A chooser cap is similar to the flexible cap except that the exercises are chosen by the buyer, who will choose an exercise strategy that is most profitable.

Cancellable Swap

A cancellable swap is an interest-rate swap where one of the parties has the right to terminate the contract on a predetermined date at no cost. The right is equivalent to a swaption to enter a reverse swap. Owing to the right, the interest-rate payment of the counterparty will be reduced for compensation.

Resettable Swap

A resettable swap is a swap where the fixed rate paid depended on the market swap rate for the remaining maturity. The change in the swap rate is triggered by a barrier. With the right to reset, the payer can take advantage of a potential fall of the market swap rate.

Ratchet Swap

A ratchet swap is a swap where the floating leg is capped with a strike that is reset at the beginning of each interest-rate period. For example, a floating-rate payment at T_{j+1} can take the form

$$\min(f\,(T_j; T_j, T_{j+1}) + s_1,\ f\,(T_{j-1}; T_{j-1}, T_j) + s_2),$$

where s_1 and s_2 are two pre-specified spreads.

Constant Maturity Swaps

A constant maturity swap (CMS) is a variant of the vanilla swap. One leg of the swap is either fixed or reset at LIBOR. The "constant maturity" leg is indexed to a long-term interest rate, typically a long-term swap rate. When the long-term rate is a yield of a government bond, for example, a 10-year Treasury rate, then the swap is called a constant maturity treasury (CMT) swap.

Cross-Currency Swaps

A cross-currency swap is a swap of cash flows between two currencies. For example, a currency swap between USD and Euros can be structured as follows:

the term of the contract is five years, the frequency of interest-rate payment is two per year, and the principal and interest rate for each currency are listed in Table 3.6.

TABLE 3.6: Contractual Details of a Dollar/Euro Swap

	Dollar	Euro
Principal	1.25 m	1 m
Interest rate (%)	5	4.5

A major difference between a cross-currency swap and a usual vanilla swap is that, with the former, the principals are also exchanged at the maturity of the swap. Currency swaps are very sensitive to exchange-rate risk. They are perhaps the most popular OTC instruments to hedge against exchange-rate risk.

Dual Currency Basis Swap

A dual currency basis swap allows one party to switch from one currency to another on a specific date at a specified exchange rate. By giving such a right, the counterparty should be compensated with a lower interest rate.

Callable Range Accrual Swap

A callable range accrual swap is a swap where the payment on one of the legs depends on the daily fixing of an interest-rate index within a specific range. The interest payment on each payment date can be written as nK/N, where K is a fixed coupon rate, N the total number of days of the observation period, and n the number of days the index lies within the range.

Range Accruals Note

A range accrual note is an FRN with a cap and a floor on the floating rate.

Callable Range Accruals Note

A callable range accrual note is like a range accrual note, except that the issuer has the right to redeem the note after a lockout period. Owing to this call feature, the issuer needs to compensate the investor with higher interest rates.

Target Redemption Note

A target redemption note is an inverse floater structure that pays coupons equal to $\Delta T_j \max[K - \alpha f(T_j; T_j, T_{j+1}), K_{\text{floor}}]$ at time T_{j+1}, as long as the sum of all previous coupons is below a predetermined barrier. Here K_{floor} is the floor and α is the gearing factor.

3.8 Repurchasing Agreement

A repurchasing agreement (repo) is a contract to sell an asset with the promise to buy it back later (for a higher price). It is actually a short-term loan secured by the asset, which plays the role of collateral, and the difference between the sell price and the buy price is the interest of the loan. U.S. Treasury bonds are most often used as collaterals for repo trades. Thanks to the use of collaterals, the loan is safe to the lender, and the implied interest rate is lower to the borrower. Note that repos were invented around 1912 in the U.S. capital market and, about 100 years later when the LIBOR market was shrinking, became pretty much the only mean for borrowing/lending between institutions. Moreover, out of the repo rates a new index, called SOFR for Secured Overnight Financing Rate, was constructed and used for the replacement of LIBOR. The shift from LIBOR derivatives to SOFR derivatives is the primary motivation for this new edition.

Exercises

1. Compute the Macaulay duration of a 30-year par bond with 6% coupons paid semiannually. If the yield decreases by one basis point, by how much will the value of the bond change?

2. Let t_j^+ be the moment immediately after a coupon payment. Prove that, at $t = t_j^+$, the price of a coupon bond is

$$B = \frac{c \cdot \mathrm{Pr}}{y} \left(1 - \frac{1}{(1 + y\Delta t)^{N-j}} \right) + \frac{\mathrm{Pr}}{(1 + y\Delta t)^{N-j}}.$$

Furthermore, do the following:

   (a) Given $c = 5\%$, $j = 0$, $N = 20$, and $\Delta t = 0.5$, plot B against y for y varying from 0 to 0.5.

   (b) Given, in addition, $B = 102\text{-}24$, calculate y.

3. Use Equation 3.19 to prove that at $t = t_j^+$, the modified duration of a coupon bond trading at par and having $\nu = N - j$ coupons remaining, is given by

$$D_{\mathrm{mod}} = \frac{1}{y} \left(1 - \frac{1}{(1 + y\Delta t)^\nu} \right).$$

4. A coupon-paying bond will mature in 15 and a quarter years and the coupons are paid semiannually. Suppose that the principal is 100, the coupon rate is 7%, and the dirty price is 102. Compute the yield, modified duration, and convexity of the bond.

5. Assume continuous compounding. Prove the yield–forward rate relation

$$f(0; T_1, T_2) = \frac{T_2 y(T_2) - T_1 y(T_1)}{T_2 - T_1}, \quad T_2 > T_1,$$

where $\Delta T = T_2 - T_1$, and the instantaneous compounding forward rate satisfies

$$f(0; T) = y(T) + T y_T(T).$$

Here, $y(T)$ is the T-maturity term rate that relates to the zero-coupon price by $P(0, T) = e^{-T y(T)}$.

Chapter 4

LIBOR Transition and SOFR Derivatives Markets

From the late 1970s to the 2008 financial crisis, the LIBOR market was the predominant market for interest rate derivatives trades across major currencies. However, the mechanism of LIBOR setting had a major flaw, which was taken advantages of by panel banks and led to the LIBOR scandal, one of the biggest scandals in the history of finance which unfolded in 2005. As a result, central banks across the Atlantic nations decided in 2018 to phase out the indexes and adopt new ones. The last day of year 2021 marked the end of the LIBOR era, when the publication of most LIBOR settings of all currencies ceased permanently, with a few exceptions which were to be ceased gradually but no later than September 2024. As a matter of fact, over the years since the 2008 financial crisis, the role of LIBOR in interbank derivatives markets had already been reduced substantially and taken over steadily by indexes for risk-free overnight rates of the respective currencies. Nowadays, derivatives trades tied to risk-free overnight rates are sufficiently liquid, both in exchanges and over-the-counter markets.

4.1 LIBOR Scandal and the Cessation of LIBOR

LIBOR came into widespread use in the 1970s as a set of reference interest rates for trades in offshore Eurodollar markets, although there was no uniform LIBOR rates at that time (Newburg, 1978). In October 1984, the British Bankers' Association (BBA) finished the construction of the BBA standard for interest rate swaps as well as the interest-settlement rates, which were the predecessors of BBA LIBOR. Commenced officially in January 1986, a BBA panel of major international banks was formed and was responsible for setting LIBOR of various terms. Since then, BBA LIBOR has become benchmarks or indexes to settle various interest rate derivatives in the global interbank markets, including swaps, Eurodollar futures, FRA, caps/floors, swaptions and a variety of exotic derivatives. Moreover, the BBA LIBOR has also been used as benchmarks for risk-free interest rates for various terms, such that lenders charged borrowers LIBOR plus a spread. Before the 2008 financial

DOI: 10.1201/9781003389101-4

crisis, 45% of prime mortgages and 80% of subprime mortgages in the U.S. had mortgage rates indexed to the LIBOR. In addition, about half of variable-rate private student loans in the U.S. were tied to the LIBOR.[1]

The LIBOR fixing was based on the submission of borrowing and lending rates from the representatives of the panel banks. The submission was supposed to reflect the actual borrowing/lending costs of the the panel banks. However, it turned out not to be the case, as there had been a number of incentives, particularly for profiting from the trading positions of the panel banks, for representatives to either inflate or deflate the interest rate indexes. The manipulation was first revealed by *The Wall Street Journal* in April 2008,[2] which raised the issue that some banks understated their borrowing rates in order to falsely uplift the credit worthiness of their banks. The manipulation was later confirmed in 2010 by two economists,[3] who however uncovered that the main purpose of these manipulations was to make profits from banks' portfolios of LIBOR instruments. In response to various reports including the aforementioned, U.S. regulators launched investigations against some panel banks in March 2011.[4]

We can get some idea of the extent of the fraud from the following numbers. In 2012, the total outstanding notional value of LIBOR derivatives was about $500 trillions.[5] As such, a 10 basis point manipulation per annum means $500 billions P&L. Fraud of such a magnitude made it the priority for scrutiny and investigations by financial regulators and departments of criminal justice.

Barclays was the first bank convicted for rate manipulations. In June 2012, multiple criminal settlements by Barclays Bank revealed significant fraud and collusion by member banks that had the privilege to make rate submissions,[6] which is when the scandal began to catch public attention. By then, the breadth of the scandal had become evident and more investigations were launched by regulators in Britain, European Union, Canada, Japan, Switzerland, and Australia.

The outcome of the investigations brought a series of penalties or fines to the participating banks. From June 2012 to April 2015, a number of banks were fined for various amounts, including Barclays ($449 millions), Lloyds Bank ($380 millions), The Royal Bank of Scotland ($612 millions), Rabobank ($1.3 billions), UBS ($1.5 billions), and Deutsche Bank ($2.5 billions). Another

---

[1] https://archive.nytimes.com/www.nytimes.com/interactive/2012/07/10/business/dealbook/behind-the-libor-scandal.html?ref=barclaysplc

[2] Mollenkamp, Carrick (16 April 2008). "Bankers Cast Doubt on Key Rate Amid Crisis". Wall Street Journal. Archived from the original on 8 October 2013.

[3] Carl Schreck (11 October 2012). "US Students Detected Libor-Fixing Years Before Banking Scandal", Archived 21 July 2013 at the Wayback Machine RIA Novosti.

[4] Enrich, David; Mollenkamp, Carrick; Eaglesham, Jean (18 March 2011). "U.S. Libor Probe Includes BofA, Citi, UBS". The Wall Street Journal. Archived from the original on 10 July 2017

[5] BIS OTC derivatives statistics at end-December 2022, https://www.bis.org/publ/otc_hy2211.htm.

[6] "CFTC Orders Barclays to pay $200 Million Penalty for Attempted Manipulation of and False Reporting concerning Libor and Euribor Benchmark Interest Rates". Archived from the original on 30 June 2012.

number of smaller banks or brokers were penalized with smaller fines. However, that was not all. As of 2015, several banks were still under investigation for their involvement in the fraud, including JP Morgan, Citigroup and Bank of America. Because of the scandal, several senior bank executives, including the CEOs of Barclays and Rabobank banks, resigned. On the persecution of criminal conducts, a number of people, mostly traders, have been convicted, some of them were given jail terms.[7]

The LIBOR scandal destroyed public confidence in LIBOR as a set of reliable indexes of interest rates for the interbank markets, making a swift and profound reform necessary. Several official bodies had joined hands in carrying out a LIBOR reform. Initially, in July 2013, the International Organization of Securities Commissions (IOSCO) published principles for financial benchmarks,[8] demanding that the benchmarks should be based on arms-length market transactions if possible, and falling back on "expert judgement" only if necessary. In July 2014, the Official Sector Steering Group (OSSG) published a report, which was based on inputs from the Market Participants Group (MPG) and endorsed by the Financial Stability Board (FSB), on how to reform major interest rate benchmarks consistently with the IOSCO principles.[9]

In the same report, however, the MPG highlighted the low volume in unsecured term-lending markets, which made it difficult to develop a transactions-based LIBOR replacement and casted doubt on the prospect of the eventual survival of LIBOR. As a matter of fact, during the 2008 financial crisis, market participants already witnessed the evaporation of term unsecured interbank lending.

The Alternative Reference Rate Committee (ARRC), a group of private market participants set up by the Federal Reserve Board and the Federal Reserve Bank of New York in 2014 with the objective to identify risk-free alternative reference rates and support an orderly adoption, had considered several choices of term unsecured bank borrowing rates (Commercial papers, CDs, Eurodollar, and term fed funds) as the basis for a reference rate. Yet, these rates were ultimately dismissed by ARRC because of "structural difficulties" – limited transactions, not robust to stress tests, and an unstable sample of banks participating.[10] On the other hand, the majority of panel submissions for LIBOR each day turned out to be based solely on "expert judgement" rather than transactions, due to the thinness of the underlying market.[11] Up to 2017, banks were funding themselves much less with unsecured term borrowing, making reference rates tied to such borrowing correspondingly less important (Duffie and Stein, 2015; Bailey, 2017). At the same time, overnight repo market appeared to have become sufficient for facilitating U.S. dollar derivatives trades.

---

[7]Tracking the Libor Scandal, March 23, 2016, New York Times

[8]https://www.iosco.org/library/pubdocs/pdf/IOSCOPD409.pdf.

[9]https://www.fsb.org/2014/07/pr_140722/.

[10]https://www.newyorkfed.org/medialibrary/microsites/arrc/files/2016/ARRC-Dec-1-2016-meeting-minutes.pdf.

[11]The Held Report, https://www.bis.org/review/r190318f.htm.

The chance for a successful LIBOR reform deteriorated in 2017 when the Financial Conduct Authority (FCA) of the United Kingdom sensed the possibility of an imminent collapse of the reference rates (Bailey, 2017). For years, the FCA had been persuading panel banks to continue submitting to LIBOR. Yet in 2016 and 2017, two large banks dropped out of the dollar-LIBOR panel consecutively, and the scarcity of underlying transactions only increased the possibility of LIBOR manipulations (Hou and Skeie, 2014; Bailey, 2018). As early as 2016, a Wall Street Journal article already indicated that banks were eager to drop out of the panel to distance themselves from greater legal risks associated with reporting (Burne and Eisen, 2016). Bailey's report indicated that all panel banks would prefer to cease as soon as possible due to the "costs and risks of submitting expert judgements." Nonetheless, the FCA convinced remaining panel banks to continue submitting until the end of 2021, as four years was judged to be the shortest period to allow for a feasible transition to alternative reference rates.

With a potential end-date for LIBOR established, efforts to reform LIBOR shifted from finding enhanced versions to finding new reference rates. In a Wall Street Journal editorial in August 2017, both Federal Reserve and Commodity Futures Trading Commission (CFTC) called for market participants to move away from LIBOR because of its matted prospects (Giancarlo and Powell, 2017). In the case of dollar-LIBOR, ARRC had selected the Secured Overnight Financing Rate (SOFR), which will be introduced in detail next section, as the preferred new reference rate index. In March 2018, the initial reform efforts to "fix" LIBOR were abandoned, and the ARRC started to implement the transition plan to SOFR, which was widely accepted as the best long-run solution. Market participants were told that the discontinuation of LIBOR was happening and were encouraged to make the conversion to SOFR without delay. Under this background, a lot of LIBOR derivatives with maturities beyond 2021 had been either closed or switched to SOFR-based derivatives through negotiations, and newly initiated derivatives have been exclusively indexed to overnight risk-free rates.

4.2 SOFR

In April 2018, the Federal Reserve Bank of New York started to publish the SOFR. The determination of SOFR is based on data collected from the repo market. In a repo trade, if the lender of cash in a repo trade is unable to return the collateral to the borrower at maturity, then the so-called "failure" occurs, which carries penalties. In order to avoid the risk of failures, tri-party repos have been developed, where a custodial bank holds the collateral as a third party to the repo transaction. In a tri-party repo, which is free of the risk of failure, no specific bonds can be selected as collateral, so the tri-party repo rate can be considered as a general collateral rate. For this reason, on any given day, the probability distribution of the tri-party repo rates is generally narrower than that of the bilateral repo rates. As such, repo rates have the following important features required by the ARRC for composing a new index to replace LIBOR.

1. The repo market is large enough, with over one trillion dollar transactions since 2017, that is hardly manipulable.

2. Repo rates are considered stable, in the sense that instability only comes from the specialty of collaterals, which can however be excluded from the computation of an aggregated index.

The input repo rates actually come from three sources:

1. Tri-party repos conducted with Bank of New York Mellon (BNYM) as a clearing bank.

2. Tri-party repos executed through General Collateral Financing (GCF) that are cleared by Fixed Income Clearing Corp (FICC).

3. Bilateral repos cleared through FICC's delivery-versus-payment (DVP) service, with transactions below the volume-weighted 25th percentile removed.

There are also some additional treatments to enhance the credibility of data. First, trades between affiliates are excluded. Second, doubtful data is manually excluded by the New York Fed. The SOFR index, then, is calculated as a volume-weighted median of admitted repo rates, which is the rate associated with transactions at the 50th percentile of transaction volume. Specifically, the volume-weighted median rate is calculated by ordering the transactions from lowest to highest rate, taking the cumulative sum of volumes of these transactions and identifying the rate associated with the trades at the 50th percentile of dollar volume.[12] [13] At publication, the volume-weighted median is rounded to the nearest basis point. Hence, SOFR can be regarded as a broad measure of the overnight Treasury repo rates.

4.3   Transition to Risk-Free Overnight Rates and LIBOR Fallbacks

The Federal Reserve Board and the New York Fed adopted SOFR to replace LIBOR because of its several key features, as are summarized in Table 4.1.

---

[12]https://www.newyorkfed.org/markets/reference-rates/additional-information-about-reference-rates#information_about_treasury_repo_reference_rates.

[13]Federal Register Vol. 82, No. 167.

TABLE 4.1: LIBOR vs. SOFR

LIBOR	SOFR
Bank-to-bank lending rate (with credit risk)	Risk-free rates (without credit risk)
Unsecured	Secures with U.S. Treasuries
Based on bank submissions incorporating actual transactions and expert judgement	Transaction based
$500 million USD of daily transactions in the 3m wholesale funding market	Over $1 trillion of daily transactions in the overnight Repo market since 2017
Forward-looking term structure	backward-looking overnight

While LIBOR is based on panel bank inputs, SOFR is a broad measure of the cost of borrowing cash overnight collateralized by U.S. Treasury securities in the repo market. The daily transaction volumes of the repos reached $1 trillion in 2017, and it was regularly between $4 to 5 trillions in 2023, as is shown in Figure 4.1.[14] The repo markets large transaction volume gives the ARRC confidence that SOFR will be reliable through a wide range of market conditions, making it a good long-term substitute of LIBOR.

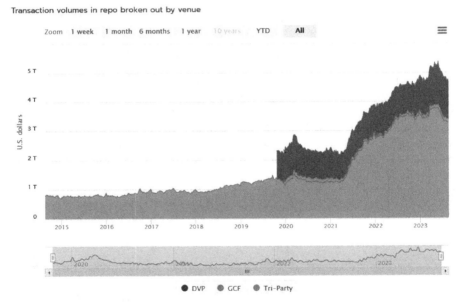

FIGURE 4.1: Repo Transaction Volumes by Venue.

The transition from LIBOR to secured overnight rates was also happening beyond the U.S. Several other markets were also transitioning away from their

---

[14]Office of Financial Research (OFR) Short-term Funding Monitor, https://www. financialresearch.gov/short-term-funding-monitor/market-digests/volume/chart-26/

relevant currency IBOR to overnight risk-free rates. By the end of 2021, the countries listed below had already transitioned to various alternatives.

- United Kingdom adopted the Sterling Over Night Indexed Average (SO-NIA).

- European Union adopted the Euro Short-Term Rate (ESTR).

- Japan adopted the Tokyo Overnight Average Rate (TONAR).

- Canada adopted the Canadian Overnight Repo Rate Average (CORRA).

Since LIBOR is referenced in many different types of financial contracts, moving away from the rate could impact a variety of businesses and individuals. This could include:

- Corporate and municipal borrowers financing operations with LIBOR-based floating rate loans and/or bonds

- End-users hedging risk with LIBOR-based derivatives

- Investment banks underwriting, issuing and making markets in LIBOR-based instruments

- Investors managing portfolios of swaps, bonds and loans tied to LIBOR

- Consumers with mortgages or student loans tied to LIBOR

- Certain credit cards that use LIBOR

For various reasons, there are still some LIBOR derivatives that would remain outstanding beyond 2021, so-called legacy contracts, and need to be serviced, thus creating the demand for credit-sensitive benchmarks to play the roles of the previous LIBOR. A 2017 survey by a LIBOR-panel bank published in the March 2019 BIS Quarterly Review found that 80 percent of respondents wanted LIBOR to continue in some form (Schrimpf and Sushko, 2019). There has been more than one proposal to adopt "synthetic LIBOR" for legacy derivatives.

On March 5, 2021, the FCA of the UK issued an announcement on the future cessation/abolishment and the fallback arrangement of the LIBOR benchmarks. It mandated that:

- all seven tenors for both Euro and Swiss franc LIBOR would be ceased permanently after December 31, 2021;

- the overnight and 12-month USD LIBOR would be ceased permanently after June 30, 2023;

The remaining LIBOR settings would, however, be published by the ICE Benchmark Administration (IBA), the LIBORs administrator, on a non-representative basis, such that

- 1-, 3- and 6-month sterling LIBOR would be published for a further period after the end of 2021;

- 1-, 3- and 6-month Yen LIBOR would be published for an additional year; and

- 1-, 3- and 6-month USD LIBOR would be published for a further period after the end of June 2023.

FCA also stressed that the use of synthetic LIBOR would not be permitted for new trades.

In an ISDA statement made on the same day,[15] March 5, 2021 was declared the "Spread Adjustment Fixing date" for all LIBOR tenors across all LIBOR currencies, and "the fallbacks (i.e., to the adjusted risk-free rate plus spread) will automatically occur for outstanding derivatives contracts" according to spreads set forth in Table 4.2.[16]

The latest development occurred on April 3, 2023, when FCA announced its decision to require IBA to continue to publish the 1-, 3- and 6-month USD LIBOR settings under an unrepresentative "synthetic" methodology until the end of September 2024, in need to serve legacy contracts. After this, publication will cease permanently.

4.4 Linear Derivatives Based on the Risk-Free Rates

4.4.1 SOFR Futures

The SOFR futures were launched in May 2018, one month after the publication of SOFR by the Federal Bank of New York. There are two kinds of SOFR futures: three-month (3m) SOFR futures and one-month (1m) SOFR futures, which yet adopt different conventions for determining settlement interest rates. For a smooth transition from unsecured rates to secured rates, the 1m SOFR futures and 3m SOFR futures have largely adopted the structures of Federal fund futures and ED futures, respectively. There is abundant liquidity for both futures contracts. As the replacement for the once almighty ED futures, the 3m SOFR futures have higher liquidity.

In terms of the structure, the 3m SOFR futures are very similar to the ED futures. One basis point of change in the expected 3m SOFR term rate will cause a P&L of $25, corresponding to a notional value of 1 million dollars for the futures contract. The tick size for the change in 3m SOFR term rate is a quarter of a basis point, corresponding to a P&L of $6.25. Similar to that of ED futures, the maturities of SOFR futures are up to 10 years. Precisely

---

[15] https://www.isda.org/2021/03/05/isda-statement-on-uk-fca-libor-announcement/.
[16] https://www.isda.org/2021/03/05/libor-cessation-and-the-impact-on-fallbacks/

TABLE 4.2: List of Impacted LIBOR Fallbacks

LIBOR	Tenor	Ticker	Spread Adjustment (%)
CHF	Spot/Next	SSF00SN Index	-0.0551
CHF	1 week	SSF0001W Index	-0.0705
CHF	1 month	SSF0001M Index	-0.0571
CHF	2 month	SSF0002M Index	-0.0231
CHF	3 month	SSF0003M Index	0.0031
CHF	6 month	SSF0006M Index	0.0741
CHF	12 month	SSF0012M Index	0.2048
EUR	Overnight	SEE00ON Index	0.0017
EUR	1 week	SEE0001W Index	0.0243
EUR	1 month	SEE0001M Index	0.0456
EUR	2 month	SEE0002M Index	0.0753
EUR	3 month	SEE0003M Index	0.0962
EUR	6 month	SEE0006M Index	0.1537
EUR	12 month	SEE0012M Index	0.2993
GBP	Overnight	SBP00ON Index	-0.0024
GBP	1 week	SBP0001W Index	0.0168
GBP	1 month	SBP0001M Index	0.0326
GBP	2 month	SBP0002M Index	0.0633
GBP	3 month	SBP0003M Index	0.1193
GBP	6 month	SBP0006M Index	0.2766
GBP	12 month	SBP0012M Index	0.4644
JPY	Spot/Next	SBP00SN Index	-0.01839
JPY	1 week	SJY0001W Index	-0.01981
JPY	1 month	SJY0001M Index	-0.02923
JPY	2 month	SJY0002M Index	-0.00449
JPY	3 month	SJY0003M Index	0.00835
JPY	6 month	SJY0006M Index	0.05809
JPY	12 month	SJY0012M Index	0.16600
USD	Overnight	SUS00ON Index	0.00644
USD	1 week	SJY0001W Index	0.03839
USD	1 month	SUS0001M Index	0.11448
USD	2 month	SUS0002M Index	0.18456
USD	3 month	SUS0003M Index	0.26161
USD	6 month	SUS0006M Index	0.42826
USD	12 month	SUS0012M Index	0.71513

https://assets.bbhub.io/professional/sites/10/IBOR-Fallbacks-LIBOR-Cessation_Announcement_20210305.pdf speaking, the maturities of 3m futures are the third Friday of the nearest quarterly months of March, June, September and December. Perhaps the only difference between the SOFR futures and the ED futures is in the calculation of the settlement rates at futures' maturity: the former uses daily compounding over the reference period which lies ahead of the maturity of the SOFR futures, while the latter adopted the 3m LIBOR rate for the reference period which was beyond the maturity of the ED futures. The formula for the settlement rate of 3m SOFR futures is[17]

$$\tilde{f}_{3m} = \frac{\prod_{i=1}^{N_b} \left(1 + \frac{SOFR_i \times d_i}{360}\right) - 1}{\frac{N_c}{360}},$$

where $SOFR_i$ is the annualized SOFR rate for business day i, d_i is the number of calendar days that $SOFR_i$ applies, N_b is the number of business days for the reference quarter, and N_c is the number of calendar days for the reference quarter. Note that the SOFR rate for a business day is published at 8am U.S. eastern time (EST) of the next business day, and the SOFR rate for weekends or holidays is taken from the last proceeding business day.

The 1m SOFR futures calculate settlement interest rates through arithmetic average of daily SOFR rates over the corresponding reference month. The formula for the settlement rate is

$$\tilde{f}_{1m} = \frac{1}{N_c} \sum_{i=1}^{N_b} SOFR_i \times d_i.$$

The available maturities of 1m SOFR futures are the ends of the nearest 13 calendar months. The 1m SOFR futures copy the structure of the Fed fund futures. One basis point of change in the expected 1m SOFR futures rate will cause a P&L of $41.67, corresponding to a notional value of 5 million dollars for the 1m SOFR futures contract. The tick size for the change in 1m SOFR futures rate is a quarter of a basis point, corresponding to a P&L of $10.4175.

Attributing to the rapidly increasing volume of the 3m SOFR futures, options have also been introduced on these futures. A partial list of SOFR products traded in Chicago Mercantile Exchange (CME) is shown in Table 4.3. The so-called mid-curve options are options that expire before the underlying futures mature, which were first launched in January 2020. On March 22, for example, options on 3m SOFR futures available to trade are shown on the options grid that is demonstrated in Figure 4.2.

The launch of 1m SOFR (SR1) futures and 3m SOFR (SR3) futures also brought several new inter-commodity spreads (ICS) to the CMEs short-term interest rate (STIR) futures offerings, as listed below:

- One-month SOFR/30-day Federal Funds (SR1/FF)

---

[17]https://www.newyorkfed.org/markets/reference-rates/additional-information-about-reference-rates#tgcr_bgcr_sofr_calculation_methodology.

TABLE 4.3: SOFR Futures and Futures Options

Globex	Product Name
SR3	Options on Three-Month SOFR Futures
SR3	Three-Month SOFR Futures
S0	One-Year Mid-Curve Options on Three-Month SOFR Futures
S2	Two-Year Mid-Curve Options on Three-Month SOFR Futures
SR1	One-Month SOFR Futures
S3	Three-Year Mid-Curve Options on Three-Month SOFR Futures

- Three-month SOFR/Three-month Eurodollar (SR3/ED)

- One-month SOFR/Three-month SOFR (SR1/SR3)

- Three-month SOFR/30-day Federal Funds (SR3/FF)

All such spread positions will be cleared by CME Clearing, and will be eligible for margin offsets/spread credits, irrespective of how they are executed. Note that the SOFR is regarded as a secured interest rate or risk-free rate, while the Fed Fund rate is not.[18] It is well known that there are bases between these secured and unsecured interest rates, and basis swaps serve the purpose of eliminating or protecting against the basis risks. The end of the publication of 3m LIBOR in June 30, 2023 inevitably brought an end to the trading of ED futures as well as the basis swaps between 3m SOFR and 3m LIBOR.

4.4.2   SOFR Swaps and SOFR Discount Curve

In an overnight index swap, the fixed leg makes periodic interest payments based on a fixed rate, while the floating leg makes interest payments accrued through daily compounding based on an overnight rate, and both legs share the same notional value. In Figure 4.3, we display the cash flow pattern of a receiver's swap, where arrows pointing upwards (downwards) represent cash inflows (outflows), and the dash-line arrows represent uncertain cash outflows. The effective Federal fund rate (EFFR) had been the only choice of the overnight rate since the late 1990s, until SOFR emerged as an alternative in 2018, and since then the volume of the SOFR swaps has grown rapidly. The structures of these two OIS swaps are almost identical and the two swap rates stay very close most of the time. Let us look at the SOFR swaps and the determination of the so-called at-the-money swap rates that nullifies the value of the SOFR swaps.

Without loss of generality, we allow different payment frequencies for the two legs[19] and take the notional value of $1. Note that both legs take the actual/360 day-count convention (DCC) to calculate payments. Let the fixed

---

[18]Yet SOFR is actually more volatile than the fed fund rate.

[19]Typical SOFR swaps, so-called SOFR OIS swaps, adopt annual payments for both leg.

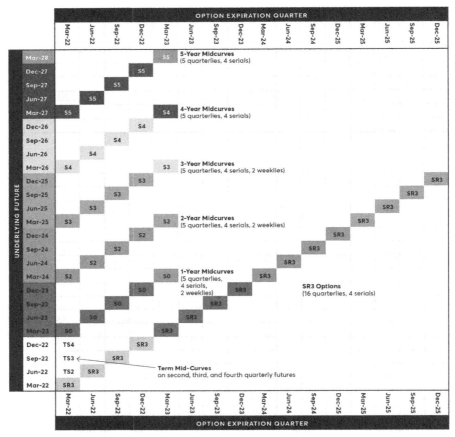

Underlying futures contract months identified by Month-Year and Strip color in column on vertical axis

FIGURE 4.2: Mid-Curve Options Grid on SOFR futures as of March 2022.

rate be K and the swap tenor be $(\tilde{T}_m, \tilde{T}_n)$. Then the value of the fixed leg is simply

$$OIS_{fixed}(t; \tilde{T}_m, \tilde{T}_n, k) = K \sum_{j=m+1}^{n} \Delta \tilde{T}_j P(t, \tilde{T}_j), \qquad (4.1)$$

where $\Delta \tilde{T}_j = \tilde{T}_j - \tilde{T}_{j-1}$, and $P(t, \tilde{T}_j)$ is the SOFR discount factor or SOFR zero-coupon bond, defined by

$$
\begin{aligned}
P(t, T) &= E_t^Q \left[\prod_{i=1}^{N_{t,T}} \left(1 + \frac{SOFR_i \times d_i}{360} \right)^{-1} \right] \\
&\approx E_t^Q \left[e^{-\int_t^T r_u du} \right].
\end{aligned}
\qquad (4.2)
$$

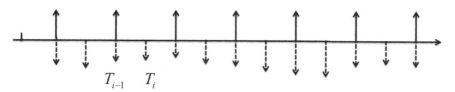

FIGURE 4.3: Cash flows of a receiver's swap.

Here, $\mathbb{Q}$ is the martingale measure corresponding to the SOFR saving account as the numeraire asset, $N_{t,T}$ is the number of business days between t and T, r_t is the continuous version of the SOFR rate, and daily compounding is approximated by continuous compounding in order to pay the way for continuous-time modeling of SOFR derivatives.

The floating leg consists of a sequence of cash flows at $T_j, j = 1, \ldots, N$, in the form of

$$\prod_{i=1}^{N_{b,j}} \left(1 + \frac{SOFR_i^{(j)} \times d_i}{360} \right) - 1. \tag{4.3}$$

Here, $SOFR_i^{(j)}$ is the overnight SOFR rate for the ith business date in the period of $(T_{j-1}, T_j]$, d_i is the number of calendar days $SOFR_i^{(j)}$ applies, and $N_{b,j}$ is the number of business days over the period $(T_{j-1}, T_j]$. The payment frequency of the floating leg can be different from that of the fixed leg. The value of the floating leg can be figured out through cash flow replications by being a long $\tilde{T}_m$-maturity SOFR ZCB and being short a $\tilde{T}_n$-maturity SOFR ZCB, similar to the pricing of the floating leg of the corresponding LIBOR swaps.[20] At its maturity, the $\tilde{T}_m$-maturity ZCB is deposited in the saving account. In each payment day, $T_j, j = m\theta + 1, \ldots, n\theta$, where θ is the ratio of floating-leg payment frequency over the fixed-leg payment frequency, the interest is paid out to meet the liability of the swap. At T_n, the notional amount of the saving account is used to cover the liability of the T_n-maturity ZCB. As such, the value of the floating-leg is

$$OIS_{float}(t; \tilde{T}_m, \tilde{T}_n, K) = P(t, \tilde{T}_m) - P(t, \tilde{T}_n). \tag{4.4}$$

Combining Equation 4.1 and Equation 4.4, we obtain the value of the SOFR payer's swap at $t \leq \tilde{T}_m$:

$$\text{OIS}(t; \tilde{T}_m, \tilde{T}_n, K) = P(t, \tilde{T}_m) - P(t, \tilde{T}_n) - K \sum_{j=m+1}^{n} \Delta \tilde{T}_j P(t, \tilde{T}_j). \tag{4.5}$$

It is a market convention that, at initiation, the swap value is zero, which

---

[20]In reality, this can only be achieved approximately through the use of rolling reverse repos.

is achieved by taking the market-prevailing swap rate (also called the ATM swap rate), defined as

$$R_{m,n}(t) = \frac{P(t, \tilde{T}_m) - P(t, \tilde{T}_n)}{A_{m,n}(t)}, \qquad (4.6)$$

where

$$A_{m,n}(t) \overset{\triangle}{=} \sum_{j=m+1}^{n} \Delta\tilde{T}_j P(t, \tilde{T}_j) \qquad (4.7)$$

is an annuity and "$\overset{\triangle}{=}$" means "being defined by". Not surprisingly, this is the same as the formula for the market prevailing LIBOR swap rate that market participants were used to. When $t = \tilde{T}_m$, the SOFR swap is "spot-starting." Otherwise, when $t < \tilde{T}_m$, the SOFR swap is "forward-starting." By early 2022, the liquidity of SOFR swaps had caught up with that of the LIBOR swaps.[21] About half of all activity in these standard swaps occurs in just four tenors - 2Y, 5Y, 10Y and 30Y.

Here, we make comments on an outstanding and important issue: the determination of the SOFR discount curve $P(t, T)$ for $T > t$. For this purpose, we must take most liquid (linear) SOFR derivatives as inputs, which include 3m SOFR futures of maturities up to two years, 1m SOFR futures of maturities up to a year, and SOFR swaps of maturities up to ten years. For terms beyond ten years and under thirty years, we will use the SOFR swaps or, in case the liquidity is not satisfying, the basis swaps of Fed fund rates versus SOFR. Out of the price of futures, swaps and basis swaps, we will be able to extract the backward-looking term rates for the 30-year horizon, and a bootstrapping process will be necessary for the extraction. The details of a bootstrapping method for production uses is presented in Chapter 9. In addition, it is well known that the futures rates carry convexity bias over their corresponding forward rates, and the technique of convexity adjustment must be adopted to remove such bias. Yet, the technique of convexity adjustment is model dependent. In three of the following chapters, we will address convexity adjustment under three different models for SOFR derivatives.

4.4.3　Forward-Rate Agreements

Forward-rate agreements (FRA) in the SOFR derivatives market are similar to the SOFR futures, yet they are traded over the counter and no margin account is required. In a SOFR FRA, two parties bet on the realized term rate over the period from $(T - \Delta T, T)$ in the future to be above or below an agreed rate, say f_0 at time $t = 0$. At $t = T$, the maturity of the FRA, the party who takes the long position will have a P&L defined by

$$\text{P\&L}(T) = Not. \times \Delta T(f_T - f_0),$$

---

[21]Adam Peralta (2022), SOFR liquidity eclipses LIBOR. https://www.bloomberg.com/professional/blog/sofr-liquidity-eclipses-libor/

where $Not.$ stands the notional value of the FRA and f_T is the realized term rate seen at the end of the reference period,

$$f_T = \frac{\prod_{i=1}^{N_b}(1 + \frac{r_i \times d_i}{360}) - 1}{\frac{N_c}{360}}.$$

Here, r_i is used in place of $SOFR_i$ for notational efficiency, d_i is the number of days r_i applies, N_b is the number of business days over the reference period $(T - \Delta T, T)$, and N_c is the number of calendar days over the reference period.

Initially, at $t = 0$, there is no upfront payment for the FRA. Assume a notional value of \$1. The party who short the FRA can secure his payment through the following transactions:

- short $P(0, T - \Delta T)/P(0, T)$ unit of T-maturity ZCB, and

- long 1 unit of $T - \Delta T$-maturity ZCB.

Note that the short FRA plus the above transactions constitute a set of zero-net transactions. At time $T - \Delta T$, a loan of \$1 is deposited in the savings account that earns SOFR with daily compounding. At T when the FRA matures, there is the following P&L for the zero-net transactions:

$$\text{P\&L} = (1 + \Delta T f_0) - \frac{P(0, T - \Delta T)}{P(0, T)}.$$

To avoid arbitrage, the fixed forward rate must be taken to be

$$f_0 = f(0; T - \Delta T, T) \triangleq \frac{1}{\Delta T}\left(\frac{P(0, T - \Delta T)}{P(0, T)} - 1\right).$$

With identical arguments, we establish that the fixed forward rate at a later time $t \leq T - \Delta T$ is

$$f_t = f(t; T - \Delta T, T) \triangleq \frac{1}{\Delta T}\left(\frac{P(t, T - \Delta T)}{P(t, T)} - 1\right).$$

If the party who is short the FRA wants to close it out at $t \leq T - \Delta T$, he/she only needs to long the FRA at the fixed rate f_t, and thus ends up with the following P&L:

$$\text{P\&L} = P(t, T)\Delta T(f_0 - f_t),$$

which is the general P&L formula for the party who is short the FRA.

For mathematical modeling, we consider the limiting case, $\Delta T \to 0$, for the forward rate. There is

$$
\begin{aligned}
f(t,T) &\triangleq \lim_{\Delta T \to 0} f(t; T - \Delta T, T) \\
&= \lim_{\Delta T \to 0} \frac{1}{\Delta T} \left(\frac{P(t, T - \Delta T)}{P(t,T)} - 1 \right) \\
&= \frac{-1}{P(t,T)} \frac{\partial P(t,T)}{\partial T} \\
&= -\frac{\partial \ln P(t,T)}{\partial T},
\end{aligned} \tag{4.8}
$$

which is the *instantaneous forward rate*. According to Equation 4.8, we can express the price of a T-maturity zero-coupon bond in terms of $f(t,s)$, $t \le s \le T$:

$$
P(t,T) = \mathrm{e}^{-\int_t^T f(t,s)\mathrm{d}s}.
$$

Therefore, once we have a model for the forward rates, we also have a model for the discount bonds. It turned that the instantaneous forward rates had indeed been a popular choice of state variables for interest-rate modeling.

4.5   Nonlinear Derivatives Based on the Risk-Free Rates

4.5.1   Caps and Floors

Parallel to nonlinear LIBOR derivatives, nonlinear derivatives based on USD SOFR include caps, floors, and swaptions, for which the payoffs are calculated based on daily compounding using SOFR. We begin with caps. Assume a notional value of \$1, then a cap consists of a sequence of cash flows at $T_j, j = 1, \ldots, N$, in the form of

$$
c_j(T_j) = \Delta T_j (f_j(T_j) - K)^+, \tag{4.9}
$$

where $f_j(T_j)$ is the "floating rate" over the period $(T_{j-1}, T_j]$ resulted from daily compounding using SOFR:

$$
f_j(T_j) = \frac{\prod_{i=1}^{N_{b,j}} \left(1 + \frac{r_i^{(j)} \times d_i}{360} \right) - 1}{\frac{N_{c,j}}{360}}. \tag{4.10}
$$

Here, $r_i^{(j)}$, $(T_{j-1}, T_j]$, d_i and $N_{b,j}$ are already defined under SOFR swaps, and $N_{c,j}$ is the number of calendar days over the same period. Because of the actual/360 day-count convention, we set $\Delta T_j = N_{c,j}/360$. In market jargons, the pieces of a cap's cash flows are called caplets. We want to emphasize that the floating rate $f_j(T_j)$ is to be fixed at T_j, unlike the forward-looking LIBOR rate for the same period $(T_{j-1}, T_j]$ which is to be fixed at T_{j-1}.

The value of the cap is obtained by aggregating the values of caplets. According to the general pricing principle established in Chapter 2, the price of the jth caplet at any time $t \leq T_j$ can be expressed below using the risk-neutral measure:

$$c_j(t) = \Delta T_j E_t^Q \left[e^{-\int_{T_{j-1}}^{T_j} r_s ds} (f_j(T_j) - K)^+ \right], \quad j = 1, \ldots, N. \qquad (4.11)$$

Here, we have replaced daily compounding by continuous compounding for convenience, and

$$f_j(T_j) \approx \frac{e^{\int_{T_{j-1}}^{T_j} r_s ds} - 1}{\Delta T_j}.$$

A floor consists of a sequence of put options on the backward-looking forward rates, with maturity T_j and payoff

$$fl_j(T_j) = \Delta T_j (K - f_j(T_j))^+, \quad \text{for } j = 1, \ldots, N. \qquad (4.12)$$

Each of these cash flows is called a floorlet, whose time-t value can be expressed as

$$fl_j(t) = \Delta T_j E_t^Q \left[e^{-\int_{T_{j-1}}^{T_j} r_s ds} (K - f_j(T_j))^+ \right], \quad j = 1, \ldots, N.$$

In fact, once we have obtained the price of the caplet with the same strike, we can derive the price of the floorlet through call-put parity.

4.5.2 Swaptions

A swaption is an option to enter into a swap, as either a payer or receiver. Take the payer's swap for example. Let the maturity of the swaption be $\tilde{T}_m$, tenor of the underlying swap be $(\tilde{T}_m, \tilde{T}_n)$, and the strike rate be K. At time $\tilde{T}_m$, the holder of the swaption has the right to enter into a swap to pay the interests accrued based on the fixed strike rate and receive the interests accrued through daily compounding out of a pre-agreed notional value, according to perhaps different payment frequencies. If the underlying is a typical SOFR swap, then both the fixed leg and the floating leg pay annually. At the maturity of the swaption, the value of the swaption is

$$\text{swtn}(\tilde{T}_m, \tilde{T}_n, K) = \max \left(\text{swap}(\tilde{T}_m, \tilde{T}_n, K), 0 \right)$$
$$= \max \left(1 - P(\tilde{T}_m, \tilde{T}_n) - K \sum_{i=m+1}^{n} \Delta \tilde{T}_i P(\tilde{T}_m, \tilde{T}_i), 0 \right). \qquad (4.13)$$

If the swaption ends in the money at maturity, the swaption holder could enter into a receiver's swap at the ATM swap rate, thus secure a sequence of fixed cash flows with time-$\tilde{T}_m$ present value equals to

$$\text{swtn}(\tilde{T}_m, \tilde{T}_n, K) = \left(\sum_{j=m+1}^{n} \Delta \tilde{T}_j P(\tilde{T}_m, \tilde{T}_j) \right) \left(R_{m,n}(\tilde{T}_m) - K \right)^+. \quad (4.14)$$

The last expression allows us to treat the swaption as a call option on the swap rate, thus paving a way for more efficient pricing methodologies.

4.5.3   Bermudan Swaptions

Finally, we mention Bermudan swaptions. As a kind of the most popular products of the LIBOR markets, Bermudan swaptions were carried over directly to the SOFR markets. After slight adaptations, pricing methods for Bermudan swaptions on LIBOR swaps can be applied for pricing Bermudan swaptions on SOFR swaps. Bermudan swaptions are intriguing because of the right to early exercise, and typically this feature is priced through dynamic programming, which will be introduced in Chapter 9 in the context of Monte Carlo simulation methods.

Chapter 5

Forward Measures and the Black Formula

Forward measures are the martingale measures corresponding to the numeraire assets of zero-coupon bonds. Asset pricing under appropriate forward measures can avoid dealing with stochastic interest rates explicitly. Through switching from the risk-neutral measure to forward measures, we demonstrate the advantages brought forward by choosing proper pricing measures, which is at the disposal of quantitative analysts or traders, in derivatives pricing. We will discuss the change of measure corresponding to the change of numeraire assets in general.

In contrast to forward prices, we also introduce futures prices, the latter are the fair strike prices for futures contracts. The results serve as the foundation for pricing SOFR futures, which will be addressed in detail after the introduction of each interest rate model.

5.1 Lognormal Model: The Starting Point

The theoretical basis of this chapter starts from the usual assumption of lognormal asset dynamics for zero-coupon bonds of all maturities:

$$\mathrm{d}P(t,T) = P(t,T)\left[\mu(t,T)\,\mathrm{d}t + \mathbf{\Sigma}^{\mathrm{T}}(t,T)\,\mathrm{d}\tilde{\mathbf{W}}_t\right], \tag{5.1}$$

under the physical measure, $\mathbb{P}$. Here, $\mu(t,T)$ is a scalar function of t and T, $\mathbf{\Sigma}(t,T)$ is a column vector,

$$\mathbf{\Sigma}(t,T) = \left(\Sigma_1(t,T), \Sigma_2(t,T), \ldots, \Sigma_n(t,T)\right)^{\mathrm{T}},$$

and $\tilde{\mathbf{W}}_t$ is an n-dimensional $\mathbb{P}$-Brownian motion,

$$\tilde{\mathbf{W}}_t = \left(\tilde{W}_1(t), \tilde{W}_2(t), \ldots, \tilde{W}_n(t)\right)^{\mathrm{T}}.$$

In principle, the coefficients in Equation 5.1 can be estimated from time series data of zero-coupon bonds, yet it is not guaranteed that Equation 5.1 with estimated drift and volatility functions can exclude arbitrage. For the time being, we assume that both $\mu(t,T)$ and $\mathbf{\Sigma}(t,T)$ are sufficiently regular

DOI: 10.1201/9781003389101-5

deterministic functions of t, so that the stochastic differential equation (SDE) (Equation 5.1) admits a unique strong solution.

The purpose of a model like Equation 5.1 is to price derivatives depending on (a portfolio of) $P(t,T), \forall T$ and $t \leq T$. For this purpose, we need to find a martingale measure for zero-coupon bonds of all maturities. Similar to our discussions on the multiple-asset market, we define an $\mathcal{F}_t$-adaptive process, $\boldsymbol{\gamma}_t$, that satisfies the following equation:

$$\boldsymbol{\Sigma}^{\mathrm{T}}(t,T)\boldsymbol{\gamma}_t = \boldsymbol{\mu}(t,T) - r_t \mathbf{I}. \tag{5.2}$$

Suppose that such a $\boldsymbol{\gamma}_t$ exists, is independent of T, and satisfies the Novikov condition. We can define a measure, $\mathbb{Q}$, through

$$\left.\frac{\mathrm{d}\mathbb{Q}}{\mathrm{d}\mathbb{P}}\right|_{\mathcal{F}_t} = \exp\left(\int_0^t -\boldsymbol{\gamma}_s^{\mathrm{T}}\mathrm{d}\tilde{\mathbf{W}}_s - \frac{1}{2}\|\boldsymbol{\gamma}_s\|_2^2\,\mathrm{d}s\right). \tag{5.3}$$

Then, by the CMG theorem, the process

$$\mathbf{W}_t = \tilde{\mathbf{W}}_t + \int_0^t \boldsymbol{\gamma}_s\,\mathrm{d}s \tag{5.4}$$

is a $\mathbb{Q}$-Brownian motion, and, in terms of $\tilde{\mathbf{W}}_t$, we can rewrite Equation 5.1 as

$$\begin{aligned}\mathrm{d}P(t,T) &= P(t,T)\left[r_t\,\mathrm{d}t + \boldsymbol{\Sigma}^{\mathrm{T}}(t,T)\left(\mathrm{d}\tilde{\mathbf{W}}_t + \boldsymbol{\gamma}_t\,\mathrm{d}t\right)\right]\\ &= P(t,T)\left[r_t\,\mathrm{d}t + \boldsymbol{\Sigma}^{\mathrm{T}}(t,T)\,\mathrm{d}\mathbf{W}_t\right].\end{aligned} \tag{5.5}$$

It then follows that the discounted prices of all maturities, $B_t^{-1}P(t,T)$, are $\mathbb{Q}$-martingales.

Now let us address the existence of $\boldsymbol{\gamma}_t$, the solution to Equation 5.2. Without loss of generality, we assume that the market of zero-coupon bonds is non-degenerate. That is, there exist at least n distinct zero-coupon bonds such that their volatility vectors constitute a non-singular matrix. Let $\{T_i\}_{i=1}^n$ be the maturities of the n bonds such that $T_i < T_{i+1}$, and let $\{\boldsymbol{\Sigma}(t,T_i)\}_{i=1}^n$ be the column vectors of their volatilities. By introducing matrices

$$\mathbf{A} = \begin{pmatrix} \boldsymbol{\Sigma}^{\mathrm{T}}(t,T_1)\\ \boldsymbol{\Sigma}^{\mathrm{T}}(t,T_2)\\ \vdots\\ \boldsymbol{\Sigma}^{\mathrm{T}}(t,T_n) \end{pmatrix}, \quad \boldsymbol{\mu}_t = \begin{pmatrix} \mu(t,T_1)\\ \mu(t,T_2)\\ \vdots\\ \mu(t,T_n) \end{pmatrix}, \quad \text{and} \quad \mathbf{I} = \begin{pmatrix} 1\\ 1\\ \vdots\\ 1 \end{pmatrix},$$

we then define $\boldsymbol{\gamma}_t$ as the solution to the linear system

$$\mathbf{A}\boldsymbol{\gamma}_t = \boldsymbol{\mu}_t - r_t\mathbf{I}. \tag{5.6}$$

This solution is unique provided that $\mathbf{A}$ is non-singular. Such a solution, however, appears to depend on $T_i, 1 \leq i \leq n$, the input maturities. With such a

γ_t, we can define a new measure, $\mathbb{Q}$, from Equation 5.3, and under which the discounted prices of those n zero-coupon bonds, $B_t^{-1}P(t, T_i), i = 1, \ldots, n$, are martingales.

Next, we will show that the solution to Equation 5.6, γ_t, also satisfies

$$\mathbf{\Sigma}^{\mathrm{T}}(t, T)\gamma_t = \mu(t, T) - r_t, \quad \forall T \leq T_n. \tag{5.7}$$

The implication is that the new measure, $\mathbb{Q}$, defined using γ_t is a martingale measure for zero-coupon bonds of all maturities $T \leq T_n$. To show that, we define

$$N_t = E^{\mathbb{Q}}\left[B_T^{-1} \mid \mathcal{F}_t\right], \quad \forall T \leq T_n,$$

which is a $\mathbb{Q}$-martingale by definition. According to the (multi-factor version of the) martingale representation theorem, we know that N_t is the value of a self-financing portfolio consisting of the money market account and the n zero-coupon bonds of maturities, $\{T_i\}_{i=1}^n$. The value of the portfolio at time T is nothing else but one, which is identical to the value of the T-maturity zero-coupon bond at maturity. By the dominance principle, there must be $B_t^{-1}P(t, T) = N_t$, so it follows that $B_t^{-1}P(t, T)$ is also a $\mathbb{Q}$-martingale. In terms of the $\mathbf{W}_t$ defined in Equation 5.4, the price process of $P(t, T)$ can be written as

$$\mathrm{d}\left(\frac{P(t, T)}{B(t)}\right) = \frac{P(t, T)}{B(t)}\left[\left(\mu(t, T) - r_t - \mathbf{\Sigma}^{\mathrm{T}}(t, T)\gamma_t\right)\mathrm{d}t + \mathbf{\Sigma}^{\mathrm{T}}(t, T)\,\mathrm{d}\mathbf{W}_t\right].$$

$$\tag{5.8}$$

The martingale property of $B_t^{-1}P(t, T)$ dictates that the drift term must vanish, yielding Equation 5.5 again.

The equality in Equation 5.7 can also be justified without using the martingale representation theorem. Because $\mathbf{A}$ is non-singular, for any T between 0 and T_n, there exists a unique vector, $\boldsymbol{\theta} = (\theta_1, \theta_2, \ldots, \theta_n)$, such that

$$(\theta_1 P_1, \theta_2 P_2, \ldots, \theta_n P_n)\,\mathbf{A} = P(t, T)\mathbf{\Sigma}^{\mathrm{T}}(t, T), \tag{5.9}$$

where we have denoted $P_i = P(t, T_i)$ for notational simplicity. Consider the portfolio

$$V_t = P(t, T) - \sum_{i=1}^{n}\theta_i P(t, T_i). \tag{5.10}$$

The value process of the portfolio is

$$\mathrm{d}V_t = \mathrm{d}P(t, T) - \sum_{i=1}^{n}\theta_i\,\mathrm{d}P(t, T_i)$$

$$= \left[P(t, T)\mu(t, T) - \sum_{i=1}^{n}\theta_i P(t, T_i)\mu(t, T_i)\right]\mathrm{d}t$$

$$+ \left[P(t, T)\mathbf{\Sigma}^{\mathrm{T}}(t, T) - \sum_{i=1}^{n}\theta_i P(t, T_i)\mathbf{\Sigma}^{\mathrm{T}}(t, T_i)\right]\mathrm{d}\mathbf{W}_t. \tag{5.11}$$

According to the definition of θ_i, the volatility coefficient of Equation 5.11 is zero, implying that V_t is a riskless portfolio. In the absence of arbitrage, the riskless portfolio must earn a return rate equal to the risk-free rate, meaning

$$P(t,T)\mu(t,T) - \sum_{i=1}^{n} \theta_i P(t,T_i)\mu(t,T_i) = r_t \left[P(t,T) - \sum_{i=1}^{n} \theta_i P(t,T_i) \right]. \quad (5.12)$$

By making use of Equation 5.5, we can rewrite Equation 5.12 as

$$\mu(t,T) - r_t = \sum_{i=1}^{n} \theta_i \frac{P(t,T_i)}{P(t,T)} \left(\mu(t,T_i) - r_t \right)$$

$$= \sum_{i=1}^{n} \theta_i \frac{P(t,T_i)}{P(t,T)} \mathbf{\Sigma}^{\mathrm{T}}(t,T_i)\boldsymbol{\gamma}_t$$

$$= \mathbf{\Sigma}^{\mathrm{T}}(t,T)\boldsymbol{\gamma}_t,$$

which is exactly Equation 5.7.

We comment here that the components of $\boldsymbol{\gamma}_t$ are considered to be the market prices of risks for zero-coupon bonds, $P(t,T), \forall T$. Hence, if zero-coupon bonds of all maturities are driven by a finite number of Brownian motions, then they share the same set of market prices of risks.

Under the martingale measure, $\mathbb{Q}$, the price process of a zero-coupon bond becomes

$$dP(t,T) = P(t,T) \left[r_t \, dt + \mathbf{\Sigma}^{\mathrm{T}}(t,T) \, d\mathbf{W}_t \right]. \quad (5.13)$$

For the purpose of derivatives pricing, $\mathbf{\Sigma}(t,T)$ should satisfy at least three conditions: (1) $\mathbf{\Sigma}(t,t) = 0, \forall t$; (2) $P(t,t) = 1, \forall t$; and (3) to ensure that $P(t,T)$ is the monotonically decreasing function of T. The first two conditions reflect only one fact: at maturity, the price of the zero-coupon bond equals its par value and thus has no volatility. The third condition is to ensure the term structure of interest rates, like zero-coupon yields and forward rates, are kept positive. While it is actually trivial to meet the first two conditions, we are still far from an adequate understanding on how to achieve the third condition until today. Finally, instead of a direct specification of $\mathbf{\Sigma}(t,T)$, we will take advantage of the functional relationship between the volatility functions of zero-coupon bonds and forward rates, to be introduced in the next chapter. Through such a relationship, we will meet the first two conditions naturally.

5.2 Forward Prices

Let us begin with forward contracts. A forward contract is a financial contract to buy/sell a security for a price sometime in the future, when both payment and delivery will take place. Typically, in a forward contract no upfront payment is required from either party, and this is achieved by taking a special price, called the forward price, to buy/sell. The forward price, favorably, can be figured out with arbitrage arguments only. To ensure delivery, the seller must get financed now and acquire certain units of the asset. Denote the current price of the asset by S_t, the current time by t, the delivery time by T, and the unknown buy/sell price by F. Assume, at first, for simplicity that the asset pays no dividend. To be able to deliver one unit of the asset, the seller then does the following transactions:

1. Short S_t/P_t^T units of T-maturity zero-coupon bond, producing a proceed equal to the asset value.

2. Long 1 unit of the asset.

Note that 1 and 2 are a set of zero-net transactions at time t. At the delivery time, T, the seller will deliver the asset to the buyer for the price of F, and thus ends up with the following P&L value,

$$V_T = F - \frac{S_t}{P_t^T}.$$

If arbitrage is not possible, there must be $V_T = 0$, giving the fair transaction price

$$F = \frac{S_t}{P_t^T}.$$

In an economy where there is no arbitrage, this price is fair and unique.

Next, we derive the fair transaction price for an asset that pays continuous dividends with a dividend yield of $q > 0$. The dividend is calculated as follows: over a short time interval, $(t, t + \mathrm{d}t)$, the asset holder receives $q\,\mathrm{d}t$ additional units of the asset, or equivalently, $qS_t\,\mathrm{d}t$ amount of cash. The transactions of the seller will

1. short $S_t \exp(-q(T - t))/P_t^T$ units of T-maturity zero-coupon bonds,

2. long $\exp(-q(T-t))$ units of the asset and, until T, to receive the dividend asset.

This is again a set of zero-net transactions. Because of the dividend payment, the second transaction will produce exactly one unit of asset at time T. The net value after delivering the asset for payment at T is

$$V_T = F - \frac{S_t \exp(-q(T - t))}{P_t^T}.$$

In the absence of arbitrage, there must be $V_T = 0$, yielding the fair price

$$F = \frac{S_t \exp(-q(T - t))}{P_t^T}.$$

Finally, we derive the fair transaction price when the asset pays discrete dividends. Assume that, prior to T, the asset pays cash dividend q_i at time $T_i \le T, i = 1, \ldots, n$, such that $T_{i-1} < T_i$. In such a circumstance, the seller's strategy is to

1. short q_i units of T_i-maturity zero-coupon bonds $(i \le n)$;

2. short $(S_t - \sum_{t < T_i \le T} q_i P_t^{T_i})/P_t^T$ units of T-maturity zero-coupon bonds; and

3. long 1 unit of the asset.

The proceeds from shorting are just enough to purchase one unit of the asset. Hence, this is still a set of zero-net transactions. After closing out all positions at time T, the seller ends up with the net value of

$$V_T = F - \frac{\left(S_t - \sum_{t < T_i \le T} q_i P_t^{T_i} \right)}{P_t^T}.$$

Hence, the fair price for a transaction is

$$F = \frac{\left(S_t - \sum_{t < T_i \le T} q_i P_t^{T_i} \right)}{P_t^T}. \tag{5.14}$$

For all three cases, we define

$$\hat{S}_t = \begin{cases} S_t, & \text{no dividend} \\ S_t \exp(-q(T - t)), & \text{dividend yield } q \\ S_t - \sum_{t < T_i \le T} q_i P_t^{T_i}, & \text{discrete dividend } \{q_i\} \end{cases} \tag{5.15}$$

as the *stripped-dividend* price of the asset. In terms of the stripped-dividend prices, we present the following definition.

Definition 5.2.1. *The price of a stripped-dividend asset relative to the T-maturity zero-coupon bond,*

$$F_t^T = \frac{\hat{S}_t}{P_t^T},$$

is called the forward price with delivery at time T.

Like the original asset itself, a stripped-dividend asset is also tradable (or replicable, in principle). Hence, its pricing process under the risk-neutral measure $\mathbb{Q}$ is usually assumed to be

$$d\hat{S}_t = \hat{S}_t(r_t\,dt + \boldsymbol{\Sigma}_S^{\mathrm{T}}\,d\mathbf{W}_t).$$

The process for the original asset then follows. In fact, for the asset paying a continuous dividend yield, the price process is

$$dS_t = S_t\left[(r_t - q)\,dt + \boldsymbol{\Sigma}_S^{\mathrm{T}}\,d\mathbf{W}_t\right], \tag{5.16}$$

whereas for the asset paying discrete dividends, the price process is

$$dS_t = \left[r_t S_t - \sum q_i \delta(T_i - t)\right]dt$$
$$+ \left[S_t\boldsymbol{\Sigma}_S^{\mathrm{T}} - \sum q_i \mathbf{1}_{t\leq T_i} P_t^{T_i}(\boldsymbol{\Sigma}_S - \boldsymbol{\Sigma})^{\mathrm{T}}\right]d\mathbf{W}_t, \tag{5.17}$$

where $\delta(x)$ is the Dirac delta function and $\mathbf{1}_{t\leq T_i}$ the indicator function.

5.3   Forward Measure

Because the price of a zero-coupon bond equals par at maturity, a forward price equals its spot price at the delivery date, that is, $F_T^T = \hat{S}_T = S_T$. As a result, any options written on S_T can equivalently be treated as an option on F_T^T. Next, we will try to price an option on F_T^T. For this purpose, we first need to derive the dynamics that F_t^T follows.

As tradable assets, the price of a stripped-dividend asset and a zero-coupon bond are assumed to be, respectively,

$$\begin{aligned}
d\hat{S}_t &= \hat{S}_t\big(r_t\,dt + \boldsymbol{\Sigma}_S^{\mathrm{T}}(t)\,d\mathbf{W}_t\big),\\
dP_t^T &= P_t^T\big(r_t\,dt + \boldsymbol{\Sigma}^{\mathrm{T}}(t,T)\,d\mathbf{W}_t\big).
\end{aligned} \tag{5.18}$$

By the quotient rule, the forward price satisfies

$$\begin{aligned}
d\left(\frac{\hat{S}_t}{P_t^T}\right) &= \frac{d\hat{S}_t}{P} - \frac{\hat{S}_t dP}{P^2} - \frac{d\hat{S}_t dP}{P^2} + \frac{\hat{S}_t(dP)^2}{P^3}\\
&= \frac{\hat{S}_t}{P}\left(r_t dt + \boldsymbol{\Sigma}_S^{\mathrm{T}}\,d\mathbf{W}_t - r_t\,dt - \boldsymbol{\Sigma}^{\mathrm{T}}\,d\mathbf{W}_t - \boldsymbol{\Sigma}_S^{\mathrm{T}}\boldsymbol{\Sigma}\,dt + \boldsymbol{\Sigma}^{\mathrm{T}}\boldsymbol{\Sigma}\,dt\right)\\
&= \frac{\hat{S}_t}{P}\left(\boldsymbol{\Sigma}_S - \boldsymbol{\Sigma}\right)^{\mathrm{T}}(d\mathbf{W}_t - \boldsymbol{\Sigma}\,dt).
\end{aligned}$$

Here, on the right-hand side of the equation, we have omitted the sub- and sup-index of P for simplicity. Define now a new measure, $\mathbb{Q}_T$, as

$$\left.\frac{d\mathbb{Q}_T}{d\mathbb{Q}}\right|_{\mathcal{F}_t} = \exp\left(\int_0^t \mathbf{\Sigma}^T d\mathbf{W}_s - \frac{1}{2}\mathbf{\Sigma}^T\mathbf{\Sigma}\,ds\right) = \zeta_t. \tag{5.19}$$

Then, by the CMG theorem,

$$\mathbf{W}_t^{(T)} = \mathbf{W}_t - \int_0^t \mathbf{\Sigma}(s,T)\,ds \tag{5.20}$$

is a $\mathbb{Q}_T$-Brownian motion. It follows that F_t^T is also a lognormal $\mathbb{Q}_T$-martingale, such that it follows a driftless process,

$$dF_t^T = F_t^T\mathbf{\Sigma}_F^T d\mathbf{W}_t^{(T)},$$

where

$$\mathbf{\Sigma}_F(t) = \mathbf{\Sigma}_S(t) - \mathbf{\Sigma}(t,T). \tag{5.21}$$

We call $\mathbb{Q}_T$ the forward measure with delivery at T, or simply the T-forward measure. According to the definition, there is a one-to-one correspondence between the T-maturity zero-coupon bond and the T-forward measure.

Based on the understanding that the forward price is a $\mathbb{Q}_T$-martingale, we can derive a general pricing principle of options on F_T^T. Define another $\mathbb{Q}_T$-martingale using the payoff of the option, $X_T = H(S_T) = H(F_T^T)$, such that

$$N_t = E^{\mathbb{Q}_T}[X_T \mid \mathcal{F}_t]. \tag{5.22}$$

According to the martingale representation theorem, there exists an $\mathcal{F}_t$-adaptive process, φ_t, such that

$$dN_t = \varphi_t dF_t^T. \tag{5.23}$$

We now form a portfolio consisting of

- φ_t units of the underlying asset, and

- $\psi_t = N_t - \varphi_t F_t^T$ units of the T-maturity zero-coupon bond.

Let $\hat{V}_t$ be the forward price of the portfolio at time t. By definition, $\hat{V}_t = N_t$ for all $t \le T$, and the spot price of the portfolio,

$$V_t = N_t P_t^T. \tag{5.24}$$

At the maturity of the option, there is $V_T = N_T = X_T$, meaning that the portfolio replicates the payoff of the option. In addition, we have

$$\begin{aligned}
dV_t &= P_t^T dN_t + N_t dP_t^T + dN_t dP_t^T \\
&= P_t^T \varphi_t dF_t^T + (\varphi_t F_t^T + \psi_t)\,dP_t^T + \varphi_t dF_t^T dP_t^T \\
&= \varphi_t\left(P_t^T dF_t^T + F_t^T dP_t^T + dF_t^T dP_t^T\right) + \psi_t dP_t^T \\
&= \varphi_t dS_t + \psi_t dP_t^T,
\end{aligned} \tag{5.25}$$

which implies that the portfolio, (φ_t, ψ_t), is a self-financing one. In the absence of arbitrage, the value of the option should be nothing else but that of the replicating portfolio. This yields the general price formula

$$V_t = P_t^T E^{\mathbb{Q}_T}[X_T | \mathcal{F}_t] \tag{5.26}$$

for the option. Note that this is the second option price formula in addition to the one developed in Chapter 2,

$$V_t = B_t E^{\mathbb{Q}}[B_T^{-1} X_T | \mathcal{F}_t], \tag{5.27}$$

obtained under the risk-neutral measure, $\mathbb{Q}$. Both formulae are derived through arbitrage arguments, so the values of the two formulae must be identical, or else we would be in a situation of arbitrage, at odd with the efficient market hypothesis.

Mathematically, it remains interesting to verify that the two formulae give an identical price. In fact, we can derive one formula from the other by merely a change of measure. We know that under the risk-neutral measure, $\mathbb{Q}$, P_t^T follows

$$dP_t^T = P_t^T \left(r_t dt + \mathbf{\Sigma}^{\mathrm{T}}(t, T) \, d\mathbf{W}_t \right), \tag{5.28}$$

or

$$d\left(\frac{P_t^T}{B_t}\right) = \left(\frac{P_t^T}{B_t}\right) \mathbf{\Sigma}^{\mathrm{T}}(t, T) \, d\mathbf{W}_t, \tag{5.29}$$

where $\mathbf{W}_t$ is a $\mathbb{Q}$-Brownian motion. By solving the equation, we obtain

$$\frac{P_T^T}{B_T} = \frac{P_t^T}{B_t} \exp\left(\int_t^T -\frac{1}{2}\mathbf{\Sigma}^{\mathrm{T}}\mathbf{\Sigma} \, ds + \mathbf{\Sigma}^{\mathrm{T}} d\mathbf{W}_s \right). \tag{5.30}$$

From this equation, we can express B_t in terms of P_t^T:

$$\frac{B_t}{B_T} = P_t^T \exp\left(\int_t^T -\frac{1}{2}\mathbf{\Sigma}^{\mathrm{T}}\mathbf{\Sigma} \, ds + \mathbf{\Sigma}^{\mathrm{T}} d\mathbf{W}_s \right) = P_t^T \frac{\zeta_T}{\zeta_t}, \tag{5.31}$$

by making use of Equation 5.19. Hence, starting from Equation 5.27, we have

$$\begin{aligned} V_t &= E^{\mathbb{Q}} \left[\frac{B_t}{B_T} X_T \;\middle|\; \mathcal{F}_t \right] \\ &= E^{\mathbb{Q}} \left[P_t^T \frac{\zeta_T}{\zeta_t} X_T \;\middle|\; \mathcal{F}_t \right] \\ &= P_t^T E^{\mathbb{Q}_T}[X_T | \mathcal{F}_t]. \end{aligned} \tag{5.32}$$

This procedure can be reversed to derive Equation 5.27 from Equation 5.26.

For completeness, we finish this section with a lemma on the martingale property of forward rates.

5.4   Black's Formula for Call and Put Options

In this section, we derive the price formula for both call and put options using the forward price and under its corresponding forward measure. The payoff of a call option on an asset, S_t, is

$$C_T = \max(S_T - K, 0) \overset{\Delta}{=} (S_T - K)^+. \tag{5.33}$$

In terms of the forward price, $F_t^T = \hat{S}_t / P_t^T$, we also have

$$C_T = (F_T^T - K)^+. \tag{5.34}$$

Under the T-forward measure, we know that the price is given by

$$C_t = P_t^T E^{\mathbb{Q}_T}\left[\left.\left(F_T^T - K\right)^+ \right| \mathcal{F}_t\right]. \tag{5.35}$$

The good news here is that F_t^T is a lognormal martingale under $\mathbb{Q}_T$:

$$\mathrm{d}F_t^T = F_t^T \mathbf{\Sigma}_F^T \, \mathrm{d}\mathbf{W}_t^{(T)}, \tag{5.36}$$

where $\mathbf{\Sigma}_F$ is the difference between the volatilities of the asset and the T-maturity zero-coupon bond:

$$\mathbf{\Sigma}_F = \mathbf{\Sigma}_S - \mathbf{\Sigma}(t, T). \tag{5.37}$$

The expectation in Equation 5.35 can be worked out by brute force. The forward price of the option can be rewritten as

$$\begin{aligned}
\tilde{C}_t &= E_t^{\mathbb{Q}_T}\left[\left(F_T^T - K\right)^+\right] \\
&= E_t^{\mathbb{Q}_T}\left[F_T^T 1_{F_T^T > K}\right] - K E_t^{\mathbb{Q}_T}\left[1_{F_T^T > K}\right].
\end{aligned} \tag{5.38}$$

As a lognormal martingale

$$F_T^T = F_t^T \exp\left[-\frac{1}{2}\sigma_F^2 \tau + \bar{\sigma}_F \sqrt{\tau} \cdot \varepsilon\right], \quad \varepsilon \sim N(0,1), \tag{5.39}$$

where $\tau = T - t$ and $\bar{\sigma}$ is the mean volatility,

$$\bar{\sigma}_F = \sqrt{\frac{1}{\tau}\int_0^\tau \|\Sigma_F\|^2 \mathrm{d}s}, \tag{5.40}$$

we have

$$E_t^{\mathbb{Q}_T}\left[1_{F_T^T > K}\right] = \mathrm{Prob}\left(\varepsilon > -\frac{\ln\left(F_t^T / K\right) - \frac{1}{2}\bar{\sigma}_F^2 \tau}{\bar{\sigma}_F \sqrt{\tau}}\right) = \Phi(d_2), \tag{5.41}$$

with

$$d_2 = \frac{\ln(F_t^T/K) - \frac{1}{2}\bar{\sigma}_F^2 \tau}{\bar{\sigma}_F \sqrt{\tau}}. \qquad (5.42)$$

Meanwhile,

$$
\begin{aligned}
E_t^{\mathbb{Q}_T}\left[F_T^T \mathbb{1}_{F_T^T > K}\right] &= \frac{1}{\sqrt{2\pi}} \int_{-d_2}^{\infty} F_t^T \exp\left[\left(-\frac{1}{2}\bar{\sigma}_F^2\right)\tau + \bar{\sigma}_F \sqrt{\tau} x - \frac{1}{2}x^2\right] dx \\
&= \frac{F_t^T}{\sqrt{2\pi}} \int_{-d_2 - \bar{\sigma}_F \sqrt{\tau}}^{\infty} \exp\left(-\frac{1}{2}y^2\right) dy \\
&= F_t^T \Phi(d_1), \qquad (5.43)
\end{aligned}
$$

where

$$d_1 = d_2 + \bar{\sigma}_F \sqrt{\tau}. \qquad (5.44)$$

By substituting Equations 5.41 and 5.43 into 5.38, we arrive at the celebrated Black formula (1976):

$$
\begin{aligned}
C_t = P_t^T \tilde{C}_t &= P_t^T\left[F_t^T \Phi(d_1) - K\Phi(d_2)\right] \\
&= \hat{S}_t \Phi(d_1) - K P_t^T \Phi(d_2).
\end{aligned} \qquad (5.45)
$$

By direct verification, we can show that the hedge ratio, φ_t, is

$$\frac{\partial C_t}{\partial \hat{S}_t} = \Phi(d_1). \qquad (5.46)$$

In the Black formula, the only model parameter is σ_F, which is called the Black volatility of the option.

In addition to the option's price, Black's formula also offers a hedging strategy: we can hedge the option by purchasing $\varphi_t = \Phi(d_1)$ units of underlying asset. Alternatively, we may say that the option can be replicated by a portfolio consisting of

- $\varphi_t = \Phi(d_1)$ units of the underlying asset, and

- $\psi_t = -K\Phi(d_2)$ units of T-maturity zero-coupon bonds.

The price formula for put options can be derived through the *call–put parity*: for the same strike, K, the prices of a call option, a put option, and a forward contract satisfy the relation

$$C(K) - P(K) = \hat{S}_t - P(t,T)K. \qquad (5.47)$$

Thus, the formula for a put option follows:

$$
\begin{aligned}
P(K) &= C(K) - \hat{S}_t + P(t,T)K \\
&= K P_t^T(1 - \Phi(d_2)) - \hat{S}_t(1 - \Phi(d_1)) \\
&= K P_t^T \Phi(-d_2) - \hat{S}_t \Phi(-d_1).
\end{aligned} \qquad (5.48)
$$

The hedging strategy is to short $\Phi(-d_1)$ units of the underlying asset.

As a finishing note we connect the Black formula to the Black-Scholes formula (1973), a formula proceeding the Black formula for option pricing. The Black-Scholes model assumes a constant interest rate, $r_t = r$, for continuous compounding, which has three consequences:

1. $P_t^T = e^{-r(T-t)}$,

2. $\Sigma(t,T) = 0$ and $\Sigma_F = \Sigma_S$.

Let $\sigma_S = \|\Sigma_S\|$, then $\sigma_F = \sigma_S$ and the Black formula for call options becomes

$$C_t = \hat{S}_t \Phi(d_1) - K e^{-r(T-t)} \Phi(d_2), \qquad (5.49)$$

where

$$d_1 = \frac{\ln\left(\frac{\hat{S}_t}{K}\right) + \frac{1}{2}\left(r + \sigma_S^2\right)(T-t)}{\sigma_S \sqrt{T-t}}, \qquad (5.50)$$

$$d_2 = d_1 - \sigma_S \sqrt{T-t},$$

Equation 5.49-5.50 constitute the Black-Scholes formula, which earned these two researchers, together with Merton, the Noble Prize in economics in 1997.

5.4.1 Equity Options under the Forward Measure

To price either a call or a put option by the Black formula, we need to calculate the Black volatility of the forward price (Equation 5.40). In applications, asset volatilities are often given in the form of a scalar instead of a vector, together with asset correlations. In such a situation, the (square of) Black's volatility, Equation 5.40, takes a different form. In this section, we demonstrate the formula for the Black volatility in terms of scalar volatility functions.

We start with the mean variance of the forward price:

$$\sigma_F^2 = \frac{1}{T-t}\int_t^T \|\Sigma_S(u) - \Sigma(u,T)\|^2 du$$

$$= \frac{1}{T-t}\int_t^T \left(\|\Sigma_S(u)\|^2 - 2\Sigma_S^{\mathrm{T}}(u)\Sigma(u,T) + \|\Sigma(u,T)\|^2\right)du.$$

$$(5.51)$$

Let us consider the simplest volatility specification of zero-coupon bond that satisfies the condition $\Sigma(T,T) = 0$: $\|\Sigma(u,T)\| = \sigma_0(T-u)$, let the asset–correlation correlation be ρ, and then we have

$$\sigma_F^2 = \frac{1}{T-t}\int_t^T \left(\sigma_S^2 - 2\rho\sigma_S\sigma_0(T-u) + \sigma_0^2(T-u)^2\right)du$$

$$= \sigma_S^2 - \rho\sigma_S \cdot \sigma_0(T-t) + \frac{1}{3}\sigma_0^2(T-t)^2. \qquad (5.52)$$

It is obvious that a positive correlation reduces Black's volatility. Let us further witness the effect on the price of asset–interest rate correlation in the following example.

Example 5.1. *We examine the price of a call option as a function of the asset–interest rate correlation. The parameters are taken as follows.*

$$\hat{S}_0 = 1, \quad \sigma_S = 0.3,$$
$$P(0,T) = e^{-\int_0^T (0.02+0.002s)\mathrm{d}s}, \quad \forall T; \quad \sigma_0 = 0.002,$$
$$K = 1,$$
$$\rho = -1 : 0.1 : 1.$$

(5.53)

This option is called an at-the-money (ATM) option because the strike price equals the spot price. The price curve of the option against the correlation is presented in Figure 5.1, where one can see that option prices decrease gradually, from 0.1292 to 0.1282. The explanation is intuitive: a positive correlation between the asset and the zero-coupon bond leads to a smaller volatility in the forward price, and hence a lower value for the option.

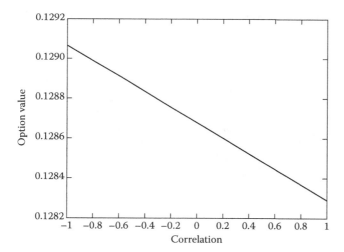

FIGURE 5.1: The price of an ATM option in relation to the correlation.

5.5 Numeraires and Change of Measures

A major achievement so far in this chapter is to take zero-coupon bonds as numeraires and price options under their corresponding forward measures. Mathematically, this is merely a technique of changing the numeraire asset, followed by taking the expectation of the option payoffs under the martingale measures of the numeraire assets. In this section, we discuss this technique in a general context.

Let $\mathbb{Q}_A$ be the martingale measure associated with reference asset A_t, meaning that, for any tradable asset V_t, its price relative to that of asset A_t,

$$\frac{V_t}{A_t},\qquad(5.54)$$

is a $\mathbb{Q}_A$-martingale. Consider another asset,[1] B_t, and its associated martingale measure, $\mathbb{Q}_B$. According to the one-price principle, the value of any traded asset at time t, V_t, satisfies

$$V_t = A_t E_t^{\mathbb{Q}_A}\left[A_T^{-1}V_T\right] = B_t E_t^{\mathbb{Q}_B}\left[B_T^{-1}V_T\right].\qquad(5.55)$$

From the above equation, we obtain

$$E_t^{\mathbb{Q}_B}\left[B_T^{-1}V_T\right] = \frac{A_t}{B_t}E_t^{\mathbb{Q}_A}\left[\frac{B_T}{A_T}\left(B_T^{-1}V_T\right)\right].\qquad(5.56)$$

Let ζ be the Radon–Nikodym derivative between $\mathbb{Q}_B$ and $\mathbb{Q}_A$:

$$\left.\frac{d\mathbb{Q}_B}{d\mathbb{Q}_A}\right|_{\mathcal{F}_t} = \zeta_t.\qquad(5.57)$$

Then,

$$E_t^{\mathbb{Q}_B}\left[B_T^{-1}V_T\right] = \zeta_t^{-1}E_t^{\mathbb{Q}_A}\left[\zeta_T(B_T^{-1}V_T)\right].\qquad(5.58)$$

Subtracting Equation 5.56 from Equation 5.58, we obtain

$$0 = E_t^{\mathbb{Q}_A}\left[\left(\frac{\zeta_T}{\zeta_t} - \frac{B_T/A_T}{B_t/A_t}\right)(B_T^{-1}V_T)\right].\qquad(5.59)$$

Note that Equation 5.59 holds for prices of any tradable assets, so we can argue that

$$\zeta_t = \left.\frac{d\mathbb{Q}_B}{d\mathbb{Q}_A}\right|_{\mathcal{F}_t} = \frac{B_t/A_t}{B_0/A_0}\quad\text{a.s.}\qquad(5.60)$$

Example 5.2. *We have already established the following correspondence between a numeraire and its martingale measure:*

$$B_t = \exp\left(\int_0^t r_s\,ds\right) \longleftrightarrow \mathbb{Q},$$
$$P_t^T = P_0^T \exp\left(\int_0^t\left(r_s - \frac{1}{2}\mathbf{\Sigma}^T\mathbf{\Sigma}\right)ds + \mathbf{\Sigma}^T d\mathbf{W}_s\right) \longleftrightarrow \mathbb{Q}_T.\qquad(5.61)$$

---

[1] A_t and B_t are either stripped divided or cum-dividend.

According to the general formula 5.60, the Radon–Nikodym derivative of
$\mathbb{Q}_T$ *with respect to* $\mathbb{Q}$ *is*

$$\left.\frac{d\mathbb{Q}_T}{d\mathbb{Q}}\right|_{\mathcal{F}_t} = \frac{P_t^T/P_0^T}{B_t/B_0}$$

$$= \exp\left(\int_0^t -\frac{1}{2}\boldsymbol{\Sigma}^{\mathrm{T}}\boldsymbol{\Sigma}\,ds + \boldsymbol{\Sigma}^{\mathrm{T}}\,d\mathbf{W}_s\right), \qquad (5.62)$$

which is exactly Equation 5.19.

5.6   Futures Price and Futures Rate

In Chapter 4, we introduced a series of SOFR derivatives. In that series, SOFR
futures and swaps of various kinds are particularly important to interest-rate
derivatives markets. This is because, initially, the value of both contracts is
zero and thus costs no capital to purchase (except the capital for maintaining
the margin account for futures), making them popular tools for either hedging
or speculation. Probably because of that, SOFR futures and swaps remain to
be the most liquid interest-rate derivatives in the post-LIBOR derivatives
market. Nonetheless, they are not necessarily easy to price. This section deals
with the determination of futures price which, due to the marking to market
of its daily P&L, has to do with the dynamics of overnight rate and thus
is model dependent. The pricing of non-vanilla swaps will be carried out in
Chapter 10, which requires additional techniques.

In this section, we introduce the general theory for determining futures
prices that is applicable to any asset categories: equity, commodity and fixed
income. The theory links up the futures prices to forward prices. Because of
the futures - forward difference in prices is caused by marking to market, the
difference is usually small, yet not negligible when the transaction volumes
are big, which is often the case for interest rate futures. To figure out a futures price or index, the existing market convention is to make the so-called
correlation adjustment to the corresponding forward price, as the latter can
be figured out without a model and is often readily available. As we shall see,
a positive asset–short rate correlation will make a futures price higher than
the forward price, whereas a negative correlation will make it lower.

5.6.1   Futures Price versus Forward Price

The value of a forward contract with maturity, T, and strike, K, on a tradable
asset is

$$V_0 = E^{\mathbb{Q}}\big[B_T^{-1}\,(S_T - K)\big|\,\mathcal{F}_0\big], \qquad (5.63)$$

where $\mathbb{Q}$ stands for the risk-neutral measure, S_T represents the time-T value of an asset, an interest rate or an interest rate index, and B_T is the balance of the saving account at T:

$$B_T = \exp\left(\int_0^T r_s \, \mathrm{d}s\right). \tag{5.64}$$

The forward price is defined as the strike price that nullifies the value of the forward contract. In view of Equation 5.63, we know that the forward price for the contract satisfies

$$F_0^T = \frac{E_0^{\mathbb{Q}}\left[B_T^{-1} S_T\right]}{E_0^{\mathbb{Q}}\left[B_T^{-1}\right]} = E_0^{\mathbb{Q}_T}[S_T], \tag{5.65}$$

that is, it is the expectation of the terminal asset price under the forward measure, $\mathbb{Q}_T$. We have just learned in this chapter that the above expectation equals

$$F_0^T = \frac{\hat{S}_0}{P(0,T)}, \tag{5.66}$$

where $\hat{S}_t$ is the stripped-dividend price of the asset at time t.

With everything else the same, a futures contract differs from a forward contract by "marking to market," meaning that the P&L from holding the contract is credited to or debited from the holder's margin account on a daily basis. The margin account, meanwhile, is accrued using risk-free interest rates. A futures price parallels the forward price and nullifies the value of a futures contract. Let $\tilde{F}_t^T$ be the futures price observed at time $t \le T$. At the maturity of the futures contract, the futures price is fixed or set to the price of the underlying security, that is, $\tilde{F}_T^T = S_T$.

Our next goal is to derive $\tilde{F}_t^T$ for $t < T$. Let us consider the price dynamics of the futures contract. At any moment $t < T$, the value of the futures contract is reflected in the balance in the margin account, V_t. The change to the futures price comes from two sources: the accrual of the margin account and the change in the futures price. Hence, in differential form, the change in the value of the futures contract is

$$\mathrm{d}V_t = r_t V_t \, \mathrm{d}t + \mathrm{d}\tilde{F}_t^T. \tag{5.67}$$

From Equation 5.67, we have the dynamics of the futures price,

$$\mathrm{d}\tilde{F}_t^T = B_t \mathrm{d}\left(B_t^{-1} V_t\right). \tag{5.68}$$

It follows that

$$\tilde{F}_t^T - \tilde{F}_0^T = \int_0^t B_s \mathrm{d}(B_s^{-1} V_s). \tag{5.69}$$

Being the value of the futures contract, a tradable security, the balance of the margin account after discounting must be a martingale under the risk-neutral measure. This gives rise to the following general formula for the futures price:

$$\tilde{F}_0^T = E^{\mathbb{Q}}\left[\tilde{F}_T^T\right] = E^{\mathbb{Q}}[S_T]. \tag{5.70}$$

We have an alternative yet intuitive approach to derive the futures price formula (Equation 5.70) that is based on the so-called dividend-yield analogy.[2] At any time, $t < T$, we will maintain $B_{t+\Delta t}$ units of the futures contract. Here,

$$B_{t+\Delta t} = (1 + r_t\,\Delta t)B_t,$$

and r_t is the backward-looking SOFR rate for yesterday which is published the next day at 8am EST.[3] This is a dynamical strategy. We start with $B_{\Delta t} > 1$ units of the futures contract at time 0, which generates a P&L in the amount of $B_{\Delta t}\left(\tilde{F}_{\Delta t}^T - \tilde{F}_0^T\right)$ at time Δt. Suppose, at time t, that we have already accumulated B_t units of the futures contract and $B_t\left(\tilde{F}_t^T - \tilde{F}_0^T\right)$ dollars in the margin account, and we add $r_t B_t \Delta t$ units of the futures contract at no cost. At time $t + \Delta t$, due to interest accrual and marking to market, the balance of the margin account will become

$$(1 + r_t\Delta t)\,B_t\left(\tilde{F}_t^T - \tilde{F}_0^T\right) + (1 + r_t\Delta t)\,B_t\left(\tilde{F}_{t+\Delta t}^T - \tilde{F}_t^T\right)$$
$$= B_{t+\Delta t}\left(\tilde{F}_{t+\Delta t}^T - \tilde{F}_0^T\right). \tag{5.71}$$

Continuing with the above strategy until $T - \Delta t$, then, at time T, when the futures contract expires, we will have a balance in the amount of

$$V_T = B_T(S_T - \tilde{F}_0^T) \tag{5.72}$$

in the margin account. The expected present value of the terminal balance is thus

$$V_0 = E^{\mathbb{Q}}\left[B_T^{-1}B_T\left(S_T - \tilde{F}_0^T\right)\right] = E^{\mathbb{Q}}[S_T] - \tilde{F}_0^T. \tag{5.73}$$

By setting $V_0 = 0$, we rederive Equation 5.70.

Our next issue of interest is the difference between the forward price and the futures price. When the interest rate is deterministic, the two prices are, apparently, equal. When the interest rate is stochastic, they are generally not the same. Let us see what causes the difference. Straightforwardly, we have

$$E^{\mathbb{Q}}[S_T] = E^{\mathbb{Q}}\left[B_T^{-1}B_T S_T\right] = P(0,T)E^{\mathbb{Q}_T}[B_T S_T]. \tag{5.74}$$

---

[2]This means the following: by receiving continuous share dividends or converting cash dividends into shares, we will increase the number of shares from one at time 0 to $\exp\left(\int_0^t q_s \mathrm{d}s\right)$ at any later time, t, where q_s is the dividend yield.

[3]We would like to use $r_{t+\Delta t}$ in place of r_t, which however is impractical.

When $S_T = 1$, Equation 5.74 yields the expectation of the money market account under the forward measure:

$$E^{\mathbb{Q}_T}[B_T] = \frac{1}{P(0,T)}. \tag{5.75}$$

It then follows that

$$
\begin{aligned}
\tilde{F}_0^T - F_0^T &= E^{\mathbb{Q}}[S_T] - E^{\mathbb{Q}_T}[S_T] \\
&= P(0,T)E^{\mathbb{Q}_T}[B_T S_T] - E^{\mathbb{Q}_T}[S_T] \\
&= P(0,T)\left(E^{\mathbb{Q}_T}[B_T S_T] - E^{\mathbb{Q}_T}[B_T]\,E^{\mathbb{Q}_T}[S_T]\right) \\
&= P(0,T) \times \mathrm{Cov}^{\mathbb{Q}_T}(B_T, S_T) \\
&= P(0,T) \times \mathrm{Cov}^{\mathbb{Q}_T}(B_T, F_T^T).
\end{aligned} \tag{5.76}
$$

Hence, if the asset is positively correlated to the money market account, then there will be $\tilde{F}_0^T > F_0^T$, or, otherwise, $\tilde{F}_0^T \leq F_0^T$. Such orders can be well explained by intuition: given a positive correlation between the asset and the short rate, a gain in the margin account will likely be accrued with a higher interest rate, whereas a loss in the account will likely be accrued with a lower interest rate. This is to the advantage of the party who is long the futures contract. To make the futures a fair game, the futures price should be set somewhat higher than the forward price.

To figure out the exact difference between the two prices, we proceed to calculate the covariance in Equation 5.76. Obviously, there are

$$
\begin{aligned}
B_T &= \frac{B_0}{P(0,T)} e^{\int_0^T -\|\Sigma(s,T)\|^2 ds - \Sigma(s,T)\cdot d\mathbf{W}_s^{(T)}}, \\
F_T^T &= F_0^T e^{\int_0^T -\|\Sigma_F(s,T)\|^2 ds + \Sigma_F(s,T)\cdot d\mathbf{W}_s^{(T)}},
\end{aligned} \tag{5.77}
$$

and both $B_t/P(t,T)$ and F_t^T are martingales under $\mathbb{Q}_T$. There is

$$
\begin{aligned}
&E^{\mathbb{Q}_T}\left[B_T F_T^T\right] \\
&= \frac{F_0^T}{P(0,T)} E^{\mathbb{Q}_T}\left[e^{\int_0^T -\frac{1}{2}(\|\Sigma(s,T)\|^2 + \|\Sigma_F(s,T)\|^2)ds - [\Sigma(s,T) - \Sigma_F(s,T)]\cdot d\mathbf{W}_s^{(T)}}\right] \\
&= \frac{F_0^T}{P(0,T)} E^{\mathbb{Q}}\left[e^{\int_0^T -\frac{1}{2}\|\Sigma(s,T)\|^2 ds + \Sigma(s,T)\cdot d\mathbf{W}_s} \right. \\
&\qquad\qquad\qquad \left. e^{\int_0^T -\frac{1}{2}(\|\Sigma(s,T)\|^2 + \|\Sigma_F(s,T)\|^2)ds - [\Sigma(s,T) - \Sigma_F(s,T)]\cdot(d\mathbf{W}_s - \Sigma(s,T)ds)}\right] \\
&= \frac{F_0^T}{P(0,T)} E^{\mathbb{Q}}\left[e^{\int_0^T -\frac{1}{2}\|\Sigma_F(s,T)\|^2 ds + \Sigma_F(s,T)\cdot d\mathbf{W}_s} e^{-\int_0^T \Sigma(s,T)\cdot\Sigma_F(s,T)ds}\right].
\end{aligned} \tag{5.78}
$$

Now define a new measure equivalent to $\mathbb{Q}$, such that

$$
\left.\frac{d\mathbb{Q}_F}{d\mathbb{Q}}\right|_{\mathcal{F}_t} = e^{\int_0^T -\frac{1}{2}\|\Sigma_F(s,T)\|^2 ds + \Sigma_F(s,T)\cdot d\mathbf{W}_s}. \tag{5.79}
$$

then in terms of $\mathbb{Q}_F$, we have

$$E^{Q_T}\left[B_T F_T^T\right] = \frac{F_0^T}{P(0,T)} E^{Q_F}\left[e^{-\int_0^T \Sigma(s,T)\cdot\Sigma_F(s,T)\mathrm{d}s}\right]. \qquad (5.80)$$

Hence the covariance between the saving account and the forward price is

$$\mathrm{Cov}^{Q_T}(B_T, S_T) = \frac{F_0^T}{P(0,T)}\left(E^{Q_F}\left[e^{-\int_0^T \Sigma(s,T)\cdot\Sigma_F(s,T)\mathrm{d}s}\right] - 1\right). \qquad (5.81)$$

Note that a positive correlation between the saving account and the forward price implied the negative correlation between the volatilities of the two assets. Putting Equation 5.81 to Equation 5.76, we obtain the formula for futures–forward adjustment. The general result for $t \leq T$ is

$$\tilde{F}_t^T - F_t^T = F_t^T\left(E^{Q_F}\left[e^{-\int_t^T \Sigma(s,T)\cdot\Sigma_F(s,T)\mathrm{d}s}\right] - 1\right). \qquad (5.82)$$

The right-hand side is usually called *correlation adjustment*. When $\Sigma(t,T)\cdot\Sigma_F(t)$ is state-independent, the above equation simplifie to

$$\tilde{F}_t^T - F_t^T = F_t^T\left(e^{-\int_t^T \Sigma(s,T)\cdot\Sigma_F(s,T)\mathrm{d}s} - 1\right). \qquad (5.83)$$

We now comment on the hedging of the futures contract. Futures are often used for hedging due to high liquidity and low cost (i.e., the funding cost for deposits in the margin account). However, when a futures price is too much off the line, futures can be arbitraged against using, for instance, the forward contract. In fact, futures can be dynamically hedged by forward contracts, using the ratio

$$\Delta_t = \frac{\partial \tilde{F}_t^T}{\partial F_t^T} = E^{Q_F}\left[e^{-\int_t^T \Sigma(s,T)\cdot\Sigma_F(s,T)\mathrm{d}s}\right], \qquad (5.84)$$

which requires information of the volatility functions.

5.6.2 Futures Rate versus Forward Rate

In SOFR swaps, caps and floors, we need to address the pricing of cash flows indexed to the SOFR term rate realized over a future period (T_{j-1}, T_j):

$$f_j(T_j) = \frac{\prod_{i=1}^{N_{b,j}}\left(1 + \frac{r_i^{(j)}\times d_i}{360}\right) - 1}{\frac{N_{c,j}}{360}} \approx \frac{e^{\int_{T_{j-1}}^{T_j} r_s \mathrm{d}s} - 1}{\Delta T_j}. \qquad (5.85)$$

As is established in Section 5.6.1, the SOFR futures rate for maturity T_j is given by

$$\tilde{f}_j(t) \triangleq E_t^Q[f_j(T_j)], \qquad (5.86)$$

and the forward rate for the same maturity is given by

$$f_j(t) \overset{\triangle}{=} E_t^{Q_j}[f_j(T_j)], \tag{5.87}$$

where Q_j stands for the T_j-forward measure, the martingale measure for the numeraire asset of $P(t, T_j)$. Note that the forward rate is used for the forward-rate agreements (FRA), and FRAs were very liquid in the LIBOR markets. Next, we will show that forward rates are completely determined by the discount curve. For this purpose, we denote the time-t value of the saving account starting with \$1 from the reference time T by

$$B(t, T) = e^{\int_T^t r_s ds}.$$

The following result regarding the backward-looking term rate $f_j(T_j)$ can be readily established.

Theorem 5.6.1. *There is*

$$E_t^{Q_j}[f_j(T_j)] = \begin{cases} \frac{1}{\Delta T_j}\left(\frac{P(t,T_{j-1})}{P(t,T_j)} - 1\right), & t \le T_{j-1}, \\ \frac{1}{\Delta T_j}\left(\frac{B(t,T_{j-1})}{P(t,T_j)} - 1\right), & T_{j-1} < t \le T_j. \end{cases} \tag{5.88}$$

Proof: First let $t \le T_{j-1}$. Using the tower law we obtain

$$\begin{aligned} & E_t^{Q_j}\left[e^{\int_{T_{j-1}}^{T_j} r_s ds}\right] \\ =& E_t^{Q_j}\left[E_{T_{j-1}}^{Q_j}\left[\frac{B(T_j,T_{j-1})}{P(T_j,T_j)} \middle/ \frac{B(T_{j-1},T_{j-1})}{P(T_{j-1},T_j)}\right] \times \frac{1}{P(T_{j-1},T_j)}\right] \\ =& E_t^{Q_j}\left[\frac{P(T_{j-1},T_{j-1})}{P(T_{j-1},T_j)}\right] = \frac{P(t,T_{j-1})}{P(t,T_j)}. \end{aligned} \tag{5.89}$$

For $t > T_{j-1}$, we have

$$E_t^{Q_j}\left[e^{\int_{T_{j-1}}^{T_j} r_s ds}\right] = E_t^{Q_j}\left[\frac{B(T_j,T_{j-1})}{P(T_j,T_j)}\right] = \frac{B(t,T_{j-1})}{P(t,T_j)}. \tag{5.90}$$

Here in both Equation 5.89 and Equation 5.90, we have made use of the martingale property of the $P(t, T_j)$-relative price under the T_j-forward measure. What remained is to insert the result of Equation 5.89 and Equation 5.90 into

$$E_t^{Q_j}[f_j(T_j)] = \frac{1}{\Delta T_j}\left(E_t^{Q_j}\left[e^{\int_{T_{j-1}}^{T_j} r_s ds}\right] - 1\right),$$

we then arrive at Equation 5.88 □

Notionally, it helps to integrate zero-coupon bonds and money market accounts into a single type of security and introduce the notion of extended zero-coupon bonds:

$$\tilde{P}(t, T) = E_t^Q\left[e^{-\int_t^T r_s ds}\right] = \begin{cases} P(t, T), & \text{if } t \le T \\ B(t, T), & \text{if } t > T. \end{cases} \tag{5.91}$$

Then, based on Theorem 5.6.1 we make the following

Definition 5.6.1. *The forward term rates based on SOFR is defined by*

$$f_j(t) = \frac{1}{\Delta T_j} \left(\frac{\tilde{P}(t, T_{j-1})}{\tilde{P}(t, T_j)} - 1 \right), \quad for \ t \leq T_j. \tag{5.92}$$

We make four remarks here. First, for $t \leq T_{j-1}$, the definition of the backward-looking SOFR forward rate is identical to that of the forward-looking LIBOR forward rate. Second, for the same tenor (T_{j-1}, T_j), both backward-looking SOFR forward rate and forward-looking LIBOR forward rates are martingales under the T_j-forward measure. Third, $f_j(t)$ is the time-t replication cost or the present value of the cash flow $f_j(T_j)$ to occur at time T_j. Finally, the evaluation of futures rates is model dependent, and can only be addressed after the introduction of models.

For notational simplicity, we drop ˜ over the notation for the extended zero-coupon bonds hereafter.

Exercises

1. Prove the equality

$$\tilde{F}_0^T - F_0^T = \frac{-1}{P(0, T)} \times \mathrm{Cov}^{\mathbb{Q}}(B_T^{-1}, S_T),$$

where $\mathbb{Q}$ stands for the risk-neutral measure.

2. Derive the pricing formula for a put option with payoff

$$V_T = (K - S_T)^+$$

under a stochastic interest rate. Assume that the volatility vector of the asset and the T-maturity zero-coupon bond are

$$\Sigma_s = \sigma_s \begin{pmatrix} \rho \\ \sqrt{1 - \rho^2} \end{pmatrix} \quad \text{and} \quad \Sigma_P = \sigma_P \begin{pmatrix} 1 \\ 0 \end{pmatrix},$$

where both σ_s and σ_P are time-dependent scalars.

   (a) Show that the correlation between $d \ln S_t$ and $d \ln P(t, T)$ is ρ.

   (b) Express the option formula in terms of σ_s, σ_P, and ρ.

3. Consider the *risk-neutralized* processes of two assets:

$$\frac{dS_1}{S_1} = r_t \, dt + \sigma_1 \, dW_1(t),$$

$$\frac{dS_2}{S_2} = r_t \, dt + \sigma_2 \, dW_2(t),$$

where $dW_1 dW_2 = \rho \, dt$ and r_t may follow some stochastic process.

    (a) What kind of process does S_1/S_2 follow under the risk-neutral measure?

    (b) Can you define a new measure under which S_1/S_2 is a martingale?

    (c) Express the Radon–Nikodym derivative in terms of the asset prices.

4. Prove that the risk-neutral dynamics of an asset with price S_t and stochastic dividend yield q_t is

$$\mathrm{d}S_t = S_t(r_t - q_t)\Delta t + \Sigma_S^T \mathrm{d}\mathbf{W}_t.$$

5. Let a *stripped-dividend price process*, $\hat{S}_t$, follow

$$\mathrm{d}\hat{S}_t = \hat{S}_t \left(r_t \, \mathrm{d}t + \Sigma_S^T \, \mathrm{d}\mathbf{W}_t \right)$$

under the risk-neutral measure. For discrete dividends $\{q_i\}$, prove that the risk-neutral process of S_t is

$$dS_t = \left[r_t S_t - \sum q_i \delta(T_i - t) \right] dt$$
$$+ \left[S_t \Sigma_S^T - \sum q_i H(T_i - t) P_t^{T_i} (\Sigma_S - \Sigma(t, T_i))^T \right] \mathrm{d}\mathbf{W}_t,$$

where $\delta(x)$ and $H(x)$ are Dirac and Heaviside functions, respectively.

6. It is understood that a SOFR term rate satisfies

$$f(0; T - \Delta T, T) = E^{\mathbb{Q}_T} \left[f(T; T - \Delta T, T) \right],$$

whereas the corresponding futures implied rate satisfies

$$\tilde{f}(0; T - \Delta T, T) = E^{\mathbb{Q}} \left[f(T; T - \Delta T, T) \right].$$

Prove that

$$f(0; T - \Delta T, T) - \tilde{f}(0; T - \Delta T, T)$$
$$= \frac{\mathrm{Cov}^{\mathbb{Q}}(f(T; T - \Delta T, T), B_T^{-1})}{E^{\mathbb{Q}}[B_T^{-1}]}.$$

Chapter 6

The Heath–Jarrow–Morton Model

Interest-rate models are necessary for pricing most interest-rate derivatives and for gauging the risk of general interest-rate instruments. The natural candidate for the state variable of interest-rate models seems to be the short rate, which in fact is the only state variable in many early models such as the well-known Vasicek (1977) model and the Cox–Ingersoll–Ross (CIR) (1985) model. Early models are largely based on macro-economical arguments, and they are thus called equilibrium models. Equilibrium models may have a sound financial economical underpinning, but they do not automatically reproduce the market price of benchmark bonds nor reproduce the yield curve unless a calibration procedure has been carried out. It is known that such a procedure may result in a model that is twisted and sometimes at odds with financial intuition. As models based on a single-state variable, short-rate models seem to lack sufficient capacity to describe the dynamics of the entire yield curve.

It was later understood that, to exclude arbitrage, an interest-rate model should take the entire yield curve as an input instead of output. This consideration has catalyzed the emergence of the so-called arbitrage pricing models. A basic feature of arbitrage pricing models of interest rates is that the prices of a whole class of basic securities, such as the prices of zero-coupon bonds of all maturities, are taken as primitive state variables. Equivalently, an arbitrage pricing model can also take some kind of yield curve, discrete or continuous, as the state variables. The zero-coupon yield curve and the par-yield curve are two of the candidates. It turns out that (the continuous or discrete compounding) forward rates are often the best choice of state variables for many applications. With the continuous compounding forward rates, Heath, Jarrow, and Morton (1992) developed an arbitrage framework for interest-rate modeling such that any specific arbitrage-free model may be fitted into the framework as a special case. In this chapter, we present the derivation, estimation, and applications of the Heath–Jarrow–Morton (HJM) model.

With historical data of U.S. Treasury yields, we demonstrate the estimation of model parameters of the HJM model using an important technique called principal component analysis (PCA). We then study in depth two classic, special cases under the HJM framework, namely, the Ho–Lee model (1986) and the Hull–White (1989) model. These two models played important roles both in the development of interest-rate models and in practical applications. We will also highlight the linear Gaussian model which, thanks to its flexibility and simplicity, has been a popular model for pricing Bermudan swaptions, one of the most important fixed-income derivatives. The remaining part of this chapter is devoted to option pricing under the HJM model.

6.1   The HJM Model

It has been established that the price process of zero-coupon bonds under the risk-neutral measure $\mathbb{Q}$ is

$$dP(t,T) = P(t,T)\left[r_t\,dt + \boldsymbol{\Sigma}^{\mathrm{T}}(t,T)\,d\mathbf{W}_t\right].\tag{6.1}$$

For the purpose of derivatives pricing, $\boldsymbol{\Sigma}(t,T)$ should satisfy at least the following additional conditions: (1) $\boldsymbol{\Sigma}(t,t) = 0, \forall t$; (2) $P(t,t) = 1, \forall t$; and (3) $P(t,T)$ is a monotonically decreasing function of T. The first two conditions reflect the fact that at maturity, the price of the zero-coupon bond equals its par value and thus has no volatility, and the last condition will ensure that forward rates are positive, otherwise an arbitrage opportunity would occur.

The specification of $\boldsymbol{\Sigma}(t,T)$ is a difficult job if we work directly with the process of $P(t,T)$. However, this job will become easier if we work with the process of forward rates. By Ito's Lemma, there is

$$d\ln P(t,T) = \left[r_t - \frac{1}{2}\boldsymbol{\Sigma}^{\mathrm{T}}(t,T)\boldsymbol{\Sigma}(t,T)\right]dt + \boldsymbol{\Sigma}^{\mathrm{T}}(t,T)\,d\mathbf{W}_t.\tag{6.2}$$

Assume, moreover, that $\boldsymbol{\Sigma}_T(t,T) = \partial\boldsymbol{\Sigma}(t,T)/\partial T$ exists and $\int_0^T \|\boldsymbol{\Sigma}_T(t,T)\|^2\,dt < \infty$. By differentiating Equation 6.2 with respect to T and recalling that

$$f(t,T) = -\frac{\partial\ln P(t,T)}{\partial T},\tag{6.3}$$

we obtain the process of forward rates under the $\mathbb{Q}$-measure,

$$df(t,T) = \boldsymbol{\Sigma}_T^{\mathrm{T}}\boldsymbol{\Sigma}\,dt - \boldsymbol{\Sigma}_T^{\mathrm{T}}\,d\mathbf{W}_t.\tag{6.4}$$

Here the arguments of $\boldsymbol{\Sigma}$ are omitted for simplicity. We consider $-\boldsymbol{\Sigma}_T(t,T)$ to be the volatility of the forward rate, and rewrite it as

$$\boldsymbol{\sigma}(t,T) = -\boldsymbol{\Sigma}_T(t,T).\tag{6.5}$$

By integrating the above equation with respect to T, we then obtain the volatility of the zero-coupon bond:

$$\boldsymbol{\Sigma}(t,T) = -\int_t^T \boldsymbol{\sigma}(t,s)\,ds.\tag{6.6}$$

Here we have made use of the condition $\mathbf{\Sigma}(t,t) = 0$. Equation 6.6 fully describes the relationship between the volatility of a forward rate and the volatility of its corresponding zero-coupon bond. In terms of $\boldsymbol{\sigma}(t,T)$, we rewrite this forward-rate process as

$$\mathrm{d}f(t,T) = \left(\boldsymbol{\sigma}^{\mathrm{T}}(t,T) \int_t^T \boldsymbol{\sigma}(t,s)\,\mathrm{d}s \right) \mathrm{d}t + \boldsymbol{\sigma}^{\mathrm{T}}(t,T)\,\mathrm{d}\mathbf{W}_t. \qquad (6.7)$$

Equation 6.7 is the famous HJM equation. An important feature of Equation 6.7 is that the drift term of the forward-rate process under the risk-neutral measure is completely determined by its volatility.

The HJM model lays down the foundation of arbitrage pricing in the context of fixed-income derivatives, and it is considered a milestone of financial derivative theory. Before 1992, fixed-income modeling was dominated by equilibrium models, which are built upon macro-economical foundations. The major limitation of the equilibrium models is that these models do not naturally reproduce the market prices of basic instruments, particularly the zero-coupon bonds, unless users go through a calibration procedure. With arbitrage pricing models, the prices of the basic instruments are taken as model inputs rather than as outputs, so their prices are naturally reproduced. The arbitrage pricing models are rooted in the efficient market hypothesis, which states that market prices of instruments do not induce any arbitrage opportunity. With an arbitrage model, derivative securities will be priced consistently with the basic instruments in the sense that no arbitrage opportunity would be induced.

There are two ways to specify the forward-rate volatility for the HJM model. The first way is to estimate $\boldsymbol{\sigma}(t,T)$ (together with its dimension) directly from time series data on forward rates of various maturities, and the estimation will be demonstrated in Section 6.2. Note that $\boldsymbol{\sigma}^{\mathrm{T}}(t,T)\boldsymbol{\sigma}(t,T')$ is the covariance between $f(t,T)$ and $f(t,T')$, and the PCA technique will enable us to figure out $\boldsymbol{\sigma}(t,T)$. The second way is to specify $\boldsymbol{\sigma}(t,T)$ exogenously using certain parametric functions of t and T. Note that different specifications of $\boldsymbol{\sigma}(t,T)$ generate different concrete models for applications. Because of that, the HJM model is treated not only as a model in its own right, but also as a framework for forward-rate models that are deemed arbitrage-free.

We should point out that the HJM Equation 6.7 is a necessary condition for no-arbitrage models, but not a sufficient one. For usual specifications of the volatility function, $\boldsymbol{\sigma}(t,T)$, the forward rate has a Gaussian distribution, so it can assume negative values with a positive possibility. It has been a major challenge to find out the necessary and sufficient conditions on $\boldsymbol{\sigma}(t,T)$ so that forward rates are guaranteed to be positive, yet this challenge remains outstanding as of today. Interestingly, it is very difficult to identify the corresponding forward-rate volatility functions for short-rate models that guarantee positive forward rates. One objective of Chapter 7 is to derive the corresponding forward-rate volatility function for short-rate models.

The HJM model also implies the dynamics of the short rate. By integrating Equation 6.4 from 0 to t, we obtain the expression of forward rates:

$$f(t,T) = f(0,T) + \int_0^t \mathbf{\Sigma}^{\mathrm{T}}(s,T)\frac{\partial \mathbf{\Sigma}(s,T)}{\partial T}\,\mathrm{d}s - \frac{\partial \mathbf{\Sigma}^{\mathrm{T}}(s,T)}{\partial T}\,\mathrm{d}\mathbf{W}_s. \qquad (6.8)$$

Then by setting $T = t$, we obtain an expression for the short rate:

$$r_t = f(t,t) = f(0,t) + \int_0^t \frac{1}{2}\frac{\partial \|\mathbf{\Sigma}(s,t)\|^2}{\partial t}\,\mathrm{d}s - \frac{\partial \mathbf{\Sigma}^{\mathrm{T}}(s,t)}{\partial t}\,\mathrm{d}\mathbf{W}_s. \qquad (6.9)$$

In differential form, Equation 6.9 becomes

$$\mathrm{d}r_t = \mathrm{drift} - \left(\int_0^t \frac{\partial^2 \mathbf{\Sigma}^{\mathrm{T}}(s,t)}{\partial t^2}\,\mathrm{d}\mathbf{W}_s\right)\mathrm{d}t - \left.\frac{\partial \mathbf{\Sigma}^{\mathrm{T}}(s,t)}{\partial t}\right|_{s=t}\,\mathrm{d}\mathbf{W}_t. \qquad (6.10)$$

By examining Equation 6.10, we understand that the short rate is in general a non-Markovian random variable under the HJM framework, unless

$$\int_0^t \frac{\partial^2 \mathbf{\Sigma}^{\mathrm{T}}(s,t)}{\partial t^2}\,\mathrm{d}\mathbf{W}_s \qquad (6.11)$$

can be expressed as a function of some Markovian variables. Only in this situation, the short-rate becomes a Markovian process. Further discussion of Markovian short-rate models is provided in Chapter 7.

For some subsequent applications, we rewrite Equation 6.9 as

$$r_t = f(0,t) + \frac{\partial}{\partial t}\int_0^t \frac{1}{2}\|\mathbf{\Sigma}(s,t)\|^2\mathrm{d}s - \mathbf{\Sigma}^{\mathrm{T}}(s,t)\,\mathrm{d}\mathbf{W}_s. \qquad (6.12)$$

Here, we have applied the following stochastic Fubini theorem (see Karatzas and Shreve, 1991) to exchange the order of differentiation and integration:

$$\frac{\partial}{\partial t}\int_0^t \theta(s,t)\,\mathrm{d}\tilde{W}_s = \theta(t,t)\frac{\mathrm{d}\tilde{W}_t}{\mathrm{d}t} + \int_0^t \frac{\partial}{\partial t}\theta(s,t)\,\mathrm{d}\tilde{W}_s,$$

and we have applied the condition $\mathbf{\Sigma}(t,t) = 0$.

Finally, we note that under the HJM model, the following price formula for zero-coupon bonds exists:

$$P(t,T) = P(0,T)\exp\left\{\int_0^t \left(r_s - \frac{1}{2}\mathbf{\Sigma}^{\mathrm{T}}(s,T)\mathbf{\Sigma}(s,T)\right)\mathrm{d}s + \mathbf{\Sigma}^{\mathrm{T}}(s,T)\,\mathrm{d}\mathbf{W}_s\right\},$$

$$\qquad (6.13)$$

which will be used repeatedly for analyses as well as computations.

6.2 Estimating the HJM Model from Yield Data

We now study the specifications for the general HJM model (Equation 6.7) based on the distributional properties of historical data of interest rates. Specifically, we need to specify the dimensions of the model, n, as well as the volatility function of the forward rates, $\sigma(t, T)$, based on time series data of bond yields of various maturities. Note that the HJM model so obtained is considered a process under the physical measure instead of the risk-neutral measure. For pricing purposes, we may have to calibrate the model to observed data of benchmark derivatives. This, however, is a very different issue and will be addressed for the market model in Chapter 9.

6.2.1 From a Yield Curve to a Forward-Rate Curve

The inputs for estimating the HJM model are historical data on the U.S. Treasury yields, demonstrated in Figure 6.1 with monthly quotes of yields for a 10-year period, from 1996 to 2006. There are seven curves in the figure, which are dot plots of yield-to-maturities for 3-month, 6-month, 2-year, 3-year, 5-year, 10-year, or 30-year maturity benchmark U.S. Treasury bonds, respectively.

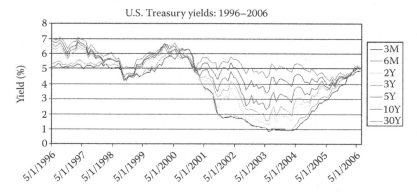

FIGURE 6.1: The monthly quotes of U.S. Treasury benchmark yields. (Adopted from Reuters.)

The first step of our model specification is to estimate the entire forward-rate curve, $f(\tau, \tau + T)$, for each day, τ, in the data set and for the 30-year horizon, $0 \leq T \leq 30$. Since we do not have any detailed information about the benchmark Treasury bonds over this 10-year period, we treat the yields in the input data set for the last five maturities, $T_3 = 2, T_4 = 3, T_5 = 5,$

$T_6 = 10$, and $T_7 = 30$, as par yields.[1] The instantaneous forward-rate curve for the day is determined by reproducing the value of the Treasury bills,

$$\frac{1}{(1 + y_i \Delta T)^{i/2}} = P(\tau, \tau + T_i), \quad \text{for } i = 1, 2, \qquad (6.14)$$

and the value of the par bonds,

$$1 = \sum_{j=1}^{n_i} y_i(\tau) \cdot \Delta T P(\tau, \tau + j\Delta T) + P(\tau, \tau + n_i \Delta T) \qquad (6.15)$$

for $i = 3, \ldots, 7$. Here $y_i(\tau)$ is the yield or par yield shown in Figure 6.1, $n_i = T_i / \Delta T$, $i = 1, 2, \ldots, 7$, $\Delta T = 0.5$, and

$$P(\tau, \tau + T) = e^{-\int_\tau^{\tau+T} f(\tau, s) \mathrm{d}s}. \qquad (6.16)$$

Because there are only a few inputs of yields, interpolation is necessary for constructing the forward rates. For clarity and notational simplicity, we present the algorithm for constructing the forward-rate curve for date $\tau = 0$ only with linear interpolation. The construction method for subsequent dates is identical. Define, in addition, $T_0 = 0$. Using $T_i, i = 0, 1, \ldots, 7$ as knot points, we assume that the forward rate is a continuous linear function between any two adjacent maturities:

$$\begin{aligned} f(0, T) &= r_0, && \text{for } T_0 \leq T < T_1, \\ f(0, T) &= f(0, T_{i-1}) + \alpha_i (T - T_{i-1}), && \text{for } T_{i-1} \leq T < T_i, \\ &&& i = 2, \ldots, 7, \end{aligned}$$

where r_0 is taken to be the three-month rate for continuous compounding, given by

$$r_0 = \frac{1}{\Delta T} \ln(1 + y_1 \Delta T), \qquad (6.17)$$

and $\{\alpha_i\}_{i=2}^7$ will be determined from the following procedure of bootstrapping.

1. Determine α_2 by matching $P(0, T_2)$ defined by Equation 6.16 to the bond price given by Equation 6.14.

2. Assume that we already have $\alpha_2, \alpha_3, \ldots, \alpha_i$.

3. To compute α_{i+1}, we use the $(i + 1)$st bond, and decompose the price into

$$1 = B_{i+1}^0 + B_{i+1}^1, \qquad (6.18)$$

where the first term represents the present value of the coupons due on or before date T_i,

$$B_{i+1}^0 = \sum_{t_j \leq T_i} \Delta T y_{i+1} P_0^{t_j}, \qquad (6.19)$$

---

[1] Otherwise we would need coupon rates and exact maturities of the bonds. The 3- and 6-month yields are for zero-coupon bonds.

where y_{i+1} is the coupon rate for the $(i+1)$st bond, and t_j is the jth coupon date of the bond. The discounted factor is calculated using the known forward rates, $f(0, T)$, for $T \leq T_i$. Assume that t_j lies between T_{k-1} and T_k, $k \leq i$, then

$$
P_0^{t_j} = P_0^{T_{k-1}} \exp \left\{ - \int_{T_{k-1}}^{t_j} f(0, s)\, \mathrm{d}s \right\}
$$

$$
= P_0^{T_{k-1}} \exp \left\{ -(t_j - T_{k-1}) \left[f(0, T_{k-1}) + \frac{1}{2} \alpha_k (t_j - T_{k-1}) \right] \right\},
$$

where

$$
P_0^{T_{k-1}} = \exp \left\{ - \int_0^{T_{k-1}} f(0, s)\, \mathrm{d}s \right\}
$$

$$
= \exp \left\{ - \sum_{j=1}^{k-1} \int_{T_{j-1}}^{T_j} \left[f(0, T_{j-1}) + \alpha_j (s - T_{j-1}) \right] \mathrm{d}s \right\}
$$

$$
= \exp \left\{ - \sum_{j=1}^{k-1} (T_j - T_{j-1}) \left[f(0, T_{j-1}) + \frac{1}{2} \alpha_j (T_j - T_{j-1}) \right] \right\}.
$$

$$(6.20)$$

The second term in Equation 6.18 is

$$
B_{i+1}^1 = P_0^{T_i} \left(\sum_{T_i < t_j \leq T_{i+1}} \Delta T y_{i+1} \mathrm{e}^{-(t_j - T_i)(f_i + (1/2)\alpha_{i+1}(t_j - T_i))} \right.
$$

$$
\left. + \mathrm{e}^{-(T_{i+1} - T_i)(f_i + (1/2)\alpha_{i+1}(T_{i+1} - T_i))} \right),
$$

$$(6.21)$$

where $f_i = f(0, T_i)$. By combining Equations 6.18, 6.19, and 6.21, we obtain an equation for α_{i+1}:

$$
\frac{1 - \sum_{t_j \leq T_i} \Delta T y_{i+1} P_0^{t_j}}{P_0^{T_i}} = \sum_{T_i < t_j \leq T_{i+1}} \Delta T y_{i+1} \mathrm{e}^{-(t_j - T_i)(f_i + (1/2)\alpha_{i+1}(t_j - T_i))}
$$

$$
+ \mathrm{e}^{-(T_{i+1} - T_i)(f_i + (1/2)\alpha_{i+1}(T_{i+1} - T_i))}.
$$

We then solve for α_{i+1} from the above equation by a root-finding algorithm. This process can continue until $i = 7$; then we will obtain all $f(0, T)$ for $T \leq 30$.

As an example, we display in Figure 6.2 the instantaneous forward-rate curve for May 31, 1996, the first date in our data, where the "$*$"s mark the forward rates of the seven benchmark maturities.

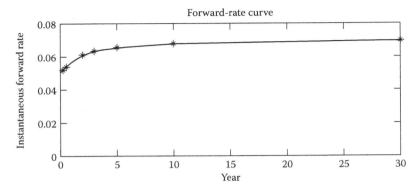

FIGURE 6.2: Instantaneous forward-rate curve of May 31, 1996, by linear interpolation.

There is a slight problem with the use of linear interpolation in the boot-strapping procedure: the curve is not differentiable at the knot points, $\{T_i\}$. If piece-wise constant interpolation was used instead, then we would have induced jumps across those maturities, worsening the problem. This kind of non-smoothness is artificial and financially not justifiable, and it can cause potential problems. According to the HJM equation, such non-smoothness propagates over time, causing non-differentiability (or jumps if piece-wise constant interpolation is used) in the short rate across certain dates in the future. When the non-smooth forward-rate curve is applied to derivatives pricing, the non-smoothness may translate into extra volatility and thus cause mispricing.

The solution for better smoothness is simple: we can adopt spline interpolation instead of linear interpolation. A spline is a piece-wise cubic polynomial that has a continuous second-order derivative. The forward-rate curve so constructed is shown in Figure 6.3.

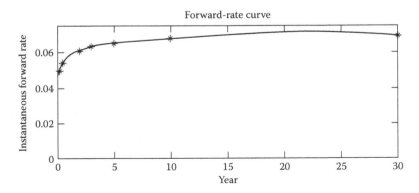

FIGURE 6.3: The instantaneous forward-rate curve for May 31, 1996, by spline interpolation.

For each date in the data set, we can construct one forward-rate curve with the above procedure. On the forward-rate curve of each date, we pick seven points, as marked in Figure 6.3, that correspond to the value of forward rates, $f(\tau, \tau + T_i)$, of the seven maturities (3-month, 6-month, 2-year, 3-year, 5-year, 10-year, or 30-year). By plotting $f(\tau, \tau + T_i), i = 1, \ldots, 7$ against τ in the data set, we obtain the time series data for the forward rates of the seven maturities, shown in Figure 6.3, where the time gap for the plots is one month, $\Delta\tau = 1/12$.

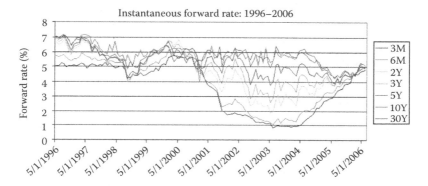

FIGURE 6.4: Forward-rate curve for the period 1996–2006.

Before we move on to analyze the covariance of the forward rates, we make a comment on the construction of the forward-rate curves. Sometimes in applications, we may need to enhance the smoothness or reduce the curvature of the forward-rate curve. These objectives can be achieved by adopting appropriate regularizations. Popular regularization functions include smoothness regularization,

$$\int_0^T \left| \frac{\partial f(0, s)}{\partial s} \right|^2 ds, \tag{6.22}$$

and curvature regularization,

$$\int_0^T \left| \frac{\partial^2 f(0, s)}{\partial s^2} \right|^2 ds. \tag{6.23}$$

Note that there also exists other parameterizations. For example, one may consider the following parameterization of the forward-rate function

$$f(0, T) = \sum_{j=0}^N \theta_j \left(\frac{T}{1 + \theta_{N+1} T} \right)^j$$

suggested in Avellaneda and Laurence (1999).

6.2.2 Principal Component Analysis

Having constructed the time series data of forward rates of the seven maturities, shown in Figure 6.4, we now proceed to the estimation of covariance among those forward rates. We then perform PCA with the covariance matrix. The results will shed light on the proper number of random factors that drive the evolution of the forward-rate curve, so that we can determine n and subsequently $\sigma(t, T)$ for the HJM equation.

Let $f(n\Delta\tau, n\Delta\tau + T_i), n = 0, 1, \ldots, N$ be the forward rates for the seven maturities, $T_i, i = 1, 2, \ldots, 7$, where $\Delta\tau = 1/12$ represents the observation interval of one month, and N the total number of months. For forward rates of each maturity, $T_i, i = 1, 2, \ldots, 7$, we calculate the change over $\Delta\tau$:

$$\Delta f_{n,i} = f((n+1)\Delta\tau, (n+1)\Delta\tau + T_i)$$
$$- f(n\Delta\tau, n\Delta\tau + T_i), \quad n = 0, 1, \ldots, N-1. \quad (6.24)$$

The empirical covariance between $\Delta f_{.,i}$ and $\Delta f_{.,j}$ is, straightforwardly,

$$\hat{c}_{ij} = \frac{1}{N} \sum_{n=0}^{N-1} (\Delta f_{n,i} - \overline{\Delta f_i})(\Delta f_{n,j} - \overline{\Delta f_j}),$$

where

$$\overline{\Delta f_i} = \frac{1}{N} \sum_{n=0}^{N-1} \Delta f_{n,i}. \quad (6.25)$$

By performing eigenvalue decomposition on the covariance matrix, $\hat{C} = (\hat{c}_{ij})$, we obtain

$$\hat{C} = V\Lambda V^T = \sum_{k=1}^{7} \lambda_k \Delta\tau \mathbf{v}_k \mathbf{v}_k^T, \quad (6.26)$$

where $\Lambda = \Delta\tau \text{diag}(\lambda_1, \lambda_2, \ldots, \lambda_7)$ and $V = (\mathbf{v}_1, \mathbf{v}_2, \ldots, \mathbf{v}_7)$ are eigenvalue and eigenvector matrices, respectively, the λ's are put in descending order, that is, $\lambda_1 \geq \lambda_2 \geq \cdots \geq \lambda_7$, and the $\mathbf{v}_k$s are normalized, $\|\mathbf{v}_k\|_2 = 1, k = 1, \ldots, 7$. These eigenvectors $\{\mathbf{v}_1, \mathbf{v}_2, \ldots, \mathbf{v}_7\}$ are also called principal components of $\hat{C}$. In terms of components, Equation 6.26 reads as

$$\hat{c}_{ij} = \sum_{k=1}^{7} \lambda_k \Delta\tau v_{ik} v_{jk}. \quad (6.27)$$

Based on the above eigenvalue decomposition, we can model the random increments of forward rates as follows:

$$\Delta \mathbf{f}_n = \overline{\Delta \mathbf{f}} + \sum_{k=1}^{7} \sqrt{\lambda_k \Delta\tau}\, \mathbf{v}_k \xi_{k,n}. \quad (6.28)$$

Here,

$$\Delta\mathbf{f}_n = \begin{pmatrix} \Delta f_{n,1} \\ \Delta f_{n,2} \\ \vdots \\ \Delta f_{n,7} \end{pmatrix}, \quad \overline{\Delta\mathbf{f}} = \begin{pmatrix} \overline{\Delta f_1} \\ \overline{\Delta f_2} \\ \vdots \\ \overline{\Delta f_7} \end{pmatrix},$$

and $\{\xi_{k,n}\}$ are independent random variables with mean and variance equal to zero and one, respectively. We say that the forward-rate curve is driven by these seven random factors. By generating seven of these random variables and applying them to Equation 6.28, we simulate the change of all forward rates over one time step, $\Delta\tau$.

The importance of each of the seven factors, however, is not the same. Since $\lambda_i \geq \lambda_{i+1}$, $\mathbf{v}_i$ is considered more important than $\mathbf{v}_{i+1}$ in shaping the forward-rate curve. Moreover, suppose that, for some $\nu < 7$, there is

$$\lambda_1 + \lambda_2 + \cdots + \lambda_\nu \gg \lambda_{\nu+1} + \cdots + \lambda_7. \tag{6.29}$$

We then can ignore those λ_k for $k > \nu$ and approximate $\hat{C}$ by

$$\hat{C} \approx (\mathbf{v}_1 \cdots \mathbf{v}_\nu) \begin{pmatrix} \lambda_1 & & \\ & \ddots & \\ & & \lambda_\nu \end{pmatrix} \begin{pmatrix} \mathbf{v}_1^T \\ \vdots \\ \mathbf{v}_\nu^T \end{pmatrix}.$$

Accordingly, we can discard the less important factors in modeling forward-rate increments:

$$\Delta\mathbf{f}_n \approx \overline{\Delta\mathbf{f}} + \sum_{k=1}^{\nu} \sqrt{\lambda_k \Delta\tau} \mathbf{v}_k \xi_{k,n}. \tag{6.30}$$

Let us take a look at the weights of various factors that drive the Treasury forward-rate curve. The eigenvalues of the covariance matrix of forward rates calculated based on the monthly observed U.S. Treasury yields are listed in Table 6.1. The second column shows their weights. We can see that the first three principal components carry an aggregated weight of over 93%.

Consider taking the first three principal components, that is, $\nu = 3$, in modeling the forward-rate increments by Equation 6.30. A random shock, $\xi_{k,n}$, will deform forward rates of the seven maturities by $\sqrt{\lambda_k \Delta\tau} v_{i,k} \xi_{k,n}$, $i = 1, \ldots, 7$. To understand the impacts of the random shocks on the forward rates, we look at the first three principal components (i.e., the eigenvectors of the first three largest eigenvalues) in Figure 6.5.

We want to highlight some stylized facts here about the PCA of yield curves. As we can see from Figure 6.5, the leading component is relatively flat, with all elements positive, which will roughly cause a parallel move of yields in response to a random shock. The second principal component tilts: its elements change from positive values to negative values. Given a positive

TABLE 6.1: Eigenvalues and Weights for the Covariance Matrix

Eigenvalue, λ_k	$\lambda_k / \sum_j \lambda_j$ (%)
0.000514	64.09
0.000152	18.91
8.32E−05	10.39
3.97E−05	4.95
1.25E−05	1.56
6.35E−07	0.08
1.78E−07	0.02

For the covariance matrix of U.S. dollar (USD) forward rates under continuous compounding.

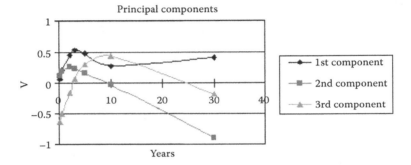

FIGURE 6.5: The first three principal components of USD forward rates.

shock, short maturity yields will increase while long maturity yields will decrease. The third principal component bends: it starts in the negative territory, goes to the positive territory, and ends up in the negative territory again. A positive shock to this component will move both short- and long-term yields lower, yet such a shock will move the median-term yields higher, thus "bending" the yield curve. Knowing that the aggregated weight of the first three eigenvalues is 93%, we may say that the deformation of the forward-rate curve is caused by parallel shifts, tilting, and bending, corresponding to the shapes of the three leading eigenvectors, 93% of the time. The eigenvalues, meanwhile, differentiate the magnitudes of the impacts on average. These stylized facts were first discovered by Litterman and Scheinkman (1991). With Treasury yield data prior to 1991, they obtained an aggregated weight for the first three eigenvalues of 98%, higher than the 93% obtained using the data between 1996 and 2006. The difference can be explained by the decorrelation of the Treasury yields that took place during the period from 2000 to 2006. Over this period, the short rates first decreased and then increased, while the long rates did not change by much.

Another feature of the leading principal components worth mentioning is that the variability concentrates on the short-term ranges, which simply reflects the fact that we witness more variation of the yield curve in short maturities.

Now we are ready to build the HJM model using the results from PCA. The process consists of the following steps:

1. Interpolate $\{v_{i,k}, i = 1, 2, \ldots, 7\}$ and $\overline{\Delta f_i}$ to get $\{v_k(T)\}$ and $\overline{\Delta f}(T)$, respectively, using a method such as continuous spline interpolation.

2. Define $\Delta f(t, t + T) = \sum_{k=1}^{\nu} v_k(T)\sqrt{\lambda_k \Delta t}\xi_k(t) + \frac{\overline{\Delta f}(T)}{\Delta \tau}\Delta t$ for general t and T, where $\xi_k(t)$ is a standard normal random variable.

3. Through comparing to the HJM equation, we define

$$\mathrm{d}f(t, t + T) = \boldsymbol{\sigma}^{\mathrm{T}}(T)\,\mathrm{d}\tilde{\mathbf{W}}_t + \mu(T)\,\mathrm{d}t, \qquad (6.31)$$

where $\sigma_k(T) = v_k(T)\sqrt{\lambda_k}$ and $\mu(T) = \overline{\Delta f}(T)/\Delta \tau$. Note that σ_k and μ depend on the time to maturity only.

4. By replacing $t + T$ by T, we finally have

$$\mathrm{d}f(t, T) = \boldsymbol{\sigma}^{\mathrm{T}}(T - t)\,\mathrm{d}\tilde{\mathbf{W}}_t + \mu(T - t)\,\mathrm{d}t. \qquad (6.32)$$

For the risk-neutral process of the forward rate, the volatility function is all we need. The risk-neutral drift is calculated according to

$$\mu(t, T) = \boldsymbol{\sigma}^{\mathrm{T}}(T - t)\int_t^T \boldsymbol{\sigma}(s - t)\,\mathrm{d}s. \qquad (6.33)$$

For a three-factor HJM model, the forward-rate volatility components estimated using yield data from 1996 to 2006 are displayed in Figure 6.6. In terms of the magnitude, we can see that there is a descending order: $\|\sigma_1(\cdot)\|_2 > \|\sigma_2(\cdot)\|_2 > \|\sigma_3(\cdot)\|_2$.

We finish this section with additional remarks on PCA analysis and HJM model estimation. In much of the literature (e.g., Litterman and Schainkman, 1991; Avellaneda and Laurence, 1999), PCA is carried out with data on yields to maturities or zero-coupon yields. We take the PCA analysis of zero-coupon yields as an example. Out of the YTM data of 1996–2006 displayed in Figure 6.1, we can bootstrap and obtain the zero-coupon yields for the same period, which are displayed in Figure 6.7. PCA analysis yields principal components for the zero-coupon yields, displayed in Figure 6.8. Yet again, we see the stylized features of the first three principal components, namely, being flat, tilted, and bent. The principal components give rise to the volatility of the zero-coupon yields.

Some studies have suggested constructing the forward-rate volatilities from the volatilities of zero-coupon yields. This approach, however, is numerically

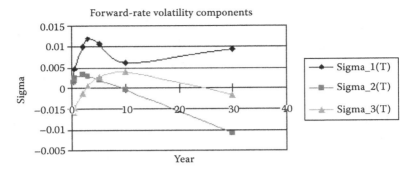

FIGURE 6.6: The three components of the forward-rate volatility, $\sigma_i(T)$, $i = 1, 2, 3$.

FIGURE 6.7: Zero-coupon yields obtained through bootstrapping.

unstable. Note that the forward-rate volatility relates to the zero-coupon yield volatility, $\boldsymbol{\sigma}_y(t, T)$, through the relationship

$$\int_t^T \boldsymbol{\sigma}(t, s)\, \mathrm{d}s = (T - t)\boldsymbol{\sigma}_y(t, T). \tag{6.34}$$

It follows that

$$\boldsymbol{\sigma}(t, T) = \boldsymbol{\sigma}_y(t, T) + (T - t)\frac{\partial \boldsymbol{\sigma}_y(t, T)}{\partial T}. \tag{6.35}$$

The above expression contains a differentiation that is then multiplied by a factor, $T - t$, which can reach 30 (years). Numerical studies have shown that forward-rate volatilities so generated are unrealistically larger than the observed volatilities. Hence, such an approach is numerically unstable and thus not feasible.

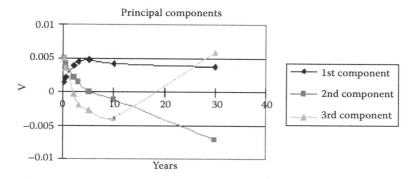

FIGURE 6.8: The first three principal components of USD zero yields.

6.3 An Example of a Two-Factor Model

Yields or forward rates of different maturities are not perfectly correlated, which was made evident in the last section with historical data. In this section, we demonstrate how to parameterize the forward-rate volatility to capture the stylized features of the forward-rate curves that are shaped by the principal components.

We consider a two-factor HJM model ($\nu = 2$) with the following forward-rate volatility components:

$$\begin{aligned} \sigma_1(T) &= ae^{-k_1 T}, \\ \sigma_2(T) &= b(1 - 2e^{-k_2 T}), \end{aligned} \tag{6.36}$$

where a, b, k_1, and k_2 are constants. To get a "flat" $\sigma_1(T)$ and a "tilted" $\sigma_2(T)$, we choose k_1 and k_2 such that $0 \le k_1 \ll 1, k_1 \ll k_2$. Similar to Avellaneda and Laurence (1999), we consider the following choice of parameters

$$a = 0.008, \quad b = 0.003, \quad k_1 = 0.0, \quad \text{and} \quad k_2 = 0.35.$$

Note that if we take $b = 0$, this two-factor model reduces to the Hull–White model, under which forward rates of all maturities are perfectly correlated.

Let us examine the correlation between the 3-month (i.e., short-term) and 30-year (long-term) forward rates. In general, the covariance between the forward rates of two maturities, T and T', is calculated according to

$$c(T, T') = \sigma_1(T)\sigma_1(T') + \sigma_2(T)\sigma_2(T'). \tag{6.37}$$

The correlation between forward rates of two maturities is thus

$$\rho(T, T') = \frac{c(T, T')}{\sqrt{c(T, T)}\sqrt{c(T', T')}}. \tag{6.38}$$

TABLE 6.2: PCA Analysis of Short and Long Rates

$\lambda_1 = 1.281 \times 10^{-4}$	$\lambda_2 = 1.51 \times 10^{-5}$
$v_1 = \begin{pmatrix} 0.6984 \\ 0.7157 \end{pmatrix}$	$v_2 = \begin{pmatrix} -0.7157 \\ 0.6984 \end{pmatrix}$
(for parallel shifts)	(for tilt moves)

Taking $T = 0.25$ (3-month) and $T' = 30$, we have

$$c(0.25, 0.25) = 7.024 \times 10^{-5},$$
$$c(0.25, 30) = 5.651 \times 10^{-5},$$
$$c(30, 30) = 7.300 \times 10^{-5}.$$

The correlation coefficient between the 3-month and 30-year forward rates is

$$\rho(0.25, 30) = \frac{c(0.25, 30)}{\sqrt{c(0.25, 0.25)}\sqrt{c(30, 30)}} = 79\%,$$

which indicates a high level of correlation between the two rates. The eigenvalues and normalized eigenvectors of the covariance matrix, $C_{2\times2}$, are listed in Table 6.2.

The relative importance of the two modes is reflected by the weights

$$\frac{\lambda_1}{\lambda_1 + \lambda_2} = 89.46\% \quad \text{and} \quad \frac{\lambda_2}{\lambda_1 + \lambda_2} = 10.54\%,$$

which are consistent with the PCA of U.S. Treasury data by Litterman and Scheinkman (1991).

Finally, we comment that the choices of the parameters correspond to reasonable short-rate volatility, which is calculated according to

$$\sigma_r = \sqrt{c(0,0)} = \sqrt{\sigma_1^2(0) + \sigma_2^2(0)} = \sqrt{a^2 + b^2} = 0.85\%.$$

Suppose the short rate is $r(0) = 5\%$, then the percentage of volatility of the short rate is $\sigma_r / r(0) = (0.85/5)\% = 17\%$. The interval of one standard deviation is

$$((5 - 0.85)\%, (5 + 0.85)\%) = (4.15\%, 5.85\%),$$

into which the short rate will fall with 67% probability.

6.4   Monte Carlo Implementations

We now consider the application of the HJM model to derivatives pricing. As a demonstration, we consider the pricing of a bond option that matures at T_0

with payoff

$$X_{T_0} = \left(\sum_{i=1}^{n} \Delta T \cdot c \cdot P_{T_0}^{T_i} + P_{T_0}^{T_n} - K \right)^+ .$$

Here, c is the coupon rate of the bond, K is the strike price of the option, and $T_i = T_0 + i\Delta T$ is the cash flow date of the ith coupon of the underlying bond. We call $T_n - T_0$, the life of the underlying bond beyond T_0, the tenor of the bond. The value of the option is given by

$$V_0 = E^{\mathbb{Q}} \left[\frac{1}{B_{T_0}} \left(\sum_{i=1}^{n} \Delta T \cdot c \cdot P_{T_0}^{T_i} + P_{T_0}^{T_n} - K \right)^+ \bigg| \mathcal{F}_0 \right]$$

$$= E^{\mathbb{Q}} \left[\left(\sum_{i=1}^{n} \Delta T c \cdot \frac{P_{T_0}^{T_i}}{B_{T_0}} + \frac{P_{T_0}^{T_n}}{B_{T_0}} - \frac{K}{B_{T_0}} \right)^+ \bigg| \mathcal{F}_0 \right], \qquad (6.39)$$

where $\mathbb{Q}$ is the risk-neutral measure. Based on Equation 6.13, we have the following expression for the discounted value of zero-coupon bonds:

$$\frac{P_{T_0}^{T_i}}{B_{T_0}} = P_0^{T_i} \exp \left(\int_0^{T_0} -\frac{1}{2} \|\boldsymbol{\Sigma}(t, T_i)\|^2 \, \mathrm{d}t + \boldsymbol{\Sigma}^{\mathrm{T}}(t, T_i) \, \mathrm{d}\mathbf{W}_t \right), \qquad (6.40)$$

for $i = 0, 1, \ldots, n$. Taking $i = 0$, in particular, we obtain the expression for the reciprocal of the money market account:

$$\frac{1}{B_{T_0}} = P_0^{T_0} \exp \left(\int_0^{T_0} -\frac{1}{2} \|\boldsymbol{\Sigma}(t, T_0)\|^2 \, \mathrm{d}t + \boldsymbol{\Sigma}^{\mathrm{T}}(t, T_0) \, \mathrm{d}\mathbf{W}_t \right). \qquad (6.41)$$

Equations 6.40 and 6.41 allow us to calculate the option's payoff. The expectation in Equation 6.39 can be evaluated by the following Monte Carlo simulation algorithm:

1. Construct the discount curve, P_0^T, $\forall T$, with the U.S. Treasury data at $t = 0$.

2. Simulate a number of Brownian paths and calculate the discounted prices according to the scheme

$$\frac{P_{t+\Delta t}^{T_i}}{B_{t+\Delta t}} = \frac{P_t^{T_i}}{B_t} \exp \left(\int_t^{t+\Delta t} -\frac{1}{2} \|\boldsymbol{\Sigma}(s, T_i)\|^2 \, \mathrm{d}s + \boldsymbol{\Sigma}^{\mathrm{T}}(s, T_i) \mathrm{d}\mathbf{W}_s \right)$$
$$(6.42)$$

   for $i = 0, 1, \ldots, n$, until $t + \Delta t$ equals T_0.

3. Average the payoffs:

$$\left(\sum_{i=1}^{n} \Delta T c \cdot \frac{P_{T_0}^{T_i}}{B_{T_0}} + \frac{P_{T_0}^{T_n}}{B_{T_0}} - \frac{K}{B_{T_0}} \right)^+ . \qquad (6.43)$$

TABLE 6.3: Yields at March 15, 2007

Maturity	Yield
3 months	4.89
6 months	4.88
2 years	4.56
3 years	4.47
5 years	4.44
10 years	4.52
30 years	4.69

We will use the Monte Carlo simulation method with the pricing of a set of options, which have the same strike price, $K = 1$, and special coupon rates defined as

$$c = \frac{\left(P_0^{T_0} - P_0^{T_n}\right)}{\sum_{i=1}^{n} \Delta T P_0^{T_i}}. \tag{6.44}$$

As we shall see later, bond options with par strike are equivalent to swaptions, the options on interest-rate swaps.

To implement the Monte Carlo method, we need to prescribe the volatility functions of zero-coupon bonds. Under the HJM framework, these functions are derived from the volatility functions of forward rates. For demonstration purposes, let us take the two-factor model discussed in the last section, which has a forward-rate volatility prescribed in Equation 6.36. We consider the following two sets of parameterizations:

1. One-factor model: $a = 0.008544$, $b = 0$, $k_1 = 0.0$, $k_2 = 0.35$.

2. Two-factor model: $a = 0.008$, $b = 0.003$, $k_1 = 0.0$, $k_2 = 0.35$.

Note that, for both models, $k_1 = 0$, and the one-factor model is exactly a Ho–Lee model. For comparison, we have chosen $a = 0.008544$ for the one-factor model so that the corresponding short-rate volatilities of the two models, $\sigma_r = \sqrt{a^2 + b^2}$, are identical. The volatility for zero-coupon bond P_t^T is obtained from an integration:

$$\mathbf{\Sigma}(t, T) = -\int_t^T \left(b\left(1 - \frac{a}{2}e^{-k_2 s}\right)\right) ds$$

$$= -\left(b\left[(T - t) + (2/k_2)\left(e^{-k_2 T} - e^{-k_2 t}\right)\right]\right). \tag{6.45}$$

The size of each time step for the Monte Carlo simulation is $\Delta t = 0.25$. To construct the discount curve, we use the yield data of March 15, 2007, listed in Table 6.3, and go through the bootstrapping procedure described earlier.

The calculated bond option prices are listed in Table 6.4. As can be seen from the table, the Ho–Lee model consistently produces higher prices, which

TABLE 6.4: Prices of Par-Bond Options in Basis Points

Bond Option		Model	
Maturity, T_0	Tenor, $T_n - T_0$	Ho–Lee	Two-factor
1	0.25	8.23	7.85
2	0.25	11.07	10.34
5	0.25	15.33	14.69
1	5	145.43	140.41
2	5	200.29	191.07
5	5	275.08	272.14
1	10	262.88	254.99
2	10	364.54	349.50
5	10	489.43	487.53

The options with tenor 0.25 are actually caplets, while the others are actually swaptions. For caplets, $\Delta T = 0.25$, whereas for swaptions, $\Delta T = 0.5$. One basis point corresponds to one cent for the notional of \$100.

can be explained as follows: under the one-factor model, the prices of all zero-coupon bonds are perfectly correlated, whereas under the two-factor model, they are not. Because a coupon bond is a portfolio of zero-coupon bonds, its volatility will be larger if all zero-coupon bonds move in the same direction, provided that the short-rate volatilities are the same. The larger volatility leads to a higher option premium.

6.4.1 Options on Coupon Bonds

Options on coupon bonds actually belong to the first generation of fixed-income derivatives. Options on Treasury bonds are liquidly traded. Earlier in this Section, we studied the pricing of coupon bonds using Monte Carlo simulations. Here, we instead introduce a methodology for approximate pricing of options on coupon bonds.

As was presented earlier, the payoffs of call options on coupon bonds take the form

$$V_{T_0} = \left(\sum_{i=1}^{N} \Delta T c P(T_0, T_i) + P(T_0, T_N) - K \right)^+ , \qquad (6.46)$$

where T_0 is the maturity of the option. Let B_t^c denote the bond price at time t. Then the T_0-forward price of the coupon bond is

$$F_t^{T_0} = \frac{B_t^c}{P(t, T_0)}$$

$$= \sum_{i=1}^{N} \Delta T c \frac{P(t, T_i)}{P(t, T_0)} + \frac{P(t, T_N)}{P(t, T_0)}$$

$$= \sum_{i=1}^{N} \Delta T c \frac{P(0, T_i)}{P(0, T_0)} M_i(t) + \frac{P(0, T_N)}{P(0, T_0)} M_N(t). \qquad (6.47)$$

Here,

$$M_i(t) = \exp\left(\int_0^t -\frac{1}{2}\|\mathbf{\Sigma}(s,T_i) - \mathbf{\Sigma}(s,T_0)\|^2 \, \mathrm{d}s\right.$$
$$\left. + (\mathbf{\Sigma}(s,T_i) - \mathbf{\Sigma}(s,T_0))^{\mathrm{T}} \, \mathrm{d}\mathbf{W}_s^{(T_0)}\right) \tag{6.48}$$

is a martingale under the T_0-forward measure. For convenience we now rewrite Equation 6.47 as

$$\frac{F_t^{T_0}}{F_0^{T_0}} = \sum_{i=1}^{N} \omega_i M_i(t), \tag{6.49}$$

where

$$\omega_i = \begin{cases} \dfrac{\Delta Tc P(0,T_i)}{B_0^c}, & i < N, \\[3mm] \dfrac{(1+\Delta Tc)P(0,T_N)}{B_0^c}, & i = N. \end{cases} \tag{6.50}$$

To price the option, we approximate the process of $F_{T_0}^{T_0}$ by a lognormal variable through moment matching:

$$\sum_{i=1}^{N} \omega_i M_i(T_0) \approx \exp\left(-\frac{1}{2}\sigma_B^2 T_0 + \sigma_B\sqrt{T_0}\cdot\varepsilon\right), \tag{6.51}$$

where $\varepsilon \sim N(0,1)$ under $\mathbb{Q}_{T_0}$, and

$$\sigma_B^2 = \frac{1}{T_0}\ln E^{\mathbb{Q}_{T_0}}\left[\left(\sum_{i=1}^{N}\omega_i M_i(T_0)\right)^2\right]$$
$$= \frac{1}{T_0}\ln\left(\sum_{i,j}\omega_i\omega_j E^{\mathbb{Q}_{T_0}}[M_i(T_0)M_j(T_0)]\right)$$
$$= \frac{1}{T_0}\ln\left(\sum_{i,j}\omega_i\omega_j e^{s_i s_j \rho_{ij} T_0}\right), \tag{6.52}$$

with

$$s_i^2 = \frac{1}{T_0}\int_0^{T_0}\|\mathbf{\Sigma}(t,T_i)-\mathbf{\Sigma}(t,T_0)\|^2\,\mathrm{d}t,$$
$$\rho_{ij} = \frac{1}{T_0 s_i s_j}\int_0^{T_0}(\mathbf{\Sigma}(t,T_i)-\mathbf{\Sigma}(t,T_0))^{\mathrm{T}}(\mathbf{\Sigma}(t,T_j)-\mathbf{\Sigma}(t,T_0))\,\mathrm{d}t. \tag{6.53}$$

Note that when $|s_i s_j \rho_{ij} T_0| \ll 1$ for all i and j, we can approximate σ_B as

$$\sigma_B \approx \sqrt{\sum_{ij}\omega_i\omega_j s_i s_j \rho_{ij}}. \tag{6.54}$$

By utilizing the lognormal approximation of the forward bond price, we have the following Black's formula for the approximate price of bond options:

$$
\begin{aligned}
V_0 &= P(0, T_0) E^{\mathbb{Q}_{T_0}} \left[\left(F_{T_0}^{T_0} - K \right)^+ \Big| \mathcal{F}_0 \right] \\
&\approx P(0, T_0) E^{\mathbb{Q}_{T_0}} \left[\left(F_0^{T_0} e^{-(1/2)\sigma_B^2 T_0 + \sigma_B \sqrt{T_0} \cdot \varepsilon} - K \right)^+ \Big| \mathcal{F}_0 \right] \\
&= P(0, T_0) \left[F_0^{T_0} \Phi(d_1) - K \Phi(d_2) \right] \\
&= B_0^c \Phi(d_1) - K P(0, T_0) \Phi(d_2),
\end{aligned}
\tag{6.55}
$$

where

$$
\begin{aligned}
d_1 &= \frac{\ln\left(F_0^{T_0}/K \right) + \sigma_B^2 T_0/2}{\sigma_B \sqrt{T_0}} = \frac{\ln\left(\frac{B_0^c}{K P(0,T_0)} \right) + \sigma_B^2 T_0/2}{\sigma_B \sqrt{T_0}}, \\
d_2 &= d_1 - \sigma_B \sqrt{T_0}.
\end{aligned}
$$

Although it appears rough, the above approximation often works quite well in the marketplace. In fact, it has become common practice to approximate an addition of lognormal random variables by a single lognormal random variable, as such an approximation seems to be rather accurate when the original random variables are reasonably correlated.

Example 6.1. *To demonstrate the accuracy of Black's formula for options on coupon bonds, we reprice the options introduced in Section 6.6, and compare the results with those from Monte Carlo simulations. The results are listed in Table 6.5. One can see that the prices from Black's formula and from Monte Carlo simulations are fairly close. In fact, the root mean squared difference for all price pairs is very close to 1%. Such closeness really does very positively support the lognormal approximation of the price distribution of coupon bonds.*

6.5 Special Cases of the HJM Model

Since the publication of the HJM model in 1992, arbitrage pricing models have quickly acquired dominant status in fixed-income modeling. Arbitrage pricing models have been generated from the HJM framework by making various specifications of forward-rate volatility. In this section, we study two specifications of the forward-rate volatility that, in terms of the short-rate dynamics, reproduce the popular one-factor models of Ho and Lee (1986) and Hull and White (1989), respectively.

TABLE 6.5: Option Prices by Black's Formula and by the Monte Carlo Simulation Method

Par-Bond Option		Ho–Lee Model		Two-Factor Model	
Maturity, T_0	Tenor $T_n - T_0$	Black	MC	Black	MC
1	0.25	8.12	8.23	7.67	7.85
2	0.25	11.01	11.07	10.31	10.34
5	0.25	15.30	15.33	14.77	14.69
1	5	147.84	145.43	139.40	140.41
2	5	200.20	200.29	190.95	191.07
5	5	276.62	275.08	271.28	272.14
1	10	264.69	262.88	253.35	254.99
2	10	357.57	364.54	345.94	349.50
5	10	492.64	489.43	486.42	487.53

6.5.1   The Ho–Lee Model

The simplest specification of the HJM model is $\sigma = $ const for $n = 1$, corresponding to the forward-rate equation

$$df(t, T) = \sigma \, dW_t + \sigma^2 (T - t) \, dt.$$

By integrating the equation over $[0, t]$, we obtain

$$f(t, T) - f(0, T) = \sigma W_t + \frac{1}{2}\sigma^2 t \, (2T - t).$$

By making $T = t$, we have the expression for the short rate:

$$r_t = f(t, t) = f(0, t) + \frac{1}{2}\sigma^2 t^2 + \sigma W_t.$$

In differential form, the last equation becomes

$$dr_t = \left(f_T(0, t) + \sigma^2 t \right) dt + \sigma \, dW_t. \tag{6.56}$$

Equation 6.56 is interpreted as the continuous-time version of the so-called Ho–Lee (1986) model, which was first developed in the context of binomial trees.

Let us take a look at a basic feature of the Ho–Lee model. It is quite obvious to see that

$$E^{\mathbb{Q}}[r_t] = f(0, t) + \frac{1}{2}\sigma^2 t^2,$$
$$\mathrm{Var}[r_t] = \sigma^2 t. \tag{6.57}$$

The two equations of Equation 6.57 suggest that the short rate will fluctuate around a quadratic function of time with increasing variance. This feature runs counter to common sense, and it has motivated alternative specifications to make the short rate behave more reasonably.

With the Ho–Lee model, we can obtain the following price formula of zero-coupon bonds:

$$P(t,T) = \exp\left\{-\int_t^T \left[f(0,s) + \sigma W_t + \frac{\sigma^2}{2}t\,(2s-t)\right]ds\right\}$$

$$= \exp\left\{-\left[\int_t^T f(0,s)\,ds + \sigma W_t\,(T-t) + \frac{\sigma^2}{2}tT\,(T-t)\right]\right\}$$

$$= \frac{P(0,T)}{P(0,t)}\exp\left\{-\left[\sigma W_t\,(T-t) + \frac{\sigma^2}{2}tT\,(T-t)\right]\right\}.$$

This formula can be used to, among other applications, price options on zero-coupon bonds.

6.5.2 The Hull–White (or Extended Vasicek) Model

It has been empirically observed that forward-rate volatility decays with time-to-maturity, $T - t$. This motivates the following specification of the volatility:

$$\sigma(t,T) = \sigma e^{-\kappa(T-t)}, \quad \kappa > 0. \tag{6.58}$$

That is, the volatility decays exponentially as time goes forward. The corresponding HJM equation now reads

$$df(t,T) = \sigma e^{-\kappa(T-t)}dW_t + \left[\sigma e^{-\kappa(T-t)}\int_t^T \sigma e^{-\kappa(s-t)}ds\right]dt$$

$$= \sigma e^{-\kappa(T-t)}dW_t + \sigma e^{-\kappa(T-t)}\frac{\sigma}{\kappa}\left[1 - e^{-\kappa(T-t)}\right]dt$$

$$= \sigma e^{-\kappa(T-t)}dW_t + \frac{\sigma^2}{\kappa}\left[e^{-\kappa(T-t)} - e^{-2\kappa(T-t)}\right]dt.$$

Integrating the above equation over $(0,t)$ yields

$$f(t,T) = f(0,T) + \sigma\int_0^t e^{-\kappa(T-s)}dW_s$$

$$+ \frac{\sigma^2}{2\kappa^2}\left[\left(1 - e^{-\kappa T}\right)^2 - \left(1 - e^{-\kappa(T-t)}\right)^2\right].$$

By making $T = t$, we obtain the expression for the short rate:

$$r_t = f(0,t) + \sigma\int_0^t e^{-\kappa(t-s)}dW_s + \frac{\sigma^2}{2\kappa^2}\left(1 - e^{-\kappa t}\right)^2. \tag{6.59}$$

Yet again, let us check the mean and variance of the short rate, which are

$$E^{\mathbb{Q}}[r_t] = f(0,t) + \frac{\sigma^2}{2\kappa^2}\left(1 - e^{-\kappa t}\right)^2,$$

$$\text{Var}[r_t] = \sigma^2 \int_0^t e^{-2\kappa(t-s)}\,ds = \sigma^2 \frac{1}{2\kappa}\left(1 - e^{-2\kappa t}\right) < \frac{\sigma^2}{2\kappa}.$$

The above equations suggest that both the mean and the variance of the short rate stay bounded, a very plausible feature for a short-rate model.

Next, let us study the differential form of Equation 6.59. Denote

$$X_t = \sigma \int_0^t e^{-\kappa(t-s)}\,dW_s,$$

which satisfies

$$dX_t = \sigma\,dW_t - \sigma\kappa \int_0^t e^{-\kappa(t-s)}\,dW_s\,dt$$

$$= \sigma\,dW_t - \kappa X_t\,dt, \tag{6.60}$$

and relates to the short rate as

$$X_t = r_t - f(0,t) - \frac{\sigma^2}{2\kappa^2}\left(1 - e^{-\kappa t}\right)^2. \tag{6.61}$$

By differentiating Equation 6.59 and making use of Equation 6.61, we then obtain

$$dr_t = f_T(0,t)\,dt + \sigma\,dW_t - \kappa X_t\,dt + \frac{\sigma^2}{\kappa}e^{-\kappa t}\left(1 - e^{-\kappa t}\right)dt$$

$$= \kappa(\theta_t - r_t)\,dt + \sigma\,dW_t, \tag{6.62}$$

for

$$\theta_t \triangleq f(0,t) + \frac{1}{\kappa}f_T(0,t) + \frac{\sigma^2}{2\kappa^2}\left(1 - e^{-2\kappa t}\right). \tag{6.63}$$

Equation 6.62 is called the Hull–White (1989) model or sometimes the extended Vasicek model, because Vasicek (1977) was the first to adopt the formalism of Equation 6.62 for short-rate modeling in an equilibrium approach. Note that when $\kappa \to 0$, the Hull–White model reduces to the Ho–Lee model (Equation 6.56).

Let us highlight the so-called mean-reverting feature of the Hull–White model: when $r_t > \theta_t$, the drift is negative; when $r_t < \theta_t$, the drift turns positive. The drift term acts like a force that pushes the short rate toward its mean level, θ_t. The contribution of Hull and White is to identify the level of mean reversion, θ_t, displayed in Equation 6.63, so that zero-coupon bond prices of all maturities are reproduced.

In terms of the short rate, we have the following formula for the zero-coupon bond price (Hull and White, 1989):

$$P(t,T) = A(t,T)e^{-B(t,T)r_t}, \tag{6.64}$$

where

$$B(t,T) = \frac{1 - e^{-\kappa(T-t)}}{\kappa}, \tag{6.65}$$

and

$$\ln A(t,T) = \ln \frac{P(0,T)}{P(0,t)} - B(t,T)\frac{\partial \ln P(0,t)}{\partial t}$$

$$- \frac{\sigma^2}{4\kappa^3} \left(e^{-\kappa T} - e^{-\kappa t}\right)^2 \left(e^{2\kappa t} - 1\right). \tag{6.66}$$

The proof is left as an exercise.

Let us consider pricing an option on a zero-coupon bond under the Hull–White model.

Example 6.2. *Consider the pricing of a zero-coupon bond option with payoff* $V_T = (P(T,\tau) - K)^+$ *for* $\tau > T$. *Let us take* $T = 2, \tau = 5$, *and* $K = P(0,\tau)/P(0,T)$. *The initial term structure of interest rates is given by*

$$f(0,T) = 0.02 + 0.002T. \tag{6.67}$$

We will price the option first under the Hull–White model, with forward-rate volatility given in Equation 6.58 for $\sigma = 0.005$ *and* $\kappa = 0.1$.

To calculate the strike, we need the price of zero-coupon bonds:

$$P(0,T) = \exp\left\{-\int_0^T f(0,u)\,du\right\}$$

$$= \exp\left\{-\int_0^T (0.02 + 0.002u)\,du\right\}$$

$$= \exp\left\{-\left(0.02T + 0.001 \times T^2\right)\right\}. \tag{6.68}$$

With the above function, we obtain

$$P(0,2) = 0.9570,$$
$$P(0,5) = 0.8825, \tag{6.69}$$

and the value of the strike,

$$K = F_0^2 = \frac{P(0,5)}{P(0,2)} = 0.9222. \tag{6.70}$$

For the Hull–White model, the volatility of a zero-coupon bonds is given by

$$\Sigma(t,T) = -\int_t^T \sigma(t,u)\,du = \frac{\sigma}{\kappa}\left[e^{-\kappa(T-t)} - 1\right]. \tag{6.71}$$

It follows that,

$$\Sigma(t,\tau) - \Sigma(t,T) = \frac{\sigma}{\kappa} \left[e^{-\kappa(\tau-t)} - e^{-\kappa(T-t)} \right]$$

$$= \frac{\sigma}{\kappa} e^{\kappa t} (e^{-\kappa\tau} - e^{-\kappa T}), \tag{6.72}$$

and the Black volatility given by

$$\sigma_F^2 = \frac{1}{T} \frac{\sigma^2}{\kappa^2} \left(e^{-\kappa\tau} - e^{-\kappa T} \right)^2 \int_0^T e^{2\kappa t}\, dt$$

$$= \frac{1}{T} \frac{\sigma^2}{\kappa^2} \left(e^{-\kappa\tau} - e^{-\kappa T} \right)^2 \frac{1}{2\kappa} \left(e^{2\kappa T} - 1 \right)$$

$$= \frac{1}{2T} \frac{\sigma^2}{\kappa^3} \left(e^{-\kappa(\tau-T)} - 1 \right)^2 \left(1 - e^{-2\kappa T} \right). \tag{6.73}$$

For the given parameters, we have

$$\sigma_F^2 = 0.00013841,$$
$$d_1 = -d_2 = 0.0083. \tag{6.74}$$

Inserting relevant numbers into the Black formula, we obtain

$$V_0 = P(0,5)\Phi(d_1) - P(0,2)K\Phi(d_2)$$
$$= P(0,5)\left(1 - 2\Phi(d_2)\right) = 0.0059, \tag{6.75}$$

or 59 basis points.

   In the limit $\kappa \to 0$, the Hull-White model reduces to the Ho–Lee model, with the Black volatility given by

$$\sigma_F^2 = \sigma^2(\tau - T)^2$$
$$= (0.005)^2 \times 9 = 0.000225. \tag{6.76}$$

It follows that

$$d_1 = 0.0106,$$
$$d_2 = -d_1 = -0.0106, \tag{6.77}$$

and the value of the option is

$$V_0 = 0.0075 \tag{6.78}$$

or 75 basis points. The bigger price under the Ho–Lee model reflects the fact of the bigger forward-rate volatility as compared to that of the Hull–White model.

   We have some additional comments on the Hull–White model from the perspective of applications. Under the Hull–White model, the short rate is a Gaussian random variable and thus can take a negative value with a positive probability, which is unrealistic and is considered a major disadvantage of the model. There are two ways to avoid negative interest rates. First, we impose a floor at zero level on interest rate. Second, we impose a reflecting boundary condition at $r_t = 0$. Moreover, since the short rate is a Gaussian variable, Hull–White model can be implemented through a lattice tree and thus has become a popular choice of model to price path-dependent options.

6.6 Linear Gaussian Models

For the purpose of either pricing or hedging, it is not necessary to have a numeraire that is tradable. As it is pointed out in Hagan and Woodward (1999), a "modern interest rate model consists of three parts: a numeraire, a set of random evolution equations in the risk-neutral world, and the martingale pricing formula." The so-called Linear Gaussian Markov Model (LGM) is such an artificial device for interest-rate pricing, which is defined as follows.

1. The numeraire:

$$N_t = \frac{1}{P(0,t)} \exp \left\{ h(t)z_t + \frac{1}{2}h^2(t)\zeta_t \right\}. \tag{6.79}$$

2. The random evolution equation:

$$dz_t = \alpha_t dW_t, \quad z_0 = 0, \quad \text{and} \quad \zeta_t = \int_0^t \alpha^2(s)ds. \tag{6.80}$$

3. The martingale pricing formula:

$$\frac{V_t}{N_t} = E_t^N \left[\frac{V_T}{N_T} \right]. \tag{6.81}$$

where the sup-index N means the martingale measure corresponding to N_t as the numeraire.

Under the LGM model, we have a closed-form formula for zero-coupon bonds. By putting $V_T = 1$, we can derive the formula for zero-coupon bonds as

$$P(t,T;z_t) = \frac{P(0,T)}{P(0,t)} \exp \left\{ -(h(T) - h(t))z_t - \frac{1}{2} \left(h^2(T) - h^2(t) \right) \zeta_t \right\}. \tag{6.82}$$

One can easily verify that the LGM automatically calibrates to the spot discount curve. Since both the numeraire and the zero-coupon bond prices are driven by the same normal random factor, the model has a high level of analytical tractability. For instance, under LGM swap rates can be expressed in closed form. Under the LGM, $h(t)$ is at our disposal, and different choices or features of $h(t)$ will yield different models. In fact, it can be verified that if we take

$$h(t) = \int_0^t e^{-\kappa s} ds \quad \text{for} \quad \kappa > 0, \tag{6.83}$$

we will reproduce the Hull-White model with "mean reversion speed" κ. LGM can be generalized to multiple driving factors. Before pricing interest rate derivatives, we ought to calibrate α_t to vanilla interest rate derivatives like caps, floors and swaptions.

6.7 Pricing SOFR Derivatives under HJM Model

Now we are ready to price SOFR derivatives, including SOFR futures, SOFR caps/floors and SOFR swaptions under the HJM model. The same derivatives will be later priced again under other popular models.

6.7.1 SOFR Futures

First in our list for pricing is the 3m SOFR futures. Let $t \leq T_{j-1}$. According to the general formula established in Chapter 5, the 3m futures rate is given by

$$
\begin{aligned}
\tilde{f}_j(t) &= E_t^Q \left[\tilde{f}_j(T_j) \right] \\
&= \frac{1}{\Delta T_j} \left(E_t^Q \left[\frac{P(T_j, T_{j-1})}{P(T_j, T_j)} \right] - 1 \right) \\
&= \frac{1}{\Delta T_j} \left(\frac{P(t, T_{j-1})}{P(t, T_j)} e^{\int_t^{T_j} -\Sigma(s, T_j) \cdot [\Sigma(s, T_{j-1}) - \Sigma(s, T_j)] ds} - 1 \right).
\end{aligned}
\tag{6.84}
$$

For notational simplicity, we denote $\Sigma(t, T_{j-1}, T_j) \triangleq \Sigma(t, T_{j-1}) - \Sigma(t, T_j)$ from now on. The futures–forward difference is thus

$$
\begin{aligned}
\tilde{f}_j(t) - f_j(t) &= \frac{1}{\Delta T_j} \frac{P(t, T_{j-1})}{P(t, T_j)} \left(e^{\int_t^{T_j} -\Sigma(s, T_j) \cdot \Sigma(s, T_{j-1}, T_j) ds} - 1 \right) \\
&= \frac{1 + \Delta T_j f_j(t)}{\Delta T_j} \left(e^{\int_t^{T_j} -\Sigma(s, T_j) \cdot \Sigma(s, T_{j-1}, T_j) ds} - 1 \right).
\end{aligned}
\tag{6.85}
$$

The right-hand side of the above equation is a general formula for the difference between a futures rate and its corresponding forward rate, so-called correlation adjustment. In general, we have the inequality $\|\Sigma(s, T_j)\| \geq \|\Sigma(s, T_{j-1})\|$

for the volatility functions of ZCB, which simply reflects the fact that the longer the maturity, the bigger the volatility. Using Cauchy-Schwarz inequality, we readily obtain $\Sigma(s, T_j) \cdot (\Sigma(s, T_j) - \Sigma(s, T_{j-1})) \geq 0$, implying the futures rate is bigger than its corresponding forward rate. Finally, we take the limit $T_{j-1} \to T_j = T$, and then we will end up with the correlation adjustment for the pair of futures rate and forward rate under continuous compounding:

$$
\begin{aligned}
\tilde{f}(t,T) &= f(t,T) + \int_t^T \Sigma(u,T) \frac{\partial \Sigma(u,T)}{\partial T} du \\
&= f(t,T) - \int_t^T \Sigma(u,T) \cdot \sigma(u,T) du,
\end{aligned}
\tag{6.86}
$$

where $\sigma(t,T)$ is the volatility function of $f(t,T)$, and this result can be obtained directly from the HJM model.

6.7.2 Correlation Adjustment under the Hull–White Model

Now let us consider correlation adjustment under the Hull–White model (1989) when the zero-coupon bond volatilities are of the form

$$
\Sigma(t, T_j) = -\sigma B(t, T_j),
$$

where $B(t, T_j)$ is defined according to Equation 6.65, and it follows that

$$
\Sigma(t, T_{j-1}, T_j) = \sigma \frac{e^{-\kappa T_{j-1}} - e^{-\kappa T_j}}{\kappa} e^{\kappa t}.
$$

We then have

$$
-\int_t^{T_j} \Sigma(s, T_j) \cdot \Sigma(s, T_{j-1}, T_j) ds = -\frac{\sigma^2}{2} B(\Delta T_j, 0) B^2(t, T_j).
$$

The correlation adjustment between futures–forward rates is thus

$$
\tilde{f}_j(t) - f_j(t) = \frac{1 + \Delta T_j f_j(t)}{\Delta T_j} \left(e^{-\frac{\sigma^2}{2} B(\Delta T_j, 0) B^2(t, T_j)} - 1 \right).
\tag{6.87}
$$

At the limit $\kappa \to 0$,

$$
\begin{aligned}
B(\Delta T_j, 0) &\to -\Delta T_j, \\
B(t, T_j) &\to T_j - t,
\end{aligned}
$$

we thus have

$$
\begin{aligned}
\tilde{f}_j(t) - f_j(t) &= (1 + \Delta T_j f_j(t)) \frac{e^{\frac{\sigma^2}{2}(T_j - t)^2 \Delta T_j} - 1}{\Delta T_j} \\
&\approx (1 + \Delta T_j f_j(t)) \times \frac{1}{2} \sigma^2 (T_j - t)^2,
\end{aligned}
\tag{6.88}
$$

which is the correlation adjustment formula under the Ho–Lee model (1986). The last approximation is rather accurate because $\sigma^2 (T_j - t)^2$ is, typically, small.

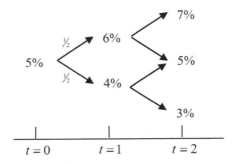

FIGURE 6.9: Two-year risk-neutral interest rate tree.

6.7.3 An Example of Arbitrage

What happens if a futures rate is taken to be a forward rate in a market? The answer is arbitrage. In the following example, we will use futures contracts to arbitrage against a wrongly priced forward rate agreement.

Assume that the two-year back-looking rate tree evolves according to a binomial tree, like the one in Figure 6.9, under the risk-neutral measure. The price of the one- and two-year maturity zero-coupon bonds can be calculated through backward induction as demonstrated in Figure 6.10 and Figure 6.11.

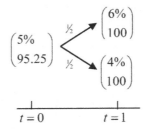

FIGURE 6.10: Price tree of the one-year maturity ZCB.

With these bond prices, we can calculate the in-1-to-1 backward-looking forward rate (for the term from the end of the first year to the end of the second year) as

$$f(0;1,2) = \frac{1}{2-1}\left(\frac{P(0,1)}{P(0,2)}-1\right) = 4.9714\%, \qquad (6.89)$$

The expected in-1-to-1 futures rate for the same period is obviously

$$\tilde{f}(0;1,2) \overset{\triangle}{=} E^{\mathbb{Q}}[r_{\cdot,2}] = 5\% > f(0;1,2). \qquad (6.90)$$

If we buy one unit of FRA at the futures rate at no cost, we then generate an arbitrage opportunity to the market. To see this, we first look at the price

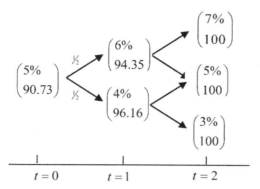

FIGURE 6.11: Price tree of the two-year maturity ZCB.

		7%
		0.02
	6%	
	0.009346	
5%		5%
-0.00026		0
	4%	
	-0.009709	
		3%
		-0.02

FIGURE 6.12: MtM values of the long FRA on $\tilde{f}(0; 1, 2)$.

tree of the FRA with the forward or strike rate $\tilde{f}(0; 1, 2) = 5\%$, shown in Figure 6.12, which is out of the money due to the unfavorable forward rate.

We can trade against the FRA by being short futures with a dynamical strategy and earn a profit at no cost. The price tree of the dynamical strategy is demonstrated in Figure 6.13, where the numbers in parenthesis indicate the unit of SOFR futures being short. At the maturity of the FRA, the cash flows of the short futures dominate those of the FRA in all states and thus constitute an arbitrage trade. Note that the delta at the (i, j) state is chosen to dominate the cash flows of the FRA at all final states, such that

$$\delta_{0,0} = \frac{0.02}{(r_{0,1} - r_{0,0})(1 + r_{0,2}) + (1 + r_{1,2})(r_{0,2} - r_{0,1})} = -0.9615;$$
$$\delta_{0,1} = \delta_{1,1} = \delta_{0,0}(1 + r_{1,2}) = -1.01.$$

		7%
		0.02038
	(-1.01)	
	6%	
	0.009615	
(-0.9615)		
5%		5%
0		0
	(-1.01)	
	4%	
	-0.009615	
		3%
		-0.02

FIGURE 6.13: MtM values of the long SOFR futures on $\tilde{f}(0;1,2)$.

6.7.4　SOFR Futures Options

SOFR futures options can be treated as options on futures rates. Let us start with options on 3m SOFR futures. Similar to options on ED futures, options on SOFR futures expire a week before the reference period. According to Equation 6.84, the formula for 3m SOFR futures rates, the price of a call option on the SOFR futures with reference period (T_{j-1}, T_j) is given by

$$
V_t = \Delta T_j P(t, T_{j-1}) E_t^{Q_{j-1}} \left[\left(\tilde{f}_j(T_{j-1}) - k \right)^+ \right]
$$

$$
= \Delta T_j P(t, T_{j-1}) E_t^{Q_{j-1}} \left[\left(\frac{1}{\Delta T_j} \left(\frac{P(T_{j-1}, T_{j-1})}{P(T_{j-1}, T_j)} e^{\int_{T_{j-1}}^{T_j} \|\Sigma(s, T_j)\|^2 ds} - 1 \right) - k \right)^+ \right]
$$

$$
= P(t, T_{j-1}) E_t^{Q_{j-1}} \left[\left(\frac{P(T_{j-1}, T_{j-1})}{P(T_{j-1}, T_j)} e^{\int_{T_{j-1}}^{T_j} \|\Sigma(s, T_j)\|^2 ds} - (1 + \Delta T_j k) \right)^+ \right].
$$

Here, we have taken into account that $\Sigma(s, T_{j-1}) = 0$ for $s \geq T_{j-1}$. Under the Heath-Jarrow-Mortan model, T_j-forward price of the T_{j-1}-maturity ZCB follows a lognormal process, such that

$$
\frac{P(T_{j-1}, T_{j-1})}{P(T_{j-1}, T_j)} = \frac{P(t, T_{j-1})}{P(t, T_j)} e^{\int_t^{T_{j-1}} -\frac{1}{2} \|\Sigma(s, T_{j-1}, T_j)\|^2 ds}
$$

$$
\times e^{\int_t^{T_{j-1}} \Sigma(s, T_{j-1}, T_j) \cdot d\mathbf{W}_s^{(j-1)}} e^{\int_t^{T_{j-1}} \|\Sigma(s, T_{j-1}, T_j)\|^2 ds},
$$

where $\mathbf{W}_t^{(j-1)} = \mathbf{W}_t - \int_0^t \Sigma(s,T_{j-1})ds$, which is the Brownian motion under the T_{j-1}-forward measure. We then make use of the above equation as well as Equation 6.84 to obtain

$$E_t^{Q_{j-1}}\left[\frac{P(T_{j-1},T_{j-1})}{P(T_{j-1},T_j)}e^{\int_{T_{j-1}}^{T_j}\|\Sigma(s,T_j)\|^2ds}\right]$$

$$=\frac{P(t,T_{j-1})}{P(t,T_j)}e^{\int_t^{T_j}\|\Sigma(s,T_{j-1},T_j)\|^2ds}$$

$$=(1+\Delta T_j\tilde{f}_j(t))e^{\int_t^{T_{j-1}}\Sigma(s,T_{j-1})\cdot\Sigma(s,T_{j-1},T_j)ds}.$$

The exponent term above can be worked out to be

$$e^{\int_t^{T_{j-1}}\Sigma(s,T_{j-1})\cdot\Sigma(s,T_{j-1},T_j)ds}=e^{-\frac{\sigma^2}{2}B(0,\Delta T_j)B^2(t,T_{j-1})}\triangleq e^{C(t,T_{j-1},\Delta T_j)}.$$

Hence we have the following Black's formula for the SOFR futures options:

$$V_t=P(t,T_{j-1})\left[(1+\Delta T_j\tilde{f}_j(t))e^{C(t,T_{j-1},\Delta T_j)}\Phi(d_1^{(j)})-(1+k\Delta T_j)\Phi(d_2^{(j-1)})\right],$$

with

$$d_1^{(j)}=\frac{\ln\frac{(1+\Delta T_j\tilde{f}_j(t))}{1+k\Delta T_j}+C(t,T_{j-1},\Delta T_j)+\frac{1}{2}\int_t^{T_{j-1}}\|\Sigma(s,T_{j-1},T_j)\|^2ds}{\sqrt{\int_t^{T_{j-1}}\|\Sigma(s,T_{j-1},T_j)\|^2ds}}$$

$$d_2^{(j)}=d_1^{(j)}-\sqrt{\int_t^{T_{j-1}}\|\Sigma(s,T_{j-1},T_j)\|^2ds}.$$

Finally,

$$\int_t^{T_{j-1}}\|\Sigma(s,T_{j-1},T_j)\|^2ds=\sigma^2B^2(0,\Delta T_j)\frac{1-e^{2\kappa(T_{j-1}-t)}}{2\kappa}.$$

The value of the corresponding put option on SOFR futures can be obtained via put-call parity, such that

$$V_t=P(t,T_{j-1})\left[(1+k\Delta T_j)\Phi(-d_2^{(j-1)})-(1+\Delta T_j\tilde{f}_j(t))e^{C(t,T_{j-1},\Delta T_j)}\Phi(-d_1^{(j)})\right].$$

(6.91)

6.7.5 SOFR Caps/Floors

SOFR caps/floors are similar to LIBOR caps/floors, except that the forward rates are set in arrears using either compound average rate or simple average rate. The value of a caplet is given by

$$
\begin{aligned}
c_j(t) &= \Delta T_j P(t,T_j) E_t^{Q_j} \left[(f_j(T_j) - k)^+ \right] \\
&= \Delta T_j P(t,T_j) E_t^{Q_j} \left[\left(\frac{B(t,T_j)/B(t,T_{j-1}) - 1}{\Delta T_j} - k \right)^+ \right] \\
&= P(t,T_j) E_t^{Q_j} \left[\left(\frac{B(t,T_j)}{B(t,T_{j-1})} - (1 + \Delta T_j k) \right)^+ \right].
\end{aligned}
\tag{6.92}
$$

According to Equation 6.41,

$$
\begin{aligned}
\frac{B(t,T_j)}{B(t,T_{j-1})} = \frac{P(t,T_{j-1})}{P(t,T_j)} e^{\int_t^{T_j} -\frac{1}{2}\|\Sigma(s,T_{j-1},T_j)\|^2 ds} \\
\times\, e^{\Sigma(u,T_{j-1},T_j)\cdot(d\mathbf{W}_s - \Sigma(s,T_j)ds)}
\end{aligned}
\tag{6.93}
$$

is an martingale under the T_j-forward measure, so it follows that

$$
\begin{aligned}
c_j(t) &= P(t,T_j) \left[\frac{P(t,T_{j-1})}{P(t,T_j)} \Phi(d_1^{(j)}) - (1 + \Delta T_j k)\Phi(d_2^{(j)}) \right] \\
&= P(t,T_j) \left[(1 + \Delta T_j f_j(t))\Phi(d_1^{(j)}) - (1 + \Delta T_j k)\Phi(d_2^{(j)}) \right],
\end{aligned}
\tag{6.94}
$$

where

$$
d_1^{(j)} = \frac{\ln \frac{1+\Delta T_j f_j(t)}{1+\Delta T_j k} + \frac{1}{2}\int_t^{T_j} \|\Sigma(s,T_{j-1},T_j)\|^2 ds}{\sqrt{\int_t^{T_j} \|\Sigma(s,T_{j-1},T_j)\|^2 ds}}
$$

$$
d_2^{(j)} = d_1^{(j)} - \sqrt{\int_t^{T_j} \|\Sigma(s,T_{j-1},T_j)\|^2 ds}.
$$

By aggregating the values of caplets, we obtain the value of caps.

The price formula for floors can be derived through call-put parity.

6.8   Notes

We finish this section with comments on the role of the HJM model as a framework of arbitrage-free models by choosing specific forward-rate volatilities. As we have seen, by choosing constant and exponentially decaying volatilities, we reproduce Ho–Lee and Hull–White models. However, we are unable to identify corresponding forward-rate volatilities of any Gaussian short-rate models, like the well-known Cox-Ingersoll-Rubinstein (1985) model and Black-Karasinski (1991) model which feature positive short rate. Put in another way, we do not yet know how to specify the forward-rate volatility so as to ensure that forward rates of all maturities will stay non-negative. For this objective, it has been tempted to adopt a forward-rate volatility that is a function of the forward rate itself, so-called state-dependent volatility. This attempt, however, has failed because the resulted HJM equation typically blows up in finite time. The details are provided in the appendix to this chapter.

The prospect of negative interest rates under the HJM framework had motivated studies on alternative frameworks of interest-rate models that maintain positive interest rates. Rogers (1997) takes a fundamental approach and works on a pricing kernel, which is the Radon–Nikodym derivative of the risk-neutral measure with respect to the physical measure. The key is to treat the pricing kernel as a potential,[2] which will result in monotonically decreasing zero-coupon bond price with respect to maturity, and thus ensure positive forward rates. Rogers casts the pricing kernel as

$$\zeta_t = E_t^{\mathbb{P}}\left[\int_t^\infty \mu_\zeta(\tau)\, \mathrm{d}\tau\right]. \tag{6.95}$$

Then, what is left to do is to prescribe $\mu_\zeta(\tau)$, the expected rate of change of the pricing kernel.

Rogers' approach includes a slightly earlier model by Flesaker and Hughston (1996) as a special case. Jin and Glasserman (2001) later proved that these two approaches are equivalent for positive interest-rate models, in the sense that a model under one specification can be converted to that under the other.

There are, however, outstanding issues in both the theory and application of positive interest-rate models. On one hand, the positive interest-rate framework by either Rogers or Flesaker and Hughston is not convenient for option pricing. On the other hand, the connection between these frameworks and other existing positive interest-rate models, including the CIR (1985) model, the Black–Karasinski (1991) model, and the general affine term structure models (ATSMs), is far from clear. These issues await further research.

Exercises

1. It is known that zero-coupon bonds relate to forward rates by

$$P(t,T) = \exp\left(-\int_t^T f(t,s)\, \mathrm{d}s\right) \triangleq \mathrm{e}^{M_t}.$$

Given the HJM equation under the physical measure,

$$\mathrm{d}f(t,T) = \mu(t,T)\, \mathrm{d}t + \boldsymbol{\sigma}^{\mathrm{T}}(t,T)\, \mathrm{d}\mathbf{W}_t,$$

---

[2]A potential, say Z_t, is a right-continuous non-negative supermartingale satisfying $E^{\mathbb{P}}[Z_t] \overset{t\to\infty}{\longrightarrow} 0$.

prove that the price processes of zero-coupon bonds are

$$dP(t,T) = P(t,T) \left(r_t \, dt + \mathbf{\Sigma}^{\mathrm{T}}(t,T) \left(d\mathbf{W}_t + \boldsymbol{\gamma}_t \, dt \right) \right),$$

where $\mathbf{\Sigma}(t,T) = -\int_t^T \boldsymbol{\sigma}(t,s) \, ds$ and $\boldsymbol{\gamma}_t$ satisfies

$$\mu(t,T) + \boldsymbol{\sigma}^{\mathrm{T}}(t,T) \left(\mathbf{\Sigma}(t,T) - \boldsymbol{\gamma}_t \right) = 0.$$

2. With Ho–Lee's model, prove that

$$P_0^t = E^{\mathbb{Q}} \left[e^{-\int_0^t r_s \, ds} \right],$$

where $\mathbb{Q}$ stands for the risk-neutral measure.

3. The risk-neutral process for zero-coupon bonds under Ho–Lee's model is

$$\frac{dP_t^T}{P_t^T} = -(T-t)\sigma \, dW_t + \left(f(0;t) + \sigma W_t + \frac{1}{2}\sigma^2 t^2 \right) dt.$$

Use the above process to show that $P_T^T = 1$, that is, the price equals par at maturity.

4. Prove that the Hull–White model is mean-reverting by showing that the short rate follows

$$dr_t = \kappa(\theta_t - r_t) \, dt + \sigma \, dW_t,$$

with

$$\theta(t) = f(0,t) + \frac{1}{\kappa} f_t(0,t) + \frac{\sigma^2}{2\kappa^2} \left(1 - e^{-2\kappa t} \right).$$

5. For the HJM model, prove the following formulae for (discounted) zero-coupon bonds and money market account:

$$\frac{P_{T_0}^{T_j}}{B_{T_0}} = P_0^{T_j} \exp \left(\int_0^{T_0} -\frac{1}{2} \|\mathbf{\Sigma}(t,T_j)\|^2 \, dt + \mathbf{\Sigma}^{\mathrm{T}}(t,T_j) \, d\mathbf{W}_t \right),$$

$$T_j \geq T_0,$$

$$B_{T_0} = \frac{1}{P_0^{T_0}} \exp \left(\int_0^{T_0} \frac{1}{2} \|\mathbf{\Sigma}(t,T_0)\|^2 \, dt - \mathbf{\Sigma}^{\mathrm{T}}(t,T_0) \, d\mathbf{W}_t \right),$$

where $\mathbf{W}_t$ is a Brownian motion under the risk-neutral measure.

6. Suppose that the risk-neutral process of the short rate follows the *Vasicek* model:

$$dr_t = \kappa(\theta - r_t) \, dt + \sigma \, dW_t,$$

where $\kappa, \theta,$ and σ are constants. Prove the following price formula for zero-coupon bonds:

$$P(t,T) = A(t,T)e^{-B(t,T)r_t},$$

where

$$B(t,T) = \frac{1 - e^{-\kappa(T-t)}}{\kappa},$$

$$A(t,T) = \exp\left\{ \frac{(B(t,T) - T + t)\left(\kappa^2\theta - \sigma^2/2\right)}{\kappa^2} - \frac{\sigma^2 B^2(t,T)}{4\kappa} \right\}.$$

7. Price a five-year maturity option on a 10-year zero-coupon bond under the one-factor Ho–Lee model. Let the spot forward-rate term structure be

$$f(0,T) = 0.02 + 0.001T,$$

the forward-rate volatility be $\sigma = 0.015$, and the strike price be the forward price of the 10-year zero-coupon bond, $K = P(0,10)/P(0,5)$.

8. Redo the above problem using the Hull–White model: all inputs remain the same except that the forward-rate volatility becomes

$$\sigma(t,T) = 0.015e^{-0.1(T-t)}.$$

9. Derive the price formula for zero-coupon bonds, Equations 6.64 through 6.66, for the Hull–White model.

10. Price a five-year maturity option on the post-dividend price of a coupon bond under the one-factor Ho-Lee model. The coupon rate and the maturity of the bond are 5.25% and nine and a half years, respectively. The strike price of the option is $100. Let the spot forward-rate term structure be flat at 5% and the forward-rate volatility be $\sigma = 0.015$.

11. Redo the pricing of the par-bond options in Table 6.4 using T_0-forward measure. Prove first the following formulae for the T_0-forward prices of zero-coupon bonds:

$$\frac{P_t^{T_j}}{P_t^{T_0}} = \frac{P_0^{T_j}}{P_0^{T_0}} \exp\left\{ \int_0^t -\frac{1}{2}\|\Sigma(s,T_j) - \Sigma(s,T_0)\|^2 ds \right.$$

$$\left. + (\Sigma(s,T_j) - \Sigma(s,T_0))^T d\mathbf{W}_t^{(T_0)} \right\}, \quad T_j \geq T_0,$$

where $\mathbf{W}_t^{(T_0)}$ is a Brownian motion under the T_0-forward measure. Then, propose a scheme for Monte Carlo simulations.

Appendix: On the Lognormal Specification of Forward Rates

We now explore the possibility of using the state-dependent volatility function in the HJM model. Without loss of generality, we consider the forward-rate volatility function of the form

$$\sigma(t, T) = \sigma_0(t, T) f^\alpha(t, T), \tag{6.96}$$

where $\sigma_0(t, T)$ is a deterministic function and α a positive exponent. In the special case, $\alpha = 0$, we obtain a Gaussian model.

Similar to Avellaneda and Laurence (1999), we show that the "lognormal" model, corresponding to $\alpha = 1$, blows up in finite time in the sense that a forward rate reaches infinity. This result was first obtained by Morton (1988). One can imagine that similar results may apply to the case of $\alpha > 0$. Hence, volatility specification in the form of Equation 6.96 is denied.

It suffices to show the result with a one-factor model. The no-arbitrage condition dictates that the drift must be

$$\mu(t, T) = f(t, T)\sigma_0(t, T) \int_t^T f(t, s)\sigma_0(t, s) \, \mathrm{d}s, \tag{6.97}$$

which depends on the entire curve of $f(t, s)$, $t \le s \le T$. Consider the simplest specification of $\sigma_0(t, T)$: $\sigma_0(t, T) = \sigma_0 = \text{constant}$. The HJM equation then becomes

$$\frac{\mathrm{d}f(t, T)}{f(t, T)} = \sigma_0 \, \mathrm{d}W_t + \left(\sigma_0^2 \int_t^T f(t, s) \, \mathrm{d}s \right) \mathrm{d}t.$$

The formal solution to the above equation is

$$f(t, T) = f(0, T) \exp\left(\sigma_0 W_t - \frac{\sigma_0^2}{2}t + \sigma_0^2 \int_0^t \left(\int_s^T f(s, u) \, \mathrm{d}u \right) \mathrm{d}s \right)$$

$$= f(0, T) M(t) \exp\left(\sigma_0^2 \int_0^t \left(\int_s^T f(s, u) \, \mathrm{d}u \right) \mathrm{d}s \right), \tag{6.98}$$

where $M(t) = \exp(\sigma_0 W_t - (\sigma_0^2/2)t)$. Assume for simplicity that the initial term structure is flat, that is, $f(0, T) = f_0 = \text{constant}$. Differentiating both sides of Equation 6.98 with respect to T, we obtain

$$\frac{\partial}{\partial T} f(t,T) = f_0 M(t) \exp\left(\sigma_0^2 \int_0^t \left(\int_s^T f(s,u)\,du\right) ds\right) \sigma_0^2 \int_0^t f(s,T)\,ds$$

$$= \sigma_0^2 f(t,T) \int_0^t f(s,T)\,ds$$

$$= \frac{\sigma_0^2}{2} \frac{\partial}{\partial t} \left(\int_0^t f(s,T)\,ds\right)^2.$$

Integrating the above equation with respect to t, we then have

$$\frac{\partial}{\partial T} \int_0^t f(s,T)\,ds = \frac{\sigma_0^2}{2} \left(\int_0^t f(s,T)\,ds\right)^2. \qquad (6.99)$$

Now setting

$$X_t(T) = \int_0^t f(s,T)\,ds$$

and solving for $X_t(t)$ from Equation 6.99, we obtain

$$X_t(T) = \frac{2X_t(t)}{2 - \sigma_0^2 X_t(t)(T-t)}.$$

It is not hard to see that for any given $X_t(t) > 0$, $X_t(T)$ blows up at

$$T_0 = t + \frac{2}{\sigma_0^2 X_t(t)}.$$

This implies that for any $T \geq T_0$, there will be $f(s,T) = \infty$ for some $s \leq t$, that is, the forward rate blows up in finite time.

Through a proper transformation, we can show that the forward rate also blows up for the volatility specification (Equation 6.96) with $\alpha > 0$. As a result, level-dependent volatilities are ruled out for the HJM model.

Chapter 7

Short-Rate Models and Lattice Implementation

Short-rate models hold a special place in fixed-income modeling: they are the first generation of interest-rate models, and some of them still play active roles in today's markets. Short-rate models remain attractive, thanks to two distinct advantages. First, they are intuitive, as many of them were established based on financial economics theories. Second, as Markovian models of a single state variable, they can be implemented by lattice trees, and thus are often adopted for pricing path-dependent options. In this chapter, we address three important issues about the short-rate models. The first is their relationship with the HJM framework, or whether a short-rate model belongs to the framework. The second is to find out, contrary to the first issue, under what conditions an HJM model implies a Markovian short-rate model. The third is the implementations of short-rate models through interest rate trees, for the purpose of derivatives pricing.

In Section 7.1, we first derive the equations governing the forward-rate volatility corresponding to a general short-rate model. We then focus on Cox-Ingersoll-Ross model (Cox *et al.*, 1985), one of the most popular models that ensure positive short rate, to learn the reality that if the short rate volatility is state dependent, the forward rate volatility, if exists, will depend on the short rate as well. The implication is that, in general, it is nontrivial to know if a short rate model with state dependent volatility belongs to the HJM framework.

In Section 7.2, we establish a sufficient condition under which an HJM model implies a Markovian short-rate model. We will see that the Cox-Ingersoll-Ross model does not satisfy the sufficient condition and, by itself, the short rate process consistent with the current term structure is non-Markovian. Yet, by adding another state variable, the joint process can be turned Markovian. This result can be generalized to other short rate models with state dependent volatility and thus facilitates the efficient implementation of these models for derivatives pricing though simulation methods.

In Section 7.3, we present an efficient methodology to construct interest-rate lattice or interest-rate tree, by utilizing the price of Arrow-Debrew securities. For replication pricing, an interest-rate tree must price the underlying securities, namely, zero-coupon bonds, correctly. Along traditional approaches, a lattice tree is built first and then fitted or calibrated to the discount curve afterwards, which is not always feasible, before being used for pricing derivatives. We instead make the calibration part of the construction procedure, so that the tree is calibrated once it is built. The calibration requirement affects both branching and branching probabilities, and the resulted tree will then evolve around the mean value of interest rate, and will automatically truncate itself if the underlying short-rate model carries the mean-reverting feature. The new calibration technology is illustrated with the Hull–White model and it can be applied to a very general class of short-rate models.

7.1   From Short-Rate Models to Forward-Rate Models

Short-rate models dominated fixed-income modeling before the emergence of the no-arbitrage framework of Heath, Jarrow, and Morton (1992), which is in terms of forward rates. Short-rate models are arbitrage free if they can be identified as the special cases of the HJM framework, as are Ho–Lee model and the Hull–White model. The focus of this section is to find out the forward-rate volatility, if exists, corresponding to a general short-rate model, with which we can then claim that the short-rate model also belongs to the HJM framework.

Consider in general an Ito's process for the short rate under the risk-neutral measure, $\mathbb{Q}$,

$$\mathrm{d}r_t = \nu(r_t, t)\,\mathrm{d}t + \rho(r_t, t)\,\mathrm{d}W_t, \tag{7.1}$$

where the drift, $\nu(r_t, t)$, and volatility, $\rho(r_t, t)$, are deterministic functions of their arguments. Define an auxiliary function

$$g(x, t, T) = -\ln E^{\mathbb{Q}}\left[\exp\left(-\int_t^T r_s\,\mathrm{d}s\right)\Big|\, r_t = x\right], \tag{7.2}$$

in term of which we have the following result (Baxter and Rennie, 1996).

Theorem 7.1.1. *An arbitrage-free short-rate model is an HJM model with forward-rate volatility given by*

$$\sigma(t, T) = \rho(r_t, t)\frac{\partial^2 g}{\partial x \partial T}(r_t, t, T). \tag{7.3}$$

Proof. According to its definition, $g(r_t, t, T) = -\ln P(t, T)$. It follows that

$$f(t, T) = -\frac{\partial \ln P(t, T)}{\partial T} = \frac{\partial g}{\partial T}(r_t, t, T).$$

The process for the instantaneous forward rate is then

$$\begin{aligned} \mathrm{d}f(t, T) &= \frac{\partial f}{\partial x}\mathrm{d}r_t + \left(\frac{\partial f}{\partial t} + \frac{1}{2}\rho^2\frac{\partial^2 f}{\partial x^2}\right)\mathrm{d}t \\ &= \rho\frac{\partial^2 g}{\partial x \partial T}\mathrm{d}W_t + \text{drift term}. \end{aligned} \tag{7.4}$$

This ends the proof.   □

We now try to derive the forward-rate volatility, based on Theorem 7.1.1, for two popular short-rate models. We start with the Hull–White model for the short rate,

$$dr_t = \kappa(\theta_t - r_t)\,dt + \sigma\,dW_t. \tag{7.5}$$

As is shown in Equations 6.64 through 6.66, the expectation in Equation 7.2 can be worked out and is equal to

$$g(r_t, t, T) = B(t,T)r_t - \ln A(t,T), \tag{7.6}$$

for

$$B(t,T) = \frac{1 - e^{-\kappa(T-t)}}{\kappa}. \tag{7.7}$$

There is

$$\frac{\partial^2 g}{\partial x \partial T}(r_t, t, T) = \frac{\partial B(t,T)}{\partial T} = e^{-\kappa(T-t)}. \tag{7.8}$$

According to Theorem 7.1.1, the volatility of the forward rate is

$$\sigma(t,T) = \sigma\frac{\partial^2 g}{\partial x \partial T}(r_t, t, T) = \sigma e^{-\kappa(T-t)}, \tag{7.9}$$

which is a known result for the Hull–White model. Putting $\kappa = 0$, we also obtain the forward-rate volatility of the Ho–Lee model.

The application of Theorem 7.1.1 to short-rate models other than Hull–White, however, is often less trivial. Let us consider another short-rate model of the form

$$dr_t = \sigma_t\sqrt{r_t}\,dW_t + \kappa_t(\theta_t - r_t)\,dt, \tag{7.10}$$

where σ_t and $\kappa_t > 0$ are deterministic functions of time, and θ_t the subject to the term structure of interest rates. This is the famous Cox-Ingersoll-Ross model, which was proposed as a modification of the Vasicek (1977) model so as to endure that the spot interest rate remains positive. Intuitively, due to the factor of $\sqrt{r_t}$ in the volatility term, dr_t is positive at $r_t = 0$, provided that there is $\kappa_t\theta_t > 0$, which pulls the interest rate back to the positive territory.

To derive the corresponding forward-rate volatility, we need to evaluate $g(r_t, t, T)$. It will be shown in Chapter 8 that function $g(r_t, t, T)$ takes the following form:

$$g(r,t,T) = rB(t,T) + \int_t^T \theta_s B(s,T)\,ds, \tag{7.11}$$

where $B(t,T)$ satisfies the Riccarti equation,

$$\frac{\partial B}{\partial t} = \frac{1}{2}\sigma_t^2 B^2(t,T) + \kappa_t B(t,T) - 1, \quad B(T,T) = 0. \tag{7.12}$$

For simplicity, we assume constant volatility and reversion strength, that is,

$\sigma_t = \sigma$ and $\kappa_t = \kappa$. Then the above Riccarti equation can be solved analytically, such that

$$B(t,T) = \frac{d-\kappa}{\sigma^2} \frac{1 - e^{d(T-t)}}{1 - h e^{d(T-t)}}, \tag{7.13}$$

with

$$d = \sqrt{\kappa^2 + 2\sigma^2} \quad \text{and} \quad h = \frac{d-\kappa}{d+\kappa}. \tag{7.14}$$

Meanwhile, θ_t, the mean of the short rate, is implied by Equation 7.11, an integral equation. In the context of the HJM framework, the CIR process is represented by the following volatility function of the forward rate,

$$\sigma(t,T) = \sigma_t \sqrt{r_t} \left(\frac{\partial B(t,T)}{\partial T} + \int_t^T \frac{\partial \theta_s}{\partial r_t} \frac{\partial B(s,T)}{\partial T} \, ds \right). \tag{7.15}$$

There are undesirable features in Equation 7.15. First, θ_t is implied by the integral Equation 7.11, for which a reliable solution is not easily obtainable. Second, $\sigma(t,T)$ depends on r_t, another stochastic process, so the HJM equation is not self-contained. This feature will be echoed in the next section where we are force to adopt an alternative formulation of the CIR model for production use. From both analytic and computational point of view, Equation 7.15 is not very useful because we must solve for θ_t numerically in advance, a procedure equivalent to calibration, which should not be necessary in the HJM context. To some extent, the undesirable features in Equation 7.15 demonstrate the limitation of the HJM framework for interest-rate modeling.

7.2 General Markovian Models

In this section, we address the opposite question: under what specifications of forward-rate volatility should the resulting short-rate process be Markovian? Answers to this question will help us to calibrate and implement a short-rate model more efficiently.

According to Equation 6.9, the short rate can be expressed as

$$r_t = f(t,t) = f(0,t) + \int_0^t \left[-\sigma^{\mathrm{T}}(s,t)\Sigma(s,t) \, ds + \sigma^{\mathrm{T}}(s,t) \, d\mathbf{W}_s \right], \tag{7.16}$$

where $\mathbf{W}_t$ is the n-dimensional Brownian motion under the risk-neutral measure, $\sigma(t,T)$ the forward-rate volatility, and $\Sigma(t,T)$ the volatility of the T-maturity zero-coupon bond, given by $\Sigma(t,T) = -\int_t^T \sigma(t,u) du$. The stochastic differentiation of the short rate is

$$
\begin{aligned}
\mathrm{d}r_t &= \left[f_t(0,t) + \int_0^t \left(-\frac{\partial}{\partial t}(\boldsymbol{\sigma}^{\mathrm{T}}(s,t)\boldsymbol{\Sigma}(s,t))\,\mathrm{d}s + \frac{\partial \boldsymbol{\sigma}^{\mathrm{T}}(s,t)}{\partial t}\,\mathrm{d}\mathbf{W}_s \right) \right] \mathrm{d}t \\
&\quad + \boldsymbol{\sigma}^{\mathrm{T}}(t,t)\,\mathrm{d}\mathbf{W}_t \\
&= [f_T(t,T)]_{T=t}\,\mathrm{d}t + \boldsymbol{\sigma}^{\mathrm{T}}(t,t)\,\mathrm{d}\mathbf{W}_t.
\end{aligned} \tag{7.17}
$$

Based on Equation 7.17 we can make the following statements: for the short-rate model to be a Markovian process, we need the drift term, $[f_T(t,T)]_{T=t}$, to be a function of a finite set of state variables that are jointly Markovian in their evolution.

To write the short rate as a function of several state variables, we introduce auxiliary functions

$$
b_i(t,T) = \sigma_i(t,T) \int_t^T \sigma_i(t,s)\,\mathrm{d}s, \quad i = 1, 2, \ldots, n, \tag{7.18}
$$

and define

$$
\chi_i(t) = \int_0^t b_i(s,t)\,\mathrm{d}s + \sigma_i(s,t)\,\mathrm{d}W_i(s), \quad i = 1, 2, \ldots, n, \tag{7.19}
$$

we can then write

$$
r_t = f(0,t) + \sum_{i=1}^n \chi_i(t). \tag{7.20}
$$

If we can find the conditions for $\chi_i(t)$ to be Markovian variables, then, under the same conditions, r_t will also be a Markovian variable.

Define, in addition,

$$
\varphi_i(t) = \int_0^t \sigma_i^2(s,t)\,\mathrm{d}s, \quad i = 1, 2, \ldots, n. \tag{7.21}
$$

The following theorem presents a sufficient condition under which the pair of functions $\{\chi_i, \varphi_i\}$ are jointly Markovian variables (Ritchken and Sankarasubramanian, 1995; Inui and Kijima, 1998).

Theorem 7.2.1. *Suppose that the forward-rate volatility satisfies*

$$
\frac{\partial \sigma_i(t,T)}{\partial T} = -\kappa_i(T)\sigma_i(t,T), \quad i = 1, 2, \ldots, n, \tag{7.22}
$$

for some deterministic functions $\kappa_i(T)$. Then,

$$
\begin{aligned}
d\varphi_i(t) &= \left(\sigma_i^2(t,t) - 2\kappa_i(t)\varphi_i(t) \right) dt, \\
d\chi_i(t) &= \left(\varphi_i(t) - \kappa_i(t)\chi_i(t) \right) dt + \sigma_i(t,t)\,dW_i(t),
\end{aligned} \tag{7.23}
$$

for $i = 1, \ldots, n$.

Proof. For φ_i, there is

$$
\begin{aligned}
\mathrm{d}\varphi_i(t) &= \sigma_i^2(t,t)\,\mathrm{d}t + \left(\int_0^t 2\sigma_i(s,t)\frac{\partial \sigma_i(s,t)}{\partial t}\,\mathrm{d}s\right)\mathrm{d}t \\
&= \sigma_i^2(t,t)\,\mathrm{d}t - 2\kappa_i(t)\left(\int_0^t \sigma_i^2(s,t)\,\mathrm{d}s\right)\mathrm{d}t \\
&= \left(\sigma_i^2(t,t) - 2\kappa_i(t)\varphi_i(t)\right)\mathrm{d}t,
\end{aligned}
\tag{7.24}
$$

while for $\chi_i(t)$, we have, noticing that $b_i(t,t) = 0$,

$$
\mathrm{d}\chi_i(t) = \left(\int_0^t \frac{\partial b_i(s,t)}{\partial t}\,\mathrm{d}s + \frac{\partial \sigma_i(s,t)}{\partial t}\,\mathrm{d}W_i(s)\right)\mathrm{d}t + \sigma_i(t,t)\,\mathrm{d}W_i(t). \tag{7.25}
$$

Because of Equation 7.22 and

$$
\begin{aligned}
\frac{\partial b_i(s,t)}{\partial t} &= \frac{\partial \sigma_i(s,t)}{\partial t}\int_s^t \sigma_i(s,u)\,\mathrm{d}u + \sigma_i^2(s,t) \\
&= -\kappa_i(t)b_i(s,t) + \sigma_i^2(s,t),
\end{aligned}
\tag{7.26}
$$

we can rewrite Equation 7.25 as

$$
\begin{aligned}
\mathrm{d}\chi_i(t) &= -\kappa_i(t)\left(\int_0^t b_i(s,t)\,\mathrm{d}s + \sigma_i(s,t)\,\mathrm{d}W_i(s)\right)\mathrm{d}t \\
&\quad + \left(\int_0^t \sigma_i^2(s,t)\,\mathrm{d}s\right)\mathrm{d}t + \sigma_i(t,t)\,\mathrm{d}W_i(t) \\
&= (\varphi_i(t) - \kappa_i(t)\chi_i(t))\,\mathrm{d}t + \sigma_i(t,t)\,\mathrm{d}W_i(t).
\end{aligned}
\tag{7.27}
$$

This completes the proof. $\qquad\square$

The implication of the above theorem is that, under condition 7.22, the HJM model implies a Markovian short-rate process, which is given by

$$
\begin{aligned}
\mathrm{d}r_t &= f_t(0,t)\,\mathrm{d}t + \sum_{i=1}^n \mathrm{d}\chi_i(t) \\
&= f_t(0,t)\,\mathrm{d}t + \sum_{i=1}^n (\varphi_i(t) - \kappa_i(t)\chi_i(t))\,\mathrm{d}t + \sum_{i=1}^n \sigma_i(t,t)\,\mathrm{d}W_i(t) \\
&= \left(f_t(0,t) + \sum_{i=1}^n \left(\varphi_i(t) + (\kappa_n(t) - \kappa_i(t))\chi_i(t)\right)\right)\mathrm{d}t \\
&\quad - \kappa_n(t)\left(\sum_{i=1}^n \chi_i(t)\right)\mathrm{d}t + \sum_{i=1}^n \sigma_i(t,t)\,\mathrm{d}W_i(t).
\end{aligned}
\tag{7.28}
$$

Denote

$$
\Phi(t) = \sum_{i=1}^n (\varphi_i(t) + (\kappa_n - \kappa_i)\chi_i(t)), \tag{7.29}
$$

then Equation 7.28 is simplified to

$$\mathrm{d}r_t = [\kappa_n(t)\,(\,f(0,t) - r_t) + f_t(0,t) + \Phi(t)]\,\mathrm{d}t + \sum_{i=1}^n \sigma_i(t,t)\,\mathrm{d}W_i(t). \quad (7.30)$$

For $\kappa_n(t) > 0$, Equation 7.30 demonstrates the mean-reverting feature for the short-rate process.

Next, we will show that zero-coupon bonds across all maturities can be expressed in terms of the Markovian state variables, $\{\varphi_i(t), \chi_i(t)\}$. Under the HJM model, the formula for the price of a zero-coupon bond is

$$P(t,T) = \frac{P(0,T)}{P(0,t)} \exp\left\{ -\int_t^T \left(\sum_{i=1}^n \int_0^t b_i(s,u)\,\mathrm{d}s + \sigma_i(s,u)\,\mathrm{d}W_i(s) \right) \mathrm{d}u \right\}. \quad (7.31)$$

For the first term in the exponent, we have

$$\int_t^T \int_0^t b_i(s,u)\,\mathrm{d}s\,\mathrm{d}u = \int_0^t \left(\int_t^T b_i(s,u)\,\mathrm{d}u \right) \mathrm{d}s$$

$$= \int_0^t \left(\int_t^T \sigma_i(s,u) \int_s^u \sigma_i(s,\nu)\,\mathrm{d}\nu\,\mathrm{d}u \right) \mathrm{d}s, \quad (7.32)$$

while the integrand of Equation 7.32 can be written as

$$\int_t^T \sigma_i(s,u) \int_s^u \sigma_i(s,\nu)\,\mathrm{d}\nu\,\mathrm{d}u$$

$$= \sigma_i(s,t) \int_t^T \mathrm{e}^{-\int_t^u \kappa_i(x)\,\mathrm{d}x} \left[\int_s^t + \int_t^u \sigma_i(s,\nu)\,\mathrm{d}\nu \right] \mathrm{d}u$$

$$= b_i(s,t) \int_t^T \mathrm{e}^{-\int_t^u \kappa_i(x)\,\mathrm{d}x}\,\mathrm{d}u$$

$$+ \sigma_i^2(s,t) \int_t^T \mathrm{e}^{-\int_t^u \kappa_i(x)\,\mathrm{d}x} \int_t^u \mathrm{e}^{-\int_t^\nu \kappa_i(x)\,\mathrm{d}x}\,\mathrm{d}\nu\,\mathrm{d}u. \quad (7.33)$$

Defining

$$\beta_i(t,T) = \int_t^T \mathrm{e}^{-\int_t^u \kappa_i(x)\,\mathrm{d}x}\,\mathrm{d}u, \quad t \le T, \quad (7.34)$$

and noticing $\beta_i(t,t) = 0$, we have

$$\int_t^T \mathrm{e}^{-\int_t^u \kappa_i(x)\,\mathrm{d}x} \int_t^u \mathrm{e}^{-\int_t^\nu \kappa_i(x)\,\mathrm{d}x}\,\mathrm{d}\nu\,\mathrm{d}u = \int_t^T \frac{\partial \beta_i(t,u)}{\partial u}\beta_i(t,u)\,\mathrm{d}u$$

$$= \frac{1}{2}\beta_i^2(t,T). \quad (7.35)$$

By combining Equations 7.32 through 7.35, we obtain

$$\int_t^T b_i(s,u)\,\mathrm{d}u = \beta_i(t,T)b_i(s,t) + \frac{1}{2}\beta_i^2(t,T)\sigma_i^2(s,t). \quad (7.36)$$

Next, we consider the second term in the exponent of Equation 7.31. Equation 7.22 implies that

$$\int_t^T \sigma_i(s, u)\, du = \beta_i(t, T)\sigma_i(s, t), \tag{7.37}$$

and it follows that

$$\int_t^T \int_0^t \left(b_i(s, t)\, ds + \sigma_i(s, t)\, dW_i(s) \right) du$$

$$= \beta_i(t, T) \int_0^t \left(b_i(s, t)\, ds + \sigma_i(s, t)\, dW_i(s) \right) + \frac{1}{2}\beta_i^2(t, T) \int_0^t \sigma_i^2(s, t)\, ds$$

$$= \beta_i(t, T)\chi_i(t) + \frac{1}{2}\beta_i^2(t, T)\varphi_i(t). \tag{7.38}$$

The price formula for zero-coupon bonds is thus

$$P(t, T) = \frac{P(0, T)}{P(0, t)} \exp\left(-\sum_{i=1}^n \beta_i(t, T)\chi_i(t) - \frac{1}{2}\sum_{i=1}^n \beta_i^2(t, T)\varphi_i(t) \right), \quad t \le T. \tag{7.39}$$

Note that in order to price options on coupon bonds, we only need the distribution of $\varphi_i(t)$ and $\chi_i(t)$.

Condition 7.22 is sufficient for the short rate to be a Markovian process. We now show that it is also a necessary condition if the short-rate volatility is state independent.

Theorem 7.2.2. *Suppose that the short-rate volatility, $\boldsymbol{\sigma}(t, t)$, is a deterministic function of time. Then a necessary condition for the short rate to be Markovian is*

$$\frac{\partial \sigma_i(t, T)}{\partial T} = -\kappa_i(T)\sigma_i(t, T), \tag{7.40}$$

for some scalar function, $\kappa_i(T), 1 \le i \le n$.

Proof. For the short rate to be Markovian, we require that $r_T - r_t$ depends only on r_t and $\{dW_s, s \in (t, T)\}$. In fact, we have

$$r_T - r_t = f(0, T) - f(0, t) + \int_0^T \boldsymbol{\sigma}^T(s, T)\boldsymbol{\Sigma}(s, T)\, ds$$

$$- \int_0^t \boldsymbol{\sigma}^T(s, t)\boldsymbol{\Sigma}(s, t)\, ds + \int_0^T \boldsymbol{\sigma}^T(s, T)\, dW_s - \int_0^t \boldsymbol{\sigma}^T(s, t)\, dW_s$$

$$= f(0, T) - f(0, t) + \int_0^T \boldsymbol{\sigma}^T(s, T)\boldsymbol{\Sigma}(s, T)\, ds$$

$$- \int_0^t \boldsymbol{\sigma}^T(s, t)\boldsymbol{\Sigma}(s, t)\, ds + \int_t^T \boldsymbol{\sigma}^T(s, T)\, dW_s$$

$$+ \int_0^t (\boldsymbol{\sigma}(s, T) - \boldsymbol{\sigma}(s, t))^T\, dW_s. \tag{7.41}$$

The last term in the above equation cannot depend on $\{\mathrm{d}\mathbf{W}_s, s \in (t, T)\}$, so it can depend only on r_t. Because

$$\int_0^t \boldsymbol{\sigma}^{\mathrm{T}}(s, t) \, \mathrm{d}\mathbf{W}_s = r_t + \text{deterministic function}, \qquad (7.42)$$

we conclude that

$$\int_0^t \boldsymbol{\sigma}^{\mathrm{T}}(s, T) \, \mathrm{d}\mathbf{W}_s \qquad (7.43)$$

is also a deterministic function of r_t. Hence, there is

$$\text{Correlation}\left[\int_0^t \boldsymbol{\sigma}^{\mathrm{T}}(s, T) \, \mathrm{d}\mathbf{W}_s, \int_0^t \boldsymbol{\sigma}^{\mathrm{T}}(s, t) \, \mathrm{d}\mathbf{W}_s\right] = 1. \qquad (7.44)$$

The last equality can be rewritten into

$$E^{\mathbb{Q}}\left[\left(\int_0^t \boldsymbol{\sigma}^{\mathrm{T}}(s, T) \, \mathrm{d}\mathbf{W}_s\right) \times \left(\int_0^t \boldsymbol{\sigma}^{\mathrm{T}}(s, t) \, \mathrm{d}\mathbf{W}_s\right)\right]$$

$$= E^{\mathbb{Q}}\left[\left(\int_0^t \boldsymbol{\sigma}^{\mathrm{T}}(s, T) \, \mathrm{d}\mathbf{W}_s\right)^2\right]^{1/2} \times E^{\mathbb{Q}}\left[\left(\int_0^t \boldsymbol{\sigma}^{\mathrm{T}}(s, t) \, \mathrm{d}\mathbf{W}_s\right)^2\right]^{1/2}.$$

$$(7.45)$$

By Ito's isometry, Equation 7.45 can be recast to

$$\left|\int_0^t \boldsymbol{\sigma}^{\mathrm{T}}(s, T)\boldsymbol{\sigma}(s, t) \, \mathrm{d}s\right| = \left(\int_0^t \|\boldsymbol{\sigma}(s, T)\|^2 \, \mathrm{d}s\right)^{1/2} \times \left(\int_0^t \|\boldsymbol{\sigma}(s, t)\|^2 \, \mathrm{d}s\right)^{1/2},$$

$$(7.46)$$

that is, the equality is achieved in a Cauchy–Schwartz inequality (Rudin, 1976), which is possible if and only if

$$\boldsymbol{\sigma}(s, t) = \alpha(t, T)\boldsymbol{\sigma}(s, T), \quad 0 \le s \le t \qquad (7.47)$$

for some deterministic scalar function, α. Similarly, we also have

$$\boldsymbol{\sigma}(s, t) = \alpha(t, T')\boldsymbol{\sigma}(s, T'), \quad 0 \le s \le t \qquad (7.48)$$

for any other T'. Assume that $\sigma_i(s, t) \neq 0$, we then have

$$\frac{\sigma_i(s, T)}{\sigma_i(s, T')} = \frac{\alpha(t, T')}{\alpha(t, T)} = \frac{\alpha(0, T')}{\alpha(0, T)}, \quad i = 1, \ldots, n. \qquad (7.49)$$

Making $T' = s$, we have then proved that $\sigma_i(s, T)$ can be factorized:

$$\sigma_i(s, T) = x_i(s)y_i(T). \qquad (7.50)$$

By differentiating the above equation with respect to T, we obtain

$$\frac{\partial \sigma_i(s, T)}{\partial T} = x_i(s)y_i(T)\frac{\partial \ln y_i(T)}{\partial T}. \qquad (7.51)$$

Denote $\partial \ln y_i(T)/\partial T$ by $-\kappa_i(T)$, we then arrive at Equation 7.40. $\qquad \square$

7.2.1 One-Factor Models

A one-factor model, for $n = 1$, can be cast as

$$
\begin{aligned}
\mathrm{d}r_t &= \left(\kappa(t)\left(f(0,t) - r_t\right) + f_t(0,t) + \varphi_t\right)\mathrm{d}t + \sigma(t,t)\,\mathrm{d}W_t, \\
\mathrm{d}\varphi_t &= \left(\sigma^2(t,t) - 2\kappa(t)\varphi_t\right)\mathrm{d}t,
\end{aligned}
\tag{7.52}
$$

based on Equations 7.23 and 7.30. When the forward-rate volatility takes $\sigma(t,t) = \sigma_0$ and $\kappa > 0$, we have

$$
\varphi_t = \frac{\sigma_0^2}{2\kappa}\left(1 - \mathrm{e}^{-2kt}\right).
\tag{7.53}
$$

By substituting Equation 7.53 back to the short-rate equation in Equation 7.52, then, unsurprisingly, we reproduce the Hull–White model in terms of the short rate.

When the short-rate volatility takes the form

$$
\sigma(t,t) = \sigma_0\sqrt{r_t},
\tag{7.54}
$$

the governing equations of the short rate become

$$
\begin{aligned}
\mathrm{d}r_t &= \left(\kappa\left(f(0,t) - r_t\right) + f_t(0,t) + \varphi(t)\right)\mathrm{d}t + \sigma_0\sqrt{r_t}\,\mathrm{d}W_t, \\
\mathrm{d}\varphi(t) &= \left(\sigma_0^2 r_t - 2\kappa\varphi(t)\right)\mathrm{d}t.
\end{aligned}
\tag{7.55}
$$

Solving for φ_t, we obtain

$$
\varphi_t = \sigma_0^2 \int_0^t r_s \mathrm{e}^{-2\kappa(t-s)}\,\mathrm{d}s.
\tag{7.56}
$$

Eliminating φ_t in the equation for the short rate, we finally obtain

$$
\mathrm{d}r_t = \left(\kappa\left(f(0,t) - r_t\right) + f_t(0,t) + \sigma_0^2 \int_0^t r_s \mathrm{e}^{-2\kappa(t-s)}\,\mathrm{d}s\right)\mathrm{d}t + \sigma_0\sqrt{r_t}\,\mathrm{d}W_t.
\tag{7.57}
$$

The insight here is that the CIR model of the form (Equation 7.10) that is consistent with the term structure has a path-dependent mean level, θ_t. We thus expect that, in general, one-factor short-rate models that have a state-dependent volatility,

$$
\mathrm{d}r_t = \kappa(t)(\theta_t - r_t)\,\mathrm{d}t + \sigma(r_t,t)\,\mathrm{d}W_t,
\tag{7.58}
$$

have a path-dependent drift term.

Finally, we present a result of Inui and Kijima (1998) on the positiveness of forward rates under Equation 7.58, a general short-rate model with mean reversion.

Theorem 7.2.3. *Suppose that the short-rate volatility, $\sigma(r,t)$, is Lipschitz continuous with $\sigma(0,t) = 0$. If $r_0 \geq 0$ and*

$$\kappa(t)f(0,t) + f_T(0,t) > 0, \quad 0 \leq t, \tag{7.59}$$

then the short rate and forward rates are positive almost surely.

Proof. Under the conditions of the theorem, the strong solution of r_t exists. Moreover, since zero is an unattainable boundary (Karlin and Taylor, 1981) and r_t has continuous sample paths, we know from Equation 7.52 that r_t is positive for all $t > 0$. On the other hand, from the bond price formula

$$\begin{aligned}
P(t,T) &= \frac{P(0,T)}{P(0,t)} \exp\left\{ -\left(\beta(t,T)\chi(t) + \frac{1}{2}\beta^2(t,T)\varphi(t) \right) \right\} \\
&= \frac{P(0,T)}{P(0,t)} \exp\left\{ -\left(\beta(t,T)(r_t - f(0,t)) + \frac{1}{2}\beta^2(t,T)\varphi(t) \right) \right\}, \quad (7.60)
\end{aligned}$$

we have

$$f(t,T) = f(0,T) - e^{-\int_t^T \kappa(x)\,\mathrm{d}x} f(0,t) + e^{-\int_t^T \kappa(x)\,\mathrm{d}x} \left[r_t + \beta(t,T)\varphi(t) \right]. \tag{7.61}$$

Under condition 7.59, the second term on the right-hand side of Equation 7.61 is monotonically increasing in t, implying

$$e^{-\int_t^T \kappa(x)\,\mathrm{d}x} f(0,t) < e^{-\int_t^T \kappa(x)\,\mathrm{d}x} f(0,t) \Big|_{t=T} = f(0,T). \tag{7.62}$$

According to Equation 7.61, there must be $f(t,T) > 0$.   □

7.2.2   Monte Carlo Simulations for Options Pricing

Thanks to the Markovian property of short-rate models, path simulations by Monte Carlo methods can be carried out efficiently, which is important for pricing exotic and path-dependent options. Take the pricing of the option on a zero-coupon for example. The value can be expressed as

$$V_t = E_t^{\mathbb{Q}} \left[e^{-\int_t^T r_s\,\mathrm{d}s} (P(T,\tau) - K)^+ \right], \quad t < T < \tau, \tag{7.63}$$

where $\mathbb{Q}$ stands for the risk-neutral measure, r_t is given by Equation 7.20, and the bond price is given by Equation 7.39. Both variables are expressed in terms of $\chi_i(t)$ and $\varphi_i(t)$, $i = 1, \ldots, n$, which evolve according to Equation 7.23. The corresponding simulation scheme for $\chi_i(t)$ and $\varphi_i(t)$ is

$$\begin{aligned}
\varphi_i(t + \Delta t) &= \varphi_i(t) + \left(\sigma_i^2(t,t) - 2\kappa_i(t)\varphi_i(t) \right)\Delta t, \\
\chi_i(t + \Delta t) &= \chi_i(t) + \left(\varphi_i(t) - \kappa_i(t)\chi_i(t) \right)\mathrm{d}t + \sigma_i(t,t)\Delta W_i(t),
\end{aligned} \tag{7.64}$$

which is simply the so-called Euler scheme. The bond option is priced by simulating many payoffs before taking an average.

In Inui and Kijima (1998), the following example is considered:

$$\boldsymbol{\sigma}(t,T) = \begin{pmatrix} c_1 r_t^{\alpha} \\ c_2 r_t^{\beta} e^{-\kappa(T-t)} \end{pmatrix}, \tag{7.65}$$

where $c_i, i = 1, 2, \alpha, \beta$, and κ are non-negative constants. It can be verified that the components of the volatility vector satisfy

$$\frac{\partial \sigma_1(t,T)}{\partial T} = 0, \qquad \frac{\partial \sigma_2(t,T)}{\partial T} = -\kappa \sigma_2(t,T). \tag{7.66}$$

According to Theorem 7.2.2, the corresponding short-rate process is a Markovian process. For various strikes, we consider the pricing of two-year maturity European options on the seven-year maturity bond, that is, $t = 0, T = 2, \tau = 7$. Assume a flat initial term structure:

$$f(0,T) = 0.05, \quad \forall T,$$

and take the following model specifications:

$$\alpha = \beta = 0, \quad \kappa = 0.05, \quad c_1 = c_2 = 0.01.$$

The size for time stepping for Equation 7.64 is $\Delta t = 1/200$, and the number of paths is $N = 50,000$. The results by the simulation method, together with the exact results (Heath *et al.*, 1992), are listed in Table 7.1. As can be seen from the table, the accuracy of simulation pricing is rather high. The simulation scheme performs just as robustly for $\alpha \neq 0$ or $\beta \neq 0$, but there is no exact solution for these cases.

7.3 Binomial Trees of Interest Rates

We have seen that interest-rate options can be priced by Monte Carlo simulations. For Markovian short-rate models, we actually have another choice of numerical method, namely, the lattice tree method. An interest-rate tree can also be regarded as a form of path simulations of interest rates, yet the paths are connected in certain ways, so that the number of states in each time step remains small, making computation efficient. For short-rate models, lattice tree methods (also called lattice or tree methods) are preferred over Monte Carlo simulation methods because the former offer higher efficiency at much less computational cost. Lattice methods are particularly powerful for pricing American options for which optimal early exercises must be taken into account.

TABLE 7.1: Simulated Call Values Versus Exact Values

Strike	Simulated Value	Exact Value	Difference
0.500	0.252346	0.252269	7.70E−05
0.525	0.229725	0.229649	7.60E−05
0.550	0.207105	0.207029	7.60E−05
0.575	0.184491	0.184414	7.70E−05
0.600	0.161906	0.161826	8.00E−05
0.625	0.139417	0.139333	8.40E−05
0.650	0.117207	0.117112	9.50E−05
0.675	0.095628	0.095517	1.11E−04
0.700	0.075271	0.075106	1.65E−04
0.725	0.056802	0.056581	2.21E−04
0.750	0.040866	0.040619	2.47E−04
0.775	0.027960	0.027675	2.85E−04
0.800	0.018086	0.017848	2.38E−04
0.825	0.011090	0.010881	2.09E−04
0.850	0.006422	0.006272	1.50E−04
0.875	0.003504	0.003420	8.40E−05

Reprinted from Inui K and Kijima M, 1998. *Journal of Financial and Quantitative Analysis* 33(3): 423–440. With permission.

An interest-rate tree is usually created by discretizing a continuous-time short-rate model. The tree must at first fit to some chosen term structures of interest rates before being applied for derivatives pricing. This process is called calibration. The term structures, for example, can be the zero-coupon yield curve and the volatility curve of zero-coupon yields, or the swap rate curve and the volatility curve of the swap rates. Note that even if a tree is obtained from discretizing an already fitted continuous-time short-rate model, calibration may still be needed because the discretizing errors may have spoiled the original fitting. With a calibrated tree, we can price interest-rate options by calculating the expected payoff values through a procedure of backward induction, where the interest rates are also used for discounting. In the following sections, we limit ourselves to the interest rate trees by respectively discretizing the Ho–Lee and Hull–White models, two popular Gaussian models of interest rates, thus obtaining calibrated binomial and trinomial interest rate trees.

7.3.1   A Binomial Tree for the Ho–Lee Model

The Ho–Lee model was first presented with a binomial tree. For a Gaussian short-rate model with mean and variance of change over $(t, t + \Delta t)$ given by

$$E^{\mathbb{Q}}[\Delta r_t] = \theta_t \Delta t,$$
$$\text{VaR}(\Delta r_t) = \sigma^2 \Delta t,$$

$$(7.67)$$

we consider a rather natural binomial tree approximation as illustrated in Figure 7.1, where, without loss of generality, the branching probabilities are uniformly one half.

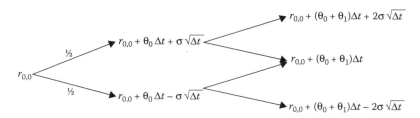

FIGURE 7.1: A binomial tree for the Ho–Lee model.

For notational efficiency, we let

$$r_{i,j} = r_{0,0} + \Delta t \sum_{k=1}^{j-1} \theta_k + (2i - j)\,\sigma\sqrt{\Delta t}, \quad i = 0, 1, \ldots, n.$$

$$j = 0.1, \ldots, n. \tag{7.68}$$

Then we have a multi-period tree as shown in Figure 7.2.

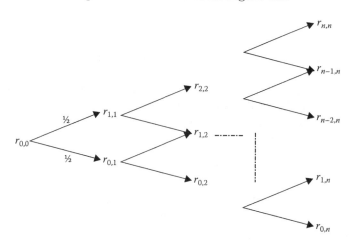

FIGURE 7.2: A general multi-period binomial interest-rate tree.

Before being applied to derivatives pricing, such a tree must first be calibrated to the current term structure of the interest rates. For the Ho–Lee model, we need to pin down the drift, $\{\theta_j\}$, by reproducing the prices of zero-coupon bonds of all maturities. This task can be efficiently achieved with the help of the so-called Arrow–Debreu security prices.

7.3.2   Arrow–Debreu Security Prices

An Arrow–Debreu (1954) security is a canonical asset that has a cash flow of \$1 if a particular state (of interest rate) is realized, or nothing otherwise. The pattern of payment is shown in Figure 7.3, where we let $Q_{i,j}$ denote the price of the security at time 0 that would pay \$1 at time j if the state i is realized, or nothing if otherwise. Let us emphasize here the interest rate $r(i,j)$ is backward-looking and used for discounting cash flows from $j\Delta t$ to $(j-1)\Delta t$.

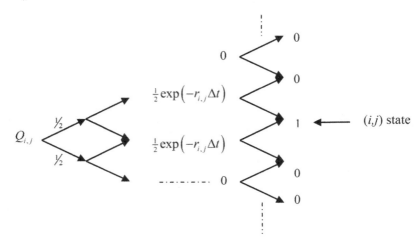

FIGURE 7.3: The payoff pattern and inductive tree for an Arrow–Debreu security.

Note that a zero-coupon bond can be regarded as a portfolio of Arrow–Debreu securities. By linearity, the price of the zero-coupon bond maturing in time $j\Delta t$ is equal to

$$P(0, j\Delta t) = \sum_{i=0}^{j} Q_{i,j}. \tag{7.69}$$

Given an interest-rate tree as in Figure 7.2, we can construct the Arrow–Debreu price tree through a forward induction process, starting with

$$Q_{0,0} = 1. \tag{7.70}$$

The calculations of $Q_{1,1}$ and $Q_{0,1}$ are done by "expectation pricing" using one-period trees in Figure 7.4.

Intuitively, the prices of the two Arrow–Debreu securities are given by

$$Q_{0,1} = \frac{1}{2}e^{-r_{0,1}\Delta t} \quad \text{and} \quad Q_{1,1} = \frac{1}{2}e^{-r_{1,1}\Delta t}. \tag{7.71}$$

We emphasize here the backward-looking term rates are used for discounting.

FIGURE 7.4: One-period trees for Arrow–Debreu prices.

Suppose that we have already obtained

$$Q_{i,k}, \quad i = 0, 1, \ldots, k, \text{ and } k = 0, 1, \ldots, j-1.$$

We proceed to the calculations of $Q_{.,j}$ using the formula

$$
\begin{aligned}
Q_{0,j} &= \frac{e^{-r_{0,j}\Delta t}}{2} Q_{0,j-1}, \\
Q_{i,j} &= \frac{e^{-r_{i,j}\Delta t}}{2} \left(Q_{i-1,j-1} + Q_{i,j-1} \right), \quad i = 1 : j, \\
Q_{j,j} &= \frac{e^{-r_{j,j}\Delta t}}{2} Q_{j-1,j-1}.
\end{aligned}
\tag{7.72}
$$

The rationale of the above inductive scheme can be seen in Figure 7.3. After one step of backward induction from time j for calculating $Q_{i,j}$, there are only two non-zero nodal values, at nodes $(i-1, j-1)$ and $(i, j-1)$. We then take advantage of the Arrow–Debreu security prices $Q_{.,j-1}$, already obtained, to calculate $Q_{i,j}$, which is reflected by the second formula in Equation 7.72. The first and the third formula in Equation 7.72 are for calculating Arrow–Debreu prices corresponding to the first and the last node at time j.

Eventually, the above process generates the Arrow–Debreu price tree, shown in Figure 7.5.

7.3.3 A Calibrated Tree for the Ho–Lee Model

In reality, Arrow–Debreu prices must be determined as a part of the calibration procedure, when we must pin down $\{\theta_j\}$ using price information from discount bonds. The procedure is described below.

The first step is to determine θ_0 by using the price of $P(0, \Delta t)$, as

$$
\begin{aligned}
P(0, \Delta t) &= Q_{0,1} + Q_{1,1} = \frac{1}{2} e^{-r_{0,1}\Delta t} + \frac{1}{2} e^{-r_{1,1}\Delta t} \\
&= \frac{1}{2} \left(e^{-\left(r_{0,0} + \theta_0 \Delta t - \sigma\sqrt{\Delta t}\right)\Delta t} + e^{-\left(r_{0,0} + \theta_0 \Delta t + \sigma\sqrt{\Delta t}\right)\Delta t} \right) \\
&= \frac{1}{2} e^{-\theta_0(\Delta t)^2 + \sigma \Delta t^{3/2}} \left(e^{-r_{0,0}\Delta t} + e^{-\left(r_{0,0} + 2\sigma\sqrt{\Delta t}\right)\Delta t} \right).
\end{aligned}
\tag{7.73}
$$

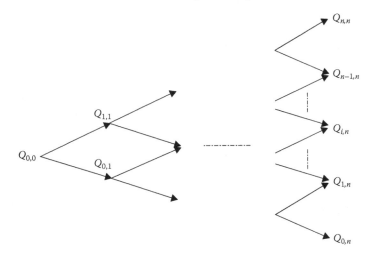

FIGURE 7.5: An Arrow–Debreu price tree.

It follows that

$$\theta_0 = \frac{1}{(\Delta t)^2}\left[\ln\frac{\left(e^{-r_{0,0}\Delta t} + e^{-\left(r_{0,0}+2\sigma\sqrt{\Delta t}\right)\Delta t}\right)}{2P(0,\Delta t)}\right] + \frac{\sigma}{\sqrt{\Delta t}}. \qquad (7.74)$$

With the newly obtained θ_0, we can calculate the nodal values of the interest rates at time 1 by

$$\begin{aligned}
r_{1,1} &= r_{0,0} + \theta_0\Delta t + \sigma\sqrt{\Delta t},\\
r_{0,1} &= r_{0,0} + \theta_0\Delta t - \sigma\sqrt{\Delta t},
\end{aligned} \qquad (7.75)$$

and, finally, calculate $Q_{i,1}$, $i = 0, 1$ using Equation 7.71.

The calibration procedure can be accomplished through induction. Assume now that we have found

$$\begin{aligned}
\theta_k &\quad \text{for } k = 0 : (j-1),\\
r_{i,k} &\quad \text{for } i = 0 : k, \ k = 0 : j,\\
Q_{i,k} &\quad \text{for } i = 0 : k, \ k = 0 : j
\end{aligned}$$

for $j \geq 1$, we can proceed to find θ_j, $r_{\cdot,j+1}$, and $Q_{\cdot,j+1}$ by matching to the price of $P(0, (j+1)\Delta t)$ as follows:

$$
\begin{aligned}
P(0, (j+1)\Delta t) &= \frac{1}{2} Q_{0,j} e^{-(r_{0,j} + \theta_j \Delta t - \sigma\sqrt{\Delta t})\Delta t} \\
&\quad + \sum_{i=1}^{j} \frac{Q_{i-1,j} + Q_{i,j}}{2} e^{-(r_{i,j} + \theta_j \Delta t - \sigma\sqrt{\Delta t})\Delta t} \\
&\quad + \frac{1}{2} Q_{j,j} e^{-(r_{j,j} + \theta_j \Delta t + \sigma\sqrt{\Delta t})\Delta t} \\
&\triangleq e^{-\theta_j(\Delta t)^2 + \sigma(\Delta t)^{3/2}} \frac{f(\{r_{.,j}\}, \{Q_{.,j}\})}{2},
\end{aligned} \tag{7.76}
$$

where

$$
\begin{aligned}
f(\{r_{.,j}\}, \{Q_{.,j}\}) &= Q_{0,j} e^{-r_{0,j}\Delta t} + \sum_{i=1}^{j} (Q_{i-1,j} + Q_{i,j}) e^{-r_{i,j}\Delta t} \\
&\quad + Q_{j,j} e^{-(r_{j,j} + 2\sigma\sqrt{\Delta t})\Delta t}.
\end{aligned} \tag{7.77}
$$

We again have the explicit solution for θ_j:

$$
\theta_j = \frac{1}{(\Delta t)^2} \ln\left(\frac{f(\{r_{.,j}\}, \{Q_{.,j}\})}{2P(0, (j+1)\Delta t)} \right) + \frac{\sigma}{\sqrt{\Delta t}}. \tag{7.78}
$$

Next, we calculate $r_{.,j+1}$ according to

$$
\begin{aligned}
r_{i,j+1} &= r_{i,j} + \theta_j \Delta t - \sigma\sqrt{\Delta t}, \quad i = 0 : j \\
r_{j+1,j+1} &= r_{j,j} + \theta_j \Delta t + \sigma\sqrt{\Delta t}.
\end{aligned} \tag{7.79}
$$

Finally, we calculate the Arrow–Debreu prices, $Q_{i,j+1}, i = 0, 1, \ldots, j+1$, using Equation 7.72.

A succinct algorithm is described below.

1. Let $r_{0,0} = r_0$ and $Q_{0,0} = 1$.

2. For $j = 0 : n - 1$,

    (a) compute $f(\{r_{.,j}\}, \{Q_{.,j}\})$ according to Equation 7.77;

    (b) compute θ_j according to Equation 7.78;

    (c) calculate $r_{i,j+1}, i = 0, \ldots, j+1$ according to Equation 7.79;

    (d) compute $Q_{i,j+1}, i = 0, \ldots, j+1$ using Equation 7.72.

End of the loop.

We now calibrate the Ho–Lee model to the term structure of March 23, 2007. The discount bond prices are first bootstrapped and listed in Table 7.2. We take the volatility of the short rate, the only exogenous parameter of the

TABLE 7.2: The Discount Curve on March 23, 2007

Year	Discount Factor	Year	Discount Factor	Year	Discount Factor
0.5	0.97584	10.5	0.61896	20.5	0.38537
1	0.95223	11	0.60497	21	0.37604
1.5	0.92914	11.5	0.59125	21.5	0.3669
2	0.90712	12	0.5778	22	0.35798
2.5	0.88629	12.5	0.5646	22.5	0.34925
3	0.86643	13	0.55165	23	0.34073
3.5	0.84724	13.5	0.53895	23.5	0.33241
4	0.82856	14	0.52649	24	0.32428
4.5	0.81032	14.5	0.51427	24.5	0.31635
5	0.7925	15	0.50229	25	0.30862
5.5	0.77506	15.5	0.49055	25.5	0.30107
6	0.75799	16	0.47903	26	0.29372
6.5	0.74127	16.5	0.46774	26.5	0.28655
7	0.72489	17	0.45668	27	0.27957
7.5	0.70884	17.5	0.44584	27.5	0.27277
8	0.69312	18	0.43523	28	0.26615
8.5	0.6777	18.5	0.42483	28.5	0.25971
9	0.66258	19	0.41464	29	0.25345
9.5	0.64776	19.5	0.40467	29.5	0.24736
10	0.63322	20	0.39492	30	0.24145

Note: Constructed using Treasury yield data from Reuters.

model, to be $\sigma = 0.005$, meaning a 50 basis points standard deviation of the short rate over one year. The time step is $\Delta t = 0.5$.

The $\{\theta_j\}$ calculated using the algorithm above is plotted in Figure 7.6. In this example, the drift term for the Ho–Lee model is rather small, and it can be negative. The surface plot for the Arrow–Debreu prices are shown in Figure 7.7 (note that the price only exists for the subdiagonal part of the rectangular area), and a plot for the nodal values of the interest-rate tree is given in Figure 7.8, where we can see that while the interest rates span from -15% to 30%, the Arrow–Debreu prices are very small for interest rates that are either too high or too low.

7.4 A General Tree-Building Procedure

The procedure for tree building presented in the last section can be generalized to other diffusion models for short rates. The major disadvantage of the resulted tree is that the rate span can be too wide and fall too deep in the

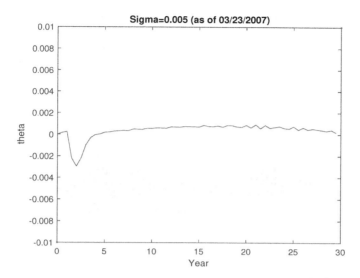

FIGURE 7.6: The Ho–Lee drift, $\theta(t)$, calibrated using the discount curve of March 23, 2007, and a short-rate volatility of $\sigma = 0.005$.

negative territory. In this section, we introduce a more sophisticated method of tree building for short-rate models with mean reversion feature. Although the resulted tree does not guarantee that the interest rate stays in the positive territory, it does provide a tree with a much narrower rate span around the level of the mean interest rate.

7.4.1 A Truncated Tree for the Hull–White Model

Without loss of generality, we consider a trinomial tree approximation of a general short-rate model with a state-dependent drift term and time-dependent volatility,

$$\mathrm{d}r_t = \mu(r_t, t; \theta_t)\,\mathrm{d}t + \sigma_t\,\mathrm{d}W_t. \tag{7.80}$$

An immediate example of μ is the drift term for the Hull–White model, $\kappa(\theta_t - r_t)$. Literally, the discrete version of Equation 7.80 is

$$\Delta r_t = \mu(r_t, t; \theta_t)\Delta t + \sigma_t \Delta W_t. \tag{7.81}$$

Its first two moments are

$$E^{\mathbb{Q}}\left[\Delta r_t \middle| r_t\right] = \mu(r_t, t; \theta_t)\Delta t,$$
$$E^{\mathbb{Q}}\left[(\Delta r_t)^2 \middle| r_t\right] = \sigma_t^2 \Delta t + \mu^2(r_t, t; \theta_t)\Delta t^2. \tag{7.82}$$

Consider an element with trinomial branching as depicted in Figure 7.9, where the interest rate at state (i, j) will evolve into one of the three possible values,

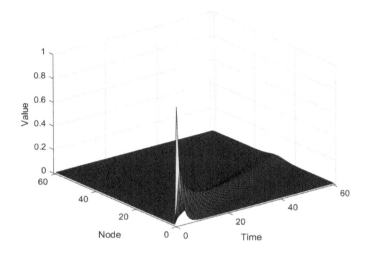

FIGURE 7.7: A surface plot of the Arrow–Debreu prices.

$r_{m(i)+k,j+1}$, $k = -1, 0, 1$, at the next time step. The index for the middle branch, $m(i)$, will be chosen so that $r_{m(i),j+1}$ is closest to the expected value of the short-rate conditional on $r_{i,j}$,

$$m(i) = \arg\min_k \left| E^{\mathbb{Q}}\left[r_{\cdot,j+1}|r_{i,j}\right] - (r_{0,j+1} + k\delta r)\right|,$$
$$r_{0,j+1} = r_{0,j} + \mu(r_{0,j}, t_j; \theta_j)\Delta t. \tag{7.83}$$

where $t_j = j\Delta t$ and $\theta_j = \theta(t_j)$. Through moment matching, we obtain the following governing equations for the branching probabilities:

$$E_t^{\mathbb{Q}}\left[\Delta r_{i,j} - \xi_{i,j}|r_{i,j}\right] = \mu(r_{i,j}, t_j; \theta_j)\Delta t - \xi_{i,j},$$
$$E_t^{\mathbb{Q}}\left[(\Delta r_{i,j} - \xi_{i,j})^2|r_{i,j}\right] = \sigma_j^2\Delta t + (\mu(r_{i,j}, t_j; \theta_j)\Delta t - \xi_{i,j})^2, \tag{7.84}$$

where $\sigma_j = \sigma(j\Delta t)$ and

$$\xi_{i,j} = (m(i) - i)\delta r + \mu(r_{0,j}, t_j; \theta_j)\Delta t. \tag{7.85}$$

As a random variable, $\Delta r_{i,j} - \xi_{i,j}$ takes the following possible values with corresponding probabilities,

$$\Delta r_{i,j} - \xi_{i,j} = \begin{cases} \delta r, & p_{i,j}^{(+1)}, \\ 0, & p_{i,j}^{(0)}, \\ -\delta r, & p_{i,j}^{(-1)}. \end{cases} \tag{7.86}$$

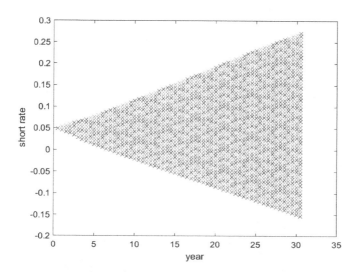

FIGURE 7.8: A surface plot of the interest-rate tree.

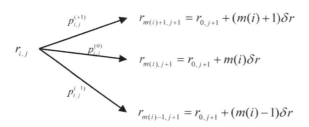

FIGURE 7.9: Trinomial tree branching of the short rate.

Then, Equation 7.84 becomes

$$
\begin{aligned}
p_{i,j}^{(+1)}\delta r - p_{i,j}^{(-1)}\delta r &= \mu(r_{i,j}, t_j; \theta_j)\Delta t - \xi_{i,j}, \\
p_{i,j}^{(+1)}\delta r^2 + p_{i,j}^{(-1)}\delta r^2 &= \sigma_j^2 \Delta t + (\mu(r_{i,j}, t_j; \theta_j)\Delta t - \xi_{i,j})^2,
\end{aligned}
\tag{7.87}
$$

subject to the following natural conditions for probabilities,

$$
\begin{aligned}
p_{i,j}^{(+1)} + p_{i,j}^{(0)} + p_{i,j}^{(-1)} &= 1, \\
p_{i,j}^{(+1)}, \ p_{i,j}^{(0)}, \ p_{i,j}^{(-1)} &\geq 0.
\end{aligned}
\tag{7.88}
$$

Let

$$
\eta_{i,j} = \mu(r_{i,j}, t_j; \theta_j)\Delta t - \xi_{i,j}.
\tag{7.89}
$$

Then the solutions to the probabilities can be solved and expressed as

$$p_{i,j}^{(+1)} = \frac{1}{2} \left(\frac{\sigma_j^2 \Delta t + \eta_{i,j}^2}{\delta r^2} + \frac{\eta_{i,j}}{\delta r} \right),$$

$$p_{i,j}^{(-1)} = \frac{1}{2} \left(\frac{\sigma_j^2 \Delta t + \eta_{i,j}^2}{\delta r^2} - \frac{\eta_{i,j}}{\delta r} \right), \tag{7.90}$$

$$p_{i,j}^{(0)} = 1 - \frac{\sigma_j^2 \Delta t + \eta_{i,j}^2}{\delta r^2}.$$

A good choice for the ratio $\Delta t / \delta r^2$ is to set

$$\frac{\sigma_j^2 \Delta t}{\delta r^2} = \frac{1}{3}, \tag{7.91}$$

which will match the third moments of the continuous and discrete processes (and also implies a varying time step if σ_j is not a constant).

When the short rate is mean-reverting, the above discretization method will automatically produce a truncated tree. To see that we take, for example, $\mu(r, t; \theta) = \kappa(\theta_t - r)$. We can justify that the nodes that can be reached at time t_n are within the band

$$r_{0,j} - \left(\left\lfloor \frac{1}{2\kappa\Delta t} \right\rfloor + 1 \right) \delta r \leq r_{i,j} \leq r_{0,j} + \left(\left\lfloor \frac{1}{2\kappa\Delta t} \right\rfloor + 1 \right) \delta r, \tag{7.92}$$

where $\lfloor x \rfloor$ stands for the integer part of x. The maximum number of nodes at any time step is

$$n_{\max} = 2 \left\lfloor \frac{1}{2\kappa\Delta t} \right\rfloor + 3. \tag{7.93}$$

The resulted multi-period trinomial tree will be truncated, as is shown in Figure 7.10.

Yet again, for the sake of arbitrage pricing, we must calibrate the trinomial tree to the current discount curve by taking proper θ_j's. When $\eta_{i,j}$ of Equation 7.89 is independent of θ_j, the calibration can be achieved very efficiently. This is the case for the Hull–White model, which has the drift term

$$\mu(r_t, t; \theta_t) = \kappa(\theta_t - r), \tag{7.94}$$

and the corresponding $\eta_{i,j}$ is

$$\begin{aligned} \eta_{i,j} &= \mu(r_{i,j}, t_j; \theta_j)\Delta t - \xi_{i,j} \\ &= ((1 - \kappa\Delta t)\, i - m(i))\, \delta r, \end{aligned} \tag{7.95}$$

which contains no θ_j. Furthermore, using Equation 7.95, we can determine the range for the first index for states of all time steps:

$$|i| \leq \left\lfloor \frac{1}{2\kappa\Delta t} \right\rfloor + 1 \overset{\Delta}{=} I. \tag{7.96}$$

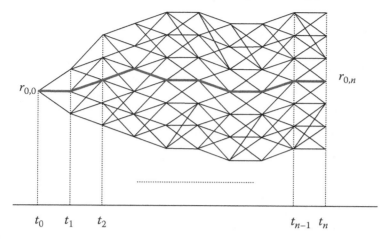

FIGURE 7.10: A truncated trinomial tree.

Note that $m(i)$ minimizes $|\eta_{i,j}|$ and $n_{\max} = 2I + 1$. In addition, the central node of branching from a node j is

$$m(i) = \begin{cases} -I+1, & \text{if } i = -I, \\ i, & \text{if } |i| < I, \\ I-1, & \text{if } i = I. \end{cases} \tag{7.97}$$

Starting from $Q_{0,0} = 1$, we determine $Q_{i,j}$ through the following procedure:

1. Taking $Q_{0,0} = 1$, $r_{0,0} = r_0$.

2. For $j = 0 : n - 1$, repeat the following steps.

   (a) Define

   $$r_{i,j+1} = (1 - \kappa\Delta t)\,r_{0,j+1} + i\delta r, \quad -\min(j,I) \le i \le \min(j,I).$$

   (b) Calculate θ_j from the equation

   $$P\left(0,\,(j+1)\Delta t\right) = e^{-\kappa\theta_j(\Delta t)^2} \sum_{i=-\min(j,I)}^{\min(j,I)} Q_{i,j} \sum_{k=-1}^{+1} p_{i,j}^{(k)} e^{-r_{m(i)+k,j+1}\Delta t},$$

   $$\tag{7.98}$$

   where $p_{i,j}^{(k)}$, $k = -1 : 1$ are calculated using Equations 7.90. Then,

   $$\theta_j = \frac{1}{\kappa(\Delta t)^2} \ln\left(\frac{\displaystyle\sum_{i=-\min(j,I)}^{\min(j,I)} Q_{i,j} \sum_{k=-1}^{+1} p_{i,j}^{(k)} e^{-r_{m(i)+k,j+1}\Delta t}}{P(0,(j+1)\Delta t)} \right). \tag{7.99}$$

(c) For $i = -\min(j, I) : \min(j, I)$ update $r_{i,j}$ as

$$r_{i,j+1} = r_{i,j+1} + \kappa\theta_j\Delta t.$$

(d) For $i = -\min(j, I) : \min(j, I)$ calculate $Q_{i,j+1}$ by

$$Q_{m(i)-1,j+1} = Q_{m(i)-1,j+1} + Q_{i,j}p_{i,j}^{(-1)}\exp(-r_{m(i)-1,j+1}\Delta t),$$
$$Q_{m(i),j+1} = Q_{m(i),j+1} + Q_{i,j}p_{i,j}^{(0)}\exp(-r_{m(i),j+1}\Delta t),$$
$$Q_{m(i)+1,j+1} = Q_{m(i)+1,j+1} + Q_{i,j}p_{i,j}^{(+1)}\exp(-r_{m(i)+1,j+1}\Delta t).$$

End of the loop.

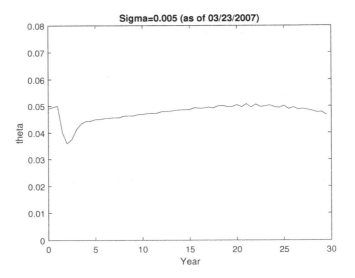

FIGURE 7.11: Theta for the Hull–White model.

We now take
$$\kappa = 0.25 \quad \text{and} \quad \sigma = 0.5\%,$$
and calibrate the tree for the Hull–White model for the discount curve in Table 7.2. Note that this choice of κ corresponds to the half-life of mean reversion equal to
$$T = \frac{\ln 2}{\kappa} = 2.8 \text{ years}$$
and a one-year standard deviation of 50 basis point for the short rate. The algorithm performs robustly, and the results are displayed in Figures 7.11 through 7.13. Figure 7.11 shows the θ_t obtained by calibration. Interestingly, the shape of the curve is very similar to that of θ_t for the Ho–Lee model, which

is shown in Figure 7.6. As the level for mean reversion, θ_t appears in the very reasonable range of interest rates, from about 3.5% to 5%. The Arrow–Debreu prices also look nice. What is perhaps most interesting is the interest-rate tree itself, displayed in Figure 7.13, which has a small span and evolves around the mean level represented by the "$*$" plots. Computationally, such a naturally truncated tree only renders more efficiency.

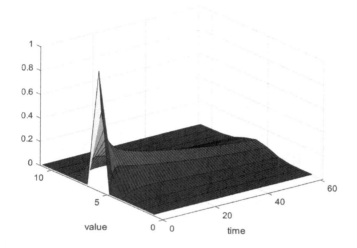

FIGURE 7.12: Arrow–Debreu prices for the Hull–White model.

7.4.2 Trinomial Trees with Adaptive Time Steps

In applications, we often need to value a portfolio of interest-rate derivatives. The cash-flow dates of these derivatives will not necessarily be Δt period apart, unless Δt is very small, which, however, would result in a dense tree and extensive calculations. The trick we introduce here is to use a tree with a variable size of time step so that node points are positioned at the cash-flow dates of the portfolio, whenever necessary. The size of the time step has to observe the following constraint,

$$\frac{\sigma_j^2 \Delta t}{\delta r^2} \leq 1,$$

to ensure that branching probabilities are non-negative. With the adoption of variable time step, the changes to the algorithm above are very limited. Potentially, trees with adaptive time steps are very useful. In Figure 7.14, we draw one such tree with variable time steps.

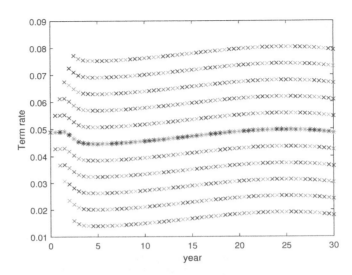

FIGURE 7.13: A short-rate tree for the Hull–White model.

7.4.3 The Black–Karasinski Model

The discussions of tree methods for short-rate models have been surrounding the Ho–Lee and Hull–White models. In this section, we discuss a more general class of short-rate models and their lattice implementations. This class of models can be cast in the form

$$df(r_t) = \kappa(\theta_t - f(r_t))\,dt + \sigma_t\,dW_t, \tag{7.100}$$

for some monotonically increasing function, $f(x)$, such that $f^{-1}(y)$ exists. For a given function of θ_t, we can apply the procedure from Equations 7.83 through 7.90 to build a tree for $R_t = f(r_t)$. The interest-rate tree for r_t then results from the one-to-one correspondence between R_t and r_t. For the calibration procedure, however, there is a major difference from the one for the Hull–White model: due to the non-linearity of $f(x)$, θ_n will be solved through a root-finding procedure, instead of being obtained explicitly. Specifically, with r_t taken over by R_t everywhere in the algorithm, we need to replace Equations 7.98 and 7.99 with a single equation

$$P(0, (j+1)\Delta t) = \sum_{i=-\min(j,I)}^{\min(j,I)} Q_{i,j} \sum_{k=-1}^{+1} p_{i,j}^{(k)} e^{-f^{-1}\left(R_{m(i)+k,j+1}+\kappa\theta_j\Delta t\right)\Delta t}, \tag{7.101}$$

solve for θ_j iteratively, and then update $R_{.,j+1}$.

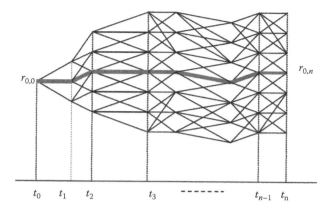

FIGURE 7.14: An interest-rate tree with adaptive time steps.

As a major model in the class of Equation 7.100, we now introduce the Black and Karasinski (1991) model that corresponds to $f(x) = \ln x$. Alternatively, we can write the short rate as an exponential function,

$$r_t = e^{X_t},$$

where X_t follows a generalized Vasicek process,

$$dX_t = \sigma_t \, dW_t + \kappa_t(\theta_t - X_t) \, dt.$$

Here σ_t and κ_t can be deterministic functions of time, and θ_t is subject to the term structure. The Black and Karasinski model is partly motivated by the objective of ensuring positive interest rates, and it also has the desired feature of mean reversion. Similar to the CIR model, it is not consistent with the HJM framework.

The Black–Karasinski model has three parameters. In applications, users of the model have managed to gain extra freedom by turning these parameters into time-dependent functions. These time-dependent functions can be determined by fitting the model to the following three term structures:

- The zero-coupon yield curve.

- The volatility curve of the zero-coupon yields.

- The local volatility of $\ln r_t$, $\forall t$.

For the details of the algorithm, we refer to Black and Karasinski (1991). Note that the stylized algorithm introduced in the last section can also be modified for such a comprehensive calibration task.

We complete this section with a comment. Originally, the Black–Karasinski model did not start from a continuous time as in Equation 7.100. Instead, it started from simple random walks of the interest rate. Black, Derman, and Toy (1990) managed to fit a simple random walk to

- the zero-coupon yield curve; and

- the volatility curve of the zero-coupon yields.

It was realized later that under the limit $\Delta t \to 0$, the simple random walk converges to the continuous-time process,

$$dX_t = \frac{\sigma'_t}{\sigma_t} (\theta_t + X_t)\, dt + \sigma_t\, dW_t, \qquad (7.102)$$

which is called the Black–Derman–Toy model (1990). This model is not necessarily mean-reverting unless there is $\sigma_t < 0$, $\forall t$.

Exercises

1. Derive the drift term of a short-rate model whose volatility is proportional to the short rate. Take the approach of Section 7.2.1.

2. Let $\sigma(s, t)$ be a deterministic function. Given Equation 7.45, prove Equation 7.46.

3. Price a European option on a bond using a trinomial tree under the Hull–White model. The strike price and maturity of the option is $100 and two years, respectively. The underlying bond has a coupon rate of 5%. At the maturity of the option, the bond has a remaining life of two years. The parameters for the model are $\kappa = 0.1$, $\sigma = 0.08$, and the yield curve is flat at 5%. Take $\Delta t = 1/2$ year.

4. Price an American option on a bond using the trinomial tree under the Hull–White model, using the same input parameters as in Problem 3.

5. Redo the pricing of the European option in Problem 3 with the Black–Karasinski model. Take the same model input except $\sigma = 0.2$.

6. Use a binomial tree for the one-factor Ho–Lee model to price the following bond options with payoff at T_0 given by

$$\left(\sum_{i=1}^{N} \Delta T \cdot c \cdot P_{T_0}^{T_i} + P_{T_0}^{T_N} - 1 \right)^+,$$

where $T_i = T_0 + i\Delta T$, $\Delta T = 0.25$ or 0.5, and c is the swap rate for the period $[T_0, T_N]$ seen at time $t = 0$, defined by

$$c = \frac{\left(P_0^{T_0} - P_0^{T_N} \right)}{\sum_{i=1}^{N} \Delta T P_0^{T_i}}.$$

The forward-rate volatility of the model takes the value $\sigma_0 = 0.008544$. You need to do the following:

   (a) Take the discount factors given in Table 7.2.

   (b) Fill in the column under Ho–Lee in Table 7.3, using both $\Delta t = 0.25$ and $\Delta t = 0.5$ in your calculations.

TABLE 7.3: Prices of Par-Bond Options in Basis Points

Maturity, T_0	Tenor, $T_n - T_0$	Ho–Lee
Bond Options		
1	0.25	
2	0.25	
5	0.25	
1	5	
2	5	
5	5	

One basis point corresponds to one cent for the notional of $100.

Chapter 8

Affine Term Structure Models

The class of ATSMs (Duffie and Kan, 1996; Duffie and Singleton, 1999; Dai and Singleton, 2000; etc.) holds an important position in the literature of interest-rate derivatives models. After the establishment of the HJM framework for arbitrage-free interest-rate models, research on interest-rate models had moved forward largely in two directions—represented by the LIBOR market model and the ATSMs. The basic feature of ATSMs is that the short rate is an affine function of some Markov state variables; the latter can follow diffusion, jump-diffusion, or Lévy dynamics (Sato, 1999). There are several reasons that make the ATSMs attractive. First, an ATSM is usually parsimonious in its formulary. One only needs to specify a few parameters of the model, yet the model still has a rich capacity to describe the dynamics of the term structure of interest rates. Second, the class of ATSMs has a relatively high degree of analytical tractability, so that vanilla options on bonds or interest rates can often be priced semi-analytically (through Laplace transforms). Third, the class of ATSMs serves as a rather natural framework for models that keep interest rates positive.[1] Last but not least, owing to their Markov property, affine models are well suited for pricing path-dependent options, through either Monte Carlo simulation methods or, when the number of factors is small, lattice or finite difference methods. In the community of academic finance, ATSMs are often the choice of models for interest rates. There is also a huge literature on empirical studies with ATSMs.

In this chapter, we limit our discussions to affine models that are driven by Brownian diffusions only. We go over a cycle of model construction and model applications, from fitting a model to current term structures of interest rates, to pricing bond options and swaptions under either one-factor or multi-factor ATSMs. Moreover, we address the numerical implementation of the models, which makes use of fast Fourier transforms (FFTs) and thus represents another dimension of the numerical methodology for option pricing. For ATSMs based on more general dynamics, we refer readers to Duffie, Pan, and Singleton (2000), Filipovic (2001), and Duffie, Filipovic, and Schachermayer (2002), among others.

---

[1] In the context of credit risk modeling, affine models allow users to generate models for positive credit spreads.

8.1   An Exposition with One-Factor Models

Instead of starting with an affine function for the short rate, we start, equivalently, with an "exponential affine" function for zero-coupon bonds. Under the risk-neutral measure, the price of zero-coupon bonds can be written as an exponential affine function of an Ito's process, X_t, such that

$$P(t,T) = e^{\alpha(t,T) + \beta(t,T)X_t}, \tag{8.1}$$

where $\alpha(t,T)$ and $\beta(t,T)$ are deterministic functions satisfying

$$\alpha(t,t) = \beta(t,t) = 0, \quad \forall t,$$

and X_t follows

$$dX_t = \mu_X(X_t,t)\, dt + \sigma_X(X_t,t)\, dW_t. \tag{8.2}$$

Here W_t is a Brownian motion under the risk-neutral measure.

In an exponential affine model with zero-coupon bonds, yields, forward rates, and the short rate are all affine functions. In fact, from the price–yield relationship of zero-coupon bonds, we readily have

$$y(t,T) = -\frac{1}{T-t}\ln P(t,T) = -\frac{1}{T-t}\left(\alpha(t,T) + \beta(t,T)X_t\right). \tag{8.3}$$

For forward rates, $f(t,T)$ of any T, we have

$$\int_t^T f(t,s)\, ds = -\ln P(t,T) = -\alpha(t,T) - \beta(t,T)X_t. \tag{8.4}$$

Thus,

$$f(t,T) = -\alpha_T(t,T) - \beta_T(t,T)X_t, \tag{8.5}$$

where the subindex, "T," stands for the partial derivative with respect to T. Putting $T = t$, we obtain a linear functional for the short rate as well:

$$r_t = f(t,t) = -\alpha_T(t,t) - \beta_T(t,t)X_t. \tag{8.6}$$

For notational simplicity, we denote $\delta_0(t) = -\alpha_T(t,t)$, $\delta(t) = -\beta_T(t,t)$ hereafter. If we start from an affine function, Equation 8.6, for the short rate, we can also show, through Lemma 8.1.1, that the zero-coupon bond price is an exponential affine function.

Next, we derive conditions on $\alpha(t,T)$ and $\beta(t,T)$ so that the model is arbitrage-free. The risk-neutral process of the zero-coupon bond is

$$dP(t,T) = P(t,T)\left(r_t\, dt + \Sigma(t,T)\, dW_t\right). \tag{8.7}$$

By Ito's Lemma, we can obtain the following expressions for the short rate and the volatility of the zero-coupon bond in terms of $P(t,T)$:

$$r_t = \frac{1}{P(t,T)}\left[\frac{\partial P}{\partial t} + \mu_X \frac{\partial P}{\partial X} + \frac{1}{2}\sigma_X^2 \frac{\partial^2 P}{\partial X^2}\right],$$

$$\Sigma(t,T) = \frac{1}{P(t,T)}\frac{\partial P}{\partial X}\sigma_X. \tag{8.8}$$

Using the functional form of $P(t,T)$ in Equation 8.1, we can make Equation 8.8 more specific:

$$r_t = \alpha_t + \beta_t X_t + \mu_X \beta + \frac{1}{2}\sigma_X^2 \beta^2,$$

$$\Sigma(t,T) = \beta(t,T)\sigma_X(X_t,t). \tag{8.9}$$

By comparing the first equation of Equation 8.9 with Equation 8.6, we obtain the equality

$$\alpha_t + \beta_t X_t + \mu_X \beta + \frac{1}{2}\sigma_X^2 \beta^2 = \delta_0(t) + \delta(t)X_t, \tag{8.10}$$

or

$$\frac{1}{2}\sigma_X^2 \beta^2 + \mu_X \beta = (\delta(t) - \beta_t)X_t + (\delta_0(t) - \alpha_t). \tag{8.11}$$

The above equality suggests that σ_X^2 and μ_X should be affine functions as well. In fact, this is indeed the case. To see that, we take two distinct maturities, T_1 and T_2, for Equation 8.11, resulting in two joint equations for (μ_X, σ_X^2):

$$\begin{pmatrix} \beta(t,T_1) & (1/2)\beta^2(t,T_1) \\ \beta(t,T_2) & (1/2)\beta^2(t,T_2) \end{pmatrix} \begin{pmatrix} \mu_X \\ \sigma_X^2 \end{pmatrix}$$

$$= \begin{pmatrix} \delta(t) - \beta_t(t,T_1) \\ \delta(t) - \beta_t(t,T_2) \end{pmatrix} X_t + \begin{pmatrix} \delta_0(t) - \alpha_t(t,T_1) \\ \delta_0(t) - \alpha_t(t,T_2) \end{pmatrix}. \tag{8.12}$$

If the coefficient matrix is non-singular, we can solve for (μ_X, σ_X^2) and obtain a solution also in affine form. It is obvious that the coefficient matrix is singular only for the case when $\beta(t,T_1) = \beta(t,T_2)$ or the case when one of the betas equals zero. Suppose that the dependence of $\beta(t,T)$ on T is genuine. Then, the matrix is in general non-singular, so (μ_X, σ_X^2) must be affine functions. Hence, we can write

$$\mu_X(x,t) = K_0(t) + K_1(t)x,$$

$$\sigma_X^2(x,t) = H_0(t) + H_1(t)x. \tag{8.13}$$

Plugging in the above functional forms into Equation 8.11, we then obtain the equation

$$\left(\frac{1}{2}H_1(t)\beta^2 + K_1(t)\beta\right)X_t + \frac{1}{2}H_0(t)\beta^2 + K_0(t)\beta$$

$$= (\delta(t) - \beta_t)X_t + (\delta_0(t) - \alpha_t). \tag{8.14}$$

By equating the corresponding coefficients of the terms,[2] $\{1, X_t\}$, on both sides of Equation 8.14, we obtain the following equations for α and β:

$$\beta_t = \delta(t) - K_1(t)\beta - \frac{1}{2}H_1(t)\beta^2, \quad \beta(T,T) = 0,$$

$$\alpha_t = \delta_0(t) - K_0(t)\beta - \frac{1}{2}H_0(t)\beta^2, \quad \alpha(T,T) = 0. \tag{8.15}$$

The first equation in Equation 8.15 is a Riccarti equation. Its solution is finite under certain technical conditions on H_1 and K_1. The solution of α follows from that of β:

$$\alpha(t,T) = -\int_t^T \delta_0(s)\,\mathrm{d}s + \bar{\alpha}(t,T), \tag{8.16}$$

where

$$\bar{\alpha}(t,T) = \int_t^T \left[K_0(s)\beta(s,T) + \frac{1}{2}H_0(s)\beta^2(s,T) \right]\mathrm{d}s.$$

Note that we have decomposed $\alpha(t,T)$ in such a way that the first term will depend on the term structure of interest rates, whereas the second term will not.

In applications, δ is often fixed as a positive constant. Because of that, both $\beta(t,T)$ and $\bar{\alpha}(t,T)$ do not depend on the term structure of interest rates. Function $\delta_0(t)$, meanwhile, is subject to the term structure of interest rates. In fact, given the term structure of $P(0,T), \forall T, \delta_0(t)$ satisfies the equation

$$P(0,T) = \exp\left(-\int_0^T \delta_0(s)\,\mathrm{d}s + \bar{\alpha}(0,T) + \beta(0,T)X_0 \right), \tag{8.17}$$

or

$$\int_0^T \delta_0(s)\,\mathrm{d}s = -\ln P(0,T) + \bar{\alpha}(0,T) + \beta(0,T)X_0. \tag{8.18}$$

Differentiating Equation 8.18 with respect to T, we obtain

$$\delta_0(T) = f(0,T) + \bar{\alpha}_T(0,T) + \beta_T(0,T)X_0. \tag{8.19}$$

Hence, by taking $\delta_0(T)$ as in Equation 8.19, we ensure that the affine model is consistent with the term structure of interest rates, which is a premise for arbitrage-free pricing of derivatives.

We make a comment Here. When the function of δ's, K's, and H's are constants, we can argue that α and β are functions of time to maturity, that is,

$$\alpha(t,T) = \alpha_0(T-t), \quad \alpha_0(0) = 0,$$
$$\beta(t,T) = \beta_0(T-t), \quad \beta_0(0) = 0,$$

---

[2]This is called the "matching principle" in the finance literature.

for some functions, α_0 and β_0.

Next, we consider option pricing under the ATSM. We define the moment-generating function with discounting as

$$\varphi(u; X_t, t, T) = E^{\mathbb{Q}}\left[\exp\left(-\int_t^T r(X_s, s)\,\mathrm{d}s\right) e^{uX_T} \mid \mathcal{F}_t\right], \qquad (8.20)$$

where $\mathbb{Q}$ stands for the risk-neutral measure. For the function φ, we have:

Lemma 8.1.1. *Suppose that both H_0 and H_1 are bounded and there is*

$$E^{\mathbb{Q}}\left[\left(\int_0^T |X_t|\,dt\right)\right] < \infty. \qquad (8.21)$$

Then, we have
$$\varphi(u; X_t, t, T) = e^{\alpha_u(t,T)+\beta_u(t,T)X_t}, \qquad (8.22)$$

where α_u and β_u are bounded solutions to the following ordinary differential equations (ODE):

$$\begin{aligned}(\beta_u)_t &= \delta - K_1\beta_u - \frac{1}{2}H_1\,\beta_u^2, \quad \beta_u(T,T) = u, \\ (\alpha_u)_t &= \delta_0 - K_0\beta_u - \frac{1}{2}H_0\,\beta_u^2, \quad \alpha_u(T,T) = 0.\end{aligned} \qquad (8.23)$$

Proof. Define an auxiliary function

$$\tilde{\varphi}_t(u) = \exp\left(-\int_0^t r(X_s, s)\,\mathrm{d}s\right) e^{\alpha_u(t,T)+\beta_u(t,T)X_t}. \qquad (8.24)$$

It suffices to show that $\tilde{\varphi}_t$ is a martingale. Then, we have

$$\begin{aligned}E_t^{\mathbb{Q}}[\tilde{\varphi}_T(u)] &= E_t^{\mathbb{Q}}\left[\exp\left(-\int_0^T r(X_s, s)\,\mathrm{d}s\right) e^{uX_T}\right] \\ &= \tilde{\varphi}_t(u) \\ &= \exp\left(-\int_0^t r_s\,\mathrm{d}s\right)\varphi(u; X_t, t, T),\end{aligned} \qquad (8.25)$$

and Equation 8.22 then follows. To prove that $\tilde{\varphi}_t$ is a martingale, we look at the differential for $\tilde{\varphi}_t$:

$$\mathrm{d}\tilde{\varphi}_t = \mu_\varphi(t)\,\mathrm{d}t + \eta_\varphi(t)\,\mathrm{d}W_t, \qquad (8.26)$$

where, by Ito's Lemma,

$$
\begin{aligned}
\mu_\varphi(t) &= \frac{\partial \tilde{\varphi}_t}{\partial t} + \mu_X \frac{\partial \tilde{\varphi}_t}{\partial X} + \frac{1}{2}\sigma_X^2 \frac{\partial^2 \tilde{\varphi}_t}{\partial X^2} \\
&= e^{-\int_0^t r_s ds} \left[(\alpha_u)_t + (\beta_u)_t X_t - r_t + \mu_X \beta_u + \frac{1}{2}\sigma_X^2 \beta_u^2 \right] \\
&= e^{-\int_0^t r_s ds} \left[(\alpha_u)_t + (\beta_u)_t X_t - (\delta_0 + \delta X_t) \right. \\
&\qquad\qquad \left. + (K_0 + K_1 X_t)\beta_u + \frac{1}{2}(H_0 + H_1 X_t)\beta_u^2 \right] \\
&= e^{-\int_0^t r_s ds} \left[\left((\alpha_u)_t - \delta_0 + K_0\beta_u + \frac{1}{2}H_0\beta_u^2 \right) \right. \\
&\qquad\qquad \left. + \left((\beta_u)_t - \delta + K_1\beta_u + \frac{1}{2}H_1\beta_u^2 \right) X_t \right] \\
&= 0.
\end{aligned}
\tag{8.27}
$$

Meanwhile,

$$
\eta_\varphi = \sigma_X \frac{\partial \tilde{\varphi}_t}{\partial X} = \sigma_X \beta_u(t, T) \tilde{\varphi}_t.
\tag{8.28}
$$

Because of Equation 8.21 and the boundedness of H_0 and H_1, we also have

$$
E^{\mathbb{Q}} \left[\int_0^T (\sigma_X \beta_u(t, T))^2 \, dt \right] < \infty.
\tag{8.29}
$$

Hence, $\tilde{\varphi}_t$ is a lognormal martingale. $\square$

Remark 8.1.1. *When $u = 0$, $\varphi(u; X_t, t, T)$ gives the price of the T-maturity zero-coupon bond. By comparing Equation 8.15 with 8.23, we know that $\beta_0 = \beta$ and $\alpha_0 = \alpha$.*

Next, we consider general option pricing under the affine model following an approach pioneered by Carr and Madan (1998). In this approach, the value of an option is regarded, whenever possible, as a convolution between the density function of the state variable(s) and the payoff function. By the Fourier convolution theorem (Rudin, 1976), the Laplace transform of the option is the product of the Laplace transforms of the density function and the payoff function. If we have the Laplace transforms of both the moment-generating function and the option payoff in closed form, we also have the Laplace transform of the option in closed form. The option's value, then, can be obtained by performing an inverse Laplace transform, which, in addition, can be implemented using the technology of FFT, thus making the method very fast.

Without loss of generality, we demonstrate Carr–Madan's approach with the pricing of a call option on a zero-coupon bond. The maturity of the bond is T', while the maturity of the option is $T < T'$. The payoff of the option is

$$V_T = \left(e^{\alpha+\beta X_T} - K\right)^+. \tag{8.30}$$

Here, $\alpha = \alpha_0$ and $\beta = \beta_0$, and we have suppressed the indices of α and β for notational simplicity. Let $k = \ln K$. Under the risk-neutral measure, $\mathbb{Q}$, we have

$$\begin{aligned}
V_t &= E_t^{\mathbb{Q}}\left[e^{-\int_t^T r_s\,ds}\left(e^{\alpha+\beta X_T} - K\right)^+\right] \\
&= E_t^{\mathbb{Q}}\left[e^{-\int_t^T r_s\,ds}\left(e^{\alpha+\beta X_T} - e^k\right)^+\right] \\
&\triangleq G(k).
\end{aligned} \tag{8.31}$$

Noticing that when $k \to -\infty$, $G(k)$ does not tend to zero, and thus the Fourier transform does not exist. For this reason, we pick a positive number, $a > 0$, and consider the damped option price, $g(k) = e^{ak}G(k)$. It can be shown (Lee, 2004) that $g(k)$ is bounded and decays exponentially for $|k| \to \infty$, its Fourier transform then exists and, moreover, we can apply the Fourier transform across the expectation in Equation 8.31:

$$\begin{aligned}
&\int_{-\infty}^{\infty} e^{(a+iu)k}G(k)dk \\
&= \int_{-\infty}^{\infty} e^{iuk}e^{ak}E_t^{\mathbb{Q}}\left[e^{-\int_t^T r_s ds}\left(e^{\alpha+\beta X_T} - e^k\right)^+\right] dk \\
&= E_t^{\mathbb{Q}}\left[e^{-\int_t^T r_s ds}\int_{-\infty}^{\infty} e^{(iu+a)k}\left(e^{\alpha+\beta X_T} - e^k\right)^+ dk\right] \\
&= E_t^{\mathbb{Q}}\left[e^{-\int_t^T r_s ds}\int_{-\infty}^{\alpha+\beta X_T} e^{(iu+a)k}\left(e^{\alpha+\beta X_T} - e^k\right) dk\right] \\
&= E_t^{\mathbb{Q}}\left[e^{-\int_t^T r_s ds}\left(\frac{e^{(iu+a+1)(\alpha+\beta X_T)}}{iu+a} - \frac{e^{(iu+a+1)(\alpha+\beta X_T)}}{iu+a+1}\right)\right] \\
&= \frac{e^{(iu+a+1)\alpha}}{(iu+a)(iu+a+1)}\varphi\left((1+a+iu)\beta; X_t, t, T\right) \\
&\triangleq \psi(u).
\end{aligned} \tag{8.32}$$

Note that the Fourier transform on the damped function is identical to the Laplace transform on the original function. With Equation 8.32, we show that the Laplace transform of $G(k)$ can be obtained explicitly in terms of the (discounted) moment-generating function of X_T. Once we have $\psi(u)$, we can calculate $G(k)$ through an inverse Laplace transform:

$$G(k) = \frac{e^{-ak}}{\pi}\int_0^{\infty} e^{-iuk}\psi(u)\,du. \tag{8.33}$$

We remark here that the above approach works as long as one has the Laplace transforms of both the moment-generating function and the payoff function. The fifth line of Equation 8.32 contains the Laplace transform of the payoff of a call option. For the Laplace transforms of other types of option payoffs, we refer readers to Lee (2004), where rigorous analysis is made on the feasibility to exchange the order of the Laplace transform and the expectation.

The inverse Laplace transform will be evaluated numerically. For that purpose, we need to truncate the infinite domain at a finite number. To choose this number, say, A, let us estimate the error of the numerical integration caused by the truncation. Let $z = (1 + a)\beta$ and assume $X_t = 0$. According to the definition of $\varphi(\cdot)$, we have

$$|\varphi\left((1 + a + iu)\beta; X_t, t, T\right)| \leq |\varphi\left((1 + a)\beta; X_t, t, T\right)| = e^{\alpha_z(t,T)}.$$

It follows that

$$|\psi(u)| \leq \left| \frac{e^{(a+1)\alpha + \alpha_z(t,T)}}{(iu + a)(iu + a + 1)} \right| \leq \frac{\left|e^{(a+1)\alpha + \alpha_z(t,T)}\right|}{u^2 + a^2},$$

and

$$\left| \int_A^\infty e^{-iuk} \psi(u)\, du \right| \leq \int_A^\infty \frac{\left|e^{(a+1)\alpha + \alpha_z(t,T)}\right|}{u^2 + a^2}\, du \leq \frac{\left|e^{(a+1)\alpha + \alpha_z(t,T)}\right|}{A}.$$

Hence, to ensure accuracy on the order of one basis point, we may truncate the integral at $A = 10^4$. Our computational experiences, however, suggest that such a truncation bound is excessively large.

A direct numerical evaluation of the truncated integral of Equation 8.33 is easy but not efficient. To make the evaluation fast, Carr and Madan (1998) introduce a very delicate technique that converts the numerical integration into a discrete FFT. The technique is described below. After a truncation is adopted, we consider the *composite trapezoidal rule* for the numerical integration:

$$H(k) = \frac{1}{\pi} \left(\frac{\psi(0)}{2} + \sum_{m=1}^{N-1} e^{-iu_m k} \psi(u_m) + \frac{e^{-iu_N k} \psi(u_N)}{2} \right) \Delta u, \qquad (8.34)$$

where $u_m = m\Delta u$ and $\Delta u = A/N$. The composite trapezoidal rule has an order of accuracy of $O(\Delta u^2)$. Since we are interested mainly in the around-the-money options, we take k around zero:

$$k_n = -b + n\Delta k, \quad \text{for some } b > 0 \text{ and } n = 0, 1, \dots, N - 1,$$

with

$$\Delta k = \frac{2b}{N}.$$

Hence, for $n = 0, 1, \ldots, N - 1$, we have

$$
H(k_n) = \frac{1}{\pi} \left(\frac{\psi(0)}{2} + \sum_{m=1}^{N-1} e^{-i\Delta u \Delta k m n} \left[e^{i b u_m} \psi(u_m) \right] \right.
$$
$$
\left. + \frac{e^{-i\Delta u \Delta k N n} \psi(u_N)}{2} \right) \Delta u.
$$

We now choose, in particular

$$
\Delta u \Delta k = \frac{2\pi}{N}, \quad \text{or} \quad b = \frac{\pi N}{A},
$$

which will then result in

$$
H(k_n) = \frac{1}{\pi} \left(\frac{\psi(0)}{2} + \sum_{m=1}^{N-1} e^{-i\Delta u \Delta k m n} \left[e^{i b u_m} \psi(u_m) \right] \right.
$$
$$
\left. + \frac{e^{-i\Delta u \Delta k N n} \psi(u_N)}{2} \right) \Delta u, \quad n = 0, 1, \ldots, N - 1.
$$

The expression of $H(k_n)$ fits the definition of discrete Fourier transform, and it can be valued via FFT (see e.g., Press et al., 1992). For later references, we call the Fourier option pricing method the FFT method.

One can, of course, consider a more accurate numerical integration scheme. Yet, our experiences suggest that the composite trapezoidal rule is accurate enough for most applications.

8.2 Analytical Solution of Riccarti Equations

We now focus on the best-known special case of the ATSM that has the following specification of the drift and volatility functions:

$$
\begin{aligned}
\mu_X(x, t) &= \kappa(\theta - x), \\
\sigma_X^2(x, t) &= \sigma_0^2 x,
\end{aligned}
\tag{8.35}
$$

where $\kappa, \theta \geq 0$. This results in the so-called square-root process

$$
\begin{cases}
dX_t = \kappa(\theta - X_t)\, dt + \sigma_0 \sqrt{X_t}\, dW_t, \\
X_0 \geq 0.
\end{cases}
\tag{8.36}
$$

In Equation 8.36, θ and κ are called the mean level and the strength of the mean reversion, respectively, and X_t remains non-negative and is expected to evolve up and down around the mean level. When we take the following special affine specification of the short rate, $r_t = X_t$, we reproduce the famous CIR model for interest rates. One of the most important motivations of the CIR model is to keep the interest-rate positive. Using the square-root processes as building blocks, we can generate a variety of models that ensure interest rates to stay positive.

For the purpose of options pricing, we need the moment-generating function of X_t. According to Lemma 8.1.1, the moment-generating function is available in explicit form:

$$\psi(u; X_t, t, T) = e^{\alpha_u(t,T) + \beta_u(t,T) X_t}, \tag{8.37}$$

where α_u and β_u satisfy

$$\begin{cases} (\beta_u)_t = \delta + \kappa \beta_u - \frac{1}{2}\sigma_0^2 \beta_u^2, & \beta_u(T,T) = u, \\ (\alpha_u)_t = \delta_0 - \kappa\theta\beta_u, & \alpha_u(T,T) = 0. \end{cases} \tag{8.38}$$

Note that δ is a constant while δ_0 is subject to the term structure of interest rates at time t. Having non-negative interest rates in the future requires $\beta_u(t,T)$ to be a decreasing function in T. The solution for α_u follows from that of β_u.

Next, let us focus on the solution of the Riccarti equation. We have

Lemma 8.2.1. *For constant coefficients, the solution to*

$$(\beta_u)_t = b_2 \beta_u^2 + b_1 \beta_u + b_0, \quad \beta_u(T,T) = u. \tag{8.39}$$

is

$$\beta_u(t,T) = -\frac{(b_1 + c)}{2b_2} \frac{\left(e^{c(T-t)} - 1\right)}{\left(g e^{c(T-t)} - 1\right)},$$

where

$$c = \sqrt{b_1^2 - 4b_0 b_2}, \quad g = \frac{2b_2 u + b_1 + c}{2b_2 u + b_1 - c}. \tag{8.40}$$

Proof. The roots for

$$0 = b_2 \beta^2 + b_1 \beta + b_0 \tag{8.41}$$

are

$$\beta_{\pm} = \frac{-b_1 \pm \sqrt{b_1^2 - 4b_0 b_1}}{2b_2} \triangleq \frac{-b_1 \pm c}{2b_2}. \tag{8.42}$$

Consider the difference, $Y(t,T) = \beta_u(t,T) - \beta_+$, which satisfies

$$\begin{aligned} Y_t &= b_2(Y + \beta_+)^2 + b_1(Y + \beta_+) + b_0 \\ &= b_2 Y^2 + (2b_2\beta_+ + b_1)Y \\ &= b_2 Y^2 + cY, \end{aligned} \tag{8.43}$$

with a final condition, $Y(T,T) = u - \beta_+$. The above ODE for Y can be rewritten into

$$\left(\frac{1}{Y}\right)_t = -b_2 - c\left(\frac{1}{Y}\right). \tag{8.44}$$

200 Interest Rate Modeling: Theory and Practice

Let $z = 1/Y$. Then, we have

$$z_t = -b_2 - cz, \quad z(T) = \frac{1}{u - \beta_+} = \frac{2b_2}{2b_2 u + b_1 - c}. \tag{8.45}$$

Solving Equation 8.45, we obtain

$$
\begin{aligned}
z(t) &= e^{c(T-t)} z(T) + \frac{b_2}{c} \left[e^{c(T-t)} - 1 \right] \\
&= \frac{b_2}{c} \left[\left(\frac{c}{b_2} z(T) + 1 \right) e^{c(T-t)} - 1 \right] \\
&= \frac{b_2}{c} \left[\left(\frac{c}{b_2} \frac{2b_2}{2b_2 u + b_1 - c} + 1 \right) e^{c(T-t)} - 1 \right] \\
&= \frac{b_2}{c} \left[\left(\frac{2b_2 u + b_1 + c}{2b_2 u + b_1 - c} \right) e^{c(T-t)} - 1 \right] \\
&= \frac{b_2}{c} \left[g e^{c(T-t)} - 1 \right],
\end{aligned} \tag{8.46}
$$

where g is defined in Equation 8.40. The solution of $z(t)$ gives rise to

$$Y(t,T) = \frac{1}{z(t)} = \frac{c}{b_2} \frac{1}{\left(g e^{c(T-t)} - 1 \right)}. \tag{8.47}$$

We thus obtain the solution to Equation 8.39:

$$
\begin{aligned}
\beta_u(t,T) &= Y(t,T) + \beta_+ \\
&= \frac{c}{b_2 \left(g e^{c(T-t)} - 1 \right)} - \frac{b_1 - c}{2b_2} \\
&= \frac{2c + (c - b_1) \left(g e^{c(T-t)} - 1 \right)}{2b_2 \left(g e^{c(T-t)} - 1 \right)} \\
&= \frac{(b_1 + c) - (b_1 + c) e^{c(T-t)}}{2b_2 \left(g e^{c(T-t)} - 1 \right)} \\
&= -\frac{(b_1 + c)}{2b_2} \frac{\left(e^{c(T-t)} - 1 \right)}{\left(g e^{c(T-t)} - 1 \right)}.
\end{aligned} \tag{8.48}
$$

$\square$

Once β_u is obtained, α_u is calculated by integrations:

$$\alpha_u(t,T) = -\int_t^T \delta_0 \, ds + \kappa\theta \int_t^T \beta_u \, ds, \tag{8.49}$$

where

$$
\begin{aligned}
\int_t^T \beta_u \, ds &= -\frac{b_1 + c}{2b_2} \int_t^T \frac{e^{c(T-s)} - 1}{g e^{c(T-s)} - 1} \, ds \\
&= -\frac{b_1 + c}{2b_2} \int_0^{T-t} \frac{e^{c\tau} - 1}{g e^{c\tau} - 1} \, d\tau \\
&= -\frac{b_1 + c}{2b_2} \left[(T - t) - \int_0^{T-t} \frac{(1 - g)e^{c\tau}}{1 - g e^{c\tau}} \, d\tau \right],
\end{aligned}
\tag{8.50}
$$

while

$$
\begin{aligned}
\int_0^{T-t} \frac{(1 - g) \, e^{c\tau}}{1 - g e^{c\tau}} \, d\tau &= \frac{1}{c} \int_1^{e^{c(T-t)}} \frac{1 - g}{1 - gu} \, du \\
&= \frac{1}{c} \left(\frac{g - 1}{g} \right) \ln \left(\frac{1 - g e^{c(T-t)}}{1 - g} \right) \\
&= \frac{2}{b_1 + c} \ln \left(\frac{1 - g e^{c(T-t)}}{1 - g} \right).
\end{aligned}
\tag{8.51}
$$

Putting the terms in place, we have

$$
\begin{aligned}
\alpha_u(t, T) &= -\int_t^T \delta_0 \, ds - \kappa\theta \left(\frac{b_1 + c}{2b_2} \right) \left[(T - t) - \frac{2}{b_1 + c} \right. \\
&\quad \left. \times \ln \left(\frac{1 - g e^{c(T-t)}}{1 - g} \right) \right] \\
&= -\int_t^T \delta_0 \, ds - \kappa\theta \left[\left(\frac{b_1 + c}{2b_2} \right) (T - t) - \frac{1}{b_2} \right. \\
&\quad \left. \times \ln \left(\frac{1 - g e^{c(T-t)}}{1 - g} \right) \right].
\end{aligned}
\tag{8.52}
$$

Recall that

$$
b_2 = -\frac{1}{2}\sigma_0^2, \quad b_1 = \kappa, \quad b_0 = \delta.
\tag{8.53}
$$

When $u = 0$, $\alpha_0(t, T)$ and $\beta_0(t, T)$ give the prices of zero-coupon bonds, $P(t, T) = e^{\alpha_0 + \beta_0 X_t}$. Since $b_0 b_2 < 0$, we have

$$
c = \sqrt{b_1^2 - 4b_0 b_2} > b_1 > 0,
\tag{8.54}
$$

which results in

$$
g = \frac{b_1 + c}{b_1 - c} < 0.
$$

By examining Equation 8.48, we conclude that $\beta_0 < 0$, meaning that the zero-coupon bond is indeed a monotonically decreasing function of X_t under the square-root process for the state variable.

8.3   Pricing Options on Coupon Bonds

The payoff of an option on a coupon bond is

$$V_T = \left(\sum_{j \geq 1} \Delta T c P(T, T_j) + P(T, T_n) - K \right)^+ , \qquad (8.55)$$

where c is the coupon rate and K the strike price. Under a one-factor ATSM,

$$V_T = \left(\sum_{j \geq 1} \Delta T c e^{\alpha(T,T_i)+\beta(T,T_j)X_T} + e^{\alpha(T,T_n)+\beta(T,T_n)X_T} - K \right)^+ . \qquad (8.56)$$

Jamshidian (1989) suggests the following approach that decomposes the option on the coupon bond into a portfolio of options on zero-coupon bonds. In view of $P(T, T_j)$ being a monotonically decreasing function of X_T, one can solve for X^* such that

$$\sum_{j \geq 1} \Delta T c e^{\alpha(T,T_j)+\beta(T,T_j)X^*} + e^{\alpha(T,T_n)+\beta(T,T_n)X^*} - K = 0. \qquad (8.57)$$

Then, we define

$$K_j = e^{\alpha(T,T_j)+\beta(T,T_j)X^*}. \qquad (8.58)$$

Note that there is

$$K = \sum_{j \geq 1} \Delta T c K_j + K_n, \qquad (8.59)$$

and we substitute the right-hand side of Equation 8.59 for K in Equation 8.56. Owing to the monotonicity of $P(T, T_j)$, we have

$$V_T = \sum_{j=1}^{n-1} \Delta T c \left(e^{\alpha(T,T_j)+\beta(T,T_j)X_T} - K_i \right)^+ + (1 + \Delta T c)$$
$$\times \left(e^{\alpha(T,T_n)+\beta(T,T_n)X_T} - K_n \right)^+ . \qquad (8.60)$$

The price of the option is given by

$$V_t = \sum_{j=1}^{n-1} \Delta T c E_t^{\mathbb{Q}} \left[e^{-\int_t^T r_s \, ds} \left(e^{\alpha(T,T_j)+\beta(T,T_j)X_T} - K_i \right)^+ \right]$$
$$+ (1 + \Delta T c) E_t^{\mathbb{Q}} \left[e^{-\int_t^T r_s ds} \left(e^{\alpha(T,T_n)+\beta(T,T_n)X_T} - K_n \right)^+ \right], \qquad (8.61)$$

where $\mathbb{Q}$ stands for the risk-neutral measure. The terms under expectation are the discounted payoff of zero-coupon options, for which we already have a closed-form solution using the transform inversion formula. This approach, however, does not apply to multi-factor ATSMs.

8.4 Distributional Properties of Square-Root Processes

The square-root process, in Equation 8.36, is often used as the basic building block for one-factor or multi-factor ATSMs. It helps us to understand the distributional properties of the process, so that we can make proper parameterization in applications.

Lemma 8.4.1. *Suppose that $X_0 > 0, \kappa \geq 0$ and $\theta \geq 0$.*

1. *If*

$$\kappa\theta \geq \frac{1}{2}\sigma_0^2, \tag{8.62}$$

   then the SDE (Equation 8.36) admits a solution, X_t, that is strictly positive for all $t > 0$.

2. *If*

$$0 < \kappa\theta < \frac{1}{2}\sigma_0^2, \tag{8.63}$$

   then the SDE (Equation 8.36) admits a unique solution, which is non-negative but occasionally hits $X = 0$.

3. *If $\kappa\theta = 0$, the process, X_t, vanishes at a finite time and remains equal to zero thereafter.*

4. *If $\kappa\theta > 0$, then as $t \to \infty$, X_t has an asymptotic Gamma distribution with density function*

$$f(x) = \frac{\left(2\kappa/\sigma_0^2\right)^{\left(2\kappa\theta/\sigma_0^2\right)}}{\Gamma(2\kappa\theta/\sigma_0^2)} x^{\left(2\kappa\theta/\sigma_0^2\right)-1} \exp\left(-\frac{2\kappa}{\sigma_0^2}x\right), \tag{8.64}$$

   where

$$\Gamma(p) = \int_0^\infty x^{p-1} e^{-x}\, dx \tag{8.65}$$

   is the Gamma function.

For the proof of the above proposition, we refer to Feller (1971) or Ikeda and Watanabe (1989).

8.5 Multi-Factor Models

Under a multi-factor model, the short rate, $r(t)$, is an affine function of a vector of the unobserved state variable, $\mathbf{X}_t$, such that

$$r(t) = \delta_0 + \boldsymbol{\delta}^\mathrm{T}\mathbf{X}_t, \tag{8.66}$$

and $\mathbf{X}_t = \left(X_1(t), X_2(t), \dots, X_N(t)\right)^{\mathrm{T}}$ follows an "affine diffusion,"

$$\mathrm{d}\mathbf{X}_t = \boldsymbol{\mu}(\mathbf{X}_t, t)\,\mathrm{d}t + \boldsymbol{\sigma}\,(\mathbf{X}_t, t)\,\mathrm{d}\mathbf{W}_t, \tag{8.67}$$

with

$$\boldsymbol{\mu}(\mathbf{X}_t, t) = K_0(t) + K_1(t)\mathbf{X}_t,$$
$$\boldsymbol{\sigma}\,(\mathbf{X}_t, t)\,\boldsymbol{\sigma}^{\mathrm{T}}(\mathbf{X}_t, t) = H_0(t) + \sum_{i=1}^{N} X_i(t)H_i(t), \tag{8.68}$$

where in Equation 8.68, K_0 is an $N \times 1$ matrix, $K_1(t)$ and $H_i(t), i = 0, \dots, N$ the $N \times N$ matrices, and $\mathbf{W}_t$ an N-dimensional Brownian motion under the risk-neutral measure, $\mathbb{Q}$.

Suppose that $\boldsymbol{\sigma}\,(\mathbf{X}_t, t)$ is well defined. It is shown that (Duffie and Kan, 1996), similar to the one-factor model, zero-coupon bond prices are exponential affine functions of the form,

$$P(t, T) = \mathrm{e}^{\alpha(t,T)+\boldsymbol{\beta}^{\mathrm{T}}(t,T)\mathbf{X}_t}, \tag{8.69}$$

where $\alpha(t, T)$ is a scalar, $\boldsymbol{\beta}(t, T)$ a vector,

$$\boldsymbol{\beta}(t, T) = (\beta_1(t, T), \beta_2(t, T), \dots, \beta_N(t, T))^{\mathrm{T}}, \tag{8.70}$$

and they satisfy the following Riccarti equations

$$\boldsymbol{\beta}_t = \boldsymbol{\delta} - K_1^{\mathrm{T}}\boldsymbol{\beta} - \frac{1}{2}\boldsymbol{\beta}^{\mathrm{T}}H\boldsymbol{\beta}, \quad \boldsymbol{\beta}(T, T) = 0,$$
$$\alpha_t = \delta_0 - K_0^{\mathrm{T}}\boldsymbol{\beta} - \frac{1}{2}\boldsymbol{\beta}^{\mathrm{T}}H_0\boldsymbol{\beta}, \quad \alpha(T, T) = 0. \tag{8.71}$$

In Equation 8.71,

$$\boldsymbol{\beta}^{\mathrm{T}}H\boldsymbol{\beta} = \left(\boldsymbol{\beta}^{\mathrm{T}}H_1\boldsymbol{\beta}, \boldsymbol{\beta}^{\mathrm{T}}H_2\boldsymbol{\beta}, \dots, \boldsymbol{\beta}^{\mathrm{T}}H_N\boldsymbol{\beta}\right)^{\mathrm{T}}. \tag{8.72}$$

Equation 8.71 can at least be solved numerically by the Runge–Kutta method (Press et al., 1992). Once we have obtained $\boldsymbol{\beta}$, we calculate α through an integral:

$$\alpha(t, T) = \int_t^T \left[-\delta_0 + K_0^{\mathrm{T}}\boldsymbol{\beta}(s, T) + \frac{1}{2}\boldsymbol{\beta}^{\mathrm{T}}H_0\boldsymbol{\beta}(s, T)\right]\mathrm{d}s. \tag{8.73}$$

Under a multi-factor ATSM, the volatilities of forward rates can be state-dependent, and the forward rates can be guaranteed to be positive. In both academic literature and applications, a lot of attention has been given to the following subclass of the affine models (Duffie and Kan, 1996; Duffie and Singleton, 1999; Dai and Singleton, 2000; etc.):

$$\mathrm{d}\mathbf{X}_t = \mathcal{K}(\boldsymbol{\theta} - \mathbf{X}_t)\,\mathrm{d}t + \boldsymbol{\Sigma}\sqrt{V(t)}\,\mathrm{d}\mathbf{W}_t. \tag{8.74}$$

In Equation 8.74, $\boldsymbol{\theta}$ is an $N \times 1$ matrix, $\mathcal{K}$ and $\boldsymbol{\Sigma}$ the $N \times N$ matrices, and $V(t)$ a diagonal matrix, such that

$$V(t) = \text{diag}(V_i(t)) = \text{diag}\left(a_i + \mathbf{b}_i^{\mathrm{T}} \mathbf{X}_t\right). \tag{8.75}$$

In formalism, Equation 8.74 is a multi-dimensional version of the square-root process (1979), which carries the important features of mean reversion and state-dependent volatility. For the generalized CIR process, Equation 8.74, the coefficient matrices for the Equation 8.71 take the form

$$K_0 = \mathcal{K}\boldsymbol{\theta}, \quad K_1 = \mathcal{K}, \quad H_0 = \boldsymbol{\Sigma}\text{diag}(a_j)\boldsymbol{\Sigma}^{\mathrm{T}},$$

$$H_j = \boldsymbol{\Sigma}\text{diag}(b_{j,.})\boldsymbol{\Sigma}^{\mathrm{T}}, \quad j = 1, \ldots, N, \tag{8.76}$$

where $\text{diag}(a_i)$ stands for the diagonal matrix with diagonal elements a_j, $j = 1, \ldots, N$, and $\text{diag}(b_{j,.})$ the diagonal matrix with diagonal elements $b_{j,i}, i = 1, \ldots, N$. For later reference, we denote

$$\mathbf{a} = (a_1, a_2, \ldots, a_N)^{\mathrm{T}}, \quad \mathcal{B} = (\mathbf{b}_1, \mathbf{b}_2, \ldots, \mathbf{b}_N).$$

8.5.1 Admissible ATSMs

For a prescription of the multi-factor affine model, Equation 8.74, to be valid, one must guarantee that $V(t)$ stays non-negative all the time. Any affine model that guarantees $V(t) \geq 0$ is called an admissible model. For an admissible model, coefficient matrices of $\mathbf{X}_t$ should satisfy certain conditions. Denote the class of admissible ATSM of N state variables such that rank $(\mathcal{B}) = m$ by $\mathbb{A}_m(N)$. It is efficient to state these conditions only with some kind of standardized formalism of affine models, which is called the canonical representation and is defined below.

Definition 8.5.1. *(Canonical Representation of $\mathbb{A}_m(N)$). For each m, we partition $\mathbf{X}_t^{\mathrm{T}}$ as $\mathbf{X}_t^{\mathrm{T}} = ((\mathbf{X}_t^{\mathrm{B}})^{\mathrm{T}}, (\mathbf{X}_t^{\mathrm{D}})^{\mathrm{T}})$, where $\mathbf{X}_t^{\mathrm{B}}$ is $m \times 1$ and $\mathbf{X}_t^{\mathrm{D}}$ is $(N - m) \times 1$, and we define the canonical representation of $\mathbb{A}_m(N)$ as the special case of Equation 8.74-8.75 with*

$$\mathcal{K} = \begin{bmatrix} \mathcal{K}_{m \times m}^{BB} & 0_{m \times (N-m)} \\ \mathcal{K}_{(N-m) \times m}^{DB} & \mathcal{K}_{(N-m) \times (N-m)}^{BB} \end{bmatrix}, \tag{8.77}$$

for $m > 0$, and $\mathcal{K}$ is a lower triangular matrix for $m = 0$,

$$\boldsymbol{\theta} = \begin{pmatrix} \boldsymbol{\theta}_{m \times 1}^B \\ 0_{(N-m) \times 1} \end{pmatrix},$$

$$\boldsymbol{\Sigma} = I,$$

$$\mathbf{a} = \begin{pmatrix} 0_{m \times 1} \\ 1_{(N-m) \times 1} \end{pmatrix}, \tag{8.78}$$

$$\mathcal{B} = \begin{pmatrix} I_{m \times m} & B_{m \times (N-m)}^{BD} \\ 0_{(N-m) \times m} & 0_{(N-m) \times (N-m)} \end{pmatrix},$$

with the following parametric restrictions imposed:

$$\delta_i \geq 0, \qquad m+1 \leq i \leq N,$$

$$\mathcal{K}_i\boldsymbol{\theta} = \sum_{j=1}^{m} \mathcal{K}_{ij}\boldsymbol{\theta}_j > 0, \quad 1 \leq i \leq m,$$

$$\mathcal{K}_{ij} \leq 0, \qquad 1 \leq j \leq m, j \neq i, \tag{8.79}$$

$$\boldsymbol{\theta}_i \geq 0, \qquad 1 \leq i \leq m,$$

$$\mathcal{B}_{ij} \geq 0, \qquad 1 \leq i \leq m, \, m+1 \leq j \leq N.$$

It is shown that any ATSM in the class of $\mathbb{A}_m(N)$ can be transformed into the canonical form using the so-called invariant transformation (Dai and Singleton, 2000), which preserves admissibility and leaves the short rate unchanged. The canonical representation also allows us to produce specific models for production use. The justification for the canonical representation is found in Dai and Singleton (2000).

We remark here that the conditions for the canonical representation to be admissible are sufficient rather than necessary. But they are known to be the set of minimal conditions so far. Furthermore, the canonical representation is not unique. The representation from Equations 8.77 through 8.79 was chosen because, with it, it is relatively easier to identify some existing cases and verify their admissibility.

8.5.2   Three-Factor ATSMs

According to Litterman and Scheinkman (1991), the term structure of interest rates is essentially driven by three factors. Because of that, three-factor models have received more attention from both researchers and practitioners. A number of three-factor models have been proposed and subjected to empirical studies. In this section, we introduce four of these three-factor ATSMs, and identify their connections with the canonical representation introduced earlier.

$$\mathbb{A}_0(3)$$

If $m = 0$, then none of the $\mathbf{X}_t$'s affect the volatility of $\mathbf{X}_t$, and $\mathbf{X}_t$ follows a three-dimensional Gaussian diffusion. The coefficient sets of the canonical representation of $\mathbb{A}_0(3)$ are given by

$$\mathcal{K} = \begin{bmatrix} \kappa_{11} & & \\ \kappa_{21} & \kappa_{22} & \\ \kappa_{31} & \kappa_{32} & \kappa_{33} \end{bmatrix}, \quad \boldsymbol{\Sigma} = \begin{bmatrix} 1 & & \\ & 1 & \\ & & 1 \end{bmatrix}, \quad \boldsymbol{\theta} = \begin{bmatrix} 0 \\ 0 \\ 0 \end{bmatrix},$$

$$\mathbf{a} = \begin{pmatrix} 1 \\ 1 \\ 1 \end{pmatrix}, \quad \mathbf{b}_i = \mathbf{0}, \quad i = 1, 2, 3,$$

where $\kappa_{11} > 0, \kappa_{22} > 0$, and $\kappa_{33} > 0$.

$$\mathbb{A}_1(3)$$

One member of $\mathbb{A}_1(3)$ is the model by Balduzzi, Das, Foresi, and Sundaram (BDFS) (1996):

$$
\begin{aligned}
\mathrm{d}u(t) &= \mu(\bar{u} - u(t))\,\mathrm{d}t + \eta\sqrt{u(t)}\,\mathrm{d}W_u(t), \\
\mathrm{d}\theta(t) &= \nu(\bar{\theta} - \theta(t))\,\mathrm{d}t + \zeta\,\mathrm{d}W_\theta(t), \\
\mathrm{d}r(t) &= \kappa(\theta(t) - r(t))\,\mathrm{d}t + \sqrt{u(t)}\,\mathrm{d}W_r(t),
\end{aligned}
\tag{8.80}
$$

with only a non-zero diffusion correlation between W_u and W_r. It can be verified that Equation 8.80 is equivalent to the following affine representation:

$$
r(t) = \delta_0 + X_2(t) + X_3(t),
\tag{8.81}
$$

where $X_2(t)$ and $X_3(t)$ satisfy

$$
\mathrm{d}\begin{pmatrix} X_1(t) \\ X_2(t) \\ X_3(t) \end{pmatrix} = \begin{bmatrix} \kappa_{11} & & \\ & \kappa_{22} & \\ \kappa_{31} & & \kappa_{33} \end{bmatrix} \left[\begin{pmatrix} \theta_1 \\ 0 \\ 0 \end{pmatrix} - \begin{pmatrix} X_1(t) \\ X_2(t) \\ X_3(t) \end{pmatrix} \right] \mathrm{d}t
$$

$$
+ \begin{bmatrix} 1 & & \\ & 1 & \\ & & 1 \end{bmatrix} \begin{bmatrix} \sqrt{V_1(t)} & & \\ & \sqrt{V_2(t)} & \\ & & \sqrt{V_3(t)} \end{bmatrix} \mathrm{d}\mathbf{W}_t,
\tag{8.82}
$$

with

$$
\begin{aligned}
V_1(t) &= X_1(t), \\
V_2(t) &= b_{1,2}X_1(t), \\
V_3(t) &= a_3.
\end{aligned}
\tag{8.83}
$$

The model in Equation 8.80 is called the BDFS model for short. It is a short-rate model where the short rate is correlated with its stochastic volatility.

$$\mathbb{A}_2(3)$$

The $\mathbb{A}_2(3)$ family is characterized by the assumption that volatilities of $\mathbf{X}_t$ are determined by affine functions of two of the three X's. A member of this subfamily is the model proposed by Chen (1996):

$$
\begin{aligned}
\mathrm{d}v(t) &= \mu(v - v(t))\,\mathrm{d}t + \eta\sqrt{v(t)}\,\mathrm{d}W_1(t), \\
\mathrm{d}\theta(t) &= v(\theta - \theta(t))\,\mathrm{d}t + \zeta\sqrt{\theta(t)}\,\mathrm{d}W_2(t), \\
\mathrm{d}r(t) &= \kappa(\theta(t) - r(t))\,\mathrm{d}t + \sqrt{v(t)}\,\mathrm{d}W_3(t),
\end{aligned}
\tag{8.84}
$$

where the Brownian motions are assumed to be mutually independent. As in the BDFS model, v and θ are interpreted as the stochastic volatility and central tendency of $r(t)$, respectively. The equivalent affine representation of the model is given below:

$$r(t) = X_2(t) + X_3(t), \qquad (8.85)$$

where $X_i(t)$, $i = 1, 2, 3$ evolve according to

$$
d \begin{pmatrix} X_1(t) \\ X_2(t) \\ X_3(t) \end{pmatrix} = \begin{bmatrix} \kappa_{11} & & \\ & \kappa_{22} & \\ & & \kappa_{33} \end{bmatrix} \left[\begin{pmatrix} \theta_1 \\ \theta_2 \\ 0 \end{pmatrix} - \begin{pmatrix} X_1(t) \\ X_2(t) \\ X_3(t) \end{pmatrix} \right] dt
$$
$$
+ \begin{bmatrix} 1 & & \\ & 1 & \\ & & 1 \end{bmatrix} \begin{bmatrix} \sqrt{V_1(t)} & & \\ & \sqrt{V_2(t)} & \\ & & \sqrt{V_3(t)} \end{bmatrix} d\mathbf{W}_t, \qquad (8.86)
$$

with $\kappa_{11} > 0, \kappa_{22} > 0$ and

$$
\begin{aligned}
V_1(t) &= X_1(t), \\
V_2(t) &= X_2(t), \\
V_3(t) &= b_{1,3} X_1(t).
\end{aligned} \qquad (8.87)
$$

$$\mathbb{A}_3(3)$$

The final subfamily of the three-factor models has $m = 3$ so that all three Xs determine the volatility structure. The canonical representation of $\mathbb{A}_3(3)$ has parameters

$$
\mathcal{K} = \begin{bmatrix} \kappa_{11} & & \\ \kappa_{21} & \kappa_{22} & \\ \kappa_{21} & \kappa_{21} & \kappa_{33} \end{bmatrix}, \quad \Sigma = \begin{bmatrix} 1 & 0 & 0 \\ 0 & 1 & 0 \\ 0 & 0 & 1 \end{bmatrix}, \quad \boldsymbol{\theta} = \begin{pmatrix} \theta_1 \\ \theta_2 \\ \theta_3 \end{pmatrix},
$$
$$
a = \begin{pmatrix} 0 \\ 0 \\ 0 \end{pmatrix}, \quad \mathbf{b}_i = \mathbf{e}_i, \quad i = 1, 2, 3, \qquad (8.88)
$$

where $\kappa_{ii} > 0$ and $\kappa_{ij} \leq 0$ for $i \neq j$.

With both Σ and $\mathcal{B}$ equal to identity matrices, the diffusion term of this model is identical to that of the three-factor CIR processes:

$$
\begin{aligned}
d\mathbf{X}_t &= K(\boldsymbol{\theta} - \mathbf{X}_t)\, dt + \Sigma \sqrt{V(t)}\, d\mathbf{W}_t, \\
\mathbf{X}_0 &\geq 0,
\end{aligned} \qquad (8.89)
$$

where $\boldsymbol{\theta} = (\theta_i)_{3 \times 1}$, and

$$
K = \text{diag}(k_i), \quad \Sigma \sqrt{V(t)} = \text{diag}(\sigma_i \sqrt{X_i}), \quad i = 1, 2, 3. \qquad (8.90)
$$

We finish this section with some remarks. Frequently, the general multi-factor affine model can be viewed as a blending of the Vasicek and CIR forms.

The conditions from Equations 8.77 through 8.79 are also restrictions in multi-factor term structure modeling with ATSMs. As a matter of fact, the CIR form perhaps offers the greatest flexibility in specifying the volatility dynamics of bond prices. However, this flexibility comes at a cost. The parameter restrictions for ensuring that Equation 8.67 provides a valid description of factor variances impose substantial restrictions on the permissible correlations among the factors. As is stated in Equations 8.89 and 8.90, in the extreme case of the pure multi-factor CIR model, the factors must be uncorrelated to ensure an admissible volatility specification.

8.6 Pricing SOFR Futures under ATSMs

Under the multi-factor ATSM, the volatility function of zero-coupon bonds are

$$\Sigma(t,T) = \boldsymbol{\sigma}^{\mathrm{T}}(X_t, t)\nabla_X P(t,T) = \boldsymbol{\sigma}^{\mathrm{T}}(X_t, t)\boldsymbol{\beta},$$

where ∇_X stands for gradient with respect to X_t. Consider the pricing of futures rates for 3m SOFR futures:

$$\tilde{f}_j(t) = E_t^Q \left[f_j(T_j) \right] = E_t^Q \left[\frac{1}{\Delta T_j} \left(\frac{P(T_j, T_{j-1})}{P(T_j, T_j)} - 1 \right) \right]. \tag{8.91}$$

There is

$$\frac{P(T_j, T_{j-1})}{P(T_j, T_j)} = \frac{P(t, T_{j-1})}{P(t, T_j)} e^{\int_t^{T_j} -\frac{1}{2}\|\Sigma(s, T_{j-1}, T_j)\|^2 ds + \Sigma(s, T_{j-1}, T_j) \cdot d\mathbf{W}_s^{(j)}}. \tag{8.92}$$

Taking expectation under the risk-neutral measure conditional on $\mathcal{F}_t$, we have

$$\begin{aligned}
E_t^Q \left[\frac{P(T_j, T_{j-1})}{P(T_j, T_j)} \right] &= \frac{P(t, T_{j-1})}{P(t, T_j)} \\
&\times E_t^Q \left[e^{\int_t^{T_j} -\frac{1}{2}\|\Sigma(s, T_{j-1}, T_j)\|^2 ds + \Sigma(s, T_{j-1}, T_j) \cdot (d\mathbf{W}_s - \Sigma(s, T_j) ds)} \right].
\end{aligned} \tag{8.93}$$

Define a new measure $\mathbb{Q}_{j-1,j}$ such that

$$\left. \frac{d\mathbb{Q}_{j-1,j}}{d\mathbb{Q}} \right|_{\mathcal{F}_t} = e^{\int_0^t -\frac{1}{2}\|\Sigma(s, T_{j-1}, T_j)\|^2 ds + \Sigma(s, T_{j-1}, T_j) \cdot d\mathbf{W}_s}.$$

We then have

$$E_t^{Q_j} \left[\frac{P(T_j, T_{j-1})}{P(T_j, T_j)} \right] = \frac{P(t, T_{j-1})}{P(t, T_j)} E_t^{Q_{j-1,j}} \left[e^{\int_t^{T_j} -\Sigma(s, T_{j-1}, T_j) \cdot \Sigma(s, T_j) ds} \right]. \tag{8.94}$$

Next, we show that the integrand in Equation 8.94 is an affine function of X_t. For notational simplicity, we denote $\beta(t, T_{j-1}, T_j) = \beta(t, T_{j-1}) - \beta(t, T_j)$. We have

$$- \Sigma(t, T_{j-1}, T_j) \cdot \Sigma(t, T_j) = -\beta^{\mathrm{T}}(t, T_{j-1}, T_j)\boldsymbol{\sigma}\boldsymbol{\sigma}^{\mathrm{T}}\beta(t, T_j)$$

$$= -\beta^{\mathrm{T}}(t, T_{j-1}, T_j) \left(H_0(t) + \sum_{i=1}^{N} X_i(t)H_i(t) \right) \beta(t, T_j)$$

$$= -\beta^{\mathrm{T}}(t, T_{j-1}, T_j)H_0(t)\beta(t, T_j) - \sum_{i=1}^{N} X_i(t)\beta^{\mathrm{T}}(t, T_{j-1}, T_j)H_i(t)\beta(t, T_j)$$

$$\overset{\triangle}{=} \phi(t, T_{j-1}, T_j) + \boldsymbol{\Psi}^{\mathrm{T}}(t, T_{j-1}, T_j)\mathbf{X}_t,$$

where

$$\boldsymbol{\Psi}(t, T_{j-1}, T_j) = - \begin{pmatrix} \beta^{\mathrm{T}}(t, T_{j-1}, T_j)H_1(t)\beta(t, T_j) \\ \beta^{\mathrm{T}}(t, T_{j-1}, T_j)H_2(t)\beta(t, T_j) \\ \vdots \\ \beta^{\mathrm{T}}(t, T_{j-1}, T_j)H_N(t)\beta(t, T_j) \end{pmatrix}$$

Under $\mathbb{Q}_{j-1,j}$, the dynamics of X_t is

$$d\mathbf{X}_t = \boldsymbol{\mu}(\mathbf{X}_t, t)\,dt + \boldsymbol{\sigma}\,(\mathbf{X}_t, t)\,d\mathbf{W}_t,$$

$$= \boldsymbol{\mu}(\mathbf{X}_t, t)\,dt + \boldsymbol{\sigma}\,(\mathbf{X}_t, t)\left(d\mathbf{W}_t^{(j-1,j)} + \Sigma(t, T_{j-1}, T_j)dt\right)$$

$$= [\boldsymbol{\mu}(\mathbf{X}_t, t) + \boldsymbol{\sigma}(\mathbf{X}_t, t)\Sigma(t, T_{j-1}, T_j)]\,dt + \boldsymbol{\sigma}(\mathbf{X}_t, t)d\mathbf{W}_t^{(j-1,j)}$$

$$= [\boldsymbol{\mu}(\mathbf{X}_t, t) + \boldsymbol{\sigma}(\mathbf{X}_t, t)\boldsymbol{\sigma}^{\mathrm{T}}(\mathbf{X}_t, t)\beta(t, T_{j-1}, T_j)]\,dt + \boldsymbol{\sigma}(\mathbf{X}_t, t)d\mathbf{W}_t^{(j-1,j)}$$

$$\overset{\triangle}{=} \tilde{\boldsymbol{\mu}}(\mathbf{X}_t, t)dt + \boldsymbol{\sigma}(\mathbf{X}_t, t)d\mathbf{W}_t^{(j-1,j)}.$$

where $\mathbf{W}_t^{(j-1,j)}$ is a Brownian motion under $\mathbb{Q}_{j-1,j}$,

$$\tilde{\boldsymbol{\mu}}(\mathbf{X}_t, t) = [K_0 + H_0\beta(t, T_{j-1}, T_j)] + (H_1, H_2, \ldots, H_N) \otimes \beta(t, T_{j-1}, T_j)\mathbf{X}_t$$

$$\overset{\triangle}{=} \tilde{K}_0 + \tilde{K}_1\mathbf{X}_t,$$

and $\otimes$ is the Kronecker product. Similar to the proof of Lemma 8.1.1 we can establish the following results.

Lemma 8.6.1. *Suppose that H_0 and $H_i, i = 1, \ldots, N$ are bounded and there is*

$$E^{\mathbb{Q}}\left[\left(\int_0^T \|\mathbf{X}_t\|\,dt\right)\right] < \infty. \tag{8.95}$$

Then, we have

$$E_t^{\mathbb{Q}_{j-1,j}}\left[e^{\int_t^{T_j} -\Sigma(s, T_{j-1}, T_j) \cdot \Sigma(s, T_j)ds}\right] = e^{\tilde{\alpha}(t, T_{j-1}, T_j) + \tilde{\beta}(t, T_{j-1}, T_j)\mathbf{X}_t}, \tag{8.96}$$

where $\tilde{\alpha}$ and $\tilde{\beta}$ are bounded solutions to the following ordinary differential equations (ODE):

$$\tilde{\boldsymbol{\beta}}_t = -\boldsymbol{\Psi} - \tilde{K}_1^{\mathrm{T}}\tilde{\boldsymbol{\beta}} - \frac{1}{2}\tilde{\boldsymbol{\beta}}^{\mathrm{T}}H\tilde{\boldsymbol{\beta}}, \quad \tilde{\boldsymbol{\beta}}(T,T) = 0,$$

$$\tilde{\alpha}_t = -\phi - \tilde{K}_0^{\mathrm{T}}\tilde{\boldsymbol{\beta}} - \frac{1}{2}\tilde{\boldsymbol{\beta}}^{\mathrm{T}}H_0\tilde{\boldsymbol{\beta}}, \quad \tilde{\alpha}(T,T) = 0,$$

(8.97)

where $\tilde{\boldsymbol{\beta}}^{\mathrm{T}}H\tilde{\boldsymbol{\beta}}$ is defined according to Equation 8.72. □

Based on the last lemma, we arrive at the following correlation adjustment between futures and forward rates under general ATSM:

$$\tilde{f}_j(t) - f_j(t) = (1 + \Delta T_j f_j(t))\frac{e^{\tilde{\alpha}(t,T_{j-1},T_j)+\tilde{\beta}(t,T_{j-1},T_j)\mathbf{X}_t} - 1}{\Delta T_j}.$$

(8.98)

We finish this section with the performance of the ATSM model for futures pricing. For pricing SOFR futures and SOFR options on futures, the (parameters of the) ATSM must first be estimated using the historical futures data. With the three-factor CIR model, Equation 8.84, we estimate the model parameters from the futures' historical data in the window of the trailing twelve months (TTM), using the extended Kalman filter maximum likelihood method (see e.g., Skov and Skovmand, 2021). The initial TTM window starts from 1/6/2018 to 5/31/2019, and ends from 1/2/2020 to 12/31/2020. Starting from 6/1/2019, we use the estimated model to price 3m SOFR futures of the nearest four maturities, using Equation 8.98. The model-implied 3m futures rates vs. the realized 3m futures rates are demonstrated in Figure 8.1 (Xia and Wu, 2024). As one can see from various sub-figures, the model is able to price the futures with very high precision, thus making the ATSM model a very promising choice for production use.

8.7 Swaption Pricing under ATSMs

In Sections 8.1 and 8.2, we have studied the pricing of options on bonds under the ATSMs. In this section, we study the pricing of options on interest rates. Without loss of generality, let us focus on the pricing of swaptions, or options on swap rates, under ATSMs, as similar approaches apply to the pricing of options on other interest rates.

As was already determined in Chapter 6, the payoff of a swaption can be cast into[3]

$$V_{T_m} = A_{m,n}(T_m)\left(R_{m,n}(T_m) - k\right)^+,$$

(8.99)

---

[3]For notational simplicity, we in this chapter drop ˜ over $T_j, j = m+1, \ldots, n$.

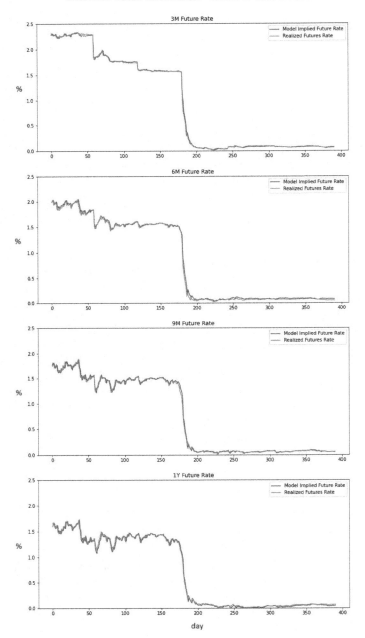

FIGURE 8.1: Model-implied 3m futures rate vs. realized 3m futures rate

where $A_{m,n}(t)$ is the annuity,

$$A_{m,n}(t) = \sum_{j=m}^{n-1} \Delta T_j P(t, T_{j+1}), \quad t \leq T_m, \tag{8.100}$$

and $R_{m,n}(t)$ is the prevailing swap rate at time t, given by

$$R_{m,n}(t) = \frac{P(t,T_m) - P(t,T_n)}{\sum_{j=m}^{n-1} \Delta T_j P(t,T_{j+1})}.$$ (8.101)

Under an ATSM, the price of a zero-coupon bond is given by

$$P(t,T) = \exp\left(\alpha(t,T) + \boldsymbol{\beta}^{\mathrm{T}}(t,T)\mathbf{X}_t\right),$$ (8.102)

where $\mathbf{X}_t$ evolves according to Equation 8.67, and $\alpha(t,T)$ and $\boldsymbol{\beta}(t,T)$ satisfy Equation 8.71.

Schrager and Pelsser (2006) take an approach parallel to the swaption pricing under the market model with a stochastic volatility approach (Wu, 2002). The price of the swaption is given by

$$V_0 = A_{m,n}(0)E^{\mathbb{Q}_{m,n}}\left[(R_{m,n}(T_m) - k)^+ \mid \mathcal{F}_0\right],$$ (8.103)

where $\mathbb{Q}_{m,n}$ stands for the forward swap measure, defined by

$$\left.\frac{\mathrm{d}\mathbb{Q}_{m,n}}{\mathrm{d}\mathbb{Q}}\right|_{\mathcal{F}_t} = \frac{A_{m,n}(t)/A_{m,n}(0)}{B(t)/B(0)}$$

$$= \frac{1}{A_{m,n}(0)} \sum_{j=m}^{n-1} \Delta T_j \frac{P(t,T_{j+1})}{B(t)} \triangleq m(t), \quad t \leq T_m.$$ (8.104)

Note that $P(t,T_{j+1})/B(t)$ is a $\mathbb{Q}$-martingale, following the process

$$\mathrm{d}\left(\frac{P(t,T_{j+1})}{B(t)}\right) = \left(\frac{P(t,T_{j+1})}{B(t)}\right)\boldsymbol{\beta}^{\mathrm{T}}(t,T_{j+1})\boldsymbol{\Sigma}\sqrt{\mathbf{V}(t)}\,\mathrm{d}\mathbf{W}_t.$$ (8.105)

The dynamics for the Radon–Nikodym derivative then is

$$\mathrm{d}m(t) = \frac{1}{A_{m,n}(0)} \sum_{j=m}^{n-1} \Delta T_j \mathrm{d}\left(\frac{P(t,T_{j+1})}{B(t)}\right)$$

$$= \frac{1}{A_{m,n}(0)} \left[\sum_{j=m}^{n-1} \Delta T_j \left(\frac{P(t,T_{j+1})}{B(t)}\right)\boldsymbol{\beta}^{\mathrm{T}}(t,T_{j+1})\right]\boldsymbol{\Sigma}\sqrt{\mathbf{V}(t)}\,\mathrm{d}\mathbf{W}_t,$$ (8.106)

and it follows that

$$\frac{\mathrm{d}m(t)}{m(t)} = \frac{B(t)}{A_{m,n}(t)} \sum_{j=m}^{n-1} \Delta T_j \mathrm{d}\left(\frac{P(t,T_{j+1})}{B(t)}\right)$$

$$= \left(\sum_{j=m}^{n-1} \Delta T_j \left(\frac{P(t,T_{j+1})}{A_{m,n}(t)}\right)\boldsymbol{\beta}^{\mathrm{T}}(t,T_{j+1})\right)\boldsymbol{\Sigma}\sqrt{\mathbf{V}(t)}\,\mathrm{d}\mathbf{W}_t$$

$$= \left(\sum_{j=m}^{n-1} \alpha_j \boldsymbol{\beta}^{\mathrm{T}}(t,T_{j+1})\right)\boldsymbol{\Sigma}\sqrt{\mathbf{V}(t)}\,\mathrm{d}\mathbf{W}_t.$$ (8.107)

Define a $\mathbb{Q}_{m,n}$-Brownian motion by

$$
\begin{aligned}
d\mathbf{W}_t^{(m,n)} &= d\mathbf{W}_t - \left\langle d\mathbf{W}_t, \frac{dm(t)}{m(t)} \right\rangle \\
&= d\mathbf{W}_t - \sqrt{\mathbf{V}(t)}\boldsymbol{\Sigma}^{\mathrm{T}} \left(\sum_{j=m}^{n-1} \alpha_j \boldsymbol{\beta}(t, T_{j+1}) \right) dt.
\end{aligned} \tag{8.108}
$$

Then, under $\mathbb{Q}_{m,n}$, the process for $\mathbf{X}_t$ becomes

$$
\begin{aligned}
d\mathbf{X}_t &= \mathcal{K}(\boldsymbol{\theta} - \mathbf{X}_t)\,dt + \boldsymbol{\Sigma}\sqrt{\mathbf{V}(t)} \left(d\mathbf{W}_t^{(m,n)} + \sqrt{\mathbf{V}(t)}\boldsymbol{\Sigma}^{\mathrm{T}} \right. \\
&\quad \times \left. \left(\sum_{j=m}^{n-1} \alpha_j(t)\boldsymbol{\beta}(t, T_{j+1}) \right) dt \right) \\
&= \left(\mathcal{K}(\boldsymbol{\theta} - \mathbf{X}_t) + \boldsymbol{\Sigma}V(t)\boldsymbol{\Sigma}^{\mathrm{T}} \left(\sum_{j=m}^{n-1} \alpha_j(t)\boldsymbol{\beta}(t, T_{j+1}) \right) \right) dt \\
&\quad + \boldsymbol{\Sigma}\sqrt{\mathbf{V}(t)}\,d\mathbf{W}_t^{(m,n)}.
\end{aligned} \tag{8.109}
$$

The process for the swap rate is

$$
\begin{aligned}
dR_{m,n}(t) &= \mathrm{drift} + \sum_{j=m}^{n} \frac{\partial R_{m,n}(t)}{\partial P(t, T_j)} dP(t, T_j) \\
&= \mathrm{drift} + \frac{dP(t, T_m)}{A_{m,n}(t)} - \frac{R_{m,n}(t)}{A_{m,n}(t)} \sum_{j=m}^{n-1} \Delta T_j dP(t, T_{j+1}) - \frac{dP(t, T_n)}{A_{m,n}(t)} \\
&= \left(\sum_{j=m}^{n} \gamma_j(t)\boldsymbol{\beta}^{\mathrm{T}}(t, T_j) \right) \boldsymbol{\Sigma}\sqrt{\mathbf{V}_t}\,d\mathbf{W}_t^{(m,n)},
\end{aligned} \tag{8.110}
$$

where

$$
\begin{aligned}
\gamma_m(t) &= \frac{P(t, T_m)}{A_{m,n}(t)}, \\
\gamma_j(t) &= R_{m,n}(t)\alpha_j, \quad j = m+1, \ldots, n-1, \\
\gamma_n(t) &= \left(R_{m,n}(t) - \frac{1}{\Delta T_{n-1}} \right) \alpha_n.
\end{aligned}
$$

Consider the frozen coefficient approximation to the processes of $R_{m,n}(t)$ and

$\mathbf{X}_t$:

$$dR_{m,n}(t) = \left(\sum_{j=m}^{n} \gamma_j(0)\boldsymbol{\beta}^{\mathrm{T}}(t, T_j) \right) \boldsymbol{\Sigma}\sqrt{\mathbf{V}_t}\, d\mathbf{W}_t^{(m,n)},$$

$$d\mathbf{X}_t = \left(\mathcal{K}\left(\boldsymbol{\theta} - \mathbf{X}_t \right) + \boldsymbol{\Sigma}V(t)\boldsymbol{\Sigma}^{\mathrm{T}} \left(\sum_{j=m}^{n-1} \alpha_j(0)\boldsymbol{\beta}(t, T_{j+1}) \right) \right) dt$$

$$+ \boldsymbol{\Sigma}\sqrt{\mathbf{V}(t)}\, d\mathbf{W}_t^{(m,n)}. \tag{8.111}$$

For notational simplicity, we let

$$\boldsymbol{\beta}_{m,n}^{\alpha}(t) = \sum_{j=m}^{n-1} \alpha_j(0)\boldsymbol{\beta}(t, T_{j+1}),$$

$$\boldsymbol{\beta}_{m,n}^{\gamma}(t) = \sum_{j=m}^{n} \gamma_j(0)\boldsymbol{\beta}(t, T_j). \tag{8.112}$$

Noticing that

$$\boldsymbol{\Sigma}V(t)\boldsymbol{\Sigma}^{\mathrm{T}}\boldsymbol{\beta}_{m,n}^{\alpha}(t)$$

$$= \boldsymbol{\Sigma}\left(\mathrm{diag}(a_i) + \mathrm{diag}(\mathbf{b}_i^{\mathrm{T}}\mathbf{X}_t) \right) \boldsymbol{\Sigma}^{\mathrm{T}}\boldsymbol{\beta}_{m,n}^{\alpha}(t)$$

$$= \boldsymbol{\Sigma}\mathrm{diag}(a_i)\boldsymbol{\Sigma}^{\mathrm{T}}\boldsymbol{\beta}_{m,n}^{\alpha}(t) + \boldsymbol{\Sigma}\mathrm{diag}(\mathbf{b}_i^{\mathrm{T}}\mathbf{X}_t)\boldsymbol{\Sigma}^{\mathrm{T}}\boldsymbol{\beta}_{m,n}^{\alpha}(t)$$

$$= \boldsymbol{\Sigma}A(t)\boldsymbol{\Sigma}^{\mathrm{T}}\boldsymbol{\beta}_{m,n}^{\alpha}(t) + \boldsymbol{\Sigma}\mathrm{diag}\big(\boldsymbol{\Sigma}^{\mathrm{T}}\boldsymbol{\beta}_{m,n}^{\alpha}(t)\big)\, B^{\mathrm{T}}(t)\mathbf{X}_t$$

$$\triangleq \boldsymbol{\xi}_t + \Phi_t\mathbf{X}_t, \tag{8.113}$$

we thus rewrite Equation 8.111 into

$$dR_{m,n}(t) = (\boldsymbol{\beta}_{m,n}^{\gamma}(t))^{\mathrm{T}}\boldsymbol{\Sigma}\sqrt{\mathbf{V}_t}\, d\mathbf{W}_t^{(m,n)},$$

$$d\mathbf{X}_t = \left(\mathcal{K}\left(\boldsymbol{\theta} - \mathbf{X}_t \right) + \boldsymbol{\xi}_t + \Phi_t\mathbf{X}_t \right) dt + \boldsymbol{\Sigma}\sqrt{\mathbf{V}(t)}\, d\mathbf{W}_t^{(m,n)}$$

$$= \left(\mathcal{K}\boldsymbol{\theta} + \boldsymbol{\xi}_t - \left(\mathcal{K} - \Phi_t \right)\mathbf{X}_t \right) dt + \boldsymbol{\Sigma}\sqrt{\mathbf{V}(t)}\, d\mathbf{W}_t^{(m,n)}. \tag{8.114}$$

Stacking the swap rate with the state variables:

$$\tilde{\mathbf{X}}_t = \begin{pmatrix} R_{m,n}(t) \\ \mathbf{X}_t \end{pmatrix},$$

and defining

$$\tilde{\mathcal{K}} = \begin{pmatrix} 0 & \\ & \mathcal{K} - \Phi \end{pmatrix}, \quad \tilde{\boldsymbol{\theta}} = \begin{pmatrix} 0 \\ (\mathcal{K} - \Phi)^{-1}(\mathcal{K}\boldsymbol{\theta} + \boldsymbol{\xi}_t) \end{pmatrix}, \quad \tilde{\boldsymbol{\Sigma}} = \begin{pmatrix} (\boldsymbol{\beta}_{m,n}^{\gamma})^{\mathrm{T}} \\ \boldsymbol{\Sigma} \end{pmatrix},$$

we finally end up with a succinct expression for the joint process

$$d\tilde{\mathbf{X}}_t = \tilde{\mathcal{K}}\left(\tilde{\boldsymbol{\theta}} - \tilde{\mathbf{X}}_t \right) dt + \tilde{\boldsymbol{\Sigma}}\sqrt{\mathbf{V}(t)}\, d\mathbf{W}_t^{(m,n)}. \tag{8.115}$$

Parallel to the proof of Lemma 8.1.1, we can show that the moment-generating function for $\tilde{\mathbf{X}}_t$ is

$$\varphi(\mathbf{u}; \tilde{\mathbf{X}}_t, t, T) = E^{\mathbb{Q}_{m,n}}\left[e^{\mathbf{u}^{\mathrm{T}}\tilde{\mathbf{X}}_T} \mid \mathcal{F}_t\right]$$

$$= \exp\left(\tilde{\alpha}_{\mathbf{u}}(t, T) + \tilde{\boldsymbol{\beta}}_{\mathbf{u}}^{\mathrm{T}}(t, T)\tilde{\mathbf{X}}_t\right), \qquad (8.116)$$

where $\tilde{\alpha}$ and $\tilde{\boldsymbol{\beta}}$ satisfy

$$\begin{aligned}
\left(\tilde{\boldsymbol{\beta}}_{\mathbf{u}}\right)_t &= \mathcal{K}^{\mathrm{T}}\tilde{\boldsymbol{\beta}}_{\mathbf{u}} - \frac{1}{2}\tilde{\boldsymbol{\beta}}_{\mathbf{u}}^{\mathrm{T}}H\tilde{\boldsymbol{\beta}}_{\mathbf{u}}, \quad \tilde{\boldsymbol{\beta}}_{\mathbf{u}}(T, T) = \mathbf{u} \\
(\tilde{\alpha}_{\mathbf{u}})_t &= -\left(\tilde{\mathcal{K}}\tilde{\boldsymbol{\theta}}\right)^{\mathrm{T}}\tilde{\boldsymbol{\beta}}_{\mathbf{u}} - \frac{1}{2}\tilde{\boldsymbol{\beta}}_{\mathbf{u}}^{\mathrm{T}}H_0\tilde{\boldsymbol{\beta}}_{\mathbf{u}}, \quad \tilde{\alpha}_{\mathbf{u}}(T, T) = 0.
\end{aligned} \qquad (8.117)$$

Here H_0 and H are defined in Equation 8.76. When we take, in particular, $\mathbf{u} = u\mathbf{e}_1$, where $\mathbf{e}_1 = (1, 0, \ldots, 0)^{\mathrm{T}}$, $\varphi(\mathbf{u}; \tilde{\mathbf{X}}_t, t, T)$ becomes the moment-generating function for $R_{m,n}(T)$ only. Denote

$$G(k) = E_t^{\mathbb{Q}_{m,n}}\left[(R_{m,n}(T) - k)^+\right], \quad \forall k. \qquad (8.118)$$

The Laplace transform of $G(k)$ with respect to k is defined by

$$\begin{aligned}
\int_{-\infty}^{\infty} e^{(a+iu)k} G(k)\, \mathrm{d}k &= \int_{-\infty}^{\infty} e^{(a+iu)k} E_t^{\mathbb{Q}_{m,n}}\left[(R_{m,n}(T) - k)^+\right] \mathrm{d}k \\
&= E_t^{\mathbb{Q}_{m,n}}\left[\int_{-\infty}^{\infty} e^{(a+iu)k}(R_{m,n}(T) - k)^+\, \mathrm{d}k\right] \\
&= E_t^{\mathbb{Q}_{m,n}}\left[\int_{-\infty}^{R_{m,n}(T)} e^{(a+iu)k}(R_{m,n}(T) - k)^+\, \mathrm{d}k\right] \\
&= E_t^{\mathbb{Q}_{m,n}}\left[\frac{e^{(a+iu)R_{m,n}(T)}}{(a+iu)^2}\right] \\
&= \frac{\varphi\left((a+iu)\mathbf{e}_1; \tilde{\mathbf{X}}_t, t, T\right)}{(a+iu)^2} \\
&\triangleq \psi(u), \qquad (8.119)
\end{aligned}$$

where a is a positive constant. The value of the swaption can then be obtained from an inverse Laplace transform:

$$V_t = A_{m,n}(t)\frac{e^{-ak}}{\pi}\int_0^{+\infty} \exp(-iuk)\,\psi(u)\, \mathrm{d}u. \qquad (8.120)$$

Schrager and Pelsser (2006) implemented the above method with a two-factor CIR model of the form

$$\begin{aligned}
r_t &= \delta + X_1(t) + X_2(t), \\
\mathrm{d}X_i(t) &= \kappa_i(\theta_i - X_i(t))\, \mathrm{d}t + \sigma_i\sqrt{X_i(t)}\, \mathrm{d}W_i(t), \qquad (8.121) \\
X_i(0) &= x_i, \quad i = 1, 2,
\end{aligned}$$

where $W_i(t)$, $i = 1, 2$ are independent Brownian motions under the risk-neutral measure, $\mathbb{Q}$, and the coefficients and initial conditions are taken as

$$\kappa_1 = 0.2, \quad \kappa_2 = 0.2,$$
$$\theta_1 = 0.03, \quad \theta_2 = 0.01,$$
$$\sigma_1 = 0.03, \quad \sigma_2 = 0.01,$$
$$\delta = 0.02,$$
$$x_1 = 0.04, \quad x_2 = 0.02.$$

For comparison, Schrager and Pelsser computed swaption prices by both the transformation method and the standard Monte Carlo simulation method. The pricing results are presented in Table 8.1, where prices (in basis points) and implied Black's volatilities are listed. Also provided in the table are differences in prices and differences in implied volatilities obtained by the two numerical methods. From the results, we can see that the approximation method is very accurate. The entire calculation takes about 0.14 s. In general, swaption pricing under ATSMs can be implemented almost instantly, which is another advantage of the model.

8.8 Notes

Whenever feasible, users of ATSMs price options by the transformation method. Numerically, this is realized through FFT. When it comes to model calibration, however, ATSMs are not easy to handle, particularly when the coefficients of the models are time-dependent. Note that for a reasonably good fit of both interest-rate caps and swaptions, we need to assume the time dependency of the model coefficients. As a matter of fact, calibration of affine models driven by the popular square-root processes is an outstanding challenge in financial engineering.

To price non-vanilla options, we may have to resort to Monte Carlo simulations. The numerical simulation of the square-root processes has attracted interest in recent years. An easy solution is to treat the square-root process as an addition of a number of squared Gaussian processes, so the time stepping of the square-root process is achieved by advancing those Gaussian processes (Roger, 1995). Several delicate new approaches were proposed recently by Andersen (2007).

Recent progress in the area of affine models is represented by the affine LIBOR model in Keller-Ressel *et al.* (2013). This model follows in the footsteps of the forward price model, but it guarantees non-negative interest rates by choosing positive driving processes, whose moment-generating function is used to define forward prices. Under the model, caplets and swaptions can be priced in closed-form – through the fast Fourier transform. Yet for the swaptions,

TABLE 8.1: Accuracy of the Approximation Method

Swap Tenor	Option Maturity			
	1		5	
ATMF				
1	25.34	9.5%	28.88	7.3%
	(−0.01)	(0.00%)	(−0.01)	(0.00%)
5	78.98	7.2%	90.61	5.3%
	(−0.02)	(0.00%)	(−0.05)	(0.00%)
10	99.82	5.6%	114.55	4.0%
	(−0.04)	(0.00%)	(−0.10)	(0.00%)
ITM				
1	101.05	(9.6%)	72.35	7.2%
	(0.03)	(0.06%)	(0.08)	(−0.04%)
5	411.03	(7.1%)	293.35	5.1%
	(0.06)	(0.08%)	(0.26)	(0.05%)
10	675.49	(5.1%)	487.83	3.6%
	(0.02)	(0.29%)	(0.24)	(0.07%)
OTM				
1	2.07	9.4%	8.43	7.3%
	(−0.05)	(−0.05%)	(−0.10)	(−0.04%)
5	2.27	7.3%	15.74	5.5%
	(−0.13)	−0.07%	(−0.41)	(−0.04%)
10	0.65	5.7%	9.73	4.2%
	(−0.10)	(−0.10%)	(−0.71)	(−0.07%)

Reprinted from Schrager DF and Pelsser AJ, 2006. *Mathematical Finance* 16(4): 673–694. With permission. The ITM and OTM are set at the levels of 85% and 115% relative to the ATM swap rate.

it encounters the curse of dimensionality: the layers of Fourier transforms increase with the underlying swap tenor, which limits its applications.

Exercises

1. Prove that if, in a one-factor ATSM, X_t follows a Gaussian process, then the affine model reduces to the Hull–White model.

2. Use the Laplace transform to price a put option on a zero-coupon bond under the ATSM (Hint: choose the damping parameter properly).

3. The value of a call option can be expressed as

$$V_0(k) = P(0, T) \int_0^\infty q(s)(s - k)^+ \mathrm{d}s,$$

where $q(s)$ is the density function of the state variable at the option's maturity, T. Let $a > 0$. Prove that the Laplace transform of the integral is

$$\int_{-\infty}^{\infty} e^{(a+iu)k} \left(\int_0^{\infty} q(s)(s - k)^+ ds \right) dk = \varphi(u) \times (a + iu)^{-2},$$

where $\varphi(u)$ and $(a+iu)^{-2}$ are the Laplace transforms of $q(s)$ and $(s-k)^+$, respectively.

4. Price an option on a zero-coupon bond with an equilibrium CIR model. The characteristics of the option are

   a. maturity of the option: two years;

   b. maturity of the bond: five years (so in two years, it will become a three-year maturity bond).

Assume, under the risk-neutral measure, $\kappa = 1, \theta = 0.05, \sigma = 0.1, \delta_0 = 0, \delta_1 = 1$, and $X_0 = \theta$. Price the option across strikes: $K = [0 : 0.1 : 1]$. You will need a numerical integration scheme to compute the transform inversions. Moreover, once you have obtained prices, calculate and plot Black's implied volatilities.

Chapter 9

Market Model for SOFR Derivatives

Market model plays a special role in both theory and practice for derivatives pricing. It has been applied to almost all asset classes: equity, forex, commodity, credit, inflation and fixed income. For a long time after its introduction in 1976 (Black, 1976), it had served as a major model for both pricing and hedging. Yet, hedging according to the Black model failed badly during the equity market crash in the Black Monday of 1987 and the emerging market debt crisis in 1998, suggesting that the Black model is seriously flawed. After each of these the crises, the relatively flat implied volatility curves consistent with the Black model have yielded to non-flat implied volatility curves consecutively in both markets. Nonetheless, the Black model has never been abandoned, and instead it has undergone a role change: from a pricing model to a quoting deviced. In this chapter, we show the analytical tractability of the Black model, and give a portrait of market reality in terms of the implied Black volatilities of SOFR futures options.

9.1 Market Model with SOFR Forward Term Rates

The Black model with the SOFR forward rates takes the following form:

$$df_j(t) = f_j(t)\gamma_j(t) \cdot d\mathbf{W}_t^{(j)}, \tag{9.1}$$

where $\mathbf{W}_t^{(j)}$ is a multi-dimensional $\mathbb{Q}_j$-Brownian motion and $\gamma_j(t)$ is a state-independent function of time. Alternatively, we can recast the Black model as

$$df_j(t) = f_j(t)\gamma_j(t) \cdot [d\mathbf{W}_t - \Sigma(t, T_j)dt], \tag{9.2}$$

where $\mathbf{W}_t$ is a $\mathbb{Q}$-Brownian motion and $\Sigma(t, T_j)$ is the percentage volatility of the T_j-maturity ZCB, so that the Black model can also be regarded as a term structure model for the spanning forward rates, which we call the market model. Next, we show that $\Sigma(t, T_j)$ is fully determined by forward rates and their volatilities. Recall the lognormal dynamics for the extended zero-coupon bonds:

$$dP(t, T_j) = P(t, T_j) [r_t dt + \Sigma(t, T_j) \cdot d\mathbf{W}_t], \tag{9.3}$$

DOI: 10.1201/9781003389101-9

from which we can derive the following dynamics for the SOFR forward rates through the use of Ito's Lemma:

$$
\begin{aligned}
df_j(t) &= \frac{1}{\Delta T_j} \left(\frac{P(t, T_{j-1})}{P(t, T_j)} \right) [\Sigma(t, T_{j-1}) - \Sigma(t, T_j)] \cdot [d\mathbf{W}_t - \Sigma(t, T_j) dt] \\
&= \frac{1 + \Delta T_j f_j(t)}{\Delta T_j} [\Sigma(t, T_{j-1}) - \Sigma(t, T_j)] \cdot d\mathbf{W}_t^{(j)}.
\end{aligned}
\tag{9.4}
$$

By comparing Equation 9.2 with Equation 9.4, we obtain the following relationship between the volatility of the SOFR forward rate and the volatility of zero-coupon bonds:

$$
\boldsymbol{\gamma}_j(t) = \frac{1 + \Delta T_j f_j(t)}{\Delta T_j f_j(t)} [\Sigma(t, T_{j-1}) - \Sigma(t, T_j)].
\tag{9.5}
$$

Equation 9.5 can be turned into an inductive scheme to derive ZCB volatilities in terms of those of the SOFR forward rates:

$$
\Sigma(t, T_j) = - \sum_{k=\eta_t+1}^{j} \frac{\Delta T_k f_k(t)}{1 + \Delta T_k f_k(t)} \boldsymbol{\gamma}_k(t),
\tag{9.6}
$$

where $\eta_t = \max\{j | T_j \leq t\}$. Note that under the market model with state-independent forward-rate volatilities, the volatilities of ZCB must be state dependent. Since there is no volatility for the SOFR forward rate at its fixing date, $\boldsymbol{\gamma}_j(t)$ must satisfy $\boldsymbol{\gamma}_j(T_j) = 0$, which is consistent with $\Sigma(T_j, T_j) = 0$. Hence, for whatever models we adopt, the diffusion volatility for the SOFR forward rates cannot be constants over their reference periods.

The market model represented by Equation 9.2 and Equation 9.6 is well-posed in the sense of the existence of a global solution, which can be proved through the approach pioneered by Brace, Gatarek, and Musiela (1997) for the LIBOR market model. Under very general conditions, they proved that a unique and strictly positive solution exists for all t. In addition, the model is found to have a number of desirable properties. First, there is mean reversion: the forward rates tend to drop when they are too high and they tend to rise when they are too low. Second, the solution is bounded from above and below by two lognormal processes. Third, if both $\boldsymbol{\gamma}_j(t)$ and $f_j(0)$ are smooth in T_j, then the solution $f_j(t)$ inherits the same degree of smoothness. For the details of their results, we refer readers to Brace, Gatarek, and Musiela (1997).

In the following subsections, we will price the major SOFR derivatives under the market model.

9.1.1 SOFR Futures

Under the market model, T_j-maturity SOFR rate at maturity is given by

$$
f_j(T_j) = f_j(t) e^{\int_t^{T_j} \left[-\boldsymbol{\gamma}_j(s) \cdot \Sigma(s, T_j) - \frac{1}{2} \|\boldsymbol{\gamma}_j(s)\|^2 \right] ds + \boldsymbol{\gamma}_j(s) \cdot d\mathbf{W}_s}.
\tag{9.7}
$$

The T_j-maturity SOFR futures rate is obtained by taking the $\mathbb{Q}$-expectation of the right-hand side (RHS) of the equation above which, however, cannot be worked out analytically. The industry convention has been to take a short cut by eliminating state dependence of the RHS of Equation 9.7 through freezing the state variables at time: for $s \geq t$, we adopt the following approximations,

$$\Sigma(s, T_j) = - \sum_{k=1}^{j} \frac{\Delta T_k f_k(s)}{1 + \Delta T_k f_k(s)} \gamma_k(s)$$

$$\approx - \sum_{k=1}^{j} \frac{\Delta T_k f_k(t)}{1 + \Delta T_k f_k(t)} \gamma_k(s) \triangleq \tilde{\Sigma}(s, T_j).$$

Then an approximation of the futures rate is easily obtained:

$$\tilde{f}_j(t) = E_t^Q \left[\tilde{f}_j(T_j) \right] \approx f_j(t) e^{\int_t^{T_j} -\gamma_j(s) \cdot \tilde{\Sigma}(s, T_j) ds}.$$

9.1.2   SOFR Futures Options

SOFR futures options are options on 3m or 1m SOFR futures. Let us focus on 3m SOFR futures. Similar to options on ED futures, options on SOFR futures expire a week before their reference periods. According to the general pricing formula under a forward measure, the value of a SOFR futures option with reference period (T_{j-1}, T_j) can be expressed as

$$V_t = \Delta T_j P(t, T_{j-1}) E_t^{Q_{j-1}} \left[\left(\tilde{f}_j^{(3m)}(T_{j-1}) - K \right)^+ \right].$$

With the SOFR market model, we can show that the futures rate is also a lognormal random variable under $\mathbb{Q}_{j-1}$, the T_{j-1}-forward measure. The implication is that the value of the futures option is given by the Black formula:

$$V_t = \Delta T_j P(t, T_{j-1}) \left(E_t^{Q_{j-1}} \left[\tilde{f}_j^{(3m)}(T_{j-1}) \right] \Phi(d_1^{(j)}) - K \Phi(d_2^{(j)}) \right),$$

with

$$d_1^{(j)} = \frac{\ln \frac{E_t^{Q_{j-1}} \left[\tilde{f}_j^{(3m)}(T_{j-1}) \right]}{k} + \frac{1}{2} \sigma_j^2 (T_{j-1} - t)}{\sigma_j \sqrt{T_{j-1} - t}} \tag{9.8}$$

$$d_2^{(j)} = d_1^{(j)} - \sigma_j \sqrt{T_{j-1} - t}.$$

and

$$\sigma_j^2 = \frac{1}{T_{j-1} - t} \int_t^{T_{j-1}} \|\gamma_j(s)\|^2 ds.$$

The value of the corresponding SOFR futures put option can be obtained via the put-call parity.

Next, we address the valuation of $E_t^{Q_{j-1}} \left[\tilde{f}_j^{(3m)}(T_{j-1}) \right]$. Because $\Sigma(s, T_{j-1}) = 0$ for $s \geq T_{j-1}$, there is

$$\tilde{f}_j^{(3m)}(T_{j-1}) = E_{T_{j-1}}^Q [f_j(T_j)] = E_{T_{j-1}}^{Q_{j-1}} [f_j(T_j)], \tag{9.9}$$

It follows that

$$E_t^{Q_{j-1}} \left[\tilde{f}_j^{(3m)}(T_{j-1}) \right] = E_t^{Q_{j-1}} [f_j(T_j)]$$

$$= f_j(t) E_t^{Q_{j-1}} \left[e^{\int_t^{T_j} \left[\boldsymbol{\gamma}_j(s) \cdot \Sigma(s;T_{j-1},T_j) - \frac{1}{2} \| \boldsymbol{\gamma}_j(s) \|^2 \right] ds + \boldsymbol{\gamma}_j(s) \cdot d\mathbf{W}_s^{(j-1)}} \right] \tag{9.10}$$

$$\approx f_j(t) e^{\int_t^{T_j} \boldsymbol{\gamma}_j(s) \cdot \tilde{\Sigma}(s;T_{j-1},T_j) ds}.$$

where $\tilde{\Sigma}(s; T_{j-1}, T_j)$ is an approximation to $\Sigma(s; T_{j-1}, T_j)$ through freezing state variables for $s \geq t$:

$$\tilde{\Sigma}(s; T_{j-1}, T_j) = -\frac{\Delta T_j f_j(t)}{1 + \Delta T_j f_j(t)} \boldsymbol{\gamma}_j(s).$$

9.1.3 Caps/Floors

Under the Black model, the value of the caplet with payoff

$$\Delta T_j \left(f_j(T_j) - K \right)^+$$

at T_j is given by

$$c_j(t) = \Delta T_j P(t, T_j) \left[f_j(t) \Phi(d_1^{(j)}(t)) - K \Phi(d_2^{(j)}(t)) \right], \tag{9.11}$$

for

$$d_1^{(j)}(t) = \frac{\ln \frac{f_j(t)}{k} + \frac{1}{2} \sigma_j^2 (T_j - t)}{\sigma_j \sqrt{T_j - t}}, \tag{9.12}$$

$$d_2^{(j)}(t) = d_1^{(j)}(t) - \sigma_j \sqrt{T_j - t},$$

and

$$\sigma_j^2 = \frac{1}{T_j - t} \int_t^{T_j} \| \boldsymbol{\gamma}_j(s) \|^2 ds.$$

Note that the formulae of Equation 9.12 are different from those in Equation 9.8.

9.1.4 Swaptions

Consider a payer's swaption of maturity $\tilde{T}_m$ and strike rate K on the underlying swap of tenor $(\tilde{T}_m, \tilde{T}_n)$. According to Equation 4.14, the payoff function of the swaption is

$$V_{\tilde{T}_m} = A_{m,n}(\tilde{T}_m) \left(R_{m,n}(\tilde{T}_m) - K \right)^+. \tag{9.13}$$

For pricing purpose, we make use of the swap measure, $\mathbb{Q}_{m,n}$, the martingale measure corresponding to the numeraire asset of $A_{m,n}(t)$. Under the swap measure $\mathbb{Q}_{m,n}$, the $A_{m,n}(t)$-relative price of any assets must be a martingale. Let V_t be the spot price of a payer's swaption with maturity $\tilde{T}_m$ and strike rate K, then it satisfies

$$
\frac{V_t}{A_{m,n}(t)} = E^{\mathbb{Q}_{m,n}} \left[\left. \frac{V_{\tilde{T}_m}}{A_{m,n}(\tilde{T}_m)} \right| \mathcal{F}_t \right]
$$

$$
= E^{\mathbb{Q}_{m,n}} \left[\left. \left(R_{m,n}(\tilde{T}_m) - K \right)^+ \right| \mathcal{F}_t \right]. \tag{9.14}
$$

The Black model for swap rates takes the following form:

$$
dR_{m,n}(t) = R_{m,n}(t)\boldsymbol{\gamma}_{m,n}(t) \cdot d\mathbf{W}_t^{(m,n)}, \tag{9.15}
$$

where $\boldsymbol{\gamma}_{m,n}(t)$ is a state-independent function and $\mathbf{W}_t^{(m,n)}$ is the Brownian motion under $\mathbb{Q}_{m,n}$. Under the Black model, the value of the payer's swaption with maturity $\tilde{T}_m$ and strike rate K is given by the Black formula:

$$
V_t = A_{m,n}(t) \left[R_{m,n}(t)\Phi(d_1^{(m,n)}(t)) - K\Phi(d_2^{(m,n)}(t)) \right], \tag{9.16}
$$

for

$$
d_1^{(m,n)}(t) = \frac{\ln \frac{R_{m,n}(t)}{k} + \frac{1}{2}\sigma_{m,n}^2(\tilde{T}_m - t)}{\sigma_{m,n}\sqrt{\tilde{T}_m - t}},
$$

$$
d_2^{(m,n)}(t) = d_1^{(m,n)}(t) - \sigma_{m,n}\sqrt{\tilde{T}_m - t},
$$

and

$$
\sigma_{m,n}^2 = \frac{1}{\tilde{T}_m - t} \int_t^{\tilde{T}_m} \|\boldsymbol{\gamma}_{m,n}(s)\|^2 ds.
$$

Given the dollar value of the prices, we can back out the volatility, $\sigma_{m,n}$, which are called the implied Black volatility. Between traders, it has been a convention to quote swaption values in terms of the implied volatilities.

We want to make comments on the Black models for forward rates and swap rates. Rigorously speaking, under the market model for forward rates, swap rates cannot be lognormal variables, and the Black model for swap rates is only an approximation to their true dynamics. Next, we derive the true dynamics of the swap rates under the market model and justify the approximation by the lognormal dynamics. Based on market reality, we regard $\tilde{\Delta T}$ as a multiple of ΔT and denote $\theta = \Delta\tilde{T}/\Delta T$, then the true dynamics of the swap rate can be derived through the use of Ito's Lemma:

$$
dR_{m,n}(t) = \sum_{j=m\theta+1}^{n\theta} \frac{\partial R_{m,n}(t)}{\partial f_j(t)} f_j(t)\boldsymbol{\gamma}_j(t) \cdot [d\mathbf{W}_t - \boldsymbol{\Sigma}_{m,n}(t)dt], \tag{9.17}
$$

where $\Sigma_{m,n}(t)$ is the volatility of $A_{m,n}(t)$, defined by

$$\mathbf{\Sigma}_{m,n}(t) \overset{\triangle}{=} \sum_{j=m+1}^{n} \alpha_j(t)\mathbf{\Sigma}(t,\tilde{T}_j) \quad \text{for} \quad \alpha_j(t) = \frac{\Delta\tilde{T}_j P(t,\tilde{T}_j)}{A_{m,n}(t)}.$$

Obviously, there is

$$\sum_{j=m+1}^{n} \alpha_j = 1,$$

meaning that $\Sigma_{m,n}(t)$ is the weighted average of $\Sigma(t,\tilde{T}_j), j = m+1,\ldots,n$. The partial derivative of the swap rate w.r.t. the forward rates are available below in closed form.

Lemma 9.1.1. *Consider swaps with fixed-leg tenor $\Delta\tilde{T}$ and floating-leg tenor ΔT such that $\Delta T \leq \Delta\tilde{T}$. There is*

$$\frac{\partial R_{m,n}(t)}{\partial f_k(t)} = \frac{\Delta T_k}{1+\Delta T_k f_k(t)}\left[\frac{1}{\Delta\tilde{T}_n}\alpha_n 1_{\{k\leq n\theta\}} + R_{m,n}\sum_{j=m+1}^{n}\alpha_j 1_{\{k\leq j\theta\}}\right]. \quad (9.18)$$

Proof. We have

$$P(t,\tilde{T}_j) = P(t,\tilde{T}_m)\prod_{l=m\theta+1}^{j\theta}[1+\Delta T_l f_l(t)]^{-1}.$$

Then, for $k \geq m\theta+1$,

$$\frac{\partial P(t,\tilde{T}_j)}{\partial f_k} = P(t,\tilde{T}_m)\prod_{l=m\theta+1}^{j\theta}[1+\Delta T_l f_l(t)]^{-1} \times \frac{-\Delta T_k}{1+\Delta T_k f_k(t)}1_{\{k\leq j\theta\}}$$

$$= \frac{-\Delta T_k}{1+\Delta T_k f_k(t)}P(t,\tilde{T}_j)1_{\{k\leq j\theta\}}.$$

According to the definition of swap rates, Equation 4.6, we have

$$\frac{\partial R_{m,n}(t)}{\partial f_k} = \frac{-\frac{\partial P(t,\tilde{T}_n)}{\partial f_k}A_{m,n}(t) - (P(t,\tilde{T}_m) - P(t,\tilde{T}_n))\frac{\partial A_{m,n}(t)}{\partial f_k}}{A_{m,n}^2(t)}$$

$$= \frac{\Delta T_k}{1+\Delta T_k f_k(t)}$$

$$\times \left[\frac{P(t,\tilde{T}_n)1_{\{k\leq n\theta\}}}{A_{m,n}(t)} + \frac{(P(t,\tilde{T}_m) - P(t,\tilde{T}_n))\sum_{j=m+1}^{n}\Delta\tilde{T}_j P(t,\tilde{T}_j)1_{\{k\leq j\theta\}}}{A_{m,n}^2(t)}\right]$$

$$= \frac{\Delta T_k}{1+\Delta T_k f_k(t)}\left[\frac{1}{\Delta\tilde{T}}\alpha_n 1_{\{k\leq n\theta\}} + R_{m,n}(t)\sum_{j=m+1}^{n}\alpha_j 1_{\{k\leq j\theta\}}\right].$$

This completes the proof of Equation 9.18. □

Since $R_{m,n}(t)$ is a martingale under $\mathbb{Q}_{m,n}$, $\mathbf{W}_t - \int_0^t \mathbf{\Sigma}_{m,n}(s)\mathrm{d}s$ must be a $\mathbb{Q}_{m,n}$-Brownian motion. This assertion can be confirmed as follows. The swap measure $\mathbb{Q}_{m,n}$ is defined by the Radon–Nikodym derivative:

$$
\begin{aligned}
\left.\frac{d\mathbb{Q}_{m,n}}{d\mathbb{Q}}\right|_{\mathcal{F}_t} &= \frac{A_{m,n}(t)/A_{m,n}(0)}{B(t)/B(0)} \\
&= \frac{1}{A_{m,n}(0)} \sum_{j=m+1}^{n} \Delta T_j \frac{P(t,\tilde{T}_j)}{B(t)} \triangleq \zeta(t), \quad t \le T_m.
\end{aligned}
\tag{9.19}
$$

Using Equation 5.29, the price process of zero-coupon bonds discounted by the money market account, we have the following equation for $\zeta(t)$, the Radon–Nikodym process:

$$
\begin{aligned}
d\zeta(t) &= \frac{1}{A_{m,n}(0)} \sum_{j=m+1}^{n} \Delta T_j\, d\left(\frac{P(t,\tilde{T}_j)}{B(t)}\right) \\
&= \frac{1}{A_{m,n}(0)} \sum_{j=m+1}^{n} \Delta T_j \left(\frac{P(t,\tilde{T}_j)}{B(t)}\right) \mathbf{\Sigma}(t,\tilde{T}_j) \cdot d\mathbf{W}_t \\
&= \frac{A_{m,n}(t)}{A_{m,n}(0)B(t)} \sum_{j=m+1}^{n} \frac{\Delta T_j P(t,\tilde{T}_j)}{A_{m,n}(t)} \mathbf{\Sigma}(t,\tilde{T}_j) \cdot d\mathbf{W}_t \\
&= \zeta(t)\mathbf{\Sigma}_{m,n}(t) \cdot d\mathbf{W}_t.
\end{aligned}
\tag{9.20}
$$

According to the CMG theorem,

$$
\mathbf{W}_t - \int_0^t \left\langle d\mathbf{W}_s, \frac{d\zeta(s)}{\zeta(s)} \right\rangle = \mathbf{W}_t - \int_0^t \mathbf{\Sigma}_{m,n}(s)\, ds
\tag{9.21}
$$

is a $\mathbb{Q}_{m,n}$-Brownian motion, which is exactly $\mathbf{W}_t^{(m,n)}$.

As it appears, the swap-rate process, Equation 9.17, has no analytical tractability and thus has limited value for swaption pricing. To reconcile market's practice of swaption pricing, we consider the following lognormal approximation to Equation 9.17: for $s \ge t$ and conditional on $\mathcal{F}_t$,

$$
\begin{aligned}
dR_{m,n}(s) &= \sum_{j=m\theta+1}^{n\theta} \frac{\partial R_{m,n}(s)}{\partial f_j} f_j(s)\boldsymbol{\gamma}_j(s) \cdot d\mathbf{W}_s^{(m,n)} \\
&= R_{m,n}(s) \sum_{j=m\theta+1}^{n\theta} \frac{\partial R_{m,n}(s)}{\partial f_j} \frac{f_j(s)}{R_{m,n}(s)} \boldsymbol{\gamma}_j(s) \cdot d\mathbf{W}_s^{(m,n)} \\
&\approx R_{m,n}(s) \sum_{j=m\theta+1}^{n\theta} \frac{\partial R_{m,n}(t)}{\partial f_j} \frac{f_j(t)}{R_{m,n}(t)} \boldsymbol{\gamma}_j(s) \cdot d\mathbf{W}_s^{(m,n)} \\
&= R_{m,n}(s) \left(\sum_{j=m\theta+1}^{n\theta} \omega_j \boldsymbol{\gamma}_j(s)\right) \cdot d\mathbf{W}_s^{(m,n)} \\
&\triangleq R_{m,n}(s)\boldsymbol{\gamma}_{m,n}(s) \cdot d\mathbf{W}_s^{(m,n)},
\end{aligned}
\tag{9.22}
$$

where

$$\gamma_{m,n}(s) \triangleq \sum_{j=m\theta+1}^{n\theta} \omega_j \gamma_j(s), \qquad (9.23)$$

with

$$\omega_j = \frac{\partial R_{m,n}(t)}{\partial f_j} \frac{f_j(t)}{R_{m,n}(t)}, \quad \text{for } j = m\theta+1, \ldots, n\theta. \qquad (9.24)$$

The approximate swap-rate process (Equation 9.22) is now a lognormal one. The approximation made in Equation 9.22 is yet another example of the "frozen coefficient" technique, often used in the industry whenever appropriate, to get rid of the dependence of model coefficients on state variables. The approximation we made here is motivated by the following facts: (1) $R_{m,n}(s)$ is a positive process; and (2) function

$$\frac{\partial R_{m,n}(s)}{\partial f_j} \frac{f_j(s)}{R_{m,n}(s)},$$

although stochastic, has very low variability. Therefore, after the approximation, the percentage volatility of the swap rate is simply a weighted average of the forward-rate volatilities.

Black's formula, Equation 9.16, also implies a replicating (or hedging) strategy for the swaption using an ATM swap and a money market account:

1. long $\Phi(d_1(t))$ units of the ATM swap; and

2. long $R_{m,n}(t)\Phi(d_1(t)) - K\Phi(d_2(t))$ units of the numeraire annuity.

The above hedging strategy is not unique. In fact, we can rewrite Black's formula as

$$V_t = \left[P(t, \tilde{T}_m) - P(t, \tilde{T}_n) \right] \Phi(d_1(t)) - KA_{m,n}(t)\Phi(d_2(t)), \qquad (9.25)$$

which suggests an alternative hedging strategy:

1. long $\Phi(d_1(t))$ unit of $\tilde{T}_m$-maturity zero-coupon bonds and short $\Phi(d_1(t))$ unit of $\tilde{T}_n$-maturity zero-coupon bonds, respectively; and

2. short $K\Phi(d_2(t))$ units of the annuity.

Note that the first hedging strategy is more practical because it uses the ATM swap, a very liquid security at no cost.

The swaption on a receiver's swap can be treated as a put option on the swap rate. Using the call–put parity, we readily obtain its price formula:

$$V_t = A_{m,n}(t) \left[K\Phi(-d_2(t)) - R_{m,n}(t)\Phi(-d_1(t)) \right]. \qquad (9.26)$$

Black's formula offers only approximate prices for European swaptions under the market model, yet they are found to be very accurate. For typical industrial applications, empirical studies (e.g., Rebonato, 1999; Sidenius, 2000) show that the pricing errors are well within one *kappa*, the usual market

TABLE 9.1: Monte Carlo Simulation and Black's Formula Prices for a Two-Factor Model

Maturity	Tenor	Strike (%)	Black	MC	STE
1	0.25	4.15	17.51	17.55	0.18
5	0.25	4.75	29.92	29.99	0.32
10	0.25	5.50	33.49	33.61	0.36
1	1	4.18	33.63	33.70	0.34
5	1	4.78	57.94	58.05	0.62
10	1	5.53	65.30	65.49	0.68
1	5	4.47	129.07	128.96	1.26
5	5	5.07	237.61	237.40	2.35
10	5	5.81	278.63	278.26	2.65
1	10	4.80	209.59	208.99	1.98
5	10	5.39	406.51	404.82	3.78
10	10	6.13	483.25	479.87	4.24

bid-ask spread defined as the change in the present value with 1% change in volatility. To demonstrate the accuracy by Black's formula, we compare the prices obtained by Black's formula and by a Monte Carlo simulation method using a two-factor model in Example 9.1.1.

Example 9.1. *We consider swaption pricing under a two-factor model, which takes the following forward-rate volatility: for $t \leq T_j$ and $1 \leq k < j$,*

$$
\gamma_j(t) = \begin{cases} \begin{pmatrix} 0.08 + 0.1e^{-0.05(j-k)} \\ 0.1 - 0.25e^{-0.1(j-k)} \end{pmatrix}, & T_{k-1} < t \leq T_k, \\ \begin{pmatrix} 0.08 + 0.1 \\ 0.1 - 0.25 \end{pmatrix} \times \frac{T_j - t}{\Delta T_j}, & T_{j-1} < t \leq T_j. \end{cases}
$$

The initial term structure of forward rates is

$$f_j(0) = 0.04 + 0.00075j, \quad \text{for all } j.$$

Here we take $\Delta T_j = 0.5$. The results for ATM swaptions are listed in Table 9.1, where the first and second column are the maturity of the options and the tenor of the underlying swaps, respectively; the third column contains the prevailing swap rates, which are taken as the strike rates of the swaptions; the fourth column lists swaption prices, in basis points, obtained by Black's formula, the fifth column is for the prices obtained by the Monte Carlo simulations, and the last column is for the standard errors of the simulation results. The simulations are made with the standard SOFR market model under the risk-neutral measure. One can see that the prices are very close across various maturities and tenors.

It helps to elaborate more on the accuracy of the lognormal approximation of the swap-rate process, Equation 9.22. Brigo and Liinev (2003) manage to calculate the Kullback–Leibler entropy distance between the original swap-rate distribution and the approximate lognormal distribution, and they find that the distance is encouragingly small. Precisely, for an implied Black's volatility of approximately 20%, the Kullback–Leibler entropy distance will be translated to a difference of about 0.1% in implied volatility, which is much smaller than 1%, the size of the bid-ask spread in swaption transactions.

For derivatives pricing, we sometimes use an alternative formula for swap rates. By making use of the martingale property of forward rates under their corresponding forward measure, we have the following formula for the floating leg of the swap:

$$
\begin{aligned}
V_{float}(t) &= \sum_{j=m\theta+1}^{n\theta} \Delta T_j P(t, T_j) E_t^{Q_j} [f_j(T_j)] \\
&= \sum_{j=m\theta+1}^{n\theta} \Delta T_j P(t, T_j) f_j(t).
\end{aligned}
\tag{9.27}
$$

By equating value of the floating leg to that of the fixed leg, we obtain

$$
R_{m,n}(t) = \sum_{j=m\theta+1}^{n\theta} \tilde{\alpha}_j f_j(t), \quad \text{for} \quad \tilde{\alpha}_j = \frac{\Delta T_j P(t, T_j)}{A_{m,n}(t)}.
\tag{9.28}
$$

The swap rate formula above will be used in swaption pricing under Lévy market model in Chapter 12.

9.2 Construction of the Initial Forward Rate Curve

In this section, we introduce the state of the art for constructing the forward rate curve, so-called *area preserving quadratic spline (APQS) interpolation* (Hagan, 2018). The objective is to find the term structure of continuous compounding forward rates, $f(0, T), 0 \leq T \leq 30$, through reproducing the prices of m instruments, such that

$$
\sum_{i=1}^{n_j} c_{i,j} e^{-\int_0^{T_{i,j}} f(0,u)du} = b_j, \quad j = 1, \ldots, m,
\tag{9.29}
$$

where $e^{-\int_0^{T_{i,j}} f(0,u)du}$ is the discount factor calculated using forward rates, b_j is the full price of the jth instrument, and $c_{i,j}$ is the ith cash flow of the jth instrument. Denote the maturities of the m instruments by

$$
0 = T_0 < T_1 < \ldots < T_m.
$$

Let $B_j = b_j, j = 1, \ldots, m$. The construction method consists of a few steps.

(i) Take constant interpolation for the forward rate such that

$$f(0,t) = A_j \quad \text{for} \quad T_{j-1} < t \leq T_j$$

and solve for $A_j, j = 1, 2, \ldots, m$, from

$$\sum_{i=1}^{n_j} c_{i,j} e^{-\sum_{k=1}^{i} \Delta T_k A_k} = B_j, \quad j = 1, \ldots, m, \tag{9.30}$$

by bootstrapping, where $\Delta T_k = T_k - T_{k-1}$.

(ii) Denote the nodal values of the forward rates as $\hat{f}_j \triangleq f(0, T_j), j = 0, \ldots, m$, and define the instantaneous forward curve through

$$f(0,t) = \hat{f}_{j-1}(1 - y_j(t)) + \hat{f}_j y_j(t) - 3\left(\hat{f}_{j-1} + \hat{f}_j - 2A_j\right) y_j(t)(1 - y_j(t)),$$
$$\text{for} \quad T_{j-1} < t < T_j, \tag{9.31}$$

with

$$y_j(t) = \frac{t - T_{j-1}}{T_j - T_{j-1}}.$$

(iii) Impose $f'(0, T_j^-) = f'(0, T_j^+)$ at T_j, thus resulting in the equations

$$\hat{f}_j = \frac{\Delta T_{j+1}}{\Delta T_{j+1} + \Delta T_j}\left(\frac{3}{2}A_j - \frac{1}{2}\hat{f}_{j-1}\right) + \frac{\Delta T_j}{\Delta T_{j+1} + \Delta T_j}\left(\frac{3}{2}A_{j+1} - \frac{1}{2}\hat{f}_{j+1}\right),$$
$$j = 1, 2, \ldots, m-1. \tag{9.32}$$

On the end points, we impose $f'(0, T_0) = f'(0, T_m) = 0$, which result in

$$\hat{f}_0 = \frac{3}{2}A_1 - \frac{1}{2}\hat{f}_1 \quad \text{and} \quad \hat{f}_m = \frac{3}{2}A_m - \frac{1}{2}\hat{f}_{m-1}. \tag{9.33}$$

Solve Equation 9.32 and Equation 9.33 for $\hat{f}_j, j = 0, 1, \ldots, m$.

(iv) Compute the fitting errors

$$\sum_{i=1}^{n_j} c_{i,j} e^{-\int_0^{T_{i,j}} f(0,u)du} - b_j = \epsilon_j, \quad \text{for} \quad j = 1, \ldots, m, \tag{9.34}$$

(vi) Stop if $|\epsilon_j| \leq$ tolerance, otherwise update B_j such that

$$B_j \longleftarrow B_j - \epsilon_j, \quad \text{for} \quad j = 1, \ldots, m,$$

and go back to step (i).

The above procedure takes only a few iterations to converge. Let us highlight the advantages of the APQS method.

1. By construction, $f(0,t)$ is continuous up to the first-order derivative.
2. The quadratic polynomial

$$g(y) = \hat{f}_{j-1}(1-y) + \hat{f}_j y - 3\left(\hat{f}_{j-1} + \hat{f}_j - 2A_j\right)y(1-y)$$

is either convex or concave, depending on the sign of $\hat{f}_{j-1} + \hat{f}_j - 2A_j$. So will be $f(0,t)$ over the interval (T_{j-1}, T_j).

3. The name of the method comes from the property that

$$\int_{T_{j-1}}^{T_j} f(0,t)dt = \Delta T_j A_j,$$

which implies

$$\int_0^{T_j} f(0,t)dt = \sum_{k=1}^{j} \Delta T_k A_k.$$

Note that the area-preserving property is close but not sufficient to ensure the price reproduction of the input instruments, but only a few iterations are needed for remedy.

An example of the forward rate curve for continuous compounding constructed by the quadratic spline is shown in Figure 9.1, using the prices of on-the-run Treasury securities of September 15, 2023.[1]

The quadratic spline is a compromise between better smoothness and lesser oscillation of the forward rate curve. Compared with cubic spline, a quadratic spline is less oscillatory and therefore has a smaller chance to generate negative forward rates, making it a much more competitive choice for constructing the forward rate curve under low-rate environment.

Once the continuous compounding forward rate curve is constructed, ZCB price curve or the discount curve follows:

$$P(0,T) = e^{-\int_0^T f(0,u)du}, \quad \text{for} \quad 0 \le T \le 30,$$

so does the initial term structure of SOFR term rates:

$$f_j(0) = \frac{1}{\Delta T_j}\left(\frac{P(0,T_{j-1})}{P(0,T_j)} - 1\right) = \frac{1}{\Delta T_j}\left(e^{-\int_{T_{j-1}}^{T_j} f(0,u)du} - 1\right),$$

for $j = 1, 2, \ldots$, which are the initial conditions to the market model.

---

[1] By the curtesy of Mr. Wencan Xia, my Ph.D. student.

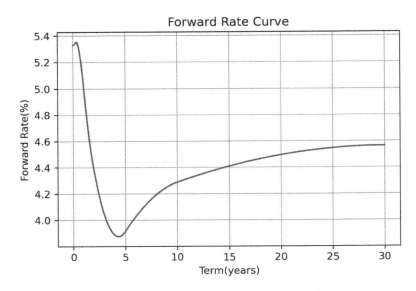

FIGURE 9.1: Forward rate curve of 9/15/2023.

9.3   Monte Carlo Simulation Method

9.3.1   The Log–Euler Scheme

Except for the few vanilla derivatives, namely, caps, floors, and swaptions, there is no closed-form formula for general SOFR derivatives. It is also not feasible to implement the market model, a multi-factor model, via a lattice tree method or a finite difference method. Hence, to price general SOFR derivatives, we have to resort to Monte Carlo simulation methods. In a Monte Carlo simulation method, we need to advance the spanning SOFR term rates in discrete time steps. The most natural scheme for the model is the log-Euler scheme, which is the result of applying the so-called Euler scheme to the stochastic equation of the logs of the forward rates under the risk-neutral measure:

$$
\mathrm{d}\ln f_j(t) = -\left(\boldsymbol{\gamma}_j(t) \cdot \tilde{\boldsymbol{\Sigma}}_j(t) + \frac{1}{2}\left\| \boldsymbol{\gamma}_j(t) \right\|^2 \right) \mathrm{d}t + \boldsymbol{\gamma}_j(t) \cdot \mathrm{d}\mathbf{W}_t, \qquad (9.35)
$$

where we write $\tilde{\boldsymbol{\Sigma}}_j(t)$ for $\tilde{\boldsymbol{\Sigma}}(t, T_j)$ for simplicity. The log-Euler scheme is

$$
\begin{aligned}
\ln f_j(t + \Delta t) = \ln f_j(t) &- \int_t^{t+\Delta t} \left(\boldsymbol{\gamma}_j(s) \cdot \tilde{\boldsymbol{\Sigma}}_j(s) + \frac{1}{2}\left\| \boldsymbol{\gamma}_j(s) \right\|^2 \right) \mathrm{d}s \\
&+ \int_t^{t+\Delta t} \boldsymbol{\gamma}_j(s) \cdot \mathrm{d}\mathbf{W}_s,
\end{aligned} \qquad (9.36)
$$

or

$$f_j(t + \Delta t) = f_j(t)e^{-\int_t^{t+\Delta t}(\gamma_j(s)\cdot\tilde{\boldsymbol{\Sigma}}_j(s)+(1/2)\|\gamma_j(s)\|^2)ds+\int_t^{t+\Delta t}\gamma_j(s)\cdot d\mathbf{W}_s}. \quad (9.37)$$

Here,

$$\int_t^{t+\Delta t} \gamma_j(s) \cdot d\mathbf{W}_s \sim N\left(0, \int_t^{t+\Delta t} \|\gamma_j(s)\|^2 ds\right).$$

The volatility of a zero-coupon bond can also be calculated from a recursive scheme,

$$\tilde{\boldsymbol{\Sigma}}_{\eta_t} = 0, \quad \text{for} \quad \eta_t = \max\{j \mid T_j \leq t\}.$$

$$\tilde{\boldsymbol{\Sigma}}_j(s) = \tilde{\boldsymbol{\Sigma}}_{j-1}(s) - \frac{\Delta T_j f_j(t)}{1 + \Delta T_j f_j(t)}\gamma_j(s), \quad j > \eta_t, \quad (9.38)$$

Let N denote the number of forward rates at time $t = 0$, and let n_f denote the number of factors. Then, for the first step in the iteration of Equation 9.37, we need to perform $2(N+1)n_f + 3$ multiplications. For each path, which typically has N steps, the total number of multiplications is

$$\sum_{j=1}^N (2(j+1)n_f + 3) = N^2 + (2n_f + 4)N,$$

which is quite manageable.

The log-Euler scheme is a first-order scheme in Δt and it is appreciated for its simplicity and intuitiveness. More sophisticated or accurate schemes for the market model have also been developed. Hunter, Jackel, and Joshi (2001) consider a predictor–corrector scheme, aiming at raising the order of accuracy to two. Pietersz, Pelsser, and Regenmortel (2004) introduce a Brownian bridge discretization for the drift. To preclude possible arbitrage arising from discretization, Glasserman and Zhao (2000) propose to simulate zero-coupon bonds first before computing the forward rates. Most of these methods achieve some improvements in pricing accuracy over the simple log-Euler method.

9.3.2 Early Exercise

The payoff of an exotic option may depend on the minimum, maximum, or average values of the state variables over the life of the option, or its cash flows can also be triggered by pre-specified barriers or a provision of early exercise. With Monte Carlo simulations, options on the minimum, maximum, or average value can be evaluated straightforwardly. But this is not the case for options with the provision of early exercise; such options are called American options. In fact, pricing the feature of early exercise in the context of Monte Carlo simulations had long been expensive and slow, due to the need for making decisions on early exercises throughout the usual backward induction procedure. This difficulty was overcome by a regression-based method developed between 1996 and 2000. In this section, we introduce this method with the pricing of Bermudan swaptions, which is also one of the most important applications of the method.

Consider a Bermudan swaption of either the first or the second kind that allows the holder to enter into a swap with a fixed rate at a pre-specified sequence of observation dates. At any of these observation dates, the holder must decide whether to exercise the option or hold on to it. These two actions result in two different values to the option. To the holder, the principle of taking either action is simple: take the one that generates higher value for the option. As is well-known, numerical option pricing is a backward induction procedure. Let $h(\mathbf{f}_i)$ denote the payoff function of a Bermudan swaption at time T_i, where $\mathbf{f}_i = \{ f_j(T_i) \}_{j=0}^m$ denotes the forward rates that are still alive. The pricing of the Bermudan swaption may be done through the following dynamic programming procedure:

$$
\begin{aligned}
V_m(\mathbf{f}_m) &= h(\mathbf{f}_m), \\
V_i(\mathbf{f}_i) &= \max\{ h(\mathbf{f}_i), E^{\mathbb{Q}}\left[D_{i,i+1} V_{i+1}(\mathbf{f}_{i+1}) \mid \mathbf{f}_i \right] \}, \\
&\quad i = m-1, m-2, \ldots, 0.
\end{aligned}
\tag{9.39}
$$

The first term in the second equation of Equation 9.39 is called the intrinsic value, the second term is called the "holding value" or "continuation value," and $D_{i,i+1}$ is the discount factor, defined by

$$
D_{i,i+1} = \frac{B(T_i)}{B(T_{i+1})} = \frac{1}{1 + \Delta T_{i+1} f_{i+1}(T_{i+1})}.
$$

Equation 9.39 is a mathematical interpretation of the holder's action: if the exercise value is higher, then exercise; otherwise, hold. The key in the implementation of the backward induction scheme lies in the valuation of the continuation value, which, however, had not been easy until a regression-based solution was first proposed by Carriere (1996) and later improved by Longstaff and Schwartz (2001). Carriere proposed to approximate the conditional expectation in Equation 9.39 by a function of the state variables:

$$
E^{\mathbb{Q}}\left[D_{i,i+1} V_{i+1}(\mathbf{f}_{i+1}) \mid \mathbf{f}_i = \mathbf{f} \right] = \sum_{j=1}^M \beta_{i,j} \psi_j(\mathbf{f}) \triangleq C_i(\mathbf{f}),
\tag{9.40}
$$

for some basis function, ψ_j, and constants, $\beta_{i,j}, j = 1, \ldots, M$. Let

$$
\begin{aligned}
\Psi(\mathbf{f}) &= (\psi_1(\mathbf{f}), \ldots, \psi_M(\mathbf{f}))^{\mathrm{T}}, \\
\boldsymbol{\beta}_i &= (\beta_{i,1}, \ldots, \beta_{i,M})^{\mathrm{T}}.
\end{aligned}
\tag{9.41}
$$

Then, by linear regression, we have the following formula for $\boldsymbol{\beta}_i$:

$$
\begin{aligned}
\boldsymbol{\beta}_i &= \left(E^{\mathbb{Q}}\left[\Psi(\mathbf{f}_i) \Psi(\mathbf{f}_i)^{\mathrm{T}} \right] \right)^{-1} E^{\mathbb{Q}}\left[\Psi(\mathbf{f}_i) D_{i,i+1} V_{i+1}(\mathbf{f}_{i+1}) \mid \mathbf{f} \right] \\
&\triangleq B_\Psi^{-1} B_{\Psi V},
\end{aligned}
\tag{9.42}
$$

where B_Ψ is an M-by-M matrix, and $B_{\Psi V}$ is an M-by-1 matrix. These two matrices can be approximated by using sample data as follows:

$$\left(\hat{B}_\Psi\right)_{q,r} = \frac{1}{N}\sum_{j=1}^{N}\psi_q\left(\mathbf{f}_i^{(j)}\right)\psi_r\left(\mathbf{f}_i^{(j)}\right), \tag{9.43}$$

and

$$\left(\hat{B}_{\Psi V}\right)_r = \frac{1}{N}\sum_{j=1}^{N}\psi_r\left(\mathbf{f}_i^{(j)}\right)D_{i,i+1}^{(j)}V_{i+1}\left(\mathbf{f}_{i+1}^{(j)}\right), \tag{9.44}$$

where

$$D_{i,i+1}^{(j)} = \frac{1}{1+\Delta T_{i+1}f_{i+1}^{(j)}(T_{i+1})}.$$

The estimate of the continuation value is then given by

$$\hat{C}_i(\mathbf{f}_i) = \hat{\beta}_i^{\mathrm{T}}\Psi(\mathbf{f}_i) \triangleq \hat{B}_\Psi^{-1}\hat{B}_{\Psi V}\Psi(\mathbf{f}_i). \tag{9.45}$$

With the continuation values, the decision about the early exercise can be made. Here is the regression-based Monte Carlo simulation algorithm.

Algorithm 9.3.1.

1. *Simulate N independent paths, $\{\mathbf{f}_0^{(j)},\ldots,\mathbf{f}_m^{(j)}\}, j = 1,\ldots,N$, of the Markov chain.*

2. *At the terminal nodes, calculate $\hat{V}_m^{(j)} = h(\mathbf{f}_m^{(j)}), j = 1,\ldots,N$.*

3. *Apply backward induction: for $i = m-1,\ldots,1,0$, do the following:*

   a. *Given estimated value $\hat{V}_{i+1}^{(j)}, j = 1,\ldots,N$, use regression formulae from Equation 9.41 to Equation 9.45 to calculate $\hat{C}_i\left(\mathbf{f}_i^{(j)}\right)$.*

   b. *Set*

   $$\hat{V}_i^{(j)} = \max\left\{h\left(\mathbf{f}_i^{(j)}\right), \hat{C}_i\left(\mathbf{f}_i^{(j)}\right)\right\}, \quad j = 1,\ldots,N. \tag{9.46}$$

4. *Take*

$$\hat{V}_0 = \max\left\{h(\mathbf{f}_0), \frac{1}{N}\sum_{j=1}^{N}D_{0,1}^{(j)}\hat{V}_1^{(j)}\right\}. \tag{9.47}$$

Although easy to comprehend, the above estimator is, however, biased upward. Here is the explanation (Glasserman, 2004). Conditional to $\mathbf{f}_i$, there is

$$E^{\mathbb{Q}}\left[\hat{V}_i(\mathbf{f}_i)\right] = E^{\mathbb{Q}}\left[\max\{h(\mathbf{f}_i),\hat{C}_i(\mathbf{f}_i)\}\right]$$

$$\geq \max\left\{h(\mathbf{f}_i), E^{\mathbb{Q}}[\hat{C}_i(\mathbf{f}_i)]\right\}$$

$$= \max\{h(\mathbf{f}_i), E^{\mathbb{Q}}[D_{i,i+1}\hat{V}_{i+1}(\mathbf{f}_{i+1})|\mathbf{f}_i]\}.$$

using Jensen's inequality. If, conditional on $\mathbf{f}_{i+1}$, we already have

$$E^{\mathbb{Q}}\left[D_{i,i+1}\hat{V}_{i+1}(\mathbf{f}_{i+1})\right] \geq D_{i,i+1}V_{i+1}(\mathbf{f}_{i+1}),$$

which is at least true for $i = m - 1$, then, by the tower law,

$$E^{\mathbb{Q}}\left[D_{i,i+1}\hat{V}_{i+1}(\mathbf{f}_{i+1})\,|\mathbf{f}_i\right] = E^{\mathbb{Q}}\left[E^{\mathbb{Q}}\left[D_{i,i+1}\hat{V}_{i+1}(\mathbf{f}_{i+1})\right]\Big|\mathbf{f}_i\right]$$
$$\geq E^{\mathbb{Q}}\left[D_{i,i+1}V_{i+1}(\mathbf{f}_{i+1})\,|\,\mathbf{f}_i\right].$$

It then follows that

$$E^{\mathbb{Q}}\left[\hat{V}_i(\mathbf{f}_i)\right] \geq \max\{h(\mathbf{f}_i), E^{\mathbb{Q}}\left[D_{i,i+1}V_{i+1}(\mathbf{f}_{i+1})\,|\,\mathbf{f}_i\right]\} = V_i(\mathbf{f}_i).$$

The upward bias of the estimator may be attributed to the use of the same information in deciding whether to exercise and the estimation of the continuation value.

To remove the bias, we must separate the exercise decision making from the valuation of continuation value. This is in fact the key to removing the bias in all Monte Carlo methods. To show the idea, we consider the problem of estimating

$$\max\{a, E\,[y]\}$$

from iid replicates $Y_1, \ldots, Y_N$. The estimator,

$$\max\{a, \bar{Y}\},$$

is upward biased since

$$E\left[\max\{a, \bar{Y}\}\right] \geq \max\{a, E\left[\bar{Y}\right]\}$$
$$= \max\{a, E\,[Y]\}.$$

To fix such a bias is nonetheless simple: separate $\{Y_i\}$ into two disjointed subsets; calculate sample means, $\bar{Y}_1$ and $\bar{Y}_2$; and set

$$\hat{V} = \begin{cases} a, & \text{if } \bar{Y}_1 \leq a, \\ \bar{Y}_2, & \text{if } \bar{Y}_1 > a. \end{cases} \tag{9.48}$$

Then,

$$E\left[\hat{V}\right] = P(\bar{Y}_1 \leq a)a + (1 - P(\bar{Y}_1 \leq a))E\left[\bar{Y}_2\right]$$
$$\leq \max\{a, E\,[Y]\}.$$

Hence, the new estimator is biased downward.

In the regression-based method, valuations like the one in Equation 9.48 ought to be made at every time step for all paths. While $\bar{Y}_1$ is obtained through regression, $\bar{Y}_2$ should be obtained by pricing a similar Bermudan option, for which we face the same problem of decision making on early exercise. It seems that the problem is not solved unless we can find an inexpensive estimation of $\bar{Y}_2$.

Longstaff and Schwartz (2001) present an alternative solution to the upward-bias problem of Algorithm 9.3.1 by figuring out cash flow date along all paths. At time step T_i, they first estimate the continuation value for all paths by the regression method of Carriere. Then, if intrinsic values are larger than the continuation values, they exercise the option and update the cash flow dates. At the end of the backward induction, they obtain a matrix of cash flow dates. The continuation value at time T_0 is defined as the average value of discounted cash flows, and the value of the American option is taken as the maximum one between the intrinsic value and the continuation value. To make the estimation of continuation values more accurate, Longstaff and Schwartz use only the paths that are in the money for regression, and the discounted values of future cash flows are taken for continuation values. The Longstaff-Schwartz algorithm is presented below.

Algorithm 9.3.2.

1. *Simulate N independent paths, $\{\mathbf{f}_0^{(j)}, \ldots, \mathbf{f}_m^{(j)}\}, j = 1, \ldots, N$, of the Markov chain.*

2. *At the terminal nodes, calculate $\hat{V}_m^{(j)} = h(\mathbf{f}_m^{(j)}), j = 1, \ldots, N$.*

3. *Apply backward induction: for $i = m - 1, \ldots, 1, 0$, do the following:*

   a. *Given estimated value $\hat{V}_{i+1}^{(j)}, j = 1, \ldots, N$, use regression formulae from Equation 9.41 to Equation 9.45 to calculate $\hat{C}_i\left(\mathbf{f}_i^{(j)}\right)$.*

   b. *Set*

   $$\hat{V}_i^{(j)} = \begin{cases} h\left(\mathbf{f}_i^{(j)}\right), & \text{if } h\left(\mathbf{f}_i^{(j)}\right) \geq \hat{C}_i\left(\mathbf{f}_i^{(j)}\right) \\ D_{i,i+1}^{(j)} \hat{V}_{i+1}^{(j)}, & \text{otherwise,} \end{cases} \quad (9.49)$$

   for $j = 1, \ldots, N$.

4. *Take*

$$\hat{V}_0 = \max\left\{ h(\mathbf{f}_0), \frac{1}{N} \sum_{j=1}^{N} D_{0,1}^{(j)} \hat{V}_1^{(j)} \right\}. \quad (9.50)$$

Let us address some technical issues in the implementation of the Longstaff–Schwartz algorithm, regarding adequate number of paths, number of basis functions, and size of time stepping for the Monte Carlo simulations. For interest rate options with early exercise provision, Longstaff and Schwartz have used the number of paths between 5,000 and 100,000. Our own experiences, meanwhile, suggest that it is safe to take the number of paths between 10,000 to 50,000. Regarding the choice and the number of the basis functions, Longstaff and Schwartz are very flexible, as they have tried monomials as well as special polynomials, and have observed that the regression is robust against the choice of basis functions. In all the examples they have considered, only low-degree monomials or polynomials are adopted for basis functions. Their view on the robustness of regression against the choice of basis functions are echoed in other researches like, for example, Pedersen (1999). For pricing Bermudan swaptions with fixed tenors, Pedersen had tried both univariate quadratic polynomial and bivariate quadratic polynomials for basis functions, such that

$$\Psi(\mathbf{f}) = \left(1, R, R^2\right)^{\mathrm{T}}, \tag{9.51}$$

or

$$\Psi(\mathbf{f}) = \left(1, R, B, R^2, RB, B^2\right)^{\mathrm{T}}, \tag{9.52}$$

where R represents the underlying swap rate and B is the value of the saving account. In Pedersen's various tests, regressions using the univariate polynomial approximation and bivariate quadratic polynomial basis functions produce rather close results. As for the size of time stepping, it varies from a month, two weeks, one week, and half a week in Longstaff and Schwartz (2001) and other researches. The search for a proper step size starts from the gap between two exercise dates which, for example, is one year for Bermudan swaptions. The proper step size also depends on the choice of models. Take the market model for example, the gap between two exercise days is taken to be the maximal step size by default, starting from which we can figure out the proper step size by redoing the Monte Carlo simulations with reduced step size, until convergence is reached.

In the Longstaff–Schwartz algorithm, the moment of optimal exercise is determined along each path, and the discounted cash flow is used for taking the average. Since any exercise strategy we can develop is suboptimal at best, the option value generated by the algorithm is biased low. The convergence of the Longstaff–Schwartz algorithm is obtained by Clement, Lamberton, and Protter (2002) for $j \to \infty$. Precisely, if the functional representation (Equation 9.40) holds exactly, they show that the limiting option value converges to the exact price.

With a lower bound estimate available, we now look at upper bound estimate. Since the discount prices $\hat{V}_i$ is a super-martingale, such that it admits the Doob-Meyer decomposition,

$$\hat{V}_i = M_i + D_i, \quad i = 1, \dots, N,$$

where M_i is a martingale with $M_0 = 0$ and D_i is a non-increasing $\mathcal{F}_{T_i}$ adaptive process. The approach to be introduced is based on a result of Chen and Glasserman (2007), so-called the duality theorem.

Proposition 9.3.1. *Let $\mathcal{M}^0$ denote the set of all martingales M with respect to $\mathcal{F}$, starting with $M_0 = 0$. The the value of the American option satisfies*

$$V_0 = \inf_{M \in \mathcal{M}^0} E^Q \left[\max_{0 \leq i \leq m} \left(\frac{h(\mathbf{f}_i)}{B(T_i)} - M_i \right) \middle| \mathcal{F}_0 \right],$$

and the infimum is achieved at the martingale obtained from the Doob-Meyer decomposition.

Based the above proposition, the option value is obtained at the martingale from the Doob-Meyer decomposition, which can expressed as

$$M_0 = 0,$$

$$M_i = \sum_{k=0}^{i} \Delta_k, \quad \text{for} \quad \Delta_i = \frac{V_i}{B(T_i)} - E^Q \left[\frac{V_i}{B(T_i)} \middle| \mathcal{F}_{T_{i-1}} \right].$$

Apparently, M_i's are not known unless we already know the value function, which leaves us in a situation of circular dependence. The only thing we can do is to choose an approximations to M_i's and thus obtain an upper bound of the option value.

There are a number of choices for the martingales, including the martingale from the approximate continuation value function and the martingales from the stopping rules. Based on their performance, we are more interested in the former. We have already known that the approximate value function satisfies

$$\hat{V}_i = \max\{h(\mathbf{f}_i), \hat{C}(\mathbf{f}_i)\},$$

where $\hat{C}_i$ is constructed by the regression method introduced earlier, we then adopt the approximation to Δ_i as

$$\hat{\Delta}_i = \frac{\hat{V}_i}{B(T_i)} - E^Q \left[\frac{\hat{V}_i}{B(T_i)} \middle| \mathcal{F}_{T_{i-1}} \right]$$

and accordingly,

$$\hat{M}_0 = 0 \quad \text{and} \quad \hat{M}_i = \sum_{k=0}^{i} \hat{\Delta}_k.$$

The algorithm for the upper bound estimate is given below.

Algorithm 9.3.3.

1. *Simulate N independent paths, $\{\mathbf{f}_0^{(j)}, \ldots, \mathbf{f}_m^{(j)}\}, j = 1, \ldots, N$, of the Markov chain.*

2. *For $j = 1, \ldots, N$ do the following*

   (a) *Take $\hat{M}_0^{(j)} = 0$.*

   (b) *For $i = 1, \ldots, m$ perform*

      i. *Calculate $\hat{C}(\mathbf{f}_i^{(j)})$ by the regression method;*

      ii. *Set $\hat{V}_i(\mathbf{f}_i^{(j)}) = \max\left(h(\mathbf{f}_i^{(j)}), \hat{C}(\mathbf{f}_i^{(j)}) \right).$*

      iii. *For $k = 1, \ldots, N_n$ perform*

A. *Simulate a path from* $\mathbf{f}_{i-1}^{(j)}$ *to time* T_i *to get* $\mathbf{f}_i^{(k,j)}$;

B. *Calculate* $\hat{V}_i^{(k,j)}/B(T_i) = \max\left(h(\mathbf{f}_i^{(k,j)}), \hat{C}_i(\mathbf{f}_i^{(k,j)})\right)/B(T_i)$;

iv. *Set* $E^Q\left[\frac{\hat{V}_i}{B(T_i)}\right] = \frac{1}{N_n}\sum_{k=1}^{N_n}\hat{V}_i^{(k,j)}/B(T_i)$;

v. *Set* $\hat{\Delta}_i = \frac{\hat{V}_i}{B(T_i)} - E^Q\left[\frac{\hat{V}_i}{B(T_i)}\Big|\mathcal{F}_{T_{i-1}}\right]$;

vi. $\hat{M}_i^{(j)} = \hat{M}_i^{(j)} + \hat{\Delta}_i^{(j)}$;

vii. $\hat{V}_0^{(j)} = \max_{0\leq i\leq m}\left\{\frac{h(\mathbf{f}_i^{(j)})}{B(T_i)} - \hat{M}_i^{(j)}\right\}$.

3. *Take* $\hat{V}_0 = \frac{1}{N}\sum_{k=1}^{N}\hat{V}_0^{(j)}$.

In the algorithm above, N_n is the number of mini-paths. It is suggested (Glasserman, 2004) that we can take $N_n \leq 100$.

For the two methods introduced above, the price difference, so-called the duality gap, is usually small. In many studies the relative difference errors are under 1%, which endorses the accuracy of both methods. Still, the Longstaff–Schwartz method is most popularly adopted for pricing American and Bermudan options.

9.4 Volatility Smiles and the Direct Adaptation of Smile Models

With every other parameter fixed, there is a one-to-one correspondence between an option and its implied Black volatility. Because of that, call and put options are often quoted in terms of the implied Black volatilities in the market place. As for traded instruments, option values are ultimately determined by supply and demand, as a result, options with the same maturity on the same underlying do not necessarily have the same implied Black volatilities, and they sometimes can be quite different. By interpolating discrete implied Black volatilities across strikes, we obtain a so-called implied volatility curve. Putting the curves for different option maturities together we obtain an implied volatility surface. The stylized features of implied volatility curves and surfaces in various sectors of financial market including, in particular, the fixed-income markets are well understood by market participants. In Figure 9.2, we present the implied volatility surface for Mid-Curve options (on SOFR futures) observed on April 22, 2024. The presence of the smile-shape implied volatility curve implies that the Black model is flawed, as otherwise the implied volatility curve should be flat. It had been a major focus for the period from 2000 to 2010 to develop alternative models to accommodate the implied volatility curves/surfaces, and very successful models have been developed that make the pricing more accurate and risk management of option portfolio more efficient. Chapter 11 is devoted to models for implied volatility smiles.

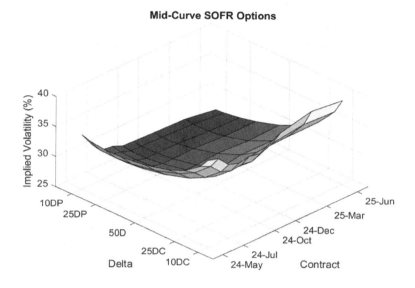

FIGURE 9.2: Implied Black volatility surface for Mid-Curve SOFR Options as of 4/22/2024 (Source of data: Bloomberg Terminal).

Exercises

1. Use the risk-neutral measure and the tower law to show that, at time $t \leq T_0$, the price of an FRN with backward-looking SOFR term rates $f_j(T_j)$ at $T_j, j = 1, \ldots, n$, is equal to $P(0, T_0)$, the price of a T_0-maturity zero-coupon bond.

2. Prove the *call–put parity* between a cap and a floor with the same strike rate and tenor:
$$\text{Cap} - \text{Floor} = \text{Swap},$$
where the swap has the same strike rate and tenor as those of the cap and floor.

3. Suggest a strategy to hedge a swap with a sequence of FRAs, and use the strategy to argue that
$$R_{m,n}(t) = \sum_{j=m+1}^{n} \alpha_j f_j(t), \ \alpha_j = \frac{\Delta T P(t, T_j)}{A_{m,n}(t)}.$$

4. Describe the swaption hedging strategy using the ATM swap and the annuity. Prove it if you think that the strategy is a self-financing one.

5. Prove the alternative version of the CMG theorem: define a new measure, $\tilde{\mathbb{Q}}$, as

$$\left.\frac{d\tilde{\mathbb{Q}}}{d\mathbb{Q}}\right|_{\mathcal{F}_t} = m(t),$$

where $m(t)$ is a $\mathbb{Q}$-martingale with $m(0) = 1$ and $m(t) > 0$. Let $\mathbf{W}_t$ be a vector of independent $\mathbb{Q}$-Brownian motions. Then, $\tilde{\mathbf{W}}_t$ defined as

$$d\tilde{\mathbf{W}}_t = d\mathbf{W}_t - \left\langle d\mathbf{W}_t, \frac{dm(t)}{m(t)} \right\rangle,$$

is a vector of independent $\tilde{\mathbb{Q}}$-Brownian motions (here $\langle .,. \rangle$ means covariance).

6. For all problems below, use the spot term structure

$$f(0, T) = 0.02 + 0.001T,$$

for the instantaneous forward rates. Pricing is done under the market model (which assumes lognormal forward rates or swap rates). The payment frequency is a quarter year for caps and half a year for swaps.

   (a) Price the 10-year cap with the strike rate to be the 10-year par yield (such a cap is called *ATM-forward cap*). Take the implied cap volatility to be 20%.

   (b) Price the in-5-to-10 ATM swaption (the maturity of the option: 5; the tenor of the underlying swap: 10). Assume a 20% swap-rate volatility.

7. Redo the pricing of swaption in Problem 8b under the SOFR market model using the method of Monte Carlo simulation. Use the dynamics of SOFR under the forward measure with delivery at the maturity day (also called terminal measure) of the swaption.

8. This problem deals with call options on coupon bonds with the par strike.

   (a) Show that any coupon-bond options with the par strike can be treated as swaptions.

   (b) Based on (a), develop a closed-form formula under the market model for coupon bond options with the par strike,

   (c) Based on (b) suggest a hedging strategy for these options using ATM swaps.

9. Provide a closed-form price formula under the market model for a SOFR

corridor, which consists of a series of cash flow, $\Delta T_j c_j$, at time $T_j, j = 1, 2, \ldots, N$, with

$$
c_j = \begin{cases} K_1, & \text{if} \quad f_j(T_j) < K_1, \\ f_j(T_j), & \text{if} \quad K_1 \leq f_j(T_j) \leq K_2, \\ K_2, & \text{if} \quad K_2 \leq f_j(T_j). \end{cases}
$$

Explain how to hedge the corridor using forward-rate agreements.

10. Consider the pricing of a cancelable swap (that exchanges SOFR for a fixed rate). The maturity of the swap is ten years, and the payer has the right to cancel the swap in five years. How to solve for the fair swap rate so that the value of the cancelable swap is zero?

Chapter 10

Convexity Adjustments

In Chapter 4, we have introduced a series of SOFR derivatives. In that series, SOFR futures and swaps of various kinds are particularly important to interest-rate derivatives markets. Unlike other derivatives, futures and swaps cost nothing to enter (except for deposits as guarantee funds), making them popular tools for either hedging or speculation. Probably because of that, SOFR futures and swaps have dominated the liquidity of the interest-rate derivatives markets since 2018. The pricing of SOFR futures has been taken care in the previous chapters. In this chapter, we consider the pricing of non-vanilla swaps relative to vanilla swaps. Specifically, we derive approximation formulae for non-vanilla swap rates in terms of available vanilla swap rates, through a technique called convexity adjustment. In addition, we also introduce cross-currency derivatives, so-called the quanto derivatives, and address their pricing, which requires the inputs of correlations between state variables and the exchange rates.

The so-called convexity adjustment essentially is about calculating the expectation of a bond yield under a forward measure. Note that the bond yield is in general not a martingale under the forward measure, but the forward price of the bond is. By utilizing the martingale property of the forward bond price, we produce an approximation formula for the expected bond yield in terms of the forward bond yield. We will see that the expected bond yield is bigger than the forward bond yield with certainty, owing to the convex feature of the bond price as a function of its yield.

From a mathematical viewpoint, what we do in this chapter can be summarized as follows: given the expectation of a random variable under one measure, compute the expectation of the variable under another equivalent measure. When the underlying of an option is an asset, it is a matter of measure change and thus is somewhat trivial. But when the underlying of an option is a yield, a change in measures does not help much because the dynamics of the yield can become too complex. For an expected yield, we produce an approximation formula by taking advantage of the martingale property of the corresponding forward bond price.

DOI: 10.1201/9781003389101-10

10.1 Pricing through Adjustments

In some FRNs and non-vanilla swaps, the cash flows depend on bond yields or swap rates of a fixed maturity. The pricing of such cash flows requires taking expectations under a series of forward measures with delivery at the cash flow days (the so-called cash flow measures). At the same time, forward bond yields and swap rates are observable (directly or indirectly) in the marketplace, which are known to be close to the expected yields or swap rates. This section presents a procedure of fine tuning that generates a good approximation of an expected yield based on its known forward yield. As we shall see, the mechanism of the adjustment (from a forward yield to an expected forward yield) lies in the convex relationship between the bond price and its yield, which leads to the name of the technique. Convexity adjustments of this kind are often needed in pricing various contingent claims on yields.

10.1.1 General Theory

We demonstrate the need of convexity adjustment with an option on a bond yield.

Example 10.1. *Consider a bond with coupon rate c, to be paid at times $T_i = T_0 + i\Delta T, i = 1, \ldots, N$. Let y_T denote the YTM of the bond at time T. We consider the pricing of the following option on the yield,*

$$V_T = 100\,(y_T - K)^+, \tag{10.1}$$

which matures at $T < T_N$. Here is the market practice. The bond yield, y_t, is assumed to follow a lognormal process under the T-forward measure, $\mathbb{Q}_T$, with a constant volatility, σ_y, and the option is priced by Black's formula:

$$V_0 = 100 P(0, T) \left(E^{\mathbb{Q}_T}[y_T]\,\Phi(d_1) - K\Phi(d_2) \right), \tag{10.2}$$

where

$$d_1 = \frac{\ln\left(E^{\mathbb{Q}_T}[y_T]/K\right) + (1/2)\sigma_y^2 T}{\sigma_y \sqrt{T}},$$
$$d_2 = d_1 - \sigma_y \sqrt{T}. \tag{10.3}$$

In the Black formula, $E^{\mathbb{Q}_T}[y_T]$ is an input and needs to be evaluated in advance. In general, the arbitrage-free drift of y_t is state dependent and can be so complicated that an exact evaluation of the expectation is very difficult. At the same time, $E^{\mathbb{Q}_T}[y_T]$ is known to be closed to the forward bond yield, and the latter is often readily available. Therefore, it is desirable to develop some approximation formulae for the expected yield by taking advantage of the available forward bond yield.

The general technique for convexity adjustment is attributed to Brotherton-Ratcliffe and Iben (1993). The key to this technique is to take advantage of the martingale property of the forward bond price under the corresponding forward measure. According to the definition of the "stripped-dividend price" in Chapter 5, the T-forward price of the bond seen at time $t \leq T$ is

$$F_t^T \triangleq \frac{B^c(t) - \sum_{T_i \leq T} \Delta T c P(t, T_i)}{P(t, T)}. \tag{10.4}$$

The forward bond yield, y_t^T, is defined to be the YTM of the forward price, that is, y_t^T solves the equation

$$\sum_{T_i > T}^{T_N} \frac{\Delta T c}{\left(1 + \Delta T y_t^T\right)^{(T_i - T)/\Delta T}} + \frac{1}{\left(1 + \Delta T y_t^T\right)^{(T_N - T)/\Delta T}} = F_t^T. \tag{10.5}$$

Note that, at time T, the forward bond yield becomes the spot bond yield, $y_T^T = y_T$. It follows that $E^{\mathbb{Q}_T}[y_T] = E^{\mathbb{Q}_T}[y_T^T]$.

Before its evaluation, we show that the convexity adjustment is positive in general. Let $B_F(y_t^T)$ denote the left-hand side of Equation 10.5. The martingale property of the forward bond price means

$$B_F(y_0^T) = E^{\mathbb{Q}_T}[B_F(y_T^T)]. \tag{10.6}$$

Since $B_F(y)$ is a monotonically decreasing and convex function of y, $y_t^T = B_F^{-1}(F_t^T)$ exists and it is also a convex function of F_t^T. By Jensen's inequality,

$$\begin{aligned} E^{\mathbb{Q}_T}\left[y_T^T\right] &= E^{\mathbb{Q}_T}\left[B_F^{-1}(F_T^T)\right] \\ &\geq B_F^{-1}\left(E^{\mathbb{Q}_T}\left[F_T^T\right]\right) = B_F^{-1}(F_0^T) = y_0^T. \end{aligned} \tag{10.7}$$

To evaluate $E^{\mathbb{Q}_T}\left[y_T^T\right]$, we consider the Taylor expansion of $B_F(y_T^T)$ around y_0^T and retain terms up to the second order:

$$B_F\left(y_T^T\right) = B_F\left(y_0^T\right) + B_F'\left(y_0^T\right)\left(y_T^T - y_0^T\right) + \frac{1}{2}B_F''\left(y_0^T\right)\left(y_T^T - y_0^T\right)^2. \tag{10.8}$$

Then, we take the $\mathbb{Q}_T$-expectation of the above equation, obtaining

$$\begin{aligned} E^{\mathbb{Q}_T}\left[B_F\left(y_T^T\right)\right] &= B_F\left(y_0^T\right) + B_F'\left(y_0^T\right)E^{\mathbb{Q}_T}\left[y_T^T - y_0^T\right] \\ &\quad + \frac{1}{2}B_F''\left(y_0^T\right)E^{\mathbb{Q}_T}\left[\left(y_T^T - y_0^T\right)^2\right] = B_F\left(y_0^T\right). \end{aligned} \tag{10.9}$$

From the equation above we obtain an approximation to the expectation of the yield:

$$E^{\mathbb{Q}_T}\left[y_T^T\right] = y_0^T - \frac{1}{2}E^{\mathbb{Q}_T}\left[\left(y_T^T - y_0^T\right)^2\right]\frac{B_F''(y_0^T)}{B_F'(y_0^T)}. \tag{10.10}$$

Because $B_F(y)$ is monotonically decreasing and convex, there are $B_F'(y) < 0$ and $B_F''(y) > 0$, implying that the order $E^{\mathbb{Q}_T}\left[y_T^T\right] \geq y_0^T$ holds with certainty.

The only problem left now is to estimate $E^{\mathbb{Q}_T}\left[\left(y_T^T - y_0^T\right)^2\right]$. Standard market practice is to assume that y_t^T follows lognormal dynamics under $\mathbb{Q}_T$,

$$dy_t^T = y_t^T \left(\mu\, dt + \sigma_y\, dW_t\right), \tag{10.11}$$

where σ_y is constant while μ is a small deterministic function. The formal solution to Equation 10.11 is

$$y_T^T = y_0^T\, e^{\left(\bar{\mu} - \frac{1}{2}\sigma_y^2\right)T + \sigma_y W_T}, \tag{10.12}$$

where $\bar{\mu} = \int_0^T \mu\, dt / T$. Furthermore, by assuming $\bar{\mu}T \ll 1$, we have

$$
\begin{aligned}
E^{\mathbb{Q}_T}\left[\left(y_T^T - y_0^T\right)^2\right] &= \left(y_0^T\right)^2 E^{\mathbb{Q}_T}\left[\left(e^{\left(\bar{\mu} - \frac{1}{2}\sigma_y^2\right)T + \sigma_y W_T} - 1\right)^2\right] \\
&= \left(y_0^T\right)^2 \left[e^{\left(2\bar{\mu} + \sigma_y^2\right)T} - 2e^{\bar{\mu}T} + 1\right] \\
&\approx \left(y_0^T\right)^2 \sigma_y^2 T.
\end{aligned} \tag{10.13}
$$

Substituting the approximation (Equation 10.13) back to Equation 10.10, we finally arrive at a general formula for convexity adjustment:

$$E^{\mathbb{Q}_T}\left[y_T^T\right] = y_0^T - \frac{1}{2}\left(y_0^T\right)^2 \sigma_y^2 T \frac{B_F''\left(y_0^T\right)}{B_F'\left(y_0^T\right)}. \tag{10.14}$$

As an example of the application of Equation 10.14, we consider a time-T cash flow linked to the yield of $(T + \Delta T)$-maturity ZCB, y_T^T. The forward price–forward yield relationship for the $(T + \Delta T)$-maturity zero-coupon bond is

$$B_F\left(y_t^T\right) = \frac{1}{1 + \Delta T y_t^T}. \tag{10.15}$$

Differentiating $B_F\left(y_t^T\right)$ repeatedly, we have

$$
\begin{aligned}
B_F'\left(y_t^T\right) &= -\frac{\Delta T}{\left(1 + \Delta T y_t^T\right)^2}, \\
B_F''\left(y_t^T\right) &= \frac{2\left(\Delta T\right)^2}{\left(1 + \Delta T y_t^T\right)^3}.
\end{aligned} \tag{10.16}
$$

Substituting Equation 10.16 into 10.14, we obtain the correction formula

$$E^{\mathbb{Q}_T}\left[y_T^T\right] = y_0^T + \left(y_0^T\right)^2 \sigma_y^2 T \frac{\Delta T}{1 + \Delta T y_0^T}, \tag{10.17}$$

where y_0^T is the forward bond yield observed at time 0.

Next, we continue with the pricing of the contingent claim on a bond yield.

Example 10.2. *Consider the valuation of the one-year bond yield three years later (i.e., in-3-to-1) using the following input parameters and term structure.*

1. *The volatility of the in-3-to-1 bond yield is 20%.*

2. *The current yield curve is flat at 5% (with annual compounding).*

Solution: *The value of the instrument is*

$$V = (1.05)^{-3} \times E^{\mathbb{Q}_3}[y], \qquad (10.18)$$

where y is the yield of the one-year bond three years forward. The relation between the yield and price of a one-year zero-coupon bond is

$$B_F(y) = \frac{1}{1+y}. \qquad (10.19)$$

There are

$$B'_F(y) = \frac{-1}{(1+y)^2} \quad and \quad B''_F(y) = \frac{2}{(1+y)^3}. \qquad (10.20)$$

Taking $y = 0.05$, we have

$$B'_F(0.05) = \frac{-1}{(1+0.05)^2} = -0.9070,$$

$$B''_F(0.05) = \frac{2}{(1+0.05)^3} = 1.7277. \qquad (10.21)$$

The convexity adjustment is

$$0.5 \times 0.05^2 \times 0.2^2 \times 3 \times \frac{1.7277}{0.9070} = 0.00057, \qquad (10.22)$$

or 5.7 basis points, so that the value of the instrument is $0.05057/(1.05)^3 = 0.043684$. Note that the price error would be 1.1% if there were no adjustment.

10.1.2   Convexity Adjustment for CMS and CMT Swaps

A major application of convexity adjustment formula, Equation 10.14, is in the evaluation of constant maturity swaps (CMS) and constant maturity Treasury swaps (CMT), which are agreements to exchange payments indexed to swap rates or Treasury yields of certain maturity for a fixed rate or SOFR. For example, the most popular CMT exchanges the payments indexed to 10-year Treasury yields for payments indexed to a fixed rate or SOFR.

To see the application of convexity adjustment to the pricing of constant maturity leg in a CMS, we consider entering a forward swap with tenor (T_m, T_n) at time $t = 0 < T_m$. The T-forward price of the fixed leg (as well as the floating leg) of the swap at $t = 0$ is

$$F_0^T = \frac{P(0, T_m)}{P(0, T)}, \qquad (10.23)$$

which is a $\mathbb{Q}_T$-martingale by definition. Let $y_t = R_{m,n}(t)$ be the market prevailing swap rate for tenor (T_m, T_n), then y_t is the uniform discount rate for coupon bonds with tenor (T_m, T_n). At time $t \leq T_m$, the T-forward price of the fixed leg of the swap initiated at $t = 0$ is given by

$$B_F(y_t) = \sum_{T_i > T, i > m} \frac{\Delta T y_0}{(1 + \Delta T y_t)^{(T_i - T)/\Delta T}} + \frac{1}{(1 + \Delta T y_t)^{(T_N - T)/\Delta T}}. \quad (10.24)$$

The convexity adjustment formula for the swap rate follows as

$$E^{\mathbb{Q}_T}[y_T] = y_0 - \frac{1}{2} y_0^2 \sigma_y^2 T \frac{B_F''(y_0)}{B_F'(y_0)}.$$

Let us apply the results above to an option on swap rates.

Example 10.3. *Evaluate the derivative on the in-3-to-3 swap rate with payoff*

$$V_3 = 100 \times (R_{3,6}(3) - R_{3,6}(0))^+. \quad (10.25)$$

Assume that the current yield curve is flat at 5%, and the volatility of the swap rate, $R_{3,6}(t)$, is 20%.
Solution: *According to Equations 10.2 and 10.3, the value of the option is given by*

$$V_0 = 100 P(0, 3) \left(E^{\mathbb{Q}_3}[R_{3,6}(3)] \, \Phi(d_1) - K \Phi(d_2) \right), \quad (10.26)$$

where

$$P(0, 3) = (1.05)^{-3},$$
$$d_1 = \frac{\ln\left(E^{\mathbb{Q}_3}[R_{3,6}(3)] / R_{3,6}(0) \right) + (1/2)(0.2)^2 \times 3}{0.2\sqrt{3}}, \quad (10.27)$$
$$d_2 = d_1 - 0.2\sqrt{3}.$$

The only input we do not yet know in Black's formula is $E^{\mathbb{Q}_3}[R_{3,6}(3)]$. For simplicity we let $y_0 = R_{3,6}(0)$ and $y = R_{3,6}(3)$. For the par bond over the period $(3, 6)$, the forward price–forward yield relation for the three-year bond is

$$B_F(y) = \frac{y_0}{1 + y} + \frac{y_0}{(1 + y)^2} + \frac{1 + y_0}{(1 + y)^3}. \quad (10.28)$$

Its first and second derivatives are

$$B_F'(y) = -\frac{y_0}{(1 + y)^2} - \frac{2y_0}{(1 + y)^3} - \frac{3(1 + y_0)}{(1 + y)^4},$$
$$B_F''(y) = \frac{2y_0}{(1 + y)^3} + \frac{6y_0}{(1 + y)^4} + \frac{12(1 + y_0)}{(1 + y)^5}. \quad (10.29)$$

Taking $y = y_0 = 5\%$ in Equation 10.29, we obtain

$$B_F'(y_0) = -2.7232, \quad B_F''(y_0) = 10.2056. \quad (10.30)$$

Hence, the adjustment is

$$\frac{1}{2} \times 0.05^2 \times 0.2^2 \times 3 \times \frac{10.2056}{2.7232} = 0.00225, \qquad (10.31)$$

and thus

$$E^{\mathbb{Q}_3}[R_{3,6}(3)] = 0.05 + 0.00225 = 0.05225. \qquad (10.32)$$

Inserting Equation 10.32 into Equations 10.26 and 10.27, we finally obtain $V_0 = 0.7093$.

Next, we proceed to the pricing of CMS and CMT. At $t = 0$, the price of the CMS or CMT leg is given by

$$V_0 = \sum_{j=m+1}^{n} \Delta T P(0, T_j) E^{\mathbb{Q}_j}[y_N(T_j, T_j)] + P(0, T_n), \qquad (10.33)$$

where $y_N(t, T_j)$ is the N-year Treasury yield beyond T_j seen at time t. The relationship between $y_N(t, T_j)$ and its corresponding T_j-forward bond price is

$$B_F(y_N) = \sum_{i=1}^{N/\Delta T} \frac{\Delta T y_N(0, T_j)}{(1 + \Delta T y_N)^i} + \frac{1}{(1 + \Delta T y_N)^{N/\Delta T}}, \qquad (10.34)$$

where, for simplicity, we let y_N stand for $y_N(t, T_j)$. Then,

$$B_F'(y_N) = \sum_{i=1}^{N/\Delta T} \frac{-i\,(\Delta T)^2\, y_N(0, T_j)}{(1 + \Delta T y_N)^{i+1}} + \frac{-N}{(1 + \Delta T y_N)^{(N/\Delta T)+1}},$$

$$B_F''(y_N) = \sum_{i=1}^{N/\Delta T} \frac{i(i+1)\,(\Delta T)^3\, y_N(0, T_j)}{(1 + \Delta T y_N)^{i+2}} + \frac{N(N + \Delta T)}{(1 + \Delta T y_N)^{(N/\Delta T)+2}}.$$

$$(10.35)$$

The convexity adjustment for the CMT swap rate is

$$E^{\mathbb{Q}_j}[y_N(T_j, T_j)] = y_N(0, T_j) - \frac{1}{2}\,(y_N(0, T_j))^2\, \sigma_{j,N}^2 T_j \frac{B_F''(y_N(0, T_j))}{B_F'(y_N(0, T_j))},$$

$$(10.36)$$

for $j = m + 1, 2, \ldots, n$, where $\sigma_{j,N}$ is the percentage volatility of the swap rate, $y_N(t, T_j)$.

10.2   Quanto Derivatives

Quanto derivatives are cross-currency derivatives where two or more currencies are involved. They can be categorized into two types. The first type is a derivative that is measured and paid in a foreign currency, before being converted to the domestic currency; and the second type is a derivative that is measured in the foreign currency but paid in the domestic currency without conversion. An example of quanto derivatives of the second type is the CME futures contract on the Nikkei futures index, for which the underlying variable is the Nikkei index but the payoff is settled directly in USD. The pricing approaches of these two types of derivatives are very different, and the pricing of quanto derivatives of the first type is a lot easier.

We let

- $\mathbb{Q}_d$ be the martingale measure corresponding to the numeraire asset of a domestic saving account, $B_d(t)$;

- $\mathbb{Q}_f$ be the martingale measure corresponding to the numeraire asset of a foreign saving account, $B_f(t)$;

- x_t be the exchange rate: the value in the domestic currency of one unit of foreign currency; and

- $y_t = 1/x_t$: the value in foreign currency of one unit of domestic currency.

Consider an option of the first type with a payoff function, $f(S_T)$, on a foreign asset, S_t. In domestic currency, the payoff is

$$V_T = x_T f(S_T). \tag{10.37}$$

To price this option, we take the foreign saving account as the numeraire and make use of the corresponding martingale measure, $\mathbb{Q}_f$. Note that the foreign saving account is a tradable asset with price $x_t B_f(t)$ in the domestic currency. Under $\mathbb{Q}_f$, we know that the price of the option is given by

$$V_0 = x_0 E^{\mathbb{Q}_f} \left[\frac{V_T}{x_T B_f(T)} \right]. \tag{10.38}$$

As an example, we consider a call option such that $f(S_T) = (S_T - K)^+$. Then,

$$V_0 = x_0 \big(S_0 \Phi(d_1) - P_f(0,T) K \Phi(d_2) \big), \tag{10.39}$$

where d_1 and d_2 take the usual values for a call option under the stochastic foreign interest rate. By examining Equations 10.37 and 10.38, we understand that general quanto options of the first type can be priced by taking expectations of the payoffs discounted in the foreign currency under the foreign risk-neutral measure, before converting the value to the domestic currency.

Quanto options of the second type are intriguing, and their pricing is less straightforward. We focus first on the pricing of quanto futures before applying relevant results to pricing quanto options.

Let S_t be a domestic asset whose price is, however, indexed to a foreign asset. As discussed in section 5.6, the futures price of the asset is given by

$$\tilde{F}_t^d = E^{\mathbb{Q}_d}[S_T \mid \mathcal{F}_t] = E_t^{\mathbb{Q}_d}[S_T]. \tag{10.40}$$

We will try to calculate $\tilde{F}_t^{\mathrm{d}}$ based on the futures price in the foreign currency,

$$\tilde{F}_t^{\mathrm{f}} = E^{\mathbb{Q}_{\mathrm{f}}}[S_T \mid \mathcal{F}_t] = E_t^{\mathbb{Q}_{\mathrm{f}}}[S_T], \qquad (10.41)$$

which is available in the foreign market. On the relationship between $\tilde{F}_t^{\mathrm{d}}$ and $\tilde{F}_t^{\mathrm{f}}$, we have

Theorem 10.2.1. *Suppose that both $\tilde{F}_t^{\mathrm{f}}$ and x_t are lognormally distributed with volatilities Σ_F and Σ_x, respectively. Then, there is*

$$\tilde{F}_t^{\mathrm{d}} = \tilde{F}_t^{\mathrm{f}} e^{-\int_t^T \Sigma_F^{\mathrm{T}} \Sigma_x \mathrm{d}t}. \qquad (10.42)$$

Proof. By a change of measure, we have

$$E_t^{\mathbb{Q}_{\mathrm{d}}}[S_T] = E_t^{\mathbb{Q}_{\mathrm{f}}}\left[S_T \frac{\zeta_T}{\zeta_t}\right], \qquad (10.43)$$

where

$$\zeta_t = \left.\frac{\mathrm{d}\mathbb{Q}_{\mathrm{d}}}{\mathrm{d}\mathbb{Q}_{\mathrm{f}}}\right|_{\mathcal{F}_t} = \frac{y_t B_{\mathrm{d}}(t)}{y_0 B_{\mathrm{f}}(t)}, \qquad (10.44)$$

and

$$B_{\mathrm{d}}(t) = e^{\int_0^t r_{\mathrm{d}}(s)\mathrm{d}s} \quad \text{and} \quad B_{\mathrm{f}}(t) = e^{\int_0^t r_{\mathrm{f}}(s)\mathrm{d}s}.$$

Here $r_{\mathrm{d}}(t)$ and $r_{\mathrm{f}}(t)$ are domestic and foreign short rates, respectively. Working with the forward price, and then making use of the futures–forward relationship, we can derive the following expression of the terminal value of asset prices under measure $\mathbb{Q}_{\mathrm{f}}$:

$$
\begin{aligned}
S_T &= F_T^T e^{\int_t^T -\frac{1}{2}\|\Sigma_F\|^2 ds + \Sigma_F^T (d\mathbf{W}_{\mathrm{f}}(s) - \Sigma_{P,\mathrm{f}} ds)} \\
&= \tilde{F}_t^T e^{\int_t^T (\Sigma_F^T \Sigma_{P,\mathrm{f}} - \frac{1}{2}\|\Sigma_F\|^2) ds + \Sigma_F^T (d\mathbf{W}_{\mathrm{f}}(s) - \Sigma_{P,\mathrm{f}} ds)} \\
&= E_t^{\mathbb{Q}_{\mathrm{f}}}[S_T] e^{\int_t^T -\frac{1}{2}\|\Sigma_F\|^2 ds + \Sigma_F^T d\mathbf{W}_{\mathrm{f}}(s)},
\end{aligned}
\qquad (10.45)
$$

where $\mathbf{W}_{\mathrm{f}}(t)$ is a $\mathbb{Q}_{\mathrm{f}}$-Brownian motion, Σ_F is the volatility of $\tilde{F}_t^{\mathrm{f}}$, and $\Sigma_{P,\mathrm{f}}$ is the volatility of the foreign ZCB with maturity T. Meanwhile, the process of y_t under $\mathbb{Q}_{\mathrm{f}}$ is

$$\mathrm{d}y_t = y_t\left((r_{\mathrm{f}} - r_{\mathrm{d}})\,\mathrm{d}t + \Sigma_y^{\mathrm{T}} \mathrm{d}\mathbf{W}_{\mathrm{f}}(t)\right). \qquad (10.46)$$

It follows that

$$\frac{\zeta_T}{\zeta_t} = \frac{y_T B_{\mathrm{d}}(T)/y_0 B_{\mathrm{f}}(T)}{y_t B_{\mathrm{d}}(t)/y_0 B_{\mathrm{f}}(t)} = e^{\int_t^T -\frac{1}{2}\|\Sigma_y\|^2 ds + \Sigma_y^T d\mathbf{W}_{\mathrm{f}}(s)}. \qquad (10.47)$$

By substituting Equations 10.45 and 10.47 into Equation 10.43, we then obtain

$$
\begin{aligned}
E_t^{\mathbb{Q}_{\mathrm{d}}}[S_T] &= E_t^{\mathbb{Q}_{\mathrm{f}}}\left[E_t^{\mathbb{Q}_{\mathrm{f}}}[S_T] e^{\int_t^T -\frac{1}{2}(\|\Sigma_F\|^2 + \|\Sigma_y\|^2) ds + (\Sigma_F + \Sigma_y)^T d\mathbf{W}_{\mathrm{f}}(s)}\right] \\
&= E_t^{\mathbb{Q}_{\mathrm{f}}}[S_T] e^{\int_t^T \Sigma_F^T \Sigma_y ds}.
\end{aligned}
\qquad (10.48)
$$

Finally, Equation 10.42 follows from the fact that $\Sigma_x = -\Sigma_y$. $\square$

Note that Equation 10.48 can be recast into the following form, which is better known:

$$E_t^{\mathbb{Q}_d}[S_T] = E_t^{\mathbb{Q}_f}[S_T]\, e^{-\int_t^T \rho\sigma_F\sigma_x dx}, \tag{10.49}$$

where $\sigma_F = \|\mathbf{\Sigma}_F\|$, $\sigma_x = \|\mathbf{\Sigma}_x\|$ and ρ is the correlation between F_t^T and x_t.

The proof for the theorem above carries useful insights on pricing quanto derivatives. Based on Equation 10.46, we can show, using the quotient rule of differentiation, that the $\mathbb{Q}_f$-dynamics of x_t is

$$dx_t = x_t\left[\left(r_d - r_f + \|\mathbf{\Sigma}_y\|^2\right)dt - \mathbf{\Sigma}_y^T d\mathbf{W}_f(t)\right]. \tag{10.50}$$

On the other hand, it is well-known that the $\mathbb{Q}_d$-dynamics of x_t is

$$dx_t = x_t\left((r_d - r_f)\,dt - \mathbf{\Sigma}_y^T d\mathbf{W}_d(t)\right), \tag{10.51}$$

where $\mathbf{W}_d(t)$ is a $\mathbb{Q}_d$-Brownian motion. The coexistence of Equations 10.50 and 10.51 had once puzzled some people of the finance community. This is the so-called Siegel's paradox (Siegel, 1972). Let us now dismiss this paradox. We can recast Equation 10.50 into

$$dx_t = x_t\left[(r_d - r_f)\,dt - \mathbf{\Sigma}_y^T(d\mathbf{W}_f(t) - \mathbf{\Sigma}_y\,dt)\right]. \tag{10.52}$$

According to Equation 10.47, ζ_t satisfies

$$d\zeta_t = \zeta_t \mathbf{\Sigma}_y^T d\mathbf{W}_f(t). \tag{10.53}$$

Therefore, we can write

$$d\mathbf{W}_f(t) - \mathbf{\Sigma}_y dt = d\mathbf{W}_f(t) - \left\langle d\mathbf{W}_f(t), \frac{d\zeta_t}{\zeta_t}\right\rangle. \tag{10.54}$$

Because ζ_t is the Radon–Nikodym derivative of $\mathbb{Q}_d$ with respect to $\mathbb{Q}_f$, according to the CMG theorem, $\mathbf{W}_d(t) = \mathbf{W}_f(t) - \int_0^t \mathbf{\Sigma}_y\,ds$ is a $\mathbb{Q}_d$-Brownian motion. Hence, Equation 10.51 is reconciled with Equation 10.50.

Next, let us try to figure out the intuition behind the adjustment formula (Equation 10.49). Let ρ be the correlation between x_t and $\tilde{F}_t^f$. To replicate the quanto futures contract, we should maintain y_t units of the foreign futures contract on the foreign asset at any time $t < T$. This is a dynamic strategy. The P&L over a small interval of time $(t, t + \Delta t)$, is

$$\Delta\tilde{F}_t^f y_t \times x_{t+\Delta t} = \Delta\tilde{F}_t^f y_t(x_t + \Delta x_t)$$
$$= \Delta\tilde{F}_t^f + \Delta\tilde{F}_t^f \frac{\Delta x_t}{x_t}$$
$$\approx \Delta\tilde{F}_t^f + \rho\sigma_F\sigma_x\tilde{F}_t^f\Delta t. \tag{10.55}$$

In Equation 10.55, we can see that the P&L comes from two sources. The first is of course the change in the futures price, and the second is the change in the exchange rate. The first term of Equation 10.55 is symmetric to upward or downward moves of the futures price, yet the second term is not. Suppose, for example, that there is a positive correlation between x_t and $\tilde{F}_t$. Then, the second term is positive regardless of the direction of movement of the futures price. If the futures price moves up, the hedger makes a profit in the contract, which is likely to be accompanied by a higher exchange rate. If, on the other hand, the futures price moves down, then the hedger suffers a loss, but it is likely to be subject to a lower exchange rate. Thus, if the price for the quanto futures is set at $\tilde{F}_t^{\mathrm{d}} = \tilde{F}_t^{\mathrm{f}}$, smart investors will short the futures and hedge this position with the foreign futures, until the $\tilde{F}_t^{\mathrm{d}}$ is driven down to the level stated in Equation 10.49.

As a general result, we now derive the price process of the foreign asset under the domestic risk-neutral measure, $\mathbb{Q}_{\mathrm{d}}$. It is well-known that, under the foreign risk-neutral measure, $\mathbb{Q}_{\mathrm{f}}$, the price of a foreign asset follows

$$\mathrm{d}S_t = S_t\big(r_{\mathrm{f}}\,\mathrm{d}t + \boldsymbol{\Sigma}_S^{\mathrm{T}}\mathrm{d}\mathbf{W}_{\mathrm{f}}(t)\big). \tag{10.56}$$

Define a $\mathbb{Q}_{\mathrm{d}}$-Brownian motion, $\mathbf{W}_{\mathrm{d}}(t)$, as

$$\begin{aligned}
\mathrm{d}\mathbf{W}_{\mathrm{d}}(t) &= \mathrm{d}\mathbf{W}_{\mathrm{f}}(t) - \left\langle \mathrm{d}\mathbf{W}_{\mathrm{f}}(t), \frac{\mathrm{d}\zeta_t}{\zeta_t} \right\rangle \\
&= \mathrm{d}\mathbf{W}_{\mathrm{f}}(t) + \boldsymbol{\Sigma}_x \mathrm{d}t.
\end{aligned} \tag{10.57}$$

Then, in terms of $\mathbf{W}_{\mathrm{d}}(t)$, we have the price process of S_t under $\mathbb{Q}_{\mathrm{d}}$ to be

$$\mathrm{d}S_t = S_t\left(\big(r_{\mathrm{f}} - \boldsymbol{\Sigma}_S^{\mathrm{T}}\boldsymbol{\Sigma}_x\big)\,\mathrm{d}t + \boldsymbol{\Sigma}_S^{\mathrm{T}}\mathrm{d}\mathbf{W}_{\mathrm{d}}(t)\right). \tag{10.58}$$

When interest rates in both currencies are deterministic, the above "risk-neutralized" process can be applied directly to pricing quanto options. Our interest, of course, is in the pricing of quanto options under stochastic interest rates.

Now, let us consider the pricing of a quanto call option of the second type. The payoff of the option,

$$V_T = (S_T - K)^+, \tag{10.59}$$

is indexed to the foreign asset in the foreign currency, but paid in the domestic currency without conversion. Because $\tilde{F}_T^{\mathrm{d}} = S_T$ by the definition of a futures price, we can rewrite the payoff as

$$V_T = \left(\tilde{F}_T^{\mathrm{d}} - K\right)^+ \text{ dollars.} \tag{10.60}$$

By the martingale property of the futures price, we have the following process of $\tilde{F}_t^{\mathrm{d}}$ under the domestic risk-neutral measure, $\mathbb{Q}_{\mathrm{d}}$:

$$\mathrm{d}\tilde{F}_t^{\mathrm{d}} = \tilde{F}_t^{\mathrm{d}}\boldsymbol{\Sigma}_F^{\mathrm{T}}\,\mathrm{d}\mathbf{W}_{\mathrm{d}}(t). \tag{10.61}$$

Furthermore, under the forward measure, $\mathbb{Q}_d^T$, Equation 10.61 becomes

$$d\tilde{F}_t^d = \tilde{F}_t^d \boldsymbol{\Sigma}_F^T \left(d\mathbf{W}_d^T(t) + \boldsymbol{\Sigma}_{P_d} dt \right), \tag{10.62}$$

where $\mathbf{W}_d^T(t)$ is a $\mathbb{Q}_d^T$-Brownian motion and $P_d = P(t,T)$ stands for the domestic zero-coupon bond. It follows that

$$V_0 = E^{\mathbb{Q}_d} \left[\frac{\left(\tilde{F}_T^d - K \right)^+}{B_d(T)} \right]$$

$$= P_d(0,T) E^{\mathbb{Q}_d^T} \left[\left(\tilde{F}_T^d - K \right)^+ \right]$$

$$= P_d(0,T) \left(E^{\mathbb{Q}_d^T} \left[\tilde{F}_T^d \right] \Phi(d_1) - K\Phi(d_2) \right), \tag{10.63}$$

with

$$d_1 = \frac{\ln\left(E^{\mathbb{Q}_d^T}\left[\tilde{F}_T^d \right]/K \right) + (1/2)\sigma_F^2 T}{\sigma_F \sqrt{T}}, \quad d_2 = d_1 - \sigma_F \sqrt{T},$$

$$\sigma_F^2 = \frac{1}{T} \int_0^T \|\boldsymbol{\Sigma}_F\|^2 \, dt, \tag{10.64}$$

and, based on Equation 10.62,

$$E^{\mathbb{Q}_d^T}\left[\tilde{F}_T^d \right] = \tilde{F}_0^d \exp\left(\int_0^T \boldsymbol{\Sigma}_F^T \boldsymbol{\Sigma}_{P_d} dt \right)$$

$$= \tilde{F}_0^d \exp\left(\int_0^T \rho_{F,P_d} \sigma_F \sigma_{P_d} dt \right). \tag{10.65}$$

When $\rho_{F,P_d} = 0$, there is apparently $E^{\mathbb{Q}_d^T}\left[\tilde{F}_T^d \right] = \tilde{F}_0^d = E^{\mathbb{Q}_d}\left[\tilde{F}_T^d \right]$. Furthermore, when there is no correlation between x_t and $\tilde{F}_t^f$, Equation 10.63 reduces to Black's formula for call options on domestic futures.

Exercises

1. Price a cash flow in three years that is equal to $100 times the three-year swap rate at the time. Let the forward-rate curve be

$$f_j(0) = 0.03 + 0.0006 \times j, \quad \Delta T_j = 0.5, \quad \forall j,$$

and the volatility of the swap rate be 30%. What will be the price if the payment is made in three and a half years?

2. (Continued from Problem 1) If the cash flow is instead from a call option on the swap rate with a strike rate of 4%, what should be the price of the option with and without payment in arrears?

3. For all problems below, we use the spot term structure

$$f_j(0) = 0.03 + 0.0003 \times j, \quad \Delta T_j = 0.25, \quad \forall j.$$

Pricing is done under the market model (which assumes lognormal forward rates or swap rates). The payment frequency is a quarter year for caps, and half a year for swaps.

a. Price the ATM caplet of one-year maturity (with the strike rate equals to $f(0; 0.75, 1)$). Take the forward-rate volatility as

$$\gamma_4(t) = \begin{cases} 20\%, & \text{for} \quad 0 \le t \le 0.75, \\ \frac{1-t}{0.25} \times 20\%, & \text{for} \quad 0.75 < t \le 1. \end{cases}$$

b. Price the five-year maturity ATM option on five-year swap rate (i.e., $R_{10,20}(5)$; the strike rate is $R_{10,20}(0)$, and $\Delta T = 0.5$). Assume a 20% swap-rate volatility. (Note that this is not a swaption and convexity adjustment will be needed.)

4. Price the Nikkei quanto futures that pay in USD. The maturity of the futures contract is three months, and the correlation between the index and the Yen-dollar exchange rate (i.e., value in USD per Yen) is -20%, and the current Nikkei index is 12,117. Describe the hedging strategy using the standard Nikkei index futures.

5. (Continued from Problem 4) Suppose that the volatility of the Nikkei index is 25%, price a quanto call option on the index that matures in three months with a strike price 12,100.

Chapter 11

Market Models with Stochastic Volatilities

After rigorous justifications were published regarding its well-posedness (Brace, Gatarek and Musiela, 1997; Jamshidian, 1997; and Miltersen, Sandmann and Sondermann, 1997), market model became the benchmark model for interest rate derivatives. One of many virtues of the market model is that it justifies the use of Black's formula for caplet and swaption prices, which has long been a standard market practice. Black's formula establishes a relationship between option prices and local volatilities of the forward rates, and such a relationship has enabled fast calibration of the standard model (Wu, 2003). Nevertheless, the standard market model is also known for generating flat implied volatility curves only, whereas the stylized patterns of implied volatility curves observed in LIBOR markets were skewed smiles. In other words, the market model cannot be calibrated to non-flat volatility curves and thus is flawed. Because of the benchmark role of caps and swaptions in the fixed-income derivatives markets, there had been great interest in extending the standard market model so as to generate the observed implied volatility smiles or skews.

It had become a major challenge to model volatility smiles in the context of the LIBOR market model for interest-rate derivatives. In modern literature of option pricing, the notion of smiles means non-flat curves of implied Black's volatilities. Over the last two decades, various solutions to smile modeling were proposed and various degrees of success were achieved. Without exception, all solutions were based on adopting at least one of the following features or risk factors: state-dependent volatilities, displaced diffusions, stochastic volatilities, and/or jumps, on top of Brownian diffusions to the driving dynamics of the forward rate curve.

Let us comment on some representative works for smile modeling. Andersen and Andreasen (2000) were among the first to adopt constant-elasticity-variance (CEV) dynamics for forward rates, a special case of state-dependent volatility models. On top of the CEV model, Andersen and Brotherton-Ratcliffe (2001) superimposed an independent square-root volatility process that serves to generate additional curvature to the otherwise monotonic volatility skews. The pricing of caps and swaptions under the CEV-type models is then done by asymptotic expansions. Andersen and Andreasen (2002) also combined displaced diffusions with stochastic volatilities. Caplet

DOI: 10.1201/9781003389101-11

pricing under such models can be achieved through Fourier transforms. In these models, there is no correlation between the forward rates and their stochastic volatilities, and the mechanism for volatility smiles or skews lies in the use of either the CEV dynamics or displaced diffusions. Around the same time, correlation-based models were also developed, represented by the stochastic alpha, beta and rho model (SABR) (Hagan *et al.*, 2002) and the Heston's type extension of the market model by Wu and Zhang (2008).[1] Under these models, the generation of volatility smiles were mainly dictated by the stochastic volatility and its correlation with a forward or swap rate, and options can be priced through asymptotic expansion, Laplace transforms, or numerical PDE methods. To model time-dependent skews, Piterbarg (2003) combines stochastic volatility with displaced diffusion for LIBOR modeling. In yet another line of research, Glasserman and Kou (2003) develop a comprehensive term structure model with the jump-diffusion dynamics. Under this model, approximate closed-form formulae for caplets and swaptions were developed (Glasserman and Merener, 2003). The Glasserman and Kou (2003) model was later extended by Eberlein and Özkan (2004), with jump-diffusion processes replaced by the general Lévy processes; the latter serves as a framework for a wide class of jump-diffusion processes. In the industry, however, it turned out that the SABR model, which combines the CEV dynamics for state variables with lognormal stochastic volatilities, achieved enormous success. The SABR model can capture various shapes of implied volatility curves, and its approximate closed-form formula for vanilla options in terms of the implied Black volatilities enables efficient calculation of Greeks, the sensitivity parameters. Empirical studies are also most supportive of the SABR model, see e.g., Wu (2012).

As is discussed in Chapter 3, FOMC sets the target range of the Fed fund rate in their routine rate-setting meetings, so that jumps or changes in the Fed fund rate occur only after the meetings. Thanks to the sufficient communications between FOMC and the financial market, jumps in the overnight Fed fund rate has not been a surprise since 2006, and the anticipated jumps in the Fed fund rate months or even years ahead are continuously updated and adequately priced in by SOFR futures. For these reasons, random jumps in the backward-looking SOFR term rates are rare, unlike the situation in the equity markets. It has become a popular belief in fixed-income derivatives markets that stochastic volatilities are the primary mechanism for the implied volatility smiles. In this chapter, we will focus on two representative models of stochastic volatility: the SABR model (Hagan *et al.*, 2002) and the Heston model in the context of SOFR term rates by Wu and Zhang (2008).

---

[1]First presented in Frontier in Finance Workshop on "Interest-rate models: theory and implementation," Ecole Polytechnique - National Center for Scientific Research (CNRS), Paris, May 2002.

11.1 SABR Model

The SABR model (Hagan *et. al.*, 2002) works with forward prices or forward rates under their corresponding forward measure,

$$
\begin{aligned}
df_t &= \alpha_t f_t^\beta dW_t, \quad f_0 = f, \\
d\alpha_t &= \nu \alpha_t dZ_t, \quad \alpha_0 = \alpha,
\end{aligned}
\tag{11.1}
$$

where $\nu \geq 0$, W_t and Z_t are Brownian motions under the forward measure of choice, and

$$
dW_t dZ_t = \rho dt \quad \text{for} \quad |\rho| \leq 1.
$$

In most applications, β is taken within the interval $(0, 1)$. The parameter ν is the volatility of the volatility, often called "vol of vol" for convenience. The model is named after three of the four parameters α, β, ρ and ν.

According to Hagan *et al.* (2002), the SABR model was motivated by a major criticism of the local volatility models, such that under the local volatility models option's implied volatility curves move in the opposite direction against the underlying securities, which is inconsistent with reality and unacceptable. Although this criticism is disputable, as local volatility models that behave otherwise can be constructed without much difficulty, the dispute has never been a concern, as it has been overshadowed by the enormously success of the SABR model.

The SABR model is the stochastic version of constant elasticity variance (CEV) model, and it has two prominent features: stochastic volatility and skewness parameter. While the former generates volatility smiles, the latter, β, affects the skewness of the smiles. At the time when the SABR model was under development, it was already widely recognized that the stochastic volatility alone could not generate volatility smiles with enough skewness, particularly for short-maturity options. A peculiar feature of the SABR model is the adoption of lognormal dynamics without mean reversion for the stochastic volatility. It is conceivable that, for a short time horizon, the drift term in the volatility dynamics is dominated by the diffusion term and thus can be ignored, resulting in higher analytical tractability to the model.

We now address the pricing of European call options under the SABR model. A European call option of maturity T has the payoff

$$
\max\left(f_T - K,\, 0\right).
$$

According to the general pricing principle, the option value satisfies

$$
P(0, T) E_0^{Q_T} \left[(f_T - K)^+ \right].
$$

In the absence of analytical tractability, there are two approaches to valuate the option: numerical approach and, often more preferably, analytical approximation approach.

Hagan *et al.* (2002) takes the approach of singular perturbation expansion, an approach of analytical approximation, for the solution of the governing PDE corresponding to a general class of stochastic volatility (SV) models:

$$df_t = \alpha_t C(f_t) dW_t, \quad f_0 = f,$$
$$d\alpha_t = \nu \alpha_t dZ_t, \quad \alpha_0 = \alpha, \tag{11.2}$$

where $C(f)$ is deterministic function which, by taking $C(f) = f^\beta$, includes the SABR model as a special case. The small parameter they identify for singular perturbation expansion is the maturity of the option, i.e., $T \ll 1$. Note that an alternative approach based on heat kernel expansion was developed later, which takes advantage of the available formula for small-time expansion of heat kernels on manifolds, coupled with the saddle-point integration method for evaluating the expectation of the option payoffs. Although the approach of heat kernel expansion is a more systematic one, it is also mathematically more demanding for comprehension. Because of that, we have chosen to present the derivation and results of Hagan *et al.* (2002). For the heat kernel expansion method, we refer readers to Henry-Labordère (2005) or Paulot (2015).

The closed-form approximate formula for call option under the general SV model is not in terms of dollar values, but in terms of the implied Black volatilities (Hagan *et al.*, 2002):

$$\sigma_{\mathrm{B}}(f, K) = \nu \frac{\log{(f/K)}}{D(\zeta)} \times \left\{ 1 + \left[\frac{2\gamma_2 - \gamma_1^2 + f_{\mathrm{av}}^{-2}}{24} \alpha^2 C^2(f_{\mathrm{av}}) \right. \right.$$
$$\left. \left. + \frac{1}{4} \rho \nu \alpha \gamma_1 C(f_{\mathrm{av}}) + \frac{2 - 3\rho^2}{24} \nu^2 \right] T \right\}. \tag{11.3}$$

The derivation is tedious and thus is put in the appendix of this chapter. Here, f_{av} is an average of f and K:

$$f_{\mathrm{av}} = \frac{f + K}{2} \quad \text{or} \quad f_{\mathrm{av}} = \sqrt{fK},$$

$$D(\zeta) = \ln\left(\frac{\sqrt{1 - 2\rho\zeta + \zeta^2} + \zeta - \rho}{1 - \rho} \right),$$

for

$$\zeta = \frac{\nu}{\alpha} \int_K^f \frac{dx}{C(x)},$$

and the two new parameters are

$$\gamma_1 = \frac{C'(f_{\mathrm{av}})}{C(f_{\mathrm{av}})} \quad \text{and} \quad \gamma_2 = \frac{C''(f_{\mathrm{av}})}{C(f_{\mathrm{av}})}.$$

The dollar value of call options are then calculated using the Black formula.

Equation 11.3 is called the HKLW formula, named using the first letters of the authors' last names.

Equation 11.3 is not valid for at-the-money (ATM) options, when $f = K$. For the implied Black volatility of the ATM options, we take the limit $K \to f$ for Equation 11.3. It is obvious that

$$\zeta = \frac{\nu}{\alpha} \frac{f - K}{C(\tilde{f})}, \quad \text{for } \tilde{f} \text{ between } f \text{ and } K.$$

When $K \to f$, there are

$$\zeta \to 0 \quad \text{and} \quad D(\zeta) \to \zeta,$$

and eventually the following limit for the implied Black volatility:

$$\sigma_{\mathrm{B}}(f, f) = \alpha \frac{C(f)}{f}$$
$$\times \left\{ 1 + \left[\frac{2\gamma_2 - \gamma_1^2 + f^{-2}}{24} \alpha^2 C^2(f) + \frac{1}{4} \rho\nu\alpha\gamma_1 C(f) + \frac{2 - 3\rho^2}{24} \nu^2 \right] T \right\}. \tag{11.4}$$

For the SABR model, $C(f) = f^\beta$, and

$$\int_K^f \frac{du}{C(u)} = \frac{1}{1 - \beta} \left[f^{1-\beta} - K^{1-\beta} \right],$$
$$\gamma_1 = \frac{\beta}{f_{\mathrm{av}}} \quad \text{and} \quad \gamma_2 = -\frac{\beta(1 - \beta)}{f_{\mathrm{av}}^2}.$$

Let us make two comments. First, the HKLW formula is accurate enough for production uses when the strike is not too close to zero and the maturity is not too big. Note that the derivation procedure for the HKLW formula is based on the assumption of a small maturity and has not involved with boundary conditions of the governing PDE. Because of that, it was anticipated that the HKLW formula would yield poorer price accuracy for options with lower strikes or longer maturities. It turns out not quite the case and the HKLW formula performs better than anticipated. As long as the strike is not too low, the formula is able to yield prices that are accurate within the bid-ask spread for options of maturities up to ten years or longer. Second, depending on market conditions, the implied normal volatility formula may be preferred over the implied Black volatility formula. As a stepping stone, the implied normal volatility formula for the normal model is first derived in Hagan *et al.* (2002):

$$\sigma_{\mathrm{N}}(f, K) = \nu \frac{f - K}{D(\zeta)} \left\{ 1 + \left[\frac{2\gamma_2 - \gamma_1^2}{24} \alpha^2 C^2(f_{\mathrm{av}}) \right. \right.$$
$$\left. \left. + \frac{\rho\nu\gamma_1}{4} \alpha C(f_{\mathrm{av}}) + \frac{2 - 3\rho^2}{24} \nu^2 \right] T \right\}.$$

Value of call options can then be calculated through the formula:

$$V_0 = (f - K)\Phi(d_1) + \frac{\sigma_N \sqrt{T}}{\sqrt{2\pi}} e^{-d_1^2/2},$$

for

$$d_1 = \frac{f - K}{\sigma_N \sqrt{T}}.$$

When $K \to f$, the implied normal volatility becomes

$$\sigma_N(f, f) = \alpha C(f) \left\{ 1 + \left[\frac{2\gamma_2 - \gamma_1^2}{24} \alpha^2 C^2(f) \right. \right.$$
$$\left. \left. + \frac{\rho\nu\gamma_1}{4} \alpha C(f) + \frac{2 - 3\rho^2}{24} \nu^2 \right] T \right\}.$$

It is worth noting that the implied normal volatility formula generally offers more accurate option values than the lognormal implied volatility formula does, and it has enjoyed certain popularity in the swaption markets. At times when the swap rates are relatively high, the swap rates may behave more like normal random variables, when it becomes more favorable to use the implied normal volatility formula.

The above formulae for the implied Black volatility and the implied normal volatility are very easy to implement. In applications, the formulae are first calibrated to a selected set of implied Black's volatilities of traded options, which only poses as a small-scale root-finding problem. Moreover, the HKLW formulae are very robust: they can be calibrated to various shapes of volatility smiles in markets, including even the hockey-stick shape, in Figure 11.1, and rarely fail.

For the SABR model, the roles played by the four model parameters have become transparent. Let us describe in detail below.

1. β, the elasticity constant or the skewness parameter.

   - β controls the backbone, known as the trace of the ATM implied volatilities, $\sigma_B(f, f)$ (represented by the dash lines of Figure 11.2).
   - With any specific choice of β, market smiles can generally be fit more or less equally well.
   - β can be estimated from the historical data of the "backbone."
   - It is popular to choose β near 0.5 (it is claimed in Hagan *et al.* (2002) that for the JPY interest rate (IR) market, $\beta = 0$; yet for the USD IR market, $\beta = 0.5$).

2. α, the level of implied volatilities.

   - α is calibrated to the level of ATM volatility.

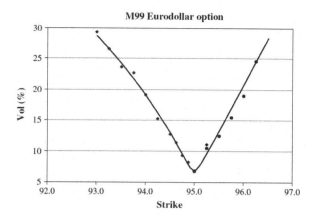

FIGURE 11.1: Implied volatility for the June 99 Eurodollar options, close-of-day values vs. predicted value (curve) by the SABR model. March 23, 1999, Bloomberg. (Courtesy of Applied. Math. Fin.)

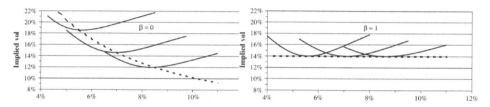

FIGURE 11.2: Backbone for different β. (Courtesy of Applied. Math. Fin.)

- One can use $\sigma_{\mathrm{ATM}} \overset{\triangle}{=} \sigma_{\mathrm{B}}(f, f)$ to replace α in the parameter set.

3. ρ and ν, the correlation and the "vol of vol."

  - ρ controls the smile and skew.
  - ν controls the curvature of the smile/skew.
  - For swaptions, the "vol of vol" ν can be big for short-dated options, and decreases as the time-to-exercise increases (Figure 11.3); where ρ starts near zero and becomes substantially more negative (Figure 11.4).

Regarding the frequency of calibration/fitting for each parameter is different, the model users have the following consensus.

- Typically, α or σ_{ATM} are updated daily or every few hours.
- ρ and ν are re-fitted every month or as needed.
- β can stay unchanged for a long time.

Finally, we discuss the calculation of Greeks and hedging with the SABR model. For this purpose, we denote the value function of a call option in terms of the implied Black volatility as

$$V_{call} = BS(f, K, \sigma_B(f, K), T).$$

- Delta

The delta risk is the risk associated with the underlying forward price/rate f. It predicts a change in option value for a side-way movement of the volatility curve caused by the change in the forward price/rate. The delta is calculated by

$$\Delta \equiv \frac{\partial V_{call}}{\partial f} = \frac{\partial BS}{\partial f} + \frac{\partial BS}{\partial \sigma_B} \frac{\partial \sigma_B(f, K)}{\partial f}.$$

Here, the first term is the ordinary delta risk that can be calculated from the Black model and the second term is the SABR model's correction to the delta risk: Black vega risk times the predicted change in the IV caused by the change in the forward f.

- Vega

Vega is defined as the sensitivity of the option value with respect to α or, more popularly, σ_{ATM}:

$$vega = \frac{\partial V_{call}}{\partial \sigma_{\text{ATM}}} = \frac{\partial V_{call}}{\partial \sigma_B} \frac{\partial \sigma_B}{\partial \sigma_{\text{ATM}}} = \frac{\partial V_{call}}{\partial \sigma_B} \frac{\frac{\partial \sigma_B}{\partial \alpha}}{\frac{\partial \sigma_{\text{ATM}}}{\partial \alpha}},$$

and

$$\frac{\partial V_{call}}{\partial \alpha} = \frac{\partial BS}{\partial \sigma_B} \frac{\partial \sigma_B(f, K; \alpha, \beta, \rho, \nu)}{\partial \alpha}.$$

According to Hagan *et al.* (2002), to the leading order, there are $\partial \sigma_B / \partial \alpha \approx \sigma_B / \alpha$ and $\partial \sigma_{\text{ATM}} / \alpha \approx \sigma_{\text{ATM}} / \alpha$, hence we have the approximation

$$vega \approx \frac{\partial BS}{\partial \sigma_B} \cdot \frac{\sigma_B(f, K)}{\sigma_{\text{ATM}}}.$$

Vega quantifies the risk of changing level of implied volatility curve.

- Vanna

The risk associated with the change in ρ is called vanna, which is calculated according to

$$vanna = \frac{\partial V_{call}}{\partial \rho} = \frac{\partial BS}{\partial \sigma_B} \cdot \frac{\partial \sigma_B(f, K; \alpha, \beta, \rho, \nu)}{\partial \rho}.$$

Vanna quantifies the risk of the changing skewness of implied volatility curve.

- Volga

The volga (vol gamma) is the risk associated with the change in ν, the "vol of vol":

$$volga = \frac{\partial V_{call}}{\partial \nu} = \frac{\partial BS}{\partial \sigma_B} \cdot \frac{\partial \sigma_B(f, K; \alpha, \beta, \rho, \nu)}{\partial \nu}.$$

If unwanted, both *vanna* and *volga* risks can be hedged by buying or selling away-from-the-money options.

The SABR model has been a great success and it has become the market standard for derivatives pricing and risk management across various asset classes, well beyond interest rate derivatives.

11.1.1 SABR Model for Interest Rate Derivatives

Initially, SABR model was not intended to be a term structure model, instead, it is a model on the volatility smile of a single forward rate or single swap rate, maturity by maturity. Taking the pricing of caplet of maturity T_{j-1} for example, the SABR model takes the following form: for $t \leq T_{j-1}$,

$$\begin{aligned} df_j(t) &= \gamma_j(t) f_j^{\beta_j}(t) dW_t^{(j)}, \\ d\gamma_j(t) &= \nu_j \gamma_j(t) dZ_t, \quad \gamma_j(0) = \alpha_j, \end{aligned} \tag{11.5}$$

where the Brownian motions are under the T_j-forward measure and correlated:

$$dW^{(j)}(t) dZ_t = \rho_j dt,$$

with $0 < \beta_j < 1, \alpha_j, \nu_j > 0$, and $|\rho_j| \leq 1$. Note that the caplet maturity is typically T_j, then we need to impose an additional condition $\gamma_j(T_j) = 0$, then the HKLW formula needs modification before being applied to caplet pricing.

For swaption pricing, the corresponding SABR model is

$$\begin{aligned} dR_{m,n}(t) &= \gamma_{m,n}(t) R_{m,n}^{\beta_{m,n}}(t) dW_t^{(m,n)}, \\ d\gamma_{m,n}(t) &= \nu_{m,n} \gamma_{m,n}(t) dZ_t, \quad \gamma_{m,n}(0) = \alpha_{m,n}(0), \end{aligned} \tag{11.6}$$

where the Brownian motions are under the forward swap measure and correlated:

$$dW_t^{(m,n)} dZ_t = \rho_{m,n} dt,$$

with $0 < \beta_{m,n} < 1, \alpha_{m,n}, \nu_{m,n} > 0$, and $|\rho_{m,n}| \leq 1$. The SABR forward rate model and the SABR swap rate model are actually inconsistent, but this has largely been harmless and thus ignored in the markets.

Now, let us focus on the performance of the SABR model on swaptions in the IR market. There are the following observations.

• There is a weak dependence of the market skew/smile on the maturity of the underlying swaps.

VOLATILITY OF VOLATILITY ν FOR EUROPEAN SWAPTIONS. ROWS ARE TIME–TO–EXERCISE; COLUMNS ARE TENOR OF THE UNDERLYING SWAP.

	1Y	2Y	3Y	4Y	5Y	7Y	10Y
1M	76.2%	75.4%	74.6%	74.1%	75.2%	73.7%	74.1%
3M	65.1%	62.0%	60.7%	60.1%	62.9%	59.7%	59.5%
6M	57.1%	52.6%	51.4%	50.8%	49.4%	50.4%	50.0%
1Y	59.8%	49.3%	47.1%	46.7%	46.0%	45.6%	44.7%
3Y	42.1%	39.1%	38.4%	38.4%	36.9%	38.0%	37.6%
5Y	33.4%	33.2%	33.1%	32.6%	31.3%	32.3%	32.2%
7Y	30.2%	29.2%	29.0%	28.2%	26.2%	27.2%	27.0%
10Y	26.7%	26.3%	26.0%	25.6%	24.8%	24.7%	24.5%

FIGURE 11.3: Fitted ν. (Courtesy of Applied. Math. Fin.)

- Both ρ and ν are fairly constant for each maturity.

Figures 11.3 and 11.4 are taken from Hagen et al. (2002), which shows the calibrated ν and ρ of swaptions for various option expiries and swap maturities.

MATRIX OF CORRELATIONS ρ BETWEEN THE UNDERLYING AND THE VOLATILITY FOR EUROPEAN SWAPTONS.

	1Y	2Y	3Y	4Y	5Y	7Y	10Y
1M	4.2%	−0.2%	−0.7%	−1.0%	−2.5%	−1.8%	−2.3%
3M	2.5%	−4.9%	−5.9%	−6.5%	−6.9%	−7.6%	−8.5%
6M	5.0%	−3.6%	−4.9%	−5.6%	−7.1%	−7.0%	−8.0%
1Y	−4.4%	−8.1%	−8.8%	−9.3%	−9.8%	−10.2%	−10.9%
3Y	−7.3%	−14.3%	−17.1%	−17.1%	−16.6%	−17.9%	−18.9%
5Y	−11.1%	−17.3%	−18.5%	−18.8%	−19.0%	−20.0%	−21.6%
7Y	−13.7%	−22.0%	−23.6%	−24.0%	−25.0%	−26.1%	−28.7%
10Y	−14.8%	−25.5%	−27.7%	−29.2%	−31.7%	−32.3%	−33.7%

FIGURE 11.4: Fitted ρ. (Courtesy of Applied. Math. Fin.)

We have the following additional remarks on the performance of the model.

1. In most markets, there is a strong smile for short-dated options that relaxes as the time-to-exercise increases.

2. Consequently, the "vol of vol" is larger for short-dated options and smaller for long-dated options, regardless of the particular underlying.

3. Correlation results are less clear: in some markets, a nearly flat skew for short-dated options develops into a strongly downward sloping skew for longer expiries.

4. In some markets, there is a strongly downward skew for all options maturities, while in other markets, the skew is close to zero for all maturities.

11.2 Wu and Zhang Model

In this section, we present a genuine correlation-based model for SOFR derivatives. The model so developed can be regarded as the SOFR term-structure version of the Heston model (1993); the latter has been one of the most popular equity option models with stochastic volatility.[2] There are two main reasons behind the adoption of Heston's model to the context of interest rate derivatives modeling. First, some quantitative analysts believe stochastic volatilities are the primary factors behind the leptokurtic feature[3] of empirical interest rate distributions, to the extent that volatility smiles can be modeled or captured solely by utilizing stochastic volatility in proper ways. This belief is somewhat reflected by the popularity of Piterbarg's (2003) model on LIBOR that couples displaced diffusion with stochastic volatility. Second, among stochastic volatility models (see e.g., Lewis, 2000), Heston's model not only has the mean reverting feature for the stochastic volatility, a feature deemed very plausible financially, but also carries good analytical tractability that renders exact closed-form pricing for call and put options.

While it is trivial to adopt Heston's model for individual forward rates or swap rates, it is not so for entire term structure of either forward or swap rates. Analytic tractability will usually be lost after a change of measure, and only to be regained after some approximations. One of the main focuses of this section is to present the approximations, adopted in Wu and Zhang (2008), to the forward rate and swap rate dynamics after measure changes for pricing purposes.

Another focus of this section is the calibration of the SOFR market model with square-root stochastic volatilities to cap prices. We try to determine the set of model parameters by optimally matching the implied caplet volatilities, not the prices themselves, in the least square sense. We identify the key parameters to be the magnitude of forward-rate volatilities, the correlation between the forward rates and stochastic volatilities, and the "vol of vol." Respectively, these three sets of parameters are responsible for the level, the skewness, and the curvature of the implied volatility surface. The calibration is achieved through a two-layer nested minimization procedure: the outer layer is for minimizing the total square error with respect to the "vol of vol," and the inner layer is for minimizing the square error in the implied volatilities for each individual maturity, with respect to the magnitude of the forward

---

[2] A lognormal process whose volatility follows a square-root process (Cox et al. 1985).
[3] Higher peak and fatter tails than that of a normal distribution.

rates and the correlations. Moreover, the inner layer consists of decoupled bi-variable minimization problems, and thus can be solved instantly. In our study, we have taken only state-independent parameters, and the procedure can be implemented in seconds. It is desirable to simultaneously calibrate to swaptions. However, there is the difficult issue of convergence that must be faced, and we thus choose to avoid the joint calibration of caps and swaptions.

Our introduction of the market model with stochastic volatility starts with the usual lognormal dynamics of the Treasury zero-coupon bonds. Let $P(t,T)$ be the price of the Treasury zero-coupon bond maturing at $T(\geq t)$ with par value \$1, and let $B(t)$ be the money market account under discrete compounding:

$$B(t) = \left(\prod_{j=0}^{\eta_t - 1} (1 + f_j(T_j)\Delta T_j) \right) (1 + f_{\eta_t}(t)(t - T_{\eta_t})),$$

where $\eta_t = \max\{j | T_j \leq t\}$. Under a risk-neutral measure, it is typical to assume dynamics of lognormal martingale for the discount price of $P(t,T)$ as

$$d \left(\frac{P(t,T)}{B(t)} \right) = \left(\frac{P(t,T)}{B(t)} \right) \Sigma(t,T) \cdot d\mathbf{W}_t. \tag{11.7}$$

The volatility function satisfies the boundedness condition, $E \left[\int_0^t \|\Sigma(s,T)\|^2 ds \right] < \infty, \forall t < T$.

Wu and Zhang (2008) adopt specifically a stochastic multiplier to the risk neutralized processes of the forward rates:

$$df_j(t) = f_j(t)\sqrt{V(t)}\gamma_j(t) \cdot \left[d\mathbf{W}_t - \sqrt{V(t)}\Sigma_j(t)dt \right],$$
$$dV(t) = \kappa(\theta - V(t))dt + \nu\sqrt{V(t)}dZ_t. \tag{11.8}$$

Here,

$$\Sigma(t,T_j) = - \sum_{k=\eta_t+1}^{j} \frac{\Delta T_k f_k(t)}{1 + \Delta T_k f_k(t)} \gamma_k(t), \tag{11.9}$$

κ, θ and ν are time-dependent variables,[4] and Z_t is an additional 1-D Brownian motion under the risk-neutral measure. As a multi-factor model, the forward rates can be correlated such that

$$\text{Cov}_{jk}^i(t) = \gamma_j(t) \cdot \gamma_k(t), \qquad i \leq j, k \leq N, \quad 1 \leq i \leq N. \tag{11.10}$$

Adoption of stochastic volatility in models similar to Equation 11.8 was first seen in Chen and Scott (2001) and Andersen and Brotherton-Ratcliffe (2001). In these models, the stochastic volatility is kept independent of interest rates, for the sake of analytical tractability. As a distinct feature of the Wu and

---

[4]The distributional properties of $V(t)$ are described in Lemma 8.4.1.

Zhang model, the correlations between the stochastic multiplier and forward rates are allowed, such as

$$E^Q \left[\left(\frac{\gamma_j(t)}{\|\gamma_j(t)\|} \cdot d\mathbf{W}_t \right) \cdot dZ_t \right] = \rho_j(t)dt, \quad \text{with} \quad |\rho_j(t)| \le 1. \qquad (11.11)$$

Here, $\left(\frac{\gamma_j(t)}{\|\gamma_j(t)\|} \cdot d\mathbf{W}_t \right)$ is equivalent to (the differential of) a single Brownian motion that drives $f_j(t)$. The correlation coefficients, $\{\rho_j(t)\}$, will play an essential role in capturing volatility smiles. For easy reference, we also call V a stochastic multiplier.

Mathematically, we can construct a market model where each component of the forward-rate volatility vector is associated with a stochastic multiplier.[5] While analytical or semi-analytical pricing of caplets and swaptions under such a model remains feasible, calibration will be extremely difficult, if not impossible. Technically, adopting a uniform volatility multiplier for all rates rather than one multiplier for each rate retains much greater analytical tractability, and it makes model calibration amenable.

11.2.1   Pricing of Caplets

We now consider caplet pricing under the extended SOFR market model, Equation 11.8. A caplet is a call option on a forward rate. Assuming that the notional value of a caplet is one dollar, then the payoff of the caplet at T_j is

$$\Delta T_j (f_j(T_j) - K)^+.$$

To price the caplet we choose $P(t, T_j)$, in particular, to be the numeraire asset and let $\mathbb{Q}_j$ denote the corresponding forward measure (i.e. the martingale measure corresponding to numeraire $P(t, T_j)$). The next proposition establishes the relationship between Brownian motions under the risk-neutral measure and under the T_j forward measure (Wu and Zhang, 2008).

Proposition 11.2.1. *Let* $\mathbf{W}_t$ *and* Z_t *be Brownian motions under* $\mathbb{Q}$, *then* $\mathbf{W}_t^{(j)}$ *and* $Z_t^{(j)}$, *defined by*

$$\begin{aligned} d\mathbf{W}_t^{(j)} &= d\mathbf{W}_t - \sqrt{V(t)} \Sigma_j(t) dt, \\ dZ_t^{(j)} &= dZ_t + \xi_j(t) \sqrt{V(t)} dt, \end{aligned} \qquad (11.12)$$

are Brownian motions under $\mathbb{Q}_j$, *where*

$$\xi_j(t) = \sum_{k=\eta_t+1}^{j} \frac{\Delta T_k f_k(t) \rho_k(t) \lambda_k(t)}{1 + \Delta T_k f_k(t)},$$

where $\lambda_k(t) = \|\gamma_k(t)\|$.

---

[5]Trolle and Schwartz (2008) develop a HJM model with such a feature.

Proof. The Radon-Nikodym derivative of $\mathbb{Q}_j$ with respect to $\mathbb{Q}$ is

$$\left.\frac{d\mathbb{Q}_j}{d\mathbb{Q}}\right|_{\mathcal{F}_t} = \frac{P(t,T_j)/P(0,T_j)}{B(t)}$$

$$= e^{\int_0^t -\frac{1}{2}V(\tau)\Sigma_j^2(\tau)d\tau + \sqrt{V(\tau)}\Sigma_j \cdot d\mathbf{W}_t}$$

$$\stackrel{\triangle}{=} m_j(t), \quad t \le T_j.$$

Clearly, we have

$$dm_j(t) = m_j(t)\sqrt{V(t)}\Sigma_j(t) \cdot d\mathbf{W}_t.$$

Let $\langle \cdot, \cdot \rangle$ denote covariance. By the CMG change of measure theorem, we obtain the Brownian motions under $\mathbb{Q}_j$:

$$d\mathbf{W}_t^{(j)} = d\mathbf{W}_t - \langle d\mathbf{W}_t, dm_j(t)/m_j(t) \rangle$$

$$= d\mathbf{W}_t - \sqrt{V(t)}\Sigma_j(t)dt,$$

$$dZ_t^{(j)} = dZ_t - \langle dZ_t, dm_j(t)/m_j(t) \rangle$$

$$= dZ_t - \langle dZ_t, \sqrt{V(t)}\Sigma_j(t) \cdot d\mathbf{W}_t \rangle$$

$$= dZ_t + \sqrt{V(t)}\sum_{k=1}^{j} \frac{\Delta T_k f_k(t)\lambda_k(t)}{1+\Delta T_k f_k(t)} \left\langle dZ_t, \frac{\gamma_k(t)}{\lambda_k(t)} \cdot d\mathbf{W}_t \right\rangle$$

$$= dZ_t + \sqrt{V(t)}\sum_{k=1}^{j} \frac{\Delta T_k f_k(t)\lambda_k(t)}{1+\Delta T_k f_k(t)}\rho_k(t)dt.$$

$\square$

In terms of $\mathbf{W}_t^{(j)}$ and $Z_t^{(j)}$, the extended market model becomes

$$df_j(t) = f_j(t)\sqrt{V(t)}\gamma_j(t) \cdot d\mathbf{W}_t^{(j)}, \tag{11.13}$$

$$dV(t) = [\kappa\theta - (\kappa + \nu\xi_j(t))V(t)]\,dt + \nu\sqrt{V(t)}dZ_t^{(j)}, \tag{11.14}$$

where

$$\gamma_j(t) = \tilde{\gamma}_j(t)\left(1_{t \le T_{j-1}} + \frac{T_j - t}{\Delta T_j}1_{t > T_{j-1}}\right),$$

so that we have $\gamma_j(T_j) = 0$. In formalism, the multiplier process remains a square-root process under $\mathbb{Q}_j$. Yet part of the coefficients, $\xi_j(t)$, depends on forward rates, and such dependence prohibits analytical option valuation. The time variability of $\xi_j(t)$, however, is small. In fact, we can write

$$\xi_j(t) = \sum_{k=1}^{j} \frac{\Delta T_k f_k(0)\rho_k(t)\lambda_k(t)}{1+\Delta T_k f_k(0)} + \frac{\rho_k(t)\lambda_k(t)\Delta T_k}{(1+\Delta T_k f_k(0))^2}(f_k(t) - f_k(0)) \tag{11.15}$$

$$+ O\left(\rho_k(t)\lambda_k(t)\Delta T_k^2(f_k(t) - f_k(0))^2\right).$$

In light of the martingale property, $E^{Q_j}[f_j(t)|\mathcal{F}_0] = f_j(0)$, we see that

$$E^{Q_j}[\xi_j(t)|\mathcal{F}_0] = \sum_{k=1}^{j} \frac{\Delta T_k f_k(0)\rho_k(t)\lambda_k(t)}{1 + \Delta T_k f_k(0)}$$

$$+ O(\rho_k(t)\lambda_k(t)\, Var(\Delta T_k f_k(t))),$$

$$Var(\xi_j(t)|\mathcal{F}_0) \approx (\rho_k(t)\lambda_k(t))^2\, Var(\Delta T_k f_k(t)).$$

According to the model, $Var(\Delta T_k f_k(t)) \sim \Delta T_k^2 f_k^2(t)\lambda_k^2(t)V(t)t$. Since $\Delta T_k f_k(t)$ is mostly under 5%, the expansion in Equation 11.15 is dominated by the first term. Hence, to remove the dependence of $V(t)$ on $f_j(t)$'s, we choose to ignore higher order terms in Equation 11.15 and consider the approximation

$$\xi_j(t) \approx \sum_{k=1}^{j} \frac{\Delta T_k f_k(0)\rho_k(t)\lambda_k(t)}{1 + \Delta T_k f_k(0)}. \tag{11.16}$$

This is close to the technique of "freezing coefficients." For notational simplicity we denote

$$\tilde{\xi}_j(t) = 1 + \frac{\nu}{\kappa}\xi_j(t),$$

and thus retain a neat equation for the process of $V(t)$:

$$dV(t) = \kappa\left[\theta - \tilde{\xi}_j(t)V(t)\right]dt + \nu\sqrt{V(t)}dZ_t^{(j)}. \tag{11.17}$$

For the processes joined by Equation 11.13 and Equation 11.17, caplets can be priced along the approach pioneered by Heston (1993). The price of the caplet on $f_j(T_j)$ can be expressed as

$$C_{let}(0) = P(0, T_j)\Delta T_j E^{Q_j}\left[(f_j(T_j) - K)^+|\mathcal{F}_0\right]$$

$$= P(0, T_j)\Delta T_j f_j(0)\left(E^{Q_j}\left[e^{X(T_j)}\mathbf{1}_{X(T_j)>k}|\mathcal{F}_0\right]\right.$$

$$\left. - e^k E^{Q_j}\left[\mathbf{1}_{X(T_j)>k}|\mathcal{F}_0\right]\right),$$

where $X(t) = \ln f_j(t)/f_j(0)$ and $k = \ln K/f_j(0)$. The two expectations above can be valuated using the moment-generating function of $X(T_j)$, defined by

$$\phi(X(t), V(t), t; z) \stackrel{\triangle}{=} E^{Q_j}\left[e^{zX(T_j)}|\mathcal{F}_t\right], \quad z \in C.$$

In terms of $\phi_{T_j}(z) \stackrel{\triangle}{=} \phi(0, V(0), 0; z)$, we have that (e.g., Kendall (1994) or Duffie, Pan and Singleton (2000))

$$E^{Q_j}\left[\mathbf{1}_{X(T_j)>k}|\mathcal{F}_0\right] = \frac{\phi_{T_j}(0)}{2} + \frac{1}{\pi}\int_0^\infty \frac{\text{Im}\{e^{-iuk}\phi_{T_j}(iu)\}}{u}du,$$

$$E^{Q_j}\left[e^{X(T_j)}\mathbf{1}_{X(T_j)>k}|\mathcal{F}_0\right] = \frac{\phi_{T_j}(1)}{2} + \frac{1}{\pi}\int_0^\infty \frac{\text{Im}\{e^{-iuk}\phi_{T_j}(1 + iu)\}}{u}du. \tag{11.18}$$

The integrals above can then be evaluated numerically. For later reference, we call this approach the Heston method.

When the Brownian motions $\mathbf{W}_t^{(j)}$ and $Z_t^{(j)}$ are independent, the moment generating function can be directly worked out. In general, one can solve for $\phi(x, V, t; z)$ from the Kolmogorov backward equation corresponding to the joint processes:

$$\frac{\partial \phi}{\partial t} + \kappa(\theta - \tilde{\xi}_j V)\frac{\partial \phi}{\partial V} - \frac{1}{2}\lambda_j^2(t)V\frac{\partial \phi}{\partial x}$$
$$+ \frac{1}{2}\nu^2 V\frac{\partial^2 \phi}{\partial V^2} + \nu\rho_j V\lambda_j(t)\frac{\partial^2 \phi}{\partial V\partial x} + \frac{1}{2}\lambda_j^2(t)V\frac{\partial^2 \phi}{\partial x^2} = 0, \quad (11.19)$$

subject to terminal condition

$$\phi(x, V, T_j; z) = e^{zx}. \tag{11.20}$$

It is known that the solution is of the form

$$\phi(x, V, t; z) = e^{A(t,z)+B(t,z)V+zx}, \tag{11.21}$$

where A and B satisfy the following equations

$$\begin{aligned} \frac{dA}{dt} + \kappa\theta B &= 0, \\ \frac{dB}{dt} + \frac{1}{2}\nu^2 B^2 + (\rho_j\nu\lambda_j z - \kappa\xi)B + \frac{1}{2}\lambda_j^2(z^2 - z) &= 0, \end{aligned} \tag{11.22}$$

subject to terminal conditions

$$A(T_j, z) = 0, \quad B(T_j, z) = 0,$$

and A and B can be solved analytically for constant coefficients (Heston, 1993). The analytical solutions can be extended to the case of piecewise constant coefficients through recursions. In the statement of the next proposition, we have suppressed the time dependence of the coefficients for simplicity. The proof of the proposition is provided in this chapter's appendix for completeness.

Proposition 11.2.2. *Suppose that all coefficients of Equation 11.22 are constants over time intervals* $T_{k-1} \le t < T_k, \quad k = 1, 2, \ldots, j,$ *and* $\nu^2 > 0$ *for all* t, *then* A *and* B *are given by the following recursive formulae*

$$\begin{cases} A(t, z) = A(T_k, z) + a_0 \left\{ \left[u_k^+ + B(T_k, z)\right](T_k - t) \right. \\ \qquad\qquad\qquad\qquad \left. - \frac{1}{b_2}\ln\left[\frac{u_k^- - u_k^+}{u_k^- - u_k^+ e^{d(T_k-t)}}\right]\right\}, \\ \\ B(t, z) = B(T_k, z) + u_k^- u_k^+ \dfrac{(1 - e^{d(T_k-t)})}{(u_k^- - u_k^+ e^{d(T_k-t)})}, \\ \qquad\qquad for \quad T_{k-1} \le t < T_k, \quad k = j, j-1, \ldots, 1, \end{cases} \tag{11.23}$$

where

$$d = \sqrt{b_1^2 - 4b_2 b_0}, \quad u_k^{\pm} = \frac{-b_1 \pm d}{2b_2} - B(\tau_k, z),$$

and

$$a_0 = \kappa\theta, \quad b_2 = \frac{1}{2}\nu^2, \quad b_1 = \rho_j \nu \lambda_j z - \kappa\xi, \quad b_0 = \frac{1}{2}\lambda_j^2(z^2 - z). \quad (11.24)$$

11.2.2 Pricing of Swaptions

The payoff of a swaption on $R_{m,n}(T_m)$ can be expressed as

$$A_{m,n}(T_m) \cdot \max(R_{m,n}(T_m) - K, 0),$$

where $R_{m,n}(t)$ is the equilibrium swap rate for the tenor (T_m, T_n), $A_{m,n}(t)$ is the annuity over the tenor (T_m, T_n), and K is the strike rate. Our next objective is to establish the dynamics of the swap rate under the forward swap measure $\mathbb{Q}_{m,n}$. For simplicity, we assume identical tenors for floating leg and fixed leg. We then have

Proposition 11.2.3. *Let $\mathbf{W}_t$ and Z_t be Brownian motions under $\mathbb{Q}$, then $\mathbf{W}_t^{(m,n)}$ and $Z_t^{(m,n)}$, defined by*

$$\begin{aligned} d\mathbf{W}_t^{(m,n)} &= d\mathbf{W}_t - \sqrt{V(t)}\Sigma_{m,n}(t)dt, \\ dZ_t^{(m,n)} &= dZ_t + \sqrt{V(t)}\xi_{m,n}(t)dt, \end{aligned} \quad (11.25)$$

are Brownian motions under $\mathbb{Q}_{m,n}$, where

$$\Sigma_{m,n}(t) = \sum_{j=m+1}^{n} \alpha_j \Sigma(t, T_j), \qquad \xi_{m,n}(t) = \sum_{j=m+1}^{n} \alpha_j \xi_j, \quad (11.26)$$

with weights

$$\alpha_j = \alpha_j(t) = \frac{\Delta T_j P(t, T_j)}{A_{m,n}(t)}.$$

$\square$

Proof: Denote the forward swap measure by $\mathbb{Q}_{m,n}$. The Radon-Nikodym derivative for $\mathbb{Q}_{m,n}$ is

$$\begin{aligned} \left. \frac{d\mathbb{Q}_{m,n}}{d\mathbb{Q}} \right|_{\mathcal{F}_t} &= \frac{A_{m,n}(t)/A_{m,n}(0)}{B(t)} \\ &= \frac{1}{A_{m,n}(0)} \sum_{j=m+1}^{n} \Delta T_j P(0, T_j) e^{\int_0^t -\frac{1}{2}V(\tau)\Sigma_j^2(\tau)d\tau + \sqrt{V(\tau)}\Sigma_j \cdot d\mathbf{W}_t} \\ &\stackrel{\triangle}{=} m_{m,n}(t), \quad t \le T_m. \end{aligned}$$

There is

$$
dm_{m,n}(t) = \frac{1}{A_{m,n}(0)} \sum_{j=m+1}^{n} \Delta T_j P(0, T_j) e^{\int_0^t -\frac{1}{2} V(\tau) \Sigma_j^2(\tau) d\tau + \sqrt{V(\tau)} \Sigma_j \cdot d\mathbf{W}_t}
$$

$$
\sqrt{V(t)} \Sigma_j(t) \cdot d\mathbf{W}_t
$$

$$
= \frac{1}{A_{m,n}(0)B(t)} \sum_{j=m+1}^{n} \Delta T_j P(t, T_j) \sqrt{V(t)} \Sigma_j(t) \cdot d\mathbf{W}_t
$$

$$
= m_{m,n}(t) \sum_{j=m+1}^{n} \alpha_j \sqrt{V(t)} \Sigma_j(t) \cdot d\mathbf{W}_t.
$$

It follows that

$$
d\mathbf{W}_t^{(m,n)} = d\mathbf{W}_t - \langle d\mathbf{W}_t, dm_{m,n}(t)/m_{m,n}(t) \rangle
$$

$$
= d\mathbf{W}_t - \sqrt{V(t)} \sum \alpha_j \Sigma_j(t) dt,
$$

$$
= d\mathbf{W}_t - \sqrt{V(t)} \Sigma_{m,n}(t) dt,
$$

$$
dZ_t^{(m,n)} = dZ_t - \langle dZ_t, dm_{m,n}(t)/m_{m,n}(t) \rangle
$$

$$
= dZ_t - \langle dZ_t, \sqrt{V(t)} \sum \alpha_j \Sigma_j(t) \cdot d\mathbf{W}_t \rangle
$$

$$
= dZ_t + \sqrt{V(t)} \sum_{j=m+1}^{n} \alpha_j \sum_{k=1}^{j} \frac{\Delta T_k f_k(t) \lambda_k(t)}{1 + \Delta T_k f_k(t)} \langle dZ_t, \frac{\gamma_k(t)}{\lambda_k(t)} \cdot d\mathbf{W}_t \rangle
$$

$$
= dZ_t + \sqrt{V(t)} \sum_{j=m+1}^{n} \alpha_j \sum_{k=1}^{j} \frac{\Delta T_k f_k(t) \lambda_k(t)}{1 + \Delta T_k f_k(t)} \rho_k(t) dt
$$

$$
= dZ_t + \sqrt{V(t)} \sum_{j=m+1}^{n} \alpha_j \xi_j(t) dt
$$

$$
= dZ_t + \sqrt{V(t)} \xi_{m,n}(t) dt.
$$

This completes the proof. □

Using Ito's Lemma, we can directly derive the swap rate process under the forward swap measure. Assume, for simplicity, the fixed leg and the floating leg of the underlying swap have the same tenor, i.e., $\Delta \tilde{T} = \Delta T$, then the swap rate process

$$
dR_{m,n}(t) = \sqrt{V(t)} \sum_{j=m+1}^{n} \frac{\partial R_{m,n}(t)}{\partial f_j(t)} f_j(t) \gamma_j(t) \cdot d\mathbf{W}_t^{(m,n)},
$$

$$
dV(t) = \kappa \left[\theta - \tilde{\xi}_{m,n}(t) V(t) \right] dt + \nu \sqrt{V(t)} dZ_t^{(m,n)}.
$$

(11.27)

Here,

$$
\tilde{\xi}_{m,n}(t) = 1 + \frac{\nu}{\kappa} \xi_{m,n}(t),
$$

$$\frac{\partial R_{m,n}(t)}{\partial f_j(t)} = \alpha_j + \frac{\Delta T_j}{1 + \Delta T_j f_j(t)} \left[\sum_{l=m+1}^{j} \alpha_l (f_l - R_{m,n}(t)) \right], \quad m+1 \leq j \leq n.$$

The proof can be found in Wu and Zhang (2008).

Similar to swaption pricing under the standard market model (e.g., Sidennius, 2000; Andersen and Andreasen, 2000), we approximate the swap rate process by a lognormal process with however a stochastic volatility:

$$
\begin{aligned}
dR_{m,n}(t) &= R_{m,n}(t)\sqrt{V(t)}\gamma_{m,n}(t) \cdot d\mathbf{W}_t^{(m,n)}, \qquad 0 \leq t < T_m, \\
dV(t) &= \kappa \left[\theta - \tilde{\xi}_{m,n}(t)V(t) \right] dt + \nu\sqrt{V(t)}dZ_t^{(m,n)},
\end{aligned}
\tag{11.28}
$$

where

$$\tilde{\xi}_{m,n}(t) = \sum_{j=m+1}^{n} \alpha_j(0)\tilde{\xi}_j(t),$$

$$\gamma_{m,n}(t) = \sum_{j=m+1}^{n} w_j(0)\gamma_j(t), \qquad w_j(t) = \frac{\partial R_{m,n}(t)}{\partial f_j}\frac{f_j(t)}{R_{m,n}(t)},$$

and $\rho_{m,n} = \sum_{j=m+1}^{n} w_j(0)\rho_j$.

In the above approximations, we have removed the dependence of $\tilde{\xi}_{m,n}(t)$ on forward rates through taking full advantage of the negligible time variability of $w_j(t)$ and $\alpha_j(t)$ (compared with that of forward rates). As a result, the approximate swap-rate process has moment generating function in closed form, and we thus retain the analytical tractability of the model under the forward swap measure. This is the key treatment in this section, which works well for market models with the square-root volatility dynamics, but may not work for those with general volatility dynamics.

Instead of following the Heston's approach for numerical pricing, we adopt a transformation method developed by Carr and Madan (1998). Under the forward swap measure, we have the following expression for swaption prices

$$SP(0) = A_{m,n}(0)R_{m,n}(0)E^{Q_{m,n}} \left[\left(e^{X(T_m)} - e^k \right)^+ | \mathcal{F}_0 \right],
\tag{11.29}$$

where

$$X(T_m) = \ln R_{m,n}(T_m)/R_{m,n}(0) \quad \text{and} \quad k = \ln K/R_{m,n}(0).$$

The moment-generating function of $X(T_m)$, $\phi_{T_m}(z) \stackrel{\triangle}{=} \phi(0, V(0), 0; z)$, which is characterized in Equation 11.21 and derived in Proposition 11.2.2, with however $\{m, \rho_{m,n}, \lambda_{m,n}\}$ taking the place of $\{j, \rho_j, \lambda_j\}$, where $\lambda_{m,n} = \|\gamma_{m,n}\|$. We treat the expectation in Equation 11.29 as a function of strike:

$$G(k) = E^{Q_{m,n}} \left[\left(e^{X(T_m)} - e^k \right)^+ | \mathcal{F}_0 \right].$$

Then, let $q(s)$ denote the density function of $X(T_m) = \ln R_{m,n}(T_m)/R_{m,n}(0)$, we write,

$$G(k) = \int_k^\infty (e^s - e^k)q(s)ds,$$

Note that $G(k)$ is not square integrable over $(-\infty, \infty)$ as it tends to 1 when k tends to $-\infty$. Take some constant $a > 0$ and consider the Laplace transform of the value function,

$$\psi(u) = \int_{-\infty}^\infty e^{(a+iu)k} G(k)dk = \frac{\phi_{T_m}(1+a+iu)}{(a+iu)(1+a+iu)}.$$

Having obtained $\psi(u)$ in closed form, the swaption price follows from an inverse Fourier transform

$$G(k) = exp(-ak) \int_0^\infty e^{iuk}\psi(u)du. \tag{11.30}$$

The FFT implementation is described with detail in Chapter 8 and will not be repeated here.

11.2.3   Model Calibration

Calibration is a procedure for determining the parameters of a model based on observed information of certain securities. This is a necessary step when the model is applied for production uses. For the market model with the square-root volatility process, specifically, we need to determine the following set of model coefficients or parameters, $\{\kappa, \theta, \nu; \tilde{\gamma}_j(t), \rho_j(t), j = 1, \ldots, N\}$, based on information on SOFR term rates, swap rates, and the prices of a set of caps, floors and swaptions. Computationally, this can be a challenging problem. It has been suggested to use some parameters estimated from time series data in order to reduce the scale (i.e., number of unknowns) of the problem. For instance, we may first estimate the multiplier process using the time series data of implied Black volatilities of at-the-money caplets, as is suggested in Chen and Scott (2001). Once the process of $V(t)$ is specified, we can proceed to determine the pair of $\{\|\tilde{\gamma}_j(t)\|, \rho_j(t)\}$ through matching to the implied volatility smile of T_j-maturity caplets. This will lead to a bivariate optimization problem, which is easily manageable. Implicitly, taking a process estimated from time series data as a risk-neutral process means that the related risk premium is treated as zero, which, at least in the case of stochastic volatility, is not justified (see for instance, Carr and Wu, 2008). Hence, using the estimated process may sometimes undermine the quality of calibration to an extent that is beyond acceptable.

In this section, we consider the simultaneous calibration of processes of forward rates and the multiplier based on caplet smiles. When there is no stochastic volatility, i.e., $\nu \equiv 0$, we want $V \equiv 1$ and thus reduce the extended model to the standard market model. For this reason, we take $V_0 = \theta = 1$. The parameter κ will also be taken fixed, but its choice is less critical as the role of κ overlaps with that of ν. Note that either a smaller κ or a bigger ν will result in stronger effect of the stochastic volatility. After V_0, θ and κ are fixed a priori, the calibration problem reduces to the determination of ν and the pairs of $\|\tilde{\gamma}_j(t)\|$ and $\rho_j(t), j = 1, \dots, N$. Since we are calibrating to caplets only, we ignore time dependence of these functions, i.e., we take $\|\tilde{\gamma}_j(t)\| = \|\tilde{\gamma}_j\|$ and $\rho_j(t) = \rho_j, j = 1, \dots, N$.

We solve for ν and $\{\|\tilde{\gamma}_j\|, \rho_j\}_{j=1}^N$ through a two-layer nested minimization:

$$\min_{\nu} \left(\sum_{j=1}^N \min_{\|\tilde{\gamma}_j\|, \rho_j} \left(\sum_{k=1}^K (v_{k,j}^{(mk)} - v_{k,j}^{(md)})^2 \right) \right). \qquad (11.31)$$

Here, $v_{k,j}^{(mk)}$ and $v_{k,j}^{(md)}$ are the implied caplet volatilities of the kth caplet of maturity T_j, and the sup-indexes "mk" and "md" stand for market and model respectively. Hence, once a ν is taken, $\|\tilde{\gamma}_j^{(\nu)}\|$ and $\rho_j^{(\nu)}$ are solved separately by matching to the implied volatility curve of T_j-maturity caplets, $j = 1, \dots, N$. Note that there is a constraint, $|\rho_j| \leq 1$, on ρ_j. This constraint is easily removed by letting $\rho_j = \cos \theta_j$ and the pair of unknowns become $\|\tilde{\gamma}_j^{(\nu)}\|$ and $\theta_j^{(\nu)}$. In general, we can parameterize $\{\|\tilde{\gamma}_j\|, \rho_j\}_{j=1}^N$ and ν as time-dependent functions. Since we are calibrating only to caplets, we have the luxury to let $\{\|\tilde{\gamma}_j\|, \rho_j\}_{j=1}^N$ be time-independent as well. Our experiences with the use of time-dependent ν, however, are not very encouraging: the resulted ν can change dramatically, and sometimes we do not even achieve convergence due to numerical instability. For this reason, we limit ourselves to finding a constant ν.

One of the key treatments adopted in our solution procedure is to make the implied Black volatility an explicit function of the state variable and the call-option value. This is achieved by interpolating the implied Black volatility over a mesh of the state variables and the call-option values. Bivariate spline functions are used for the interpolation. Note that we are matching the implied volatilities through an iteration procedure, so there are many calculations of implied volatilities that employ a root-finding procedure. Making implied volatility an explicit function avoids the root-finding procedure and thus greatly speeds up the algorithm.

Let us comment on how to determine the components of the volatility vector, $\{\tilde{\gamma}_j\}$. It is known that once $\{\|\tilde{\gamma}_j\|\}$ are obtained, the determination of $\{\tilde{\gamma}_j\}$ is subject to forward-rate correlations. Given, for instance, historical forward-rate correlations, we can solve for $\{\tilde{\gamma}_j\}$ by matching model correlations to the historical correlations. Specifically, taking the rate-multiplier correlations into account, we can derive the following equation for $\{\tilde{\gamma}_j / \|\tilde{\gamma}_j\|\}$'s,

$$(1 - \rho_j \rho_k) \left(\frac{\tilde{\gamma}_j}{\|\tilde{\gamma}_j\|} \right) \cdot \left(\frac{\tilde{\gamma}_k}{\|\tilde{\gamma}_k\|} \right) + \rho_j \rho_k = C_{jk},$$

where C_{jk} is the historical correlation between the time series data of f_j and

f_k, and it is assumed to be dependent on $T_j - T_k$. The existence of $\{\tilde{\gamma}_j / \|\tilde{\gamma}_j\|\}$ requires that the matrix with components

$$\frac{C_{jk} - \rho_j \rho_k}{1 - \rho_j \rho_k}, \quad i \leq j \wedge k, \tag{11.32}$$

be non-negative definite. Intuitively, Equation 11.32 represents the correlation between the two forward rates after the factor of stochastic volatility is removed. An eigenvalue decomposition of the matrix with the elements given in Equation 11.32 will produce $\{\tilde{\gamma}_j / \|\tilde{\gamma}_j\|\}$.

For the implementation of the FFT method, we have to fix several additional parameters: A, b, N and α. Based on the condition $Ab = \pi N$, we take

$$A = \frac{3}{2}\sqrt{\pi N}, \quad \text{and} \quad b = \frac{2}{3}\sqrt{\pi N}. \tag{11.33}$$

By taking A and b in such a way, we ensure that both A and b increase when N increases, thus a finer division in both physical and frequency spaces is coupled with broader ranges. When $N = 80$, in particular, we have

$$A = 23.78, \quad \text{and} \quad b = 10.57.$$

The dampening parameter, α, is taken to be $\alpha = 0.5$, which has performed well. Note that a big α may induce numerical instability in the inverse Laplace transform.

Example: In the following example, we calibrate the market model with stochastic volatility with market data of July 3, 2002. The LIBOR/SOFR and swap rates are shown in Table 11.1, and the cap data are listed in Table 11.2 (at the end of this section). The implied volatilities of caplets of various maturities are stripped from the cap data and are displayed later in Figure 11.3. To calibrate to this set of data, we tried using $\kappa = 0.25$ and $\kappa = 1$, two numbers that correspond to weak and strong effects of mean reversion for the stochastic volatility respectively.

Let us first present the calibration results for the smaller kappa, $\kappa = 0.25$. We take initial guess for $\{\|\tilde{\gamma}_j\|, \theta_j, \nu\}$ as

$$\{0.35, \quad \frac{\pi}{2}, \quad 1\}.$$

Note that the guideline for selecting the initial guess for $\{\|\tilde{\gamma}_j\|\}$ is simply to get close to the level of implied caplet volatilities. After the calibration procedure, we obtain

$$\nu = 1.98,$$

while $\{\|\tilde{\gamma}_j\|, \rho_j = \cos\theta_j\}$, as functions of maturity, are shown in Figures 11.5 and 11.6. From Figure 11.6, we see that $\|\tilde{\gamma}_j\|$ decays steadily with increasing maturity, consistent with the lowering implied volatility curve for increasing maturity. The ρ's obtained by calibration, meanwhile, are not as steady but remain in a reasonable range.

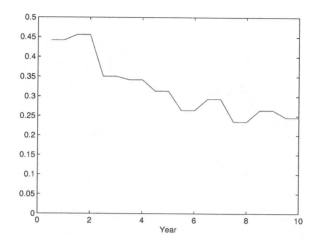

FIGURE 11.5: $\|\tilde{\gamma}_j\|$ versus T_j.

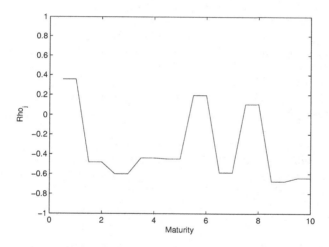

FIGURE 11.6: ρ_j versus T_j.

The implied caplet volatility surface by the calibrated model is shown in Figure 11.7, alongside the implied volatility surface for the market prices of caplets. It can seen that, although the implied caplet volatility surface by the model is not as smooth, the overall agreement between the two surfaces is quite good. More detailed comparisons are made through Figure 11.8, where the quality of fitting of implied volatility curves for all maturities is displayed. The solid curves are the implied Black volatilities of the market values of the caplets, while the dotted curves are the implied Black volatilities of the caplets calculated from the calibrated model. One can see that the level and skewness

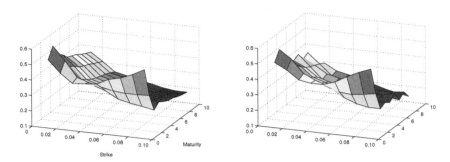

FIGURE 11.7: Implied caplet volatility surface, market (left) vs. model (right).

of all curves are well matched, but the curvature of some curves is slightly missed, probably due to the use of a constant "vol of vol," ν.

It is time to justify the choice of A, b in Equation 11.33 for $N = 80$. As demonstration, we show the Laplace transform of the 5-year caplets, $\eta_5(u)$, in Figures 11.9 and 11.10. One can see that both the real part and imaginary part are very small for $u > 5$. The Laplace transform of caplets of other maturities behaves similarly. Hence, our choice of $A = 23.78$, is thus already very conservative. Figures 11.9 and 11.10 may suggest the use of a smaller A, but, as is constrained by $Ab = \pi N$, a smaller A means a bigger b, which could cause greater errors in the calculation of option values, which is done by interpolation. We cannot reduce N either, as $\eta_5(u)$ is quite steep near the origin.

Next, we show the calibration results for the larger kappa, $\kappa = 1$. This kappa corresponds to a half-life of mean reversion of $\ln(2)/\kappa = 0.7$ years, and it represents a greater strength of mean reversion and thus a weaker effect of stochastic volatility for a longer time horizon. The initial values for $\{\|\tilde{\gamma}_j\|, \theta_j\}$, and ν are

$$\{0.35, \quad \pi/2, \quad 2\}.$$

Let us explain the results. The "vol of vol" by calibration is

$$\nu = 3.65,$$

and other results of calibration are shown in Figures 11.11 to 11.14 Figure 11.11 shows $\|\tilde{\gamma}_j\|$ versus the maturity T_j, while Figure 11.12 shows ρ_j versus the maturity T_j. The matching of implied volatility surfaces and curves is shown in Figures 11.13 and 11.14, from where we might say that the quality of calibration also looks quite good. These two calibration exercises also suggest that the result of a calibration depends on several input parameters. To gain some insight into the selection of those parameters, such as κ, we need to make more careful comparisons.

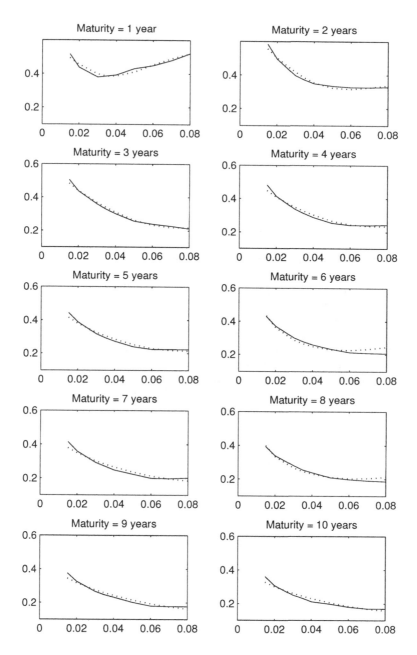

FIGURE 11.8: Implied volatility surface for the calibrated model, $\kappa = 0.25$.

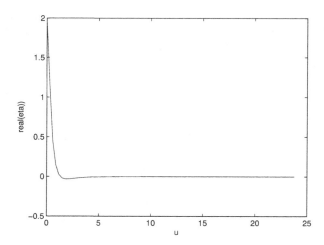

FIGURE 11.9: Real part of $\eta_T(u)$ for the 5-year caplets.

When compared with the implied volatility surface by calibration with $\kappa = 0.25$, the implied volatility surface for $\kappa = 1$ is smoother but flatter, a sign of a weak effect of stochastic volatility. In fact, we can see in Figure 11.12 that ρ_j drifts steadily toward -1 when maturity increases. This can be interpreted as that, in order to generate enough skewness, the model needs almost perfect negative correlations. The implication of this is that forward rates of long maturities would be almost perfectly correlated. Accordingly, $\|\tilde{\gamma}_j\|$ demonstrates a pattern known for calibrated standard SOFR market models: the volatilities of the forward rates first increases and then decreases,

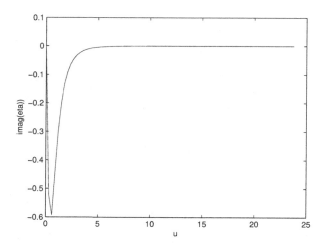

FIGURE 11.10: Imaginary part of $\eta_T(u)$ for the 5-year caplets.

in an exponential way. Note that $\|\tilde{\gamma}_j\|$ again appears to be very close to the mean level of implied caplet volatilities of maturity T_j. One can imagine that for even greater maturities, the quality of calibration would become unacceptable due to the very weak stochastic volatility. Hence, it is not advisable to adopt a kappa as large as one.

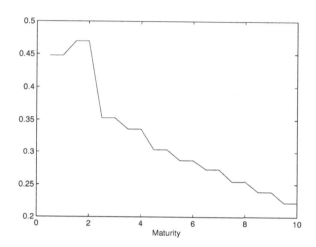

FIGURE 11.11: Real part of $\eta_T(u)$ for the 5-year caplets.

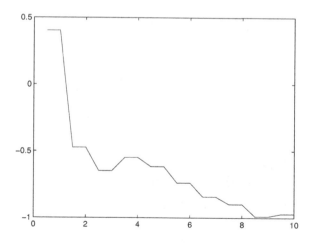

FIGURE 11.12: Imaginary part of $\eta_T(u)$ for the 5-year caplets.

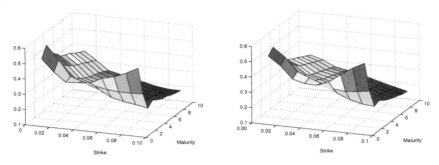

FIGURE 11.13: Implied volatility surface, market (left) vs. model (right).

TABLE 11.1: LIBOR/SOFR and Swap Rates of July 3, 2002 (Bloomberg)

	Term	Rate
LIBOR/	0.5	0.019463
SOFR	1	0.022425
Swap	2	0.031450
	3	0.037440
	4	0.041680
	5	0.044710
	6	0.047130
	7	0.049070
	8	0.050630
	9	0.051920
	10	0.052970
	15	0.056860
	20	0.058540
	30	0.059400

TABLE 11.2: Implied Cap Volatilities of July 3, 2002 (Bloomberg)

Maturity	Strikes								
	0.015	0.020	0.030	0.035	0.040	0.050	0.060	0.070	0.080
1y	51.70%	43.80%	38.00%	38.60%	39.30%	43.10%	44.90%	48.00%	52.10%
2y	55.00%	46.90%	39.50%	37.10%	35.00%	33.50%	32.70%	32.70%	33.00%
3y	52.30%	45.00%	37.30%	34.50%	32.20%	29.40%	28.30%	27.90%	27.70%
4y	50.60%	43.60%	35.90%	33.20%	30.90%	27.80%	26.40%	25.90%	25.70%
5y	48.50%	42.10%	34.60%	32.00%	29.80%	26.60%	25.00%	24.50%	24.20%
6y	46.90%	40.80%	33.50%	31.00%	28.90%	25.80%	24.00%	23.40%	23.00%
7y	45.60%	39.70%	32.60%	30.20%	28.10%	25.10%	23.10%	22.50%	22.20%
8y	44.30%	38.60%	31.80%	29.40%	27.40%	24.40%	22.50%	21.80%	21.40%
9y	43.10%	37.60%	31.00%	28.70%	26.80%	23.80%	21.80%	21.10%	20.70%
10y	42.00%	36.60%	30.20%	28.00%	26.10%	23.30%	21.30%	20.50%	20.10%

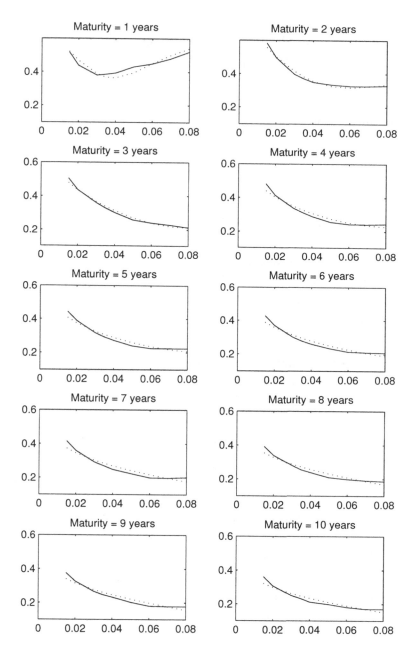

FIGURE 11.14: Implied volatility curves for the calibrated model, $\kappa = 1$.

11.3 Appendix: Derivation of the HKLW Formula

Consider a European call option with maturity T, the forward price is given by

$$V_0 = E^{Q_T}\left[\left(f_T - K\right)^+ \Big| f_0 = f, \alpha_0 = \alpha\right].$$

The governing equation of the option value is

$$\begin{cases} \dfrac{\partial V}{\partial t} = -\mathcal{L}V, & t < T, \\ V(T, x, y) = (x - K)^+ \end{cases} \tag{11.34}$$

for

$$\mathcal{L} \triangleq \frac{1}{2}y^2 C^2(x)\frac{\partial^2}{\partial x^2} + \rho\nu y^2 C(x)\frac{\partial^2}{\partial x \partial y} + \frac{1}{2}\nu^2 y^2 \frac{\partial^2}{\partial y^2}.$$

We present the approach of Hagan *et al.* (2002) to solve the PDE for an analytical approximation to V_0.

Define the transition density function p by

$$\begin{aligned} &p(0, f, \alpha; T, F, A)dFdA \\ =&\text{Prob}\left\{F < f_T < F + dF, A < \alpha_T < A + dA\,\middle|\, f_0 = f, \alpha_0 = \alpha\right\} \quad (11.35) \\ =&E^{Q_T}\left[\delta(f_T - F)\delta(\alpha_T - A)\middle|\, f_0 = f, \alpha_0 = \alpha\right] dFdA. \end{aligned}$$

Then p satisfies the terminal-value problem,

$$\begin{cases} \dfrac{\partial V}{\partial t} = -\mathcal{L}p, & t < T, \\ p(T, x, y) = \delta(x - F)\delta(y - A). \end{cases} \tag{11.36}$$

Proposition 11.3.1. *p also satisfies the forward Fokker-Planck equation:*

$$\begin{cases} \dfrac{\partial p}{\partial T} = \dfrac{1}{2}A^2\left[C^2(F)p\right]_{FF} + \rho\nu\left[A^2 C(F)p\right]_{FA} + \dfrac{1}{2}\nu^2\left[A^2 p\right]_{AA}, \\ p(0, F, A) = \delta(F - f)\delta(A - \alpha). \end{cases} \tag{11.37}$$

Proof. Consider an intermediate time θ s.t. $0 < \theta < T$. We have

$$p(0, f, \alpha; T, F, A) = \int_{R^2} p(0, f, \alpha; \theta, x, y)p(\theta, x, y; T, F, A)dX,$$

where $dX = dxdy$. Differentiate the above equation w.r.t. θ and perform integration by parts, we obtain

$$\begin{aligned} 0 =& \int_{R^2} \frac{\partial p(0, f, \alpha; \theta, x, y)}{\partial \theta}p(\theta, x, y; T, F, A)dX \\ &- p(0, f, \alpha; \theta, x, y)\mathcal{L}p(\theta, x, y; T, F, A)dX \\ =& \int_{R^2}\left[\frac{\partial p(0, f, \alpha; \theta, x, y)}{\partial \theta} - \mathcal{L}^* p(0, f, \alpha; \theta, x, y)\right]p(\theta, x, y; T, F, A)dX, \end{aligned}$$

where

$$\mathcal{L}^* p = \frac{1}{2} y^2 \left[C^2(x) p \right]_{xx} + \rho \nu \left[y^2 C(x) p \right]_{xy} + \frac{1}{2} \nu^2 \left[y^2 p \right]_{yy}.$$

Let $\theta \to T$, we then end up with

$$
\begin{aligned}
0 &= \int_{R^2} \left[\frac{\partial p(0, f, \alpha; \theta, x, y)}{\partial \theta} - \mathcal{L}^* p(0, f, \alpha; \theta, x, y) \right] \delta(x - F) \delta(y - A) dX \\
&= \frac{\partial p(0, f, \alpha; \theta, x, y)}{\partial \theta} - \mathcal{L}^* p(0, f, \alpha; \theta, x, y) \Big|_{\theta=T, x=F, y=A} \\
&= \frac{\partial p(0, f, \alpha; T, F, A)}{\partial T} - \mathcal{L}^* p(0, f, \alpha; T, F, A)
\end{aligned}
$$

$$\square$$

Next, we derive an expression of the time value of the option. Let $V(0, f, \alpha)$ be the T-forward price of the European call option at time 0, then,

$$
\begin{aligned}
V(0, f, \alpha) &= E_0^{Q_T} \left[(f_T - K)^+ \right] \\
&= \int_{R^2} (F - K)^+ p(0, f, \alpha; T, F, A) dF dA.
\end{aligned}
$$

Since

$$p(0, f, \alpha; T, F, A) = \delta(F - f) \delta(A - \alpha) + \int_0^T p_T(0, f, \alpha; T, F, A) dT.$$

So the value function of the option is

$$V(0, f, \alpha) = (f - K)^+ + \int_0^T \int_{R^2} (F - K)^+ p_T(0, f, \alpha; T, F, A) dF dA dT. \quad (11.38)$$

Substituting the first equation of Equation 11.37 into Equation 11.38, we obtain

$$V(0, f, \alpha) = (f - K)^+ + \frac{1}{2} \int_0^T \int_{R^2} A^2 (F - K)^+ \left[C^2(F) p \right]_{FF} dF dA dT. \quad (11.39)$$

Integrating by parts w.r.t. F twice, we obtain

$$
\begin{aligned}
V(0, f, \alpha) &= (f - K)^+ + \frac{1}{2} \int_0^T \int_{R^2} A^2 \delta(F - K) C^2(F) p \, dF dA dT \\
&= (f - K)^+ + \frac{1}{2} C^2(K) \int_0^T \int_{-\infty}^{+\infty} A^2 p(0, f, \alpha; T, K, A) dA dT.
\end{aligned}
$$

$$(11.40)$$

Define

$$P(0, f, \alpha; T, K) = \int_{-\infty}^{+\infty} A^2 p(0, f, \alpha; T, K, A) dA,$$

then P satisfies the backward Fokker-Planck equation:

$$\begin{cases} P_t + \frac{1}{2}\alpha^2 C^2(f)P_{ff} + \alpha^2\rho\nu C(f)P_{f\alpha} + \frac{1}{2}\nu^2\alpha^2 P_{\alpha\alpha} = 0, \quad t < T, \\ P(T, f, \alpha; T, K) = \alpha^2\delta(f - K). \end{cases} \quad (11.41)$$

Here we start using f and α in place of x and y, following the usage of the Hagen *et al.* (2002). Since t does not appear explicitly in the PDE, P depends on $T - t$ only, so we define

$$\tau = T - t,$$

then the expression for option price becomes

$$V(0, f, \alpha) = (f - K)^+ + \frac{1}{2}C^2(K)\int_0^T P(\tau, f, \alpha; T, K)d\tau, \quad (11.42)$$

where P satisfies

$$\begin{cases} P_\tau = \frac{1}{2}\alpha^2 C^2(f)P_{ff} + \alpha^2\rho\nu C(f)P_{f\alpha} + \frac{1}{2}\nu^2\alpha^2 P_{\alpha\alpha}, \quad \tau > 0, \\ P(0, f, \alpha; T, K) = \alpha^2\delta(f - K). \end{cases} \quad (11.43)$$

Once we have obtained the solution of P, we will substitute it back to Equation 11.42 to solve for $V(0, f, \alpha)$.

For a general function of $C(f)$, Equation 11.43 is nonlinear and admits no closed-form solution. When $T \ll 1$, it becomes feasible to develop an analytical approximation using the method of singular perturbation expansion. For this purpose, we make a time change:

$$\hat{\tau} = \frac{\tau}{T},$$

and write ϵ^2 for T, then initial-value problem 11.43 becomes

$$\begin{cases} P_{\hat{\tau}} = \frac{1}{2}\epsilon^2\alpha^2 C^2(f)P_{ff} + \epsilon^2\alpha^2\rho\nu C(f)P_{f\alpha} + \frac{1}{2}\epsilon^2\nu^2\alpha^2 P_{\alpha\alpha}, \quad \hat{\tau} > 0, \\ P(0, f, \alpha; T, K) = \alpha^2\delta(f - K). \end{cases} \quad (11.44)$$

In what follows, we will make a sequence of change of variables to reduce Equation 11.44 to a one-dimensional heat equation with constant coefficients and thus derive an analytic approximation to its solution. For notational simplicity we drop ˆ over τ.

11.3.1 Singular Perturbation Expansion

The singular perturbation expansion method for solving Equation 11.44 goes through seven steps.

Step 1: To simplify the leading coefficient of the PDE
Consider the Lamberti transformation for f:

$$z(f) = \frac{1}{\epsilon\alpha} \int_k^f \frac{du}{C(u)}.$$

Then $Z = z(f_t)$ satisfies

$$dZ = dW_t + \text{drift term},$$

so the transition density function of Z has the leading-order term $e^{-z^2/2\tau}/\sqrt{2\pi\tau}$, and we can find higher-order terms by using the singular perturbation method. Define

$$B(\epsilon\alpha z) = C(f).$$

We then have

$$\frac{\partial}{\partial f} \to \frac{1}{\epsilon\alpha B(\epsilon\alpha z)} \frac{\partial}{\partial z},\quad \frac{\partial}{\partial\alpha} \to \frac{\partial}{\partial\alpha} - \frac{z}{\alpha}\frac{\partial}{\partial z},$$

$$\frac{\partial^2}{\partial f^2} \to \frac{1}{\epsilon^2\alpha^2 B^2(\epsilon\alpha z)} \left\{ \frac{\partial^2}{\partial z^2} - \epsilon\alpha\frac{B'(\epsilon\alpha z)}{B(\epsilon\alpha z)}\frac{\partial}{\partial z} \right\},$$

$$\frac{\partial^2}{\partial f\partial\alpha} \to \frac{1}{\epsilon\alpha B(\epsilon\alpha z)} \left\{ \frac{\partial^2}{\partial z\partial\alpha} - \frac{z}{\alpha}\frac{\partial^2}{\partial z^2} - \frac{1}{\alpha}\frac{\partial}{\partial z} \right\},$$

$$\frac{\partial^2}{\partial\alpha^2} \to \frac{\partial^2}{\partial\alpha^2} - \frac{2z}{\alpha}\frac{\partial^2}{\partial z\partial a} + \frac{z^2}{\alpha^2}\frac{\partial^2}{\partial z^2} + \frac{2z}{\alpha^2}\frac{\partial}{\partial z},$$

and

$$\delta(f - K) = \delta(\epsilon\alpha z C(K)) = \frac{1}{\epsilon\alpha C(K)}\delta(z).$$

The expression of option value can be rewritten into

$$V(0, f, \alpha) = (f - K)^+ + \frac{1}{2}\epsilon^2 C^2(K) \int_0^1 P(\tau, z, \alpha)d\tau, \tag{11.45}$$

for $P(\tau, z, \alpha)$ that satisfies

$$\begin{cases} P_\tau = \dfrac{1}{2}\left(1 - 2\epsilon\rho\nu z + \epsilon^2\nu^2 z^2\right)P_{zz} - \dfrac{1}{2}\epsilon\alpha\dfrac{B'}{B}P_z \\[2mm] \quad + \left(\epsilon\rho\nu - \epsilon^2\nu^2 z\right)\left(\alpha P_{z\alpha} - P_z\right) + \dfrac{1}{2}\epsilon^2\nu^2\alpha^2 P_{\alpha\alpha}, \quad \tau > 0, \tag{11.46} \\[2mm] P(0, z, \alpha) = \dfrac{\alpha}{\epsilon C(K)}\delta(z). \end{cases}$$

Step 2: Turning P into a density function
Now define

$$\hat{P} = \frac{\epsilon C(K)}{\alpha}P,$$

in terms of which the expression of value becomes

$$V(0, f, \alpha) = (f - K)^+ + \frac{1}{2}\epsilon\alpha C(K) \int_0^1 \hat{P}(\tau, z, \alpha)d\tau, \tag{11.47}$$

where $\hat{P}$ solves

$$
\begin{cases}
\hat{P}_\tau = \dfrac{1}{2}\left(1 - 2\epsilon\rho\nu z + \epsilon^2\nu^2 z^2\right)\hat{P}_{zz} - \dfrac{1}{2}\epsilon\alpha\dfrac{B'}{B}\hat{P}_z \\[2mm]
\qquad + \left(\epsilon\rho\nu - \epsilon^2\nu^2 z\right)\alpha\hat{P}_{z\alpha} + \dfrac{1}{2}\epsilon^2\nu^2(\alpha^2\hat{P}_{\alpha\alpha} + 2\alpha\hat{P}_\alpha), \quad \tau > 0, \\[2mm]
\hat{P}(0, z, \alpha) = \delta(z).
\end{cases}
\tag{11.48}
$$

The asymptotic solution of the above PDE is of the form

$$
\hat{P}(\tau, z, \alpha) = \hat{P}_0(\tau, z) + \epsilon\hat{P}_1(\tau, z, \alpha) + O(\epsilon^2),
$$

and we understand that

$$
\hat{P}_{z\alpha}, \quad \hat{P}_{\alpha\alpha}, \quad \hat{P}_\alpha \sim O(\epsilon).
$$

Omitting the $O(\epsilon^3)$ terms in Equation 11.48, we arrive at

$$
\begin{cases}
\hat{P}_\tau = \dfrac{1}{2}\left(1 - 2\epsilon\rho\nu z + \epsilon^2\nu^2 z^2\right)\hat{P}_{zz} - \dfrac{1}{2}\epsilon\alpha\dfrac{B'}{B}\hat{P}_z + \epsilon\rho\nu\alpha\hat{P}_{z\alpha}, \quad \tau > 0, \\[2mm]
\hat{P}(0, z, \alpha) = \delta(z).
\end{cases}
\tag{11.49}
$$

Step 3: Removing α derivative terms
Define a function $H(\tau, z, \alpha)$ through

$$
\hat{P} = \sqrt{C(f)/C(K)}\; H = \sqrt{B(\epsilon\alpha z)/B(0)}\; H.
$$

Then,

$$
\hat{P}_z = \sqrt{B(\epsilon\alpha z)/B(0)}\left\{H_z + \dfrac{1}{2}\epsilon\alpha\dfrac{B'}{B}H\right\},
$$

$$
\hat{P}_{zz} = \sqrt{B(\epsilon\alpha z)/B(0)}\left\{H_{zz} + \epsilon\alpha\dfrac{B'}{B}H_z + \epsilon^2\alpha^2\left[\dfrac{B''}{2B} - \dfrac{B'^2}{4B^2}\right]H\right\},
$$

$$
\hat{P}_{z\alpha} = \sqrt{B(\epsilon\alpha z)/B(0)}\left\{H_{z\alpha} + \dfrac{1}{2}\epsilon z\dfrac{B'}{B}H_z + \dfrac{1}{2}\epsilon\alpha\dfrac{B'}{B}H_\alpha + \dfrac{1}{2}\epsilon\dfrac{B'}{B}H\right\}.
$$

The expression for option price becomes

$$
V(0, f, \alpha) = (f - K)^+ + \dfrac{1}{2}\epsilon\alpha\sqrt{B(0)B(\epsilon\alpha z)}\int_0^1 H(\tau, z, \alpha)d\tau,
\tag{11.50}
$$

where H satisfies

$$
\begin{cases}
H_\tau = \dfrac{1}{2}\left(1 - 2\epsilon\rho\nu z + \epsilon^2\nu^2 z^2\right)H_{zz} - \dfrac{1}{2}\epsilon^2\rho\nu\alpha\dfrac{B'}{B}(zH_z - H) \\[2mm]
\qquad + \epsilon^2\alpha^2\left(\dfrac{1}{4}\dfrac{B''}{B} - \dfrac{3}{8}\dfrac{B'^2}{B^2}\right)H + \epsilon\rho\nu\alpha\left(H_{z\alpha} + \dfrac{1}{2}\epsilon\alpha\dfrac{B'}{B}H_\alpha\right), \quad \tau > 0, \\[2mm]
H(0, z) = \delta(z).
\end{cases}
\tag{11.51}
$$

We express the solution as

$$H = H_0(\tau, z) + \epsilon H_1(\tau, z) + \epsilon^2 H_2(\tau, z, \alpha) + O(\epsilon^3),$$

with which we understand that α-derivative terms is of order $O(\epsilon^3)$ and thus ignored, yielding

$$\begin{cases} H_\tau = \dfrac{1}{2}\left(1 - 2\epsilon\rho\nu z + \epsilon^2\nu^2 z^2\right) H_{zz} - \dfrac{1}{2}\epsilon^2\rho\nu\alpha\dfrac{B'}{B}(zH_z - H) \\ \qquad\quad + \epsilon^2\alpha^2\left(\dfrac{1}{4}\dfrac{B''}{B} - \dfrac{3}{8}\dfrac{B'^2}{B^2}\right) H, \quad \tau > 0, \\ H(0, z) = \delta(z). \end{cases} \tag{11.52}$$

There is no α derivatives in the equation. As a result, we now treat α as a parameter instead of a variable in the solution.

Step 4: Removing the $\frac{B'}{B} zH_z$ term

By Taylor expansions, we have the following approximations:

$$\frac{B'(\epsilon\alpha z)}{B(\epsilon\alpha z)} = \frac{B'(\epsilon\alpha z_0)}{B(\epsilon\alpha z_0)} + O(\epsilon) \stackrel{\triangle}{=} b_1 + O(\epsilon),$$

$$\frac{B''(\epsilon\alpha z)}{B(\epsilon\alpha z)} = \frac{B''(\epsilon\alpha z_0)}{B(\epsilon\alpha z_0)} + O(\epsilon) \stackrel{\triangle}{=} b_2 + O(\epsilon),$$

where z_0 is to be determined. Define $\hat{H}$ by

$$H = e^{\epsilon^2\rho\nu\alpha b_1 z^2/4}\hat{H}.$$

The option price then becomes

$$V(0, f, \alpha) = (f - K)^+ + \frac{1}{2}\epsilon\alpha\sqrt{B(0)B(\epsilon\alpha z)}e^{\epsilon^2\rho\nu\alpha b_1 z^2/4}\int_0^1 \hat{H}(\tau, z)d\tau, \tag{11.53}$$

and $\hat{H}$ satisfies the following PDE:

$$\begin{cases} \hat{H}_\tau = \dfrac{1}{2}\left(1 - 2\epsilon\rho\nu z + \epsilon^2\nu^2 z^2\right)\hat{H}_{zz} + \epsilon^2\alpha^2\left(\dfrac{1}{4}b_2 - \dfrac{3}{8}b_1^2\right)\hat{H} + \dfrac{3}{4}\epsilon^2\rho\nu\alpha b_1\hat{H}, \\ H(0, z) = \delta(z). \end{cases}$$

$$\tag{11.54}$$

Step 5: Getting rid of the variable coefficient of $\hat{H}_{zz}$

Introduce

$$x = \frac{1}{\epsilon\nu}\int_0^{\epsilon\nu z} \frac{d\zeta}{\sqrt{1 - 2\rho\zeta + \zeta^2}} \tag{11.55}$$

or

$$\epsilon\nu z = \sinh(\epsilon\nu x) - \rho(\cosh(\epsilon\nu x) - 1),$$

and let

$$I(\zeta) = \sqrt{1 - 2\rho\zeta + \zeta^2}.$$

Then, in terms of x we change the value formula to

$$V(0, f, \alpha) = (f - K)^+ + \frac{1}{2}\epsilon\alpha\sqrt{B(0)B(\epsilon\alpha z)}e^{\epsilon^2 \rho\nu\alpha b_1 z^2/4} \int_0^1 \hat{H}(\tau, x)dx,$$

with $\hat{H}$ satisfying

$$\begin{cases} \hat{H}_\tau = \frac{1}{2}\hat{H}_{xx} - \frac{1}{2}\epsilon\nu I'(\epsilon\nu z)\hat{H}_x + \epsilon^2\alpha^2\left(\frac{1}{4}b_2 - \frac{3}{8}b_1^2\right)\hat{H} + \frac{3}{4}\epsilon^2\rho\nu\alpha b_1\hat{H}, \\ H(0, x) = \delta(x). \end{cases}$$

$$(11.56)$$

Step 6: Getting rid of the $\hat{H}_x$ term

The next step is to get rid of the $\hat{H}_x$ term by introducing function Q through

$$\hat{H} = I^{\frac{1}{2}}(\epsilon\nu z(x))Q = \left(1 - 2\epsilon\rho\nu z + \epsilon^2\nu^2 z^2\right)^{\frac{1}{4}}Q.$$

Then,

$$\hat{H}_x = I^{\frac{1}{2}}(\epsilon\nu z)\left[Q_x + \frac{1}{2}\epsilon\nu I'(\epsilon\nu z)Q\right]$$

$$\hat{H}_{xx} = I^{\frac{1}{2}}(\epsilon\nu z)\left[Q_{xx} + \epsilon\nu I'(\epsilon\nu z)Q_x + \epsilon^2\nu^2\left(\frac{1}{2}I''I + \frac{1}{4}(I')^2\right)\right],$$

$$(11.57)$$

and the value function becomes

$$V(0, f, \alpha) = (f - K)^+ + \frac{1}{2}\epsilon\alpha\sqrt{B(0)B(\epsilon\alpha z)}\ I^{\frac{1}{2}}(\epsilon\nu z)e^{\epsilon^2 \rho\nu\alpha b_1 z^2/4} \int_0^1 Q(\tau, x)d\tau,$$

$$(11.58)$$

with Q satisfying

$$\begin{cases} Q_\tau = \frac{1}{2}Q_{xx} + \epsilon^2\nu^2\left(\frac{1}{4}I''I - \frac{1}{8}(I')^2\right)Q \\ \qquad + \epsilon^2\alpha^2\left(\frac{1}{4}b_2 - \frac{3}{8}b_1^2\right)Q + \frac{3}{4}\epsilon^2\rho\nu\alpha b_1 Q, \\ Q(0, x) = \delta(x). \end{cases}$$

$$(11.59)$$

Step 7: Freezing the coefficient of Q

Put the coefficients of Q terms together and then freeze z at z_0, we commit an error of order $O(\epsilon)$. Define a constant κ by

$$\kappa = \nu^2\left(\frac{1}{4}I''(\epsilon\nu z_0)I(\epsilon\nu z_0) - \frac{1}{8}[I'(\epsilon\nu z_0)]^2\right) + \alpha^2\left(\frac{1}{4}b_2 - \frac{3}{8}b_1^2\right) + \frac{3}{4}\rho\nu\alpha b_1.$$

Through order $O(\epsilon^2)$, the equation is simplified to

$$Q_\tau = \frac{1}{2}Q_{xx} + \epsilon^2\kappa Q,$$

$$Q(0, x) = \delta(x) \quad \text{at} \quad \tau = 0.$$

$$(11.60)$$

The solution clearly is

$$Q = \frac{1}{\sqrt{2\pi\tau}} e^{-\frac{x^2}{2\tau} + \epsilon^2 \kappa \tau},$$

thus yielding the option value below:

$$V(0, f, \alpha) = (f - K)^+ + \frac{1}{2}\epsilon\alpha\sqrt{B(0)B(\epsilon\alpha z)}$$

$$\times I^{\frac{1}{2}}(\epsilon\nu z)e^{\frac{1}{4}\epsilon^2\rho\nu\alpha b_1 z^2} \int_0^1 \frac{1}{\sqrt{2\pi\tau}} e^{-\frac{x^2}{2\tau} + \epsilon^2 \kappa \tau} d\tau. \tag{11.61}$$

Introduce θ s.t.

$$\epsilon^2\theta = \ln\left(\frac{\epsilon\alpha z}{f - K}\sqrt{B(0)B(\epsilon\alpha z)}\right) + \ln\left(\frac{xI^{\frac{1}{2}}(\epsilon\nu z)}{z}\right) + \frac{1}{4}\epsilon^2\rho\nu\alpha b_1 z^2. \tag{11.62}$$

Noticing that θ/x^2 matches $\kappa/3$ up to an error of order $O(\epsilon^2)$, we have the approximation of

$$e^{\epsilon^2 \kappa \tau} = \frac{1}{\left(1 - \frac{2}{3}\kappa\epsilon^2\tau\right)^{3/2}} + O(\epsilon^4) = \frac{1}{(1 - 2\epsilon^2\tau\theta/x^2)^{3/2}} + O(\epsilon^4). \tag{11.63}$$

We thus write

$$V(0, f, \alpha) = (f - K)^+ + \frac{1}{2}\frac{f - K}{x} \int_0^1 \frac{1}{\sqrt{2\pi\tau}} \frac{e^{-\frac{x^2}{2\tau} + \epsilon^2\theta}}{\left(1 - \frac{2\tau}{x^2}\epsilon^2\theta\right)^{3/2}} d\tau. \tag{11.64}$$

Note that x and $f - K$ have the same sign. By a change of variable,

$$q = \frac{x^2}{2\tau},$$

the expression for option value becomes

$$V(0, f, \alpha) = (f - K)^+ + \frac{|f - K|}{4\sqrt{\pi}} \int_{\frac{x^2}{2}}^{\infty} \frac{e^{-q + \epsilon^2\theta}}{(q - \epsilon^2\theta)^{3/2}} dq$$

$$= (f - K)^+ + \frac{|f - K|}{4\sqrt{\pi}} \int_{\frac{x^2}{2} - \epsilon^2\theta}^{\infty} \frac{e^{-q}}{q^{3/2}} dq. \tag{11.65}$$

The option value given in Equation 11.65 can be converted to implied Black's volatility. The conversion consists of two steps: 1) the implied normal volatility, and 2) the implied Black volatility.

11.3.2 Equivalent Normal Volatility

Suppose we repeat the above analysis for the normal model,

$$df_t = \sigma_N dW_t, \quad f_0 = f,$$

where σ_N is a constant. It is a special case of the stochastic volatility model with $\alpha = \sigma_N, C(f) = 1$ and $\nu = 0$. After a time change, $t \to \hat{t} \times T$, it becomes

$$df_{\hat{t}} = \epsilon \sigma_N dW_{\hat{t}}, \quad f_0 = f.$$

According to its definition in Equation 11.62,

$$\epsilon^2 \theta = 0$$

and

$$V(t, f) = (f - K)^+ + \frac{|f - K|}{4\sqrt{\pi}} \int_{\frac{x^2}{2}}^{\infty} \frac{e^{-q}}{q^{3/2}} dq \tag{11.66}$$

for

$$x = z = \frac{f - K}{\epsilon \sigma_N}.$$

The option price under the normal model matches that of the general stochastic volatility model iff σ_N is chosen to satisfy

$$\frac{(f - K)^2}{2\epsilon^2 \sigma_N^2} = \frac{x^2}{2} - \epsilon^2 \theta,$$

that is,

$$\sigma_N(f, K) = \frac{f - K}{\epsilon x} \left\{ 1 + \epsilon^2 \frac{\theta}{x^2} + O(\epsilon^4) \right\}$$

$$= \left(\frac{f - K}{\epsilon z} \right) \left(\frac{z}{x(z)} \right) \left[1 + \epsilon^2 (\phi_1 + \phi_2 + \phi_3) + O(\epsilon^4) \right],$$

where the terms $\epsilon^2 \phi_1$, $\epsilon^2 \phi_2$ and $\epsilon^2 \phi_3$ correspond to the three terms of $\epsilon^2 \theta$ in Equation 11.62,

$$\frac{f - K}{\epsilon z} = \frac{\alpha(f - K)}{\int_K^f \frac{du}{C(u)}} = \left(\frac{1}{f - K} \int_K^f \frac{du}{\alpha C(u)} \right)^{-1}.$$

$$\frac{z}{x(z)} = \frac{\zeta}{\ln \left(\frac{I(\zeta) - \rho + \zeta}{1 - \rho} \right)} \stackrel{\triangle}{=} \frac{\zeta}{D(\zeta)},$$

for

$$\zeta = \epsilon \nu z = \frac{\nu}{\alpha} \int_K^f \frac{du}{C(u)} = \frac{\nu}{\alpha} \frac{f - K}{C(f_{av})} \left\{ 1 + O(\epsilon^2) \right\},$$

and $f_{av} = (f + K)/2$, the arithmetic average of f and K. The key to evaluate the three ϵ^2-order terms is to evaluate

$$z = \frac{1}{\epsilon \alpha} \int_K^f \frac{du}{C(u)}.$$

For this purpose we perform Taylor expansion of $1/C(u)$ at $u = (f + K)/2 = f_{av}$ and retain up to the second-order terms, obtaining

$$z = \frac{f - K}{\epsilon \alpha C(f_{av})} \left[1 + \frac{1}{6}(2\gamma_1^2 - \gamma_2) \left(\frac{f - K}{2} \right)^2 \right]$$

$$\triangleq z_0(1 + \delta),$$

for

$$\gamma_1 = \frac{C'(f_{av})}{C(f_{av})}, \quad \gamma_2 = \frac{C''(f_{av})}{C(f_{av})}.$$

Note that z_0 is chosen and defined through the equation above. Using Taylor expansions whenever needed, we obtain, with errors of order $O(\epsilon^4)$,

$$\epsilon^2 \phi_1 = \frac{1}{z^2} \ln \left(\frac{\epsilon \alpha z}{f - K} \sqrt{C(f)C(K)} \right) = \frac{2\gamma_2 - \gamma_1^2}{24} \epsilon^2 \alpha^2 C^2(f_{av}), \quad (11.67)$$

$$\epsilon^2 \phi_2 = \frac{1}{z^2} \ln \left(\frac{x}{z} \left[1 - 2\epsilon \rho \nu z + \epsilon^2 \nu^2 z^2 \right]^{1/4} \right) = \frac{2 - 3\rho^2}{24} \epsilon^2 \nu^2, \quad (11.68)$$

$$\epsilon^2 \phi_3 = \frac{1}{4} \epsilon^2 \rho \alpha \nu \frac{B'(\epsilon \alpha z_0)}{B(\epsilon \alpha z_0)} = \frac{1}{4} \epsilon^2 \rho \nu \alpha \gamma_1 C(f_{av}). \quad (11.69)$$

Putting the terms together, we obtain the normal implied volatility

$$\sigma_N(f, K) = \frac{\alpha(f - K)}{\int_K^f \frac{du}{C(u)}} \left(\frac{\zeta}{D(\zeta)} \right) \{ 1 + h_N \epsilon^2 + O(\epsilon^4) \}, \quad (11.70)$$

for

$$h_N = \frac{2\gamma_2 - \gamma_1^2}{24} \alpha^2 C^2(f_{av}) + \frac{1}{4} \rho \nu \alpha \gamma_1 C(f_{av}) + \frac{2 - 3\rho^2}{24} \nu^2. \quad (11.71)$$

In $\sigma_N(f, K)$, the $O(1)$ term provides the dominant behavior, and the $O(\epsilon^2)$ term provides a correction of 1% magnitude.

11.3.3   Equivalent Implied Black Volatility for General SV Models

After the time change, $t \to \hat{t} \times T$, the Black model becomes

$$df_{\hat{t}} = \epsilon \sigma_B f dW_{\hat{t}}, \quad f_0 = f,$$

which is another special case of the general SV model with $\nu = 0, C(f) = f$ and $\alpha = \sigma_B$. Putting $\nu = 0$ and taking the limit $\beta \to 1$ in Equation 11.70, we readily obtain

$$\sigma_N(f, K) = \frac{\sigma_B(f - K)}{\ln \frac{f}{K}} \left[1 - \frac{1}{24} \sigma_B^2 \epsilon^2 + O(\epsilon^4) \right]. \quad (11.72)$$

Note that when $\nu = 0$, then $\zeta = 0$, at which $\zeta/D(\zeta)$ is defined by its limit:

$$\frac{\zeta}{D(\zeta)} \xrightarrow{\zeta \to 0} 1.$$

Equating Equation 11.70 to Equation 11.72, we obtain

$$\sigma_B(f, K) = \frac{\alpha \ln \frac{f}{K}}{\int_K^f \frac{du}{C(u)}} \frac{\zeta}{D(\zeta)} \left[1 + \left(h_N + \frac{\sigma_B^2}{24} \right) \epsilon^2 + O(\epsilon^4) \right]$$

$$\approx \frac{\alpha \ln \frac{f}{K}}{\int_K^f \frac{du}{C(u)}} \frac{\zeta}{D(\zeta)} \left[1 + \left(h_N + \frac{1}{24} \left(\frac{\alpha \ln \frac{f}{K}}{\int_K^f \frac{du}{C(u)}} \right)^2 \right) \epsilon^2 + O(\epsilon^3) \right].$$

$$(11.73)$$

Here, we make use of the approximation $\frac{\zeta}{D(\zeta)} = 1 + O(\epsilon)$. The last term above can be further simplified. When f is near K, we write

$$\int_K^f \frac{du}{C(u)} = \frac{f - K}{C(f_{av})}$$

$$= \frac{2(Kf)^{1/2}}{C(f_{av})} \sinh \left(\ln \left(\frac{f}{K} \right) \right)$$

$$= \frac{f_{av}}{C(f_{av})} \ln \frac{f}{K} \left[1 + \frac{1}{24} \ln^2 \left(\frac{f}{K} \right) + O \left(\ln^4 \left(\frac{f}{K} \right) \right) \right].$$

Insert the leading term into Equation 11.73 and make some cancelations, we finally arrive at the final formula for the implied Black volatility:

$$\sigma_B(f, K) = \frac{\alpha \ln \frac{f}{K}}{\int_K^f \frac{du}{C(u)}} \frac{\zeta}{D(\zeta)}$$

$$\times \left\{ 1 + \left[\frac{2\gamma_2 - \gamma_1^2 + f_{av}^{-2}}{24} \alpha^2 C^2(f_{av}) + \frac{1}{4} \rho \nu \alpha \gamma_1 C(f_{av}) + \frac{2 - 3\rho^2}{24} \nu^2 \right] \epsilon^2 + O(\epsilon^3) \right\}.$$

$$(11.74)$$

This finishes the proof. □

11.4   Appendix: Proof of Proposition 11.2.2

For clarity, we let $\tau = T - t$ and $\lambda = \|\gamma_j\|$. Substituting the formal solution Equation 11.21 to Equation 12.28, we obtain the following equations for the undetermined coefficient:

$$\frac{dA}{d\tau} = a_0 B,$$

$$\frac{dB}{d\tau} = b_2 B^2 + b_1 B + b_0,$$

$$(11.75)$$

where

$$a_0 = \kappa\theta, \quad b_0 = \frac{1}{2}\lambda^2(z^2 - z), \quad b_1 = (\rho\epsilon\lambda z - \kappa\xi), \quad b_2 = \frac{1}{2}\epsilon^2.$$

Now consider Equation 11.75 with constant coefficients and general initial conditions

$$A(0) = A_0, \qquad B(0) = B_0.$$

Since B is independent of A, it is solved first. In the special case when

$$b_2 B_0^2 + b_1 B_0 + b_0 = 0,$$

we have a easy solution

$$B(\tau) = B_0,$$
$$A(\tau) = A_0 + a_0 B_0 \tau.$$

Otherwise, let Y_1 be the solution to

$$b_2 Y^2 + b_1 Y + b_0 = 0.$$

Assume $b_2 \neq 0$, then

$$Y_1 = \frac{-b_1 \pm d}{2b_2}, \quad \text{with} \quad d = \sqrt{b_1^2 - 4b_0 b_2}. \tag{11.76}$$

Without making any difference we take the "+" sign for Y_1. We then consider the difference between Y_1 and B:

$$Y_2 = B - Y_1.$$

Clearly, Y_2 satisfies

$$\begin{aligned} \frac{dY_2}{d\tau} &= \frac{d(Y_1 + Y_2)}{d\tau} \\ &= b_2(Y_1 + Y_2)^2 + b_1(Y_1 + Y_2) + b_0 \\ &= b_2 Y_2^2 + (2b_2 Y_1 + b_1)Y_2 \\ &= b_2 Y_2^2 + dY_2, \end{aligned} \tag{11.77}$$

with initial condition

$$Y_2(0) = B_0 - Y_1.$$

Note in the last equality of Equation 11.77 we have used the equation Equation 11.76. Equation 11.77 belongs to the class of Bernoulli equations which can be solved explicitly. One can verify that the solution is

$$Y_2 = \frac{d}{b_2} \frac{ge^{d\tau}}{(1 - ge^{d\tau})}, \quad \text{with} \quad g = \frac{\frac{-b_1 + d}{2b_2} - B_0}{\frac{-b_1 - d}{2b_2} - B_0}. \tag{11.78}$$

It follows that

$$B(\tau) = Y_1 + Y_2$$

$$= \frac{-b_1 + d}{2b_2} + \frac{d}{b_2}\frac{ge^{d\tau}}{(1 - ge^{d\tau})}$$

$$= B_0 + \left(\frac{-b_1 + d}{2b_2} - B_0\right)\frac{(1 - e^{d\tau})}{(1 - ge^{d\tau})}.$$

Having obtained B, we integrate the first equation of Equation 11.75 to get A:

$$A(\tau) = A_0 + a_0 \int_0^\tau B(s)ds$$

$$= A_0 + a_0 B_0 \tau + a_0 \left(\frac{-b_1 + d}{2b_2} - B_0\right)\int_0^\tau \frac{1 - e^{d\tau}}{1 - ge^{d\tau}}d\tau$$

$$= A_0 + a_0 B_0 \tau + a_0 \left(\frac{-b_1 + d}{2b_2} - B_0\right)\left[\tau - \int_0^\tau \frac{(1 - g)e^{d\tau}}{1 - ge^{d\tau}}d\tau\right]$$

$$= A_0 + a_0\frac{(-b_1 + d)\tau}{2b_2} - a_0\left(\frac{-b_1 + d}{2b_2} - B_0\right)\int_1^{e^{d\tau}}\frac{(1 - g)}{1 - gu}du$$

$$= A_0 + a_0\frac{(-b_1 + d)\tau}{2b_2} - a_0\left(\frac{-b_1 + d}{2b_2} - B_0\right)\frac{(g - 1)}{g}\ln\left(\frac{1 - ge^{d\tau}}{1 - g}\right)$$

$$= A_0 + a_0\left[\frac{(-b_1 + d)}{2b_2}\tau - \frac{1}{b_2}\ln\left(\frac{1 - ge^{d\tau}}{1 - g}\right)\right].$$

Letting

$$u^{\pm} = \frac{-b_1 \pm d}{2b_2} - B_0,$$

and then

$$A_0 = A(\tau_j, z),$$
$$B_0 = B(\tau_j, z),$$

and replacing τ by $\tau - \tau_j$, we arrive at Equation 11.23. The solution $\phi(z)$ so obtained belongs to $\mathbb{C}^1$ and hence is a weak solution to Equation 12.28.   □

Chapter 12

Lévy Market Model

The SOFR market models we have studied so far are driven by diffusions, so SOFR rates are a continuous function of time. Although less important today for interest-rate modeling than stochastic volatility as another kind of risk factor, jumps in derivatives modeling and pricing were not ignored. In the 1990s, interest rates often jumped following the routine FOMC meetings. After entering into the new millennium, FOMC had enhanced their dialogues and communications with the financial industry, such that surprises to the market upon the release of the minutes of rate-setting meetings have now become rare. Yet jumps in interest rates can be caused by other reasons, including political or economic events, see e.g., El-Jahel, Lindberg, and Perraudin (1997) and Johannes (2004). The best known example perhaps was the swift and repeated slashing of the U.S. Federal fund rate right after the event of 9/11, when the U.S. Federal Reserve lowered the interest rates by 25 basis points consecutively for several days. From an empirical point of view, volatility smiles or skews are related to the so-called leptokurtic feature of the distribution of the underlying variable, which means a higher peak and fatter tails than those of a normal distribution, and jump-diffusion processes are well-known to bear such a feature. The intention of this chapter is to take jump risks into the modeling of interest rates and present a Lévy market model for SOFR term rates, which is based on general marked point processes, a broad class of stochastic processes driven by diffusions and random jumps. The Lévy market model is an adaptation of the Lévy LIBOR model (Eberlein and Özkan, 2005) to the context of SOFR term rates. In many ways, the Lévy LIBOR model can be regarded as a generalization to the Glasserman and Kou model (2003), a pioneering piece of work that models jumps using marked point processes.

Models with jump risk have been studied extensively, and rich literatures have been produced, particularly by the academic finance community. Although intuitive and theoretically elegant, models of this type have not been popular for applications until today, due to the inconvenience or inability to hedge the jump risk. Until this situation changes, jump risk modeling perhaps will remain a topic of theoretical interests in finance.

DOI: 10.1201/9781003389101-12

12.1   Introduction to Lévy Processes

To prepare for the introduction of the Lévy market model, we first introduce the Lévy processes.

12.1.1   Infinite Divisibility

Definition 12.1.1. *A process* $\{X_t, t \geq 0\}$ *defined on a probability space* $(\Omega, \mathcal{F}, \mathbb{P})$ *is said to be a Lévy process if*

1. *The paths of* X_t *are right continuous with left limits.*

2. $X_0 = 0$ *almost surely.*

3. *For* $0 \leq s \leq t$, $X_t - X_s$ *is equal in distribution to* X_{t-s}.

4. *For* $0 \leq s \leq t$, $X_t - X_s$ *is independent of* $\{X_u, u \leq s\}$.

Define the characteristic exponent of X_t by

$$\Psi_t(\theta) = -\ln \mathbb{E}\left[e^{i\theta X_t}\right], \quad \forall \theta \in \mathbb{R}.$$

For any $t > 0$, and any positive integer n, there is

$$X_t = X_{t/n} + (X_{2t/n} - X_{t/n}) + \cdots + (X_t - X_{(n-1)t/n}).$$

Due to Properties 3 and 4 of the Lévy process, we have

$$\Psi_m(\theta) = m\Psi_1(\theta) = n\Psi_{m/n}(\theta),$$

or

$$\Psi_{m/n}(\theta) = \frac{m}{n}\Psi_1(\theta).$$

So for any rational number t,

$$\Psi_t(\theta) = t\Psi_1(\theta).$$

For any irrational number t, we find a sequence of rational numbers $\{t_n\} \downarrow t$ as $n \to \infty$, then we have

$$\Psi_{t_n}(\theta) \to \Psi_t(\theta)$$

owing to the right continuity of the Lévy process. Hence, we conclude that any Lévy process has the property of

$$\Psi_t(\theta) = t\Psi_1(\theta)$$

for any number $t > 0$. With this equality, we have just established that any infinitely divisible process has the characteristic exponent linear in t and, moreover,

$$\text{Lévy processes} \subset \text{infinitely divisible processes}.$$

From now on we denote $\Psi(\theta) = \Psi_1(\theta)$.

12.1.2 Basic Examples of the Lévy Processes

12.1.2.1 Poisson Processes

Consider a random variable N with a probability distribution

$$P(N = k) = \frac{e^{-\lambda}\lambda^k}{k!}$$

for some $\lambda > 0$. The characteristic function is

$$E\left[e^{i\theta N}\right] = \sum_{k \geq 0} e^{i\theta k}\frac{e^{-\lambda}\lambda^k}{k!} = e^{-\lambda(1-e^{i\theta})}.$$

The characteristic exponent is $\Psi(\theta) = \lambda(1 - e^{i\theta})$.

The Poisson process, $\{N_t : n \geq 0\}$, is a stationary process for which the distribution is Poisson with parameter λt, such that

$$P(N_t = k) = \frac{e^{-\lambda t}(\lambda t)^k}{k!},$$

and

$$\mathbb{E}\left[e^{i\theta N_t}\right] = e^{-\lambda t(1-e^{i\theta})} = e^{-t\Psi(\theta)}.$$

12.1.2.2 Compound Poisson Processes

Suppose $\{\xi_i : i \geq 1\}$ is a sequence of i.i.d. random variables, independent of N, with a common law F which has no atom at zero. A compound process $\{X_t, t \geq 0\}$ is defined by

$$X_t = \sum_{i=1}^{N_t} \xi_i, \quad t \geq 0.$$

By conditioning on N_t, we have the characteristic function of X_t:

$$\mathbb{E}\left[e^{i\theta \sum_{i=1}^{N_t} \xi_i}\right] = \sum_{n \geq 0} \mathbb{E}\left[e^{i\theta \sum_{k=1}^{n} \xi_i}\right] e^{-\lambda t}\frac{(\lambda t)^n}{n!}$$

$$= \sum_{n \geq 0} \left(\int_{\mathbb{R}} e^{i\theta x} F(dx)\right)^n \frac{e^{-\lambda t}(\lambda t)^n}{n!}$$

$$= e^{-\lambda t} e^{\lambda t \int_{\mathbb{R}} e^{i\theta x} F(dx)}$$

$$= e^{-\lambda t \int_{\mathbb{R}}(1-e^{i\theta x})F(dx)}.$$

The characteristic exponent of X_t is

$$\Psi_t(\theta) = \lambda t \int_{\mathbb{R}} (1 - e^{i\theta x})F(dx).$$

By direct computations, we also have

$$E\left[X_t\right] = E\left[N_t\right]E\left[\xi_1\right] = \lambda t \int_{\mathbb{R}} xF(dx).$$

Note that when $F(dx) = \delta(x-1)\,dx$, the compound Poisson process reduces to the Poisson process. Also, for $0 \le s < t < \infty$,

$$X_t = X_s + \sum_{i=N_s+1}^{N_t} \xi_i.$$

The last summation is an independent copy of X_{t-s}.

A compound Poisson with a drift is defined as

$$X_t = \sum_{i=1}^{N_t} \xi_i + ct, \quad t \ge 0,$$

with $c \in \mathbb{R}$. The characteristic exponent of X_t is

$$\Psi(\theta) = \lambda \int_{\mathbb{R}} (1 - e^{i\theta x})F(dx) - ic\theta.$$

If we take $c = \lambda \int_{\mathbb{R}} xF(dx)$, then there is

$$\mathbb{E}\left[X_t\right] = 0,$$

and the corresponding X_t is called the *centered compound Poisson process*, with characteristic exponent

$$\Psi(\theta) = \lambda \int_{\mathbb{R}} (1 - e^{i\theta x} + i\theta x)F(dx).$$

12.1.2.3 Linear Brownian Motion

Take the probability law

$$\mu_{\sigma,a}(dx) = \frac{1}{\sigma\sqrt{2\pi}} e^{-(x+a)^2/2\sigma^2}\,dx,$$

which is the normal distribution, $N(-a, \sigma^2)$, with characteristic function

$$\int_{\mathbb{R}} e^{i\theta x}\mu_{\sigma,a}(dx) = e^{-\frac{1}{2}\sigma^2\theta^2 - ia\theta},$$

so that characteristic component is

$$\Psi(\theta) = \frac{1}{2}\sigma^2\theta^2 + ia\theta,$$

which is that of a Brownian motion with a drift:

$$X_t = -at + \sigma W_t.$$

12.1.3 Introduction of the Jump Measure

Next, we introduce the jump measure to describe how jumps occur. Consider a compound Poisson process for which the arrival rate and jump size, ΔX, are described by the *jump measure*

$$\mu(dt, dx) = \text{no. of jumps over } [t, t + dt] \text{ such that } \Delta X \in [x, x + dx].$$

Define, in addition,

$$\nu(dx) = \text{expected no. of jumps over } [0, 1] \text{ such that } \Delta X \in [x, x + dx],$$

or simply

$$\nu(dx)dt = E\left[\mu(dt, dx)\right].$$

We call $\nu(dx)dt$ the compensator to $\mu(dt, dx)$. In literature, $\nu(dx)$ is more often called the Lévy measure of a Lévy process, which gives the intensity of jumps (i.e., expected number of jumps per unit time) of size in $[x, x + dx]$.

According to the definition,

$$\int_{[0,t] \times R} x\mu(ds, dx) \triangleq \int_0^t \int_R x\mu(ds, dx) = \sum_{0 \le s \le t, \Delta X_s \neq 0} \Delta X_s,$$

i.e., the aggregated jump size over $[0, t]$, while

$$\int_{[0,t] \times R} x\nu(dx)dt = E\left[\sum_{0 \le s \le t, \Delta X_s \neq 0} \Delta X_s\right],$$

i.e., the expected aggregated jump size over $[0, t]$. For the compensated jump measure,

$$\tilde{\mu}(dt, dx) = \mu(dt, dx) - \nu(dx)dt,$$

there is always

$$E\left[\int_{[0,t] \times R} f(x)\tilde{\mu}(dt, dx)\right] = 0,$$

for $f(x)$ in a rather general class of functions.

12.1.4 Characteristic Exponents for Lévy Processes

The next theorem characterizes a general Lévy process.

Theorem 12.1.1 (Lévy-Khintchine formula for Lévy processes). *Suppose $a \in \mathbb{R}$, $\sigma \ge 0$ and $\nu(x)$ is a measure on $\mathbb{R} \backslash \{0\}$ s.t.* $\displaystyle\int_\mathbb{R} (1 \wedge x^2)\nu(dx) < \infty$. *Define*

$$\Psi(\theta) = ia\theta + \frac{1}{2}\sigma^2\theta^2 + \int_\mathbb{R} \left(1 - e^{i\theta x} + i\theta x \mathbf{1}_{\{|x| < 1\}}\right) \nu(dx),$$

then $\Psi(\theta)$ is the characteristic exponent of a Lévy process. □

Here, we need to make some necessary remarks. The condition

$$\int_{\mathbb{R}} (1 \wedge x^2)\nu(dx) < \infty$$

implies

1. $\nu(|x| > 1) < \infty$. It means that the number of big jumps (with size $|x| > 1$) is finite.

2. $\int_{-1}^{1} x^2 \nu(dx) < \infty$. It means that infinite quadratic variation for small (as well as big) jumps is not allowed. Items 1 and 2 ensure that the integral in the characteristic exponent is finite.

3. There are two possibilities for $\nu(|x| < 1)$.

   (a) $\nu(|x| < 1) < \infty$. It means that the number of small jumps is finite as well. In such a case, the Lévy processes reduce to jump-diffusion processes.

   (b) $\nu(|x| < 1) = \infty$. It means that the number of small jumps is infinite. Item 2 however ensures that, $\forall \epsilon \in (0,1)$,

   $$\begin{cases} \nu(\epsilon < |x| < 1) < \infty, \\ \nu(|x| \leq \epsilon) = \infty, \end{cases}$$

   suggesting that the smaller the jump size the greater the intensity, and the discontinuity in the path is predominantly made up of arbitrarily small jumps.

A Lévy process can be represented by three parameters, a, σ, and ν. We call (a, σ, ν) the Lévy triplet, where

a — the drift coefficient,
σ — the diffusion coefficient,
ν — the Lévy measure for jump distribution in unit time.

The Lévy triplets for the three preliminary processes of Section 12.1.2 are

$$\begin{array}{rcl} \text{Poisson} & - & (0, 0, \lambda\delta(x-1)), \\ \text{Compound Poisson} & - & (0, 0, \lambda F(x)), \\ \text{Linear Brownian motion} & - & (a, \sigma, 0). \end{array}$$

We now proceed to prove Theorem 12.1.1: given $\Psi(\theta)$, there exists a Lévy process with the same characteristic exponent. What follows below is the strategy of the proof.

The first step is to rewrite the characteristic exponent as

$$\Psi(\theta) = \left\{ ia\theta + \frac{1}{2}\sigma^2\theta^2 \right\}$$

$$+ \left\{ \nu(\mathbb{R}\backslash(-1,1)) \int_{|x|\geq 1} (1 - e^{i\theta x}) \frac{\nu(dx)}{\nu(\mathbb{R}\backslash(-1,1))} \right\}$$

$$+ \left\{ \int_{0<|x|<1} (1 - e^{i\theta x} + i\theta x)\nu(dx) \right\}$$

$$\triangleq \Psi^{(1)}(\theta) + \Psi^{(2)}(\theta) + \Psi^{(3)}(\theta).$$

In case of $\nu(\mathbb{R}\backslash(-1,1)) = 0$, $\Psi^{(2)}(\theta) = 0$. We have established the following relationship between a process and its characteristic component:

$$\Psi^{(1)}(\theta) \longleftrightarrow X_t^{(1)} = \sigma W_t - at,$$

$$\Psi^{(2)}(\theta) \longleftrightarrow X_t^{(2)} = \sum_{i=1}^{N_t} \xi_i, \quad |\xi_i| > 1, \quad t \geq 0,$$

while for $\Psi^{(3)}(\theta)$, we have

$$\Psi^{(3)}(\theta) = \int_{0<|x|<1} (1 - e^{i\theta x} + i\theta x)\nu(dx)$$

$$= \sum_{n\geq 0} \left\{ \lambda_n \int_{2^{-(n+1)}\leq |x| < 2^{-n}} (1 - e^{i\theta x} + i\theta x)F_n(dx) \right\}$$

$$\triangleq \sum_{n\geq 0} \Psi_n^{(3)}(\theta),$$

where

$$\lambda_n = \nu(2^{-(n+1)} \leq |x| < 2^{-n}),$$

$$F_n(dx) = \lambda_n^{-1}\nu(dx)\mathbf{1}_{\{2^{-(n+1)}\leq |x|<2^{-n}\}}.$$

In case $\lambda_n = 0$, we let $\Psi_n^{(3)}(\theta) = 0$. There is the relationship:

$$\Psi_n^{(3)}(\theta) \longleftrightarrow M_t^{(n)} = \sum_{i=1}^{N_t^{(n)}} \xi_i - \lambda_n t \int_{\mathbb{R}} x F_n(dx),$$

where $N_t^{(n)}$ has the intensity λ_n and ξ_i has the law $F_n(dx)$. Define, in addition,

$$\mu^{(k)}(dt, dx) = \mu(dt, dx) \, \mathbf{1}_{\{2^{-(k+1)}\leq |x|<1\}},$$

$$\nu^{(k)}(dx) = \nu(dx) \, \mathbf{1}_{\{2^{-(k+1)}\leq |x|<1\}},$$

then

$$X_t^{(3,k)} = \sum_{n=0}^{k} M_i^{(n)} = \int_0^t \int_{0<|x|<1} x \left(\mu^{(k)}(ds,dx) - 1_{|x|\leq 1}\nu^{(k)}(dx)ds \right).$$

What are left to show are whether $X_t^{(3,k)}$

1. has a limit which has the characteristic exponent $\Psi^{(3)}(\theta)$ as $k \to \infty$? and

2. the limiting function is right-continuous with a left limit?

The answers to both questions are positive and are attributed to Lévy and Khintchine, yet the proof given here relies on the Lévy-Ito decomposition (Ito, 1942; Lévy, 1954).

Theorem 12.1.2 (Lévy-Ito decomposition). *Given any $a \in \mathbb{R}$, $\sigma \geq 0$ and a measure ν on $\mathbb{R}\backslash\{0\}$ satisfying*

$$\int_{\mathbb{R}} (1 \wedge x^2)\nu(dx) < \infty,$$

then (a,σ,ν) is the Lévy triplet of a Lévy process given by

$$X_t = X_t^{(1)} + X_t^{(2)} + X_t^{(3)},$$

where $X_t^{(i)}$, $i = 1,2,3$, are independent processes such that

$$X_t^{(1)} = \sigma W_t - at,$$

$$X_t^{(2)} = \sum_{i=1}^{N_t} \xi_i, \quad |\xi_i| > 1,$$

$$X_t^{(3)} = \int_0^t \int_{0<|x|<1} x \left(\mu(ds,dx) - 1_{|x|\leq 1}\nu(dx)ds \right),$$

and $X_t^{(3)}$ is a square-integrable martingale with characteristic exponent $\Psi^{(3)}(\theta)$. □

The proof is taken from Kyprianou (2008), and is put in this chapter's appendix for completeness. For notational simplification in our interest rate modeling in the subsequent sections, we do not separate big jumps from small jumps, and we reduce the three components of a Lévy process into two, such that

$$X_t = X_t^{(1)} + X_t^{(2)},$$

with

$$X_t^{(1)} = \sigma W_t - \tilde{a}t,$$

$$X_t^{(2)} = \int_0^t \int_R x \left(\mu(ds, dx) - \nu(dx)ds \right),$$

for

$$\tilde{a} = a - \int_{|x|>1} x\nu(dx).$$

We can use the new set of triplets $(\tilde{a}, \sigma, \nu)$ to represent a Lévy process.

12.1.5 The General Marked Point Processes

We now relax the restriction of time stationarity for Lévy processes, such that

$$E\left[\mu(dt, dx)\right] = \nu(t, dx)dt,$$

i.e., we allow the Lévy measures to be time dependent yet satisfy other conditions for Lévy measures, we then arrive at the general marked point processes, which were first adopted for interest rate modeling by Björk *et al.* (1997). Note that the time stationarity can sometimes become undesirable for modeling financial time series, and the relaxation only generates more flexibility. Whenever it is more convenient, we also use $\nu(dt, dx)$ for $\nu(t, dx)dt$.

12.2 Lévy HJM Model

We are now ready to develop the HJM model driven by general marked point processes. We begin with the $\mathbb{P}$ dynamics of zero-coupon bonds:

$$\frac{dP(t, T)}{P(t, T)} = m(t, T)dt + \Sigma(t, T) \cdot d\tilde{\mathbf{W}}_t$$

$$+ \int_R (H(t, x, T) - 1)(\mu_P - \nu_P)(dt, dx),$$

where $m(t, T)$ is the growth rate, Σ is the percentage volatility, $H(t, x, T) - 1$ is the percentage jump size, $\mu_P(t, x)$ is the jump measure and $\nu_P(t, x)$ is the compensator to $\mu_P(t, x)$. In the absence of arbitrage, there must exist another measure equivalent to $\mathbb{P}$ under which the discount price of a zero-coupon bond for any maturity T is a martingale. We need the following theorem for finding the martingale measure (Girsanov, 1960).

Theorem 12.2.1. *(Girsanov) If* $\mathbb{Q} \sim \mathbb{P}$, *then there must exist* $\mathbb{P}$-*measurable function* $\varphi_t(x)$ *and* $\psi(x,t)$ *such that*

$$\int_0^t \|\varphi_s\|^2 ds < \infty, \qquad \int_0^t \int_R x|\psi(x,s) - 1|\nu_P(s,dx)ds < \infty,$$

so that the Radon-Nikodyn derivative of $\mathbb{Q}$ *with respect to* $\mathbb{P}$ *can be expressed as*

$$
\begin{aligned}
\left. \frac{d\mathbb{Q}}{d\mathbb{P}} \right|_{\mathcal{F}_t} = \exp \Bigg(& \int_0^t -\frac{1}{2}\|\varphi_s\|^2 ds + \varphi_s \cdot d\tilde{\mathbf{W}}_s \\
& + \int_0^t \int_R \left[(1 - \psi(x,s)) + \ln \psi(x,s) \right] \nu_P(s,dx)ds \qquad (12.1) \\
& + \int_0^t \int_R \ln \psi(x,s)(\mu_P - \nu_P)(ds,dx) \Bigg) = \zeta_t,
\end{aligned}
$$

or equivalently

$$d\zeta_t = \zeta_t \left(\varphi_t \cdot d\tilde{\mathbf{W}}_t + \int_0^t \int_R (\psi(x,s) - 1)(\mu_P - \nu_P)(ds,dx) \right).$$

Under $\mathbb{Q}$,

$$\mathbf{W}_t = \tilde{\mathbf{W}}_t - \int_0^t \varphi_s ds$$

is a Brownian motion and $(\tau_n, \Delta X(\tau_n))$ *have the Lévy measure*

$$\nu_Q(t,dx) = \psi(x,t)\nu_P(t,dx).$$

Proof. The proof consists of two steps:

1. We first show that the results hold for the pair of

$$\mu_P^{(\epsilon)} = \mu_P \, 1_{\{|x| \geq \epsilon\}}, \qquad \nu_P^{(\epsilon)} = \nu_P \, 1_{\{|x| \geq \epsilon\}}.$$

2. We then let $\epsilon \to 0$ and show that ζ_t converges.

Let

$$U_1(t) = \int_0^t -\frac{1}{2}\|\varphi_s\|^2 ds + \varphi_s \cdot d\tilde{\mathbf{W}}_s,$$

$$U_2(t) = \int_0^t \int_R \left[(1 - \psi(x,s)) \right] \nu_P(ds,dx) + \int_0^t \int_R \ln \psi(x,s)\mu_P(ds,dx).$$

The exponential of $U_1(t)$ is clearly a $\mathbb{P}$ martingale. For the exponential of $U_2(t)$ we have (by the Ito's Lemma for jump-diffusion processes),

$$de^{U_2(t)} = e^{U_2(t-)} \left(dU_2^c(t) + \int_R [\psi(x,t) - 1] \, \mu_P(dt,dx) \right)$$

$$= e^{U_2(t-)} \int_R [\psi(x,t) - 1] \, (\mu_P - \nu_P)(dt,dx),$$

where $U_2^c(t)$ represents the continuous component of $U_2(t)$. The equation above shows that the $e^{U_2(t)}$ is also a $\mathbb{P}$ martingale. To prove the result of the first step, we need to calculate the characteristic function of $\mathbf{W}_t$ under $\mathbb{Q}$. We have

$$E^Q\left[e^{i\mathbf{u}\cdot\mathbf{W}_t}\right] = E^P\left[\zeta_t e^{i\mathbf{u}\cdot\mathbf{W}_t}\right]$$

$$= E^P\left[e^{U_1(t)} e^{i\mathbf{u}\cdot\mathbf{W}_t}\right] E^P\left[e^{U_2(t)}\right]$$

$$= E^P\left[\exp\left(\int_0^t -\frac{1}{2}\|\boldsymbol{\varphi}_s + i\mathbf{u}\|^2 ds + (\boldsymbol{\varphi}_s + i\mathbf{u})\cdot d\tilde{\mathbf{W}}_s + \frac{1}{2}\|\mathbf{u}\|^2 t\right)\right]$$

$$= \exp\left(\frac{1}{2}\|\mathbf{u}\|^2 t\right),$$

which proves that $\mathbf{W}_t$ is a $\mathbb{Q}$-Brownian motion. Meanwhile,

$$E^Q\left[e^{iu\sum_{s\leq t}\Delta X_s}\right]$$

$$= E^P\left[e^{iu\sum_{s\leq t}\Delta X_s} e^{\int_0^t \int_R (\nu_P^{(\epsilon)} - \nu_Q^{(\epsilon)})(ds,dx) + \ln\psi(x,s)\mu_P^{(\epsilon)}(ds,dx)}\right]$$

$$= e^{\int_0^t \int_R (\nu_P^{(\epsilon)} - \nu_Q^{(\epsilon)})(ds,dx)} E^{\mathbb{P}}\left[e^{\int_0^t \int_R \ln(\psi(x,s)e^{iux})\mu_P^{(\epsilon)}(ds,dx)}\right]$$

$$= e^{\int_0^t \int_R (\nu_P^{(\epsilon)} - \nu_Q^{(\epsilon)})(ds,dx)} e^{\int_0^t \int_R (\psi(x,s)e^{iux} - 1)\nu_P^{(\epsilon)}(ds,dx)}$$

$$= e^{\int_0^t \int_R (e^{iux} - 1)\nu_Q^{(\epsilon)}(ds,dx)},$$

which shows that the jump component has Lévy measure $\nu_Q^{(\epsilon)}(dt,dx)$. Finally, given the regularity condition of ψ, we can easily show that $U_2(t)$ is well defined and both the mean and variance of $U_2 - U_2^{(\epsilon)}$ converge to zero as $\epsilon \to 0$, which completes the proof. $\qquad\square$

The absence of arbitrage implies (Harrison and Pliska, 1981) the existence of $\mathbb{Q} \sim \mathbb{P}$ such that under $\mathbb{Q}$, the discount prices of zero-coupon bonds are martingales. To look for φ and ψ, we rewrite the dynamics of zero-coupon bonds to

$$\frac{dP(t,T)}{P(t,T)} = (m(t,T) + \Sigma(t,T)\cdot\boldsymbol{\varphi}_t)\,dt$$

$$+ \int_R (H(t,x,T) - 1)(\psi - 1)\nu_P(dt,dx) \qquad (12.2)$$

$$+ \Sigma(t,T)\cdot(d\tilde{\mathbf{W}}_t - \boldsymbol{\varphi}_t dt) + \int_R (H(t,x,T) - 1)(\mu - \nu_Q)(dt,dx).$$

Note that the last line of Equation 12.2 has the $\mathbb{Q}$ expectation equal to zero. For $P(t,T)/B_t$ to be a $\mathbb{Q}$ martingale, there must be

$$m(t,T) + \Sigma(t,T)\cdot\boldsymbol{\varphi}_t + \int_R (H(t,x,T) - 1)(\psi - 1)\nu_P(t,dx) = r_t,$$

where r_t is the risk-free short rate. Differentiating the above equation w.r.t. T, we then obtain an equivalent condition imposed on φ and ψ:

$$m_T(t,T) + \Sigma_T(t,T) \cdot \varphi_t + \int_R H_T(t,x,T)(\psi - 1)\nu_P(t,dx) = 0. \qquad (12.3)$$

There are usually non-unique solutions for the pair unless additional criteria are imposed. An industrial approach to determine the pair is to calibrate the model to market prices of benchmark instruments, which is a challenging task and is beyond the scope of this book .

We now derive the $\mathbb{P}$ dynamics of the instantaneous forward rates. Based on Equation 12.2, we obtain the dynamics of the log price of the zero-coupon bonds:

$$d \ln P(t,T) = \left(m(t,T) - \frac{1}{2}\|\Sigma(t,T)\|^2 \right) dt$$

$$+ \int_R \left[(1 - H(t,x,T) + \ln H(t,x,T) \right] \nu_P(t,dx)dt \qquad (12.4)$$

$$+ \Sigma(t,T) \cdot d\tilde{\mathbf{W}}_t + \int_R \ln H(t,x,T)(\mu - \nu_P)(dt,dx).$$

Differentiating Equation 12.4 with respect to T and multiply the minus sign to the resulted equation, we then obtain

$$df(t,T) = \left(-m_T(t,T) + \Sigma_T(t,T) \cdot \Sigma(t,T) \right.$$

$$\left. + \int_R \frac{H_T(t,x,T)\left[H(t,x,T) - 1\right]}{H(t,x,T)}\nu_P(t,dx) \right) dt \qquad (12.5)$$

$$-\Sigma_T(t,T) \cdot d\tilde{\mathbf{W}}_t - \int_R \frac{H_T(t,x,T)}{H(t,x,T)}(\mu_P - \nu_P)(dt,dx).$$

Define the diffusion volatility and jump size for the forward rate by

$$\sigma(t,T) = -\Sigma_T(t,T),$$

$$h(t,x,T) = -\frac{H_T(t,x,T)}{H(t,x,T)} = -\frac{\partial \ln H(t,x,T)}{\partial T}. \qquad (12.6)$$

Once $\sigma(t,T)$ and $h(t,x,T)$ are specified, we can conversely define $\Sigma(t,T)$ and $H(t,x,T)$. In fact, there are

$$\Sigma(t,T) = -\int_t^T \sigma(t,s)ds,$$

and

$$\int_t^T h(t,x,s)ds = -\ln \frac{H(t,x,T)}{H(t,x,t)}.$$

At maturity, the bond price does not jump, meaning $H(t,x,t) = 1$. It then follows that

$$H(t,x,T) = e^{-\int_t^T h(t,x,s)ds}$$

and

$$H_T(t,x,T) = -h(t,x,T)e^{-\int_t^T h(t,x,s)ds}.$$

We now rewrite Equation 12.5 into

$$df(t,T) = \alpha(t,T)dt + \sigma \cdot d\tilde{\mathbf{W}}_t + \int_R h(t,x,T)(\mu - \nu_P)(dt,dx), \qquad (12.7)$$

with

$$\alpha(t,T) = -m_T(t,T) - \sigma(t,T)\cdot\Sigma(t,T)$$
$$- \int_R h(t,x,T)\left[H(t,x,T)-1\right]\nu_P(t,dx). \qquad (12.8)$$

By making use of Equation 12.3, we further have

$$\alpha(t,T) = -\sigma(t,T)\cdot\boldsymbol{\varphi}_t - \int_R h(t,x,T)e^{-\int_t^T h(t,x,s)ds}(\psi-1)\nu_P(t,dx)$$

$$- \sigma(t,T)\cdot\Sigma(t,T) - \int_R h(t,x,T)\left[e^{-\int_t^T h(t,x,s)ds}-1\right]\nu_P(t,dx)$$

$$= -\sigma\cdot(\Sigma+\boldsymbol{\varphi}_t) - \int_R h(t,x,T)(e^{-\int_t^T h(t,x,s)ds}\psi(x,t)-1)\nu_P(t,dx),$$

$$(12.9)$$

which is the result first obtained by Eberline and Oxkan (2004) as the no-arbitrage condition. Given Equation 12.9, we obtain the risk-neutral dynamics of the forward rates:

$$df(t,T) = \sigma(t,T)\cdot\left[d\tilde{\mathbf{W}}_t - (\boldsymbol{\varphi}_t + \Sigma(t,T))dt\right]$$

$$+ \int_R h(t,x,T)(\mu - \psi(x,t)e^{-\int_t^T h(t,x,s)ds}\nu_P)(dt,dx)$$

$$= \sigma(t,T)\cdot[d\mathbf{W}_t - \Sigma(t,T)dt]$$

$$+ \int_R h(t,x,T)(\mu - e^{-\int_t^T h(t,x,s)ds}\nu_Q)(dt,dx)$$

$$(12.10)$$

Next, we will show that $f(t,T)$ is a martingale under T forward measure, $\mathbb{Q}_T$, which is yet to be defined. For this purpose, we need to solve for the price of zero-coupon bonds under the $\mathbb{Q}$ measure. We already know that under the $\mathbb{Q}$ measure, the price dynamics is

$$\frac{dP(t,T)}{P(t,T)} = r_t dt + \Sigma(t,T)\cdot d\mathbf{W}_t + \int_R (\dot{H}(t,x,T)-1)(\mu - \nu_Q)(dt,dx).$$

$$(12.11)$$

The dynamics for the log price is then

$$d\ln P(t,T) = \left(r_t + \int_R [1 - H(t,x,T) + \ln H(t,x,T)]\,\nu_Q(t,dx)\right.$$

$$\left. -\frac{1}{2}\|\Sigma(t,T)\|^2\right)dt + \Sigma(t,T)\cdot d\mathbf{W}_t + \int_R \ln H(t,x,T)(\mu - \nu_Q)(dt,dx),$$

$$(12.12)$$

from which we can solve for the solution of zero-coupon bond prices:

$$P(t, T) = P(0, T)$$

$$\exp\left\{\int_0^t \left(r_s + \int_R [1 - H(s, x, T) + \ln H(s, x, T)] \, \nu_Q(s, dx)\right.\right.$$

$$\left.-\frac{1}{2}\|\Sigma(s, T)\|^2\right) ds + \Sigma(s, T) \cdot d\mathbf{W}_s \tag{12.13}$$

$$\left.+ \int_R \ln H(s, x, T)(\mu - \nu_Q)(ds, dx)\right\}.$$

With the price formula for zero-coupon bonds, we can define the Radon-Nikodyn derivative for the forward measure to be

$$\frac{dQ_T}{dQ}\bigg|_{\mathcal{F}_t} = \frac{P(t, T)}{P(0, T)} \bigg/ \frac{B(t)}{B(0)}$$

$$= \exp\left\{\int_0^t \int_R [1 - H(s, x, T) + \ln H(s, x, T)] \, \nu_Q(ds, dx)\right.$$

$$-\frac{1}{2}\|\Sigma(s, T)\|^2 ds + \Sigma(s, T) \cdot d\mathbf{W}_s \tag{12.14}$$

$$\left.+ \int_R \ln H(s, x, T)(\mu - \nu_Q)(ds, dx)\right\} \stackrel{\triangle}{=} \zeta_t.$$

According to the Girsanov Theorem, we now understand that

$$\mathbf{W}_t^{(T)} = \mathbf{W}_t - \int_0^t \Sigma(s, T) ds$$

is a Q_T Brownian motion and

$$\nu_{Q_T} = H(t, x, T)\nu_Q$$

is the Lévy measure of jumps under Q_T. The Q_T dynamics of the forward rate is thus

$$df(t, T) = \sigma(t, T) \cdot d\mathbf{W}_t^{(T)} + \int_R h(t, x, T)(\mu - \nu_{Q_T})(dt, dx), \tag{12.15}$$

which is a martingale under the T-forward measure.

12.3 Market Model under Lévy Processes

For notational simplicity, we rewrite the SOFR forward rate for the term $(T - \Delta T, T)$ as

$$f_{\Delta T}(t, T) = \frac{1}{\Delta T}\left(\frac{P(t, T - \Delta T)}{P(t, T)} - 1\right).$$

By the Lévy-Ito's Lemma,

$$d\left(\frac{P(t, T - \Delta T)}{P(t, T)}\right) = \left(\frac{P(t-, T - \Delta T)}{P(t-, T)}\right)$$
$$\times \left([[\Sigma(t, T - \Delta T) - \Sigma(t, T)] \cdot [d\mathbf{W}_t - \Sigma(t, T)dt]\right.$$
$$+ \int_R \left[1 - \frac{H(t, x, T - \Delta T)}{H(t, x, T)}\right] (H(t, x, T) - 1)\nu_Q(dt, dx)\right]$$
$$- \int_R \left[1 - \frac{H(t, x, T - \Delta T)}{H(t, x, T)}\right] (\mu - \nu_Q)(dt, dx)\right) \tag{12.16}$$
$$= \left(\frac{P(t-, T - \Delta T)}{P(t-, T)}\right)$$
$$\times \left([[\Sigma(t, T - \Delta T) - \Sigma(t, T)] \cdot [d\mathbf{W}_t - \Sigma(t, T)dt]\right.$$
$$- \int_R \left[1 - \frac{H(t, x, T - \Delta T)}{H(t, x, T)}\right] (\mu - H(t, x, T)\nu_Q)(dt, dx)\right).$$

The dynamics of the SOFR forward rates then follows:

$$\frac{df_{\Delta T}(t, T)}{f_{\Delta T}(t-, T)} = \frac{1 + \Delta T f_{\Delta T}(t-, T)}{\Delta T f_{\Delta T}(t-, T)}$$
$$\times \left([\Sigma(t, T - \Delta T) - \Sigma(t, T)] \cdot [d\mathbf{W}_t - \Sigma(t, T)dt]\right. \tag{12.17}$$
$$+ \int_R \left[\frac{H(t, x, T - \Delta T)}{H(t, x, T)} - 1\right] (\mu - H(t, x, T)\nu_Q)(dt, dx)\right).$$

Now define

$$\gamma(t, T) = \frac{1 + \Delta T f_{\Delta T}(t-, T)}{\Delta T f_{\Delta T}(t-, T)} (\Sigma(t, T - \Delta T) - \Sigma(t, T)),$$
$$h_{\Delta T}(t, x, T) = \frac{1 + \Delta T f_{\Delta T}(t-, T)}{\Delta T f_{\Delta T}(t-, T)} \left[\frac{H(t, x, T - \Delta T)}{H(t, x, T)} - 1\right], \tag{12.18}$$

then the dynamics of the SOFR forward rates is simplified to

$$\frac{df_{\Delta T}(t, T)}{f_{\Delta T}(t-, T)} = \gamma(t, T) \cdot (d\mathbf{W}_t - \Sigma(t, T)dt)$$
$$+ \int_R h_{\Delta T}(s, x, T)(\mu - H(t, x, T)\nu_Q)(dt, dx). \tag{12.19}$$

In terms of $\gamma(t, T)$, $h_{\Delta T}$ and the forward rates, we can specify both $\Sigma(t, T)$ and $H(t, x, T)$ as follows:

$$\Sigma(t, T) = \Sigma(t, T - \Delta T) - \frac{\Delta T f_{\Delta T}(t-, T)}{1 + \Delta T f_{\Delta T}(t-, T)}\gamma(t, T)$$
$$= \ldots \tag{12.20}$$
$$= - \sum_{k=0}^{\lceil \frac{(T-t)}{\Delta T} \rceil} \frac{\Delta T f_{\Delta T}(t-, T - k\Delta T)}{1 + \Delta T f_{\Delta T}(t-, T - k\Delta T)}\gamma(t, T - k\Delta T),$$

and

$$H(t, x, T) = \frac{1 + \Delta T f_{\Delta T}(t-, T)}{1 + \Delta T f_{\Delta T}(t-, T)(1 + h_{\Delta T}(t, x, T))} H(t, x, T - \Delta T)$$

$$= \dots \qquad (12.21)$$

$$= \prod_{k=0}^{\lceil \frac{(T-t)}{\Delta T} \rceil} \frac{1 + \Delta T f_{\Delta T}(t-, T - k\Delta T)}{1 + \Delta T f_{\Delta T}(t-, T - k\Delta T)(1 + h_{\Delta T}(t, x, T - k\Delta T))},$$

where $\lceil x \rceil$ means the integer ceiling of x, $\gamma(t, T) = 0, h_{\Delta T}(t, x, T) = 0$ and $H(t, x, T) = 1$ for $t \geq T$.

We can now define the Lévy market model with the spanning forward rates:

$$f_j(t) = \frac{1}{\Delta T_j} \left(\frac{P(t, T_{j-1})}{P(t, T_j)} - 1 \right), \quad j = \eta_{t+1}, \dots,$$

where $\eta_t = \max\{j | T_j \leq t\}$. Then the dynamics of the forward rates under the risk-neutral measure are

$$\frac{df_j(t)}{f_j(t-)} = \gamma_j(t) \cdot [d\mathbf{W}_t - \Sigma(t, T_j)dt]$$

$$+ \int_R h_j(t, x)(\mu - H(t, x, T_j)\nu_Q)(dt, dx), \qquad (12.22)$$

where $h_j(t, x)$ is the percentage jump size of $f_j(t)$,

$$\Sigma(t, T_j) = - \sum_{k=\eta_t+1}^{j} \frac{\Delta T_k f_k(t-)}{1 + \Delta T_k f_k(t-)} \gamma_k(t),$$

$$H(t, x, T_j) = \prod_{k=\eta_t+1}^{j} \frac{1 + \Delta T_k f_k(t-)}{1 + \Delta T_k f_k(t-)(1 + h_j(t, x))}. \qquad (12.23)$$

Similar to the Lévy LIBOR model by Eberlein and Özkan (2005), the above model for SOFR term rates is actually based on the general marked point processes. In a retrospective of model evolution, we call it the Lévy market model.

12.3.1 Swaption Pricing

As is already known, swap rate $R_{m,n}(t)$ is a deterministic function of forward rates:

$$R_{m,n}(t) = R_{m,n}(\{f_j(t)\}),$$

for $j = m\theta + 1, \dots, n\theta$ for $\theta = \tilde{\Delta}T/\Delta T$, and the payoff function of swaptions can be expressed as

$$V_T = A_{m,n}(T)(R_{m,n}(T) - K)^+. \qquad (12.24)$$

Under the forward swap measure, $\mathbb{Q}_{m,n}$, we have the following expression for swaption value:

$$V_t = A_{m,n}(t)E_t^{Q^{m,n}}\left[(R_{m,n}(T) - K)^+\right].$$

For $t \leq T_m$, the approximate dynamics of the swap rate can be approximated as follows:

$$dR_{m,n}(t) = \left(\sum_{j=m\theta+1}^{n\theta} \frac{\partial R_{m,n}(t-)}{\partial f_j} f_j(t-)\gamma_j(t)\right) \cdot [d\mathbf{W}_t - \Sigma_{m,n}(t)dt]$$

$$+ \int_R [R_{m,n}(\{f_j(t-)(1 + h_j(t))\}) - R_{m,n}(\{f_j(t-)\})](\mu - \nu_{m,n})(dt, dx),$$

$$(12.25)$$

where $\nu_{m,n}(t, dx)$ is the Lévy measure under the swap measure $\mathbb{Q}_{m,n}$, for which we have

Proposition 12.3.1 (Glasserman and Merener, 2003). *The Lévy measure of the swap rate is*

$$\nu_{m,n}(t, dx) = \sum_{j=m}^{n-1} \alpha_j(t-) \prod_{k=m\theta+1}^{j\theta} \frac{1 + \Delta T_j f_j(t-)}{1 + \Delta T_j f_j(t-)(1 + h_j(t, x))} \nu_Q(t, dx).$$

Proof. The Radon-Nikodyn derivative of the forward swap measure is defined by

$$\left.\frac{d\mathbb{Q}_S}{d\mathbb{Q}}\right|_{\mathcal{F}_t} = \frac{A_{m,n}(t)}{A_{m,n}(0)} \Big/ \frac{B(t)}{B(0)} = \zeta(t).$$

In terms of the forward rates, we have

$$\zeta(t) = \frac{1}{A_{m,n}(0)} \sum_{k=m+1}^{n} \Delta T_k \prod_{j=m\theta+1}^{k\theta} \frac{1}{1 + \Delta T_j f_j(t)}. \qquad (12.26)$$

The jump in percentage term is

$$\frac{\zeta(\tau) - \zeta(\tau-)}{\zeta(\tau-)} = \frac{\sum_{j=m+1}^{n} \Delta T_j \prod_{k=m\theta+1}^{j\theta} \frac{1}{(1+\Delta T_k f_k(\tau-)(1+h_k(\tau,x))}}{\sum_{j=m+1}^{n} \Delta T_j \prod_{k=m\theta+1}^{j\theta} \frac{1}{1+\Delta T_k f_k(\tau-)}} - 1$$

$$= \frac{\sum_{j=m+1}^{n} \Delta T_j \prod_{k=m\theta}^{j\theta} \frac{1}{(1+\Delta T_k f_k(\tau-)} \prod_{k=\eta_t\theta+1}^{j\theta} \frac{1+\Delta T_k f_k(\tau-)}{1+\Delta T_k f_k(\tau-)(1+h_j(\tau,x)}}{\sum_{j=m+1}^{n} \Delta T_j \prod_{k=m\theta+1}^{j\theta} \frac{1}{1+\Delta T_k f_k(\tau-)}} - 1$$

$$= \sum_{j=m+1}^{n} \alpha_j(\tau-) \prod_{k=m\theta+1}^{j\theta} \frac{1 + \Delta T_k f_k(\tau-)}{1 + \Delta T_k f_k(\tau-)(1 + h_k(\tau, x))} - 1,$$

where we have made use of the definition of α_j. Using the notation of random measure, we have

$$\frac{d\zeta(t)}{\zeta(t-)} = \int_R \left(\sum_{j=m+1}^{n} \alpha_j(\tau-) \prod_{k=m\theta+1}^{j\theta} \frac{1 + \Delta T_k f_k(t-)}{1 + \Delta T_k f_k(t-)(1 + h_k(t,x))} - 1 \right)$$
$$\times (\mu_Q(dt, dx) - \nu_Q(t, dx)dt) + \dots d\mathbf{W}_t.$$

According to the Girsanov's theorem, we identify the Lévy measure under the forward swap measure to be

$$\nu_{m,n}(t, dx) = \left(\sum_{j=m+1}^{n} \alpha_j(t-) \prod_{k=m\theta+1}^{j\theta} \frac{1 + \Delta T_k f_k(t-)}{1 + \Delta T_k f_k(t-)(1 + h_k(t,x))} \right) \nu_Q(t, dx).$$

$$\square$$

To retain analytical tractability, we yet again adopt the following "freezing coefficient" treatment: we freeze the state variable $\{f_j(t-)\}$ at time zero, so as to have

$$\frac{dR_{m,n}(t)}{R_{m,n}(t-)} \approx \left(\sum_{j=m\theta+1}^{n\theta} \frac{\partial R_{m,n}(0)}{\partial f_j} \frac{f_j(0)}{R_{m,n}(0)} \gamma_j(t) \right) \cdot [d\mathbf{W}_t - \Sigma_{m,n}(t)dt]$$
$$+ \int_R \left[\frac{R_{m,n}(\{f_j(0)(1 + h_j(t))\})}{R_{m,n}(\{f_j(0)\})} - 1 \right] (\mu - \nu_{m,n})(dt, dx) \tag{12.27}$$

and

$$\nu_{m,n}(t, dx) \approx \sum_{j=m+1}^{n} \alpha_j(0) \prod_{k=m\theta+1}^{j\theta} \frac{1 + \Delta T_j f_j(0)}{1 + \Delta T_j f_j(0)(1 + h_j(t,x))} \nu_Q(t, dx).$$

Once the swap rate process is approximated by a Lévy process, swaption pricing can be done through the method of Laplace transform, as is described in Chapter 8. Here, we will instead take a different yet more stylized approach introduced by Raible (2000) for general options to price the swaptions. The value of the swaption is given by

$$V_0 = A_{m,n}(0) E_0^{Q_{m,n}} [g(R_{m,n}(T_m))]$$
$$= A_{m,n}(0) \int_{R^+} g(R_{m,n}(T_m)) dQ_{m,n}$$
$$= A_{m,n}(0) \int_{R^+} g(R_{m,n}(0)e^x) dQ_{m,n}(x)$$
$$= A_{m,n}(0) \int_{R^+} g(R_{m,n}(0)e^x) \rho(x) dx,$$

where $\rho(x)$ is the density function of $\ln R_{m,n}(T_m)$. Define $\pi(x) = g(e^{-x})$ and let $\zeta = -\ln R_{m,n}(0)$, and then we have

$$\frac{V_0}{A_{m,n}(0)} = \int_R \pi(\zeta - x)\rho(x)dx = (\pi * \rho)(\zeta), \qquad (12.28)$$

which is a convolution between π and ρ, multiplied by the discount factor. Let $\mathcal{L}_V(u)$ denote the Laplace transform of a function V at $u \in \mathbb{C}$, such that

$$\mathcal{L}_V(u) = \int_R e^{-ux}V(x)dx.$$

Applying Laplace transform to both sides of Equation 12.28 for $u \in \mathbb{C}$, we get

$$\begin{aligned}
\mathcal{L}_V(u) &= \int_R e^{-ux}(\pi * \rho)(x)dx \\
&= \int_R e^{-ux}\pi(x)dx \int_R e^{-ux}\rho(x)dx \\
&= \mathcal{L}_\pi(u)\mathcal{L}_\rho(u).
\end{aligned}$$

Finally, we perform the inverse Laplace transform to obtain the price of the swaption relative to the annuity:

$$\begin{aligned}
\frac{V_0}{A_{m,n}(0)} &= \frac{1}{2\pi} \int_R e^{\zeta(R+iu)}\mathcal{L}_V(R+iu)du \\
&= \frac{e^{\zeta R}}{2\pi} \int_R e^{i\zeta u}\mathcal{L}_\pi(R+iu)\mathcal{L}_\rho(R+iu)du \\
&= \frac{e^{\zeta R}}{2\pi} \int_R e^{i\zeta u}\mathcal{L}_\pi(R+iu)\varphi_f(R+iu)du.
\end{aligned}$$

The numerical implementation of the Laplace transform is described in Chapter 8 in detail.

Numerical Laplace transform is only one of the major computational methods for option pricing under the Lévy process, the other two major methods are the analytical approximation method and the numerical method for the so-called partial integral differential equations (PIDE). In the following section, we describe the analytical approximation method which utilizes the Merton formula (1976) under the jump-diffusion model for underlying state variables.

12.3.2 Approximate Swaption Pricing via the Merton Model

The Merton model (1976) is a combination of diffusion and Poisson jumps with a lognormal jump distribution. For the swap rate $R_{m,n}(t)$, the Merton model can be cast under the swap measure $\mathbb{Q}_{m,n}$ as follows:

$$\frac{dR_{m,n}(t)}{R_{m,n}(t-)} = -\lambda E[Y-1]dt + \sigma dW_t + [Y-1]dN_t, \qquad (12.29)$$

where W_t is a Brownian motion, λ is the intensity of Poisson jumps, and the distribution of jump size is taken to be

$$Y = e^{\mu + \sigma_Y \epsilon}, \quad \epsilon \sim N(0, 1).$$

It follows that

$$E[Y] = e^{\mu + \frac{1}{2}\sigma_Y^2} = 1 + m.$$

Let $X_t = \ln(R_{m,n}(t)/R_{m,n}(0))$. Then X_t has the following risk-neutral dynamics:

$$dX_t = \left[-\lambda m - \frac{1}{2}\sigma^2 \right] dt + \sigma dW_t + \ln Y \, dN_t.$$

Conditional to jumps, we have the following expression for the log price:

$$X_t = \left(-\lambda m - \frac{1}{2}\sigma^2 \right) t + \sigma W_t + \sum_{i=1}^{N_t} \ln Y_i.$$

The pricing of European swaptions is conditional to $0, 1, \cdots, \infty$ jumps, respectively, so we have the so-called Merton formula for swaptions:

$$V_0 = A_{m,n}(0) \sum_{n=0}^{\infty} e^{-\lambda T_m} \frac{(\lambda T_m)^n}{n!} \cdot e^{-\left[\lambda m + \frac{n}{T_m} \ln(1+m)\right]T_m}$$

$$C_B(R_{m,n}(0), K, T_m, r_n, \sigma_n)$$

$$= A_{m,n}(0) \sum_{n=0}^{\infty} e^{-\lambda(1+m)T} \frac{(\lambda T(1+m))^n}{n!} C_B(R_{m,n}(0), K, T_m, r_n, \sigma_n),$$

where C_B is the Black formula for call options:

$$C_B(F, K, T, r, \sigma) = F\Phi(d_1) - K\Phi(d_2),$$

$$d_{1,2} = \frac{\ln(F/K) \pm \sigma^2 T/2}{\sigma\sqrt{T}},$$

$$r_n = -\lambda m + \frac{n}{T}\ln(1+m),$$

$$\sigma_n^2 = \sigma^2 + \frac{n}{T}\sigma_Y^2.$$

Swaption pricing under the market model can be done by taking advantage of the Merton formula (Glasserman and Merener, 2003). For this purpose, we need to determine the mean and variance of Y by matching the first two moments of the jump size of the swap rate under the Lévy market model. For the Merton model for the swap rate, Equation 12.29, the first two moments of the jump size are

$$E\left[R_{m,n}(\tau) - R_{m,n}(\tau-)|\tau, R_{m,n}(\tau-)\right] = R_{m,n}(\tau-)m,$$

$$E\left[(R_{m,n}(\tau) - R_{m,n}(\tau-))^2|\tau, R_{m,n}(\tau-)\right]$$

$$= R_{m,n}^2(\tau-)\left[e^{\sigma_Y^2}(1+m)^2 - 2m - 1\right],$$

where σ_Y^2 is the variance of jump size of $\ln R_{m,n}(\tau)$.

By making use of the alternative formula for swap rates, Equation 9.28, we have the following expression for the first and the second moments of the jump size

$$
\begin{aligned}
I_1 &= E^{Q_{m,n}} \left[\frac{R_{m,n}(\tau) - R_{m,n}(\tau-)}{R_{m,n}(\tau-)} \Big| \tau, R_{m,n}(\tau-) \right] \\
&= \int_{R^+} h_{m,n}(\tau, x) \frac{H_{m,n}(\tau, x)}{E^Q \left[H_{m,n}(\tau, x) \right] \nu_Q(\tau, dx)}, \\
I_2 &= E^{Q_{m,n}} \left[\left(\frac{R_{m,n}(\tau) - R_{m,n}(\tau-)}{R_{m,n}(\tau-)} \right)^2 \Big| \tau, R_{m,n}(\tau-) \right] \\
&= \int_{R^+} h_{m,n}^2(\tau, x) \frac{H_{m,n}(\tau, x)}{E^Q \left[H_{m,n}(\tau, x) \right]} \nu_Q(\tau, dx),
\end{aligned}
\tag{12.30}
$$

where

$$
h_{m,n}(\tau, x) = \sum_{j=m\theta+1}^{n\theta} \tilde{\alpha}_j(\tau-) h_j(\tau, x),
$$

$$
H_{m,n}(\tau, x) = \sum_{j=m\theta+1}^{n\theta} \tilde{\alpha}_j(\tau-) H(\tau, x, T_j),
$$

for

$$
\tilde{\alpha}_j(\tau) = \frac{\Delta T_j P(\tau, T_j)}{\sum_{k=j+1}^n \Delta \tilde{T}_k P(\tau, \tilde{T}_k)},
$$

and $H(t, x, T_j)$ is given by Equation 12.23, and both $H(t, x, T_j)$ and α_j are state-dependent. For an approximation of the moments, we resort to the technique of freezing state variables such that $f_k(\tau-) \approx f_k(0), k = m\theta+1, \ldots, n\theta$. We then set the two moments of the Merton model to those of the Lévy market model, thus obtaining

$$
m = I_1,
$$
$$
e^{\sigma_Y^2} (1 + m)^2 - 2m - 1 = I_2,
\tag{12.31}
$$

for $m = e^{\mu + \sigma_Y^2/2} - 1$. With a bit more algebra, we arrive at

$$
\mu = \ln(1 + I_1) - \frac{\sigma_Y^2}{2} \quad \text{and} \quad \sigma_Y = \left[\ln \left(\frac{1 + 2I_1 + I_2}{(1 + I_1)^2} \right) \right]^{1/2}.
$$

What left is to insert μ and σ_Y to the Merton formula to obtain an approximation of the swaption value.

We finish this chapter with some remarks. We in this chapter have presented the Lévy market model along an approach simpler than the one by Eberlein and Özkan (2005). Our starting point is the exponential Lévy dynamics for the zero-coupon bond processes. The change of measure theorem by Girsanov is utilized in the derivation. The Lévy market model can be regarded as a framework for SOFR market model with both jump and diffusion risks, which includes the famous model of Glasserman and Kou (2003) as a special case. As we have demonstrated in this chapter, under the Lévy market model caplets and swaptions can be priced approximately in closed form, see also Glasserman and Merener (2003) and Eberlein and Özkan (2005). A natural question to ask is whether one can feature both stochastic volatility and jumps in a model. The answer is positive and attributed to Jarrow, Li and Zhao (2003), who have proposed a generalized market model that combines the constant elasticity variance (CEV) dynamics with stochastic volatility (Andersen and Andreasen, 2000) and jumps (Glasserman and Kou, 2003). Their empirical studies have the conclusions that the generalized model outperforms the other two models in fitting volatility smiles.

Appendix: The Lévy-Ito Decomposition

Definition 12.3.1. *Fix $T > 0$. Define $\mathcal{M}_T^2 = \mathcal{M}_T^2(\Sigma, \mathcal{F}, \mathcal{F}_t, \mathbb{P})$ to be the space of real-valued, zero mean right-continuous, square-integrable $\mathbb{P}$-martingales with respect to $\mathcal{F}$ over the finite time period $[0, T]$.*

Note that any zero mean square-integrable martingale with respect to $\{\mathcal{F}_t : t \geq 0\}$ has a right-continuous version belonging to $\mathcal{M}_T^2$.

We will show that $\mathcal{M}_T^2$ is a Hilbert space with respect to the inner product

$$\langle M, N \rangle = \mathbb{E}\left[M_T N_T\right]$$

for any M, $N \in \mathcal{M}_T^2$. It is obvious to see that, for any M, N, $Q \in \mathcal{M}_T^2$,

1. $\langle aM + bN, Q \rangle = a\langle M, Q \rangle + b\langle N, Q \rangle$ for any $a, b \in \mathbb{R}$.

2. $\langle M, N \rangle = \langle N, M \rangle$.

3. $\langle M, M \rangle \geq 0$.

4. When $\langle M, M \rangle = 0$, by Doob's maximal inequality,

$$\mathbb{E}\left[\sup_{0 \leq s \leq T} M_s^2\right] \leq 4\mathbb{E}\left[M_T^2\right] = 4\langle M, M \rangle = 0,$$

so, $\sup_{0 \leq t \leq T} M_t = 0$ almost surely.

5. Finally, we need to show any Cauchy sequences have a limit in $\mathcal{M}_T^2$.

Let $\{M^{(n)} : n = 1, 2, \ldots\}$ be a Cauchy sequence in $\mathcal{M}_T^2$ such that

$$\|M^{(m)} - M^{(n)}\| = \left(\mathbb{E}\left[(M_T^{(m)} - M_T^{(n)})^2\right]\right)^{1/2} \longrightarrow 0 \quad \text{as} \quad m, n \uparrow \infty.$$

Then, $\{M_T^{(n)} : n \geq 1\}$ must be a Cauchy sequence in the Hilbert space of zero mean, square-integrable random variables defined on $(\Omega, \mathcal{F}_t, \mathbb{P})$, which is a subspace of $L^2(\Omega, \mathcal{F}_T, \mathbb{P})$, endowed with the inner product $\langle M, N \rangle = \mathbb{E}[MN]$. Hence there exists a limiting variable $M_T \in L^2(\Omega, \mathcal{F}_T, \mathbb{P})$ with zero mean such that

$$\left(\mathbb{E}\left[(M_T^{(n)} - M_T)^2 \right] \right)^2 \longrightarrow 0, \quad \text{as } n \uparrow \infty.$$

Define

$$M_t = \mathbb{E}[M_T | \mathcal{F}_t] \quad \forall t \in [0, T],$$

which is also right-continuous, then

$$\| M_t^{(n)} - M_t \| \longrightarrow 0 \quad \text{as } n \uparrow \infty.$$

M_t is $\mathcal{F}_t$-adaptive by definition and by Jensen's inequality,

$$\begin{aligned}
\mathbb{E}\left[M_t^2\right] &= \mathbb{E}\left[(\mathbb{E}[M_T|\mathcal{F}_t])^2 \right] \\
&\leq \mathbb{E}\left[\mathbb{E}[M_T^2|\mathcal{F}_t] \right] \\
&= \mathbb{E}\left[M_T^2\right].
\end{aligned}$$

Hence $M \in \mathcal{M}_T^2$ and $\mathcal{M}_T^2$ is a Hilbert space.

Suppose that $\{\xi_i : i \geq 1\}$ is a sequence of i.i.d. random variable with a common law F (with no mass at the origin) and that $N = \{N_t : t \geq 0\}$ is a Poisson process with rate $\lambda > 0$, we have

Lemma 12.3.1. *Suppose that* $\displaystyle\int_{\mathbb{R}} |x| F(dx) < \infty.$

1. The process $M = \{M_t : t \geq 0\}$ *defined by*

$$M_t \triangleq \sum_{i=1}^{N_t} \xi_i - \lambda t \int_{\mathbb{R}} x F(dx)$$

is a zero mean martingale with respect to its natural filtration.

2. If, moreover, $\displaystyle\int_{\mathbb{R}} x^2 F(dx) < \infty$, *then*

$$\mathbb{E}\left[M_t^2\right] = \lambda t \int_{\mathbb{R}} x^2 F(dx),$$

so M is a square-integrable martingale.

Proof. The proof consists of two steps.

1. By definition, M has stationary and independent increments so it is a Lévy process. Define $\mathcal{F}_t = \sigma(M_s : s \leq t)$ then for $t \geq s \geq 0$,

$$\mathbb{E}\left[M_t | \mathcal{F}_s\right] = M_s + \mathbb{E}\left[M_t - M_s | \mathcal{F}_s\right]$$
$$= M_s + \mathbb{E}\left[M_{t-s}\right].$$

What is left to show is that

$$\mathbb{E}\left[M_u\right] = 0 \quad \text{for all } u \geq 0.$$

In fact, $\forall u \geq 0$,

$$\mathbb{E}\left[M_u\right] = \mathbb{E}\left[\sum_{i=1}^{N_u} \xi_i - \lambda u \int_{\mathbb{R}} x F(dx)\right]$$
$$= \lambda u \mathbb{E}\left[\xi_1\right] - \lambda u \int_{\mathbb{R}} x F(dx) = 0.$$

In addition

$$\mathbb{E}\left[|M_u|\right] \leq \mathbb{E}\left[\left|\sum_{i=1}^{N_u} \xi_i\right| + \lambda u \int_{\mathbb{R}} x F(dx)\right]$$
$$\leq \lambda u \mathbb{E}\left[|\xi_1|\right] + \lambda u \int_{\mathbb{R}} x F(dx)$$
$$= \lambda u \left(\int_{\mathbb{R}} (|x| + x) F(dx)\right) < \infty.$$

2. Using the i.i.d. property of $\{\xi_i, i \geq 1\}$, we have

$$\mathbb{E}\left[M_t^2\right] = \mathbb{E}\left[\left(\sum_{i=1}^{N_t} \xi_i\right)^2\right] - \lambda^2 t^2 \left(\int_{\mathbb{R}} x F(dx)\right)^2$$
$$= \mathbb{E}\left[\sum_{i=1}^{N_t} \xi_i^2\right] + \mathbb{E}\left[\sum_{i=1}^{N_t}\sum_{j=1}^{N_t} 1_{\{i \neq j\}} \xi_i \xi_j\right] - \lambda^2 t^2 \left(\int_{\mathbb{R}} x F(dx)\right)^2$$
$$= \lambda t \int_{\mathbb{R}} x^2 F(dx) + \mathbb{E}\left[N_t^2 - N_t\right] \left(\int_{\mathbb{R}} x F(dx)\right)^2$$
$$\quad - \lambda^2 t^2 \left(\int_{\mathbb{R}} x F(dx)\right)^2$$
$$= \lambda t \int_{\mathbb{R}} x^2 F(dx) + \lambda^2 t^2 \left(\int_{\mathbb{R}} x F(dx)\right)^2 - \lambda^2 t^2 \left(\int_{\mathbb{R}} x F(dx)\right)^2$$
$$= \lambda t \int_{\mathbb{R}} x^2 F(dx).$$

$\square$

Recall that

$$\lambda_n = \nu\left(2^{-(n+1)} \le |x| < 2^{-n}\right),$$
$$F_n(dx) = \lambda_n^{-1}\nu(dx)\mathbf{1}_{\{2^{-(n+1)}\le|x|<2^{-n}\}},$$

we now define

$$N^{(n)} = \{N_t^{(n)} : t \ge 0\} \quad - \quad \text{Poisson process with rate } \lambda_n,$$
$$\{\xi_i : i = 1, 2, \ldots\} \quad - \quad \text{i.i.d random variable with law } F_n,$$

and $M^{(n)} = \{M_t^{(n)} : t \ge 0\}$ such that

$$M_t^{(n)} = \sum_{i=1}^{N_t^{(n)}} \xi_i^{(n)} - \lambda_n t \int_{\mathbb{R}} x F_n(dx)$$
$$\mathcal{F}_t^{(n)} = \sigma(M_s^{(n)} : s \le t) \quad \text{for } t \ge s \ge 0.$$

Finally, we put $\{M^{(n)} : n \ge 1\}$ on the same probability space with respect to the common filtration

$$\mathcal{F}_t = \sigma\left(\bigcup_{n \ge 1} \mathcal{F}_t^{(n)}\right).$$

Theorem 12.3.1. *If*

$$\sum_{n \ge 1} \lambda_n \int_{\mathbb{R}} x^2 F_n(dx) < \infty,$$

then there is a Lévy process $X = \{X_t, t \ge 0\}$ which is also a square-integrable martingale with characteristic exponent

$$\Psi(\theta) = \int_{\mathbb{R}} (1 - e^{i\theta x} + i\theta x) \sum_{n \ge 1} \lambda_n F_n(dx)$$

for all $\theta \in \mathbb{R}$, such that for each fixed $T > 0$,

$$\lim_{k \to \infty} \mathbb{E}\left[\sup_{t \le T}\left(X_t - \sum_{n=1}^{k} M_t^{(n)}\right)^2\right] = 0.$$

Proof. We first show that $\sum_{n=1}^{k} M^{(n)}$ is a square-integrable martingale. In fact, due to independence and zero mean,

$$\mathbb{E}\left[\left(\sum_{n=1}^{k} M_t^{(n)}\right)^2\right] = \sum_{n=1}^{k} \mathbb{E}\left[\left(M_t^{(n)}\right)^2\right] = t\sum_{n=1}^{k}\lambda_n\int_{\mathbb{R}} x^2 F_n(dx) < \infty.$$

Fix $T > 0$. We now claim $X^{(k)} = \{X_t^{(k)}, 0 \leq t \leq T\}$ such that

$$X_t^{(k)} = \sum_{n=1}^{k} M_t^{(n)}$$

is a Cauchy sequence with respect to $\| \cdot \|$. Note that for $k \geq l$,

$$\|X^{(k)} - X^{(l)}\|^2 = \mathbb{E}\left[\left(X_T^{(k)} - X_T^{(l)}\right)^2\right]$$

$$= T \sum_{n=l+1}^{k} \lambda_n \int_{\mathbb{R}} x^2 F_n(dx) \longrightarrow 0 \quad \text{as} \quad k, l \uparrow \infty.$$

Then, there is X_T in $(\Omega, \mathcal{F}_T, \mathbb{P})$ such that

$$\|X_T^{(k)} - X_T\| \longrightarrow 0 \quad \text{as} \quad k \uparrow \infty, \tag{12.32}$$

owing to completeness of $L^2(\Omega, \mathcal{F}_t, \mathbb{P})$. Define

$$X_t = \mathbb{E}\left[X_T | \mathcal{F}_t\right] \tag{12.33}$$

and $X = \{X_t, 0 \leq t \leq T\}$, then there is also

$$\|X^{(k)} - X\| \longrightarrow 0 \quad \text{as} \quad k \uparrow \infty.$$

Moreover, by making use of the Doob's maximal inequality, we have

$$\lim_{k \uparrow \infty} \mathbb{E}\left[\sup_{0 \leq t \leq T}\left(X_t - X_t^{(k)}\right)^2\right] = 0.$$

The above limit also implies the convergence of (the finite dimensional) distribution. Consequently, since $X^{(k)}$ are Lévy processes,

$$\mathbb{E}\left[e^{i\theta(X_t - X_s)}\right] = \lim_{k \uparrow \infty} \mathbb{E}\left[e^{i\theta(X_t^{(k)} - X_s^{(k)})}\right]$$

$$= \lim_{k \uparrow \infty} \mathbb{E}\left[e^{i\theta X_{t-s}^{(k)}}\right]$$

$$= \mathbb{E}\left[e^{i\theta X_{t-s}}\right],$$

which shows that X has stationary and independent increments. Based on the condition of the theorem, we readily have

$$\mathbb{E}\left[e^{i\theta X_t}\right] = \lim_{k \uparrow \infty} \mathbb{E}\left[e^{i\theta X_t^{(k)}}\right]$$

$$= \lim_{k \uparrow \infty} \prod_{n=1}^{k} \mathbb{E}\left[e^{i\theta M_t^{(n)}}\right]$$

$$= \exp\left\{-\int_{\mathbb{R}}(1 - e^{i\theta x} + i\theta x)\sum_{n \geq 1} \lambda_n F_n(dx)\right\}$$

$$= \Psi^{(3)}(\theta).$$

There are two more minor issues left. The first is to show the right-continuity of X. This comes from the fact that the space of right-continuity functions over $[0, T]$ is a closed space, under the metric $d(f, g) = \sup_{0 \leq t \leq T} |f(t) - g(t)|$. The second is the dependence of X on T, which should be dismissed.

Suppose we index X by T, say X^T, using

$$\sup_n a_n^2 = \left(\sup_n |a_n| \right)^2,$$

$$\sup_n |a_n + b_n| \leq \sup_n |a_n| + \sup_n |b_n|,$$

and Minkowski's inequality, we have for $T_1 \leq T_2$,

$$\mathbb{E} \left[\sup_{t \leq T_1} \left(X_t^{T_1} - X_t^{T_2} \right)^2 \right]^{1/2}$$

$$\leq \mathbb{E} \left[\left(\sup_{t \leq T_1} \left| X_t^{T_1} - X_t^{(k)} \right| + \sup_{t \leq T_1} \left| X_t^{T_2} - X_t^{(k)} \right| \right)^2 \right]^{1/2}$$

$$\leq \mathbb{E} \left[\sup_{t \leq T_1} \left(X_t^{T_1} - X_t^{(k)} \right)^2 \right]^{1/2} + \mathbb{E} \left[\sup_{t \leq T_1} \left(X_t^{T_2} - X_t^{(k)} \right)^2 \right]^{1/2}$$

$$\longrightarrow 0 \qquad \text{as } k \uparrow \infty.$$

Note that $X_t^{T_1}$ and $X_t^{T_2}$ are defined as expectations of $X_{T_1}^{T_1}$ and $X_{T_2}^{T_2}$ in Equation 12.33, and $X_{T_1}^{T_1}$ and $X_{T_2}^{T_2}$ are defined as the limits of $X_{T_1}^{(k)}$ and $X_{T_2}^{(k)}$. So $X_t^{T_1} = X_t^{T_2}$ for any $t \in [0, T_1]$, thus X_t does not depend on T.

The limit X established in the Theorem is just $X^{(3)}$, which has a countable number of discontinuities. $\qquad \square$

Chapter 13

Market Model for Inflation Derivatives

The government inflation-indexed bond first appeared in Finland in 1945, but it only began to be treated as an asset class after the first issuance of inflation-indexed bond, known as Gilt, by the UK government in 1981. Since then, more European governments began to issue inflation-protected sovereignty debts. The U.S. Treasury joined the ranks in 1997 with the issuance of Treasury Inflation Protected Securities (TIPS), which spurred the growth of the inflation derivatives securities.

Research on pricing models for inflation-rate derivatives has also become active since 1997 (Barone and Castagna, 1997; Bezooyen *et al.*, 1997). A theoretical framework, the so-called "foreign currency analogy," first suggested by Hughston (1998) and established by Jarrow and Yildirim (2003), had became very influential. Under this framework, the real interest rate, defined as the difference between the nominal interest rate and the inflation rate, is treated as the interest rate of a foreign currency, while the Consumer Price Index (CPI) is treated as the exchange rate between the domestic and the foreign currencies. To price inflation derivatives, one needs to model the nominal (domestic) interest rate, the foreign (real) interest rate, and the exchange rate (CPI). A handy solution for modeling inflation derivatives is to adopt the Heath-Jarrow-Morton's (1992) framework for interest rates of two currencies, and bridge them using a lognormal exchange-rate process. Manning and Jones (2003) push the limit of the analytical tractability of this approach and obtain price formulae for inflation caplets and floorlets. For general inflation derivatives, one resorts to Monte Carlo simulations.

Although elegant in theory, a Heath-Jarrow-Morton-type model is inconvenient to use. The model takes the unobservable instantaneous (nominal and/or real) forward rates as state variables, while the payoffs of most inflation derivatives are written on CPI or simple compounding inflation rates, making the models hard to calibrate.

A number of different models aimed at more convenient pricing and hedging of inflation derivatives have been developed since 2003. Some researchers adopted normal or lognormal dynamics for certain observable inflation-related variables, for example, the CPI index (Belgrade and Benhamou, 2004a; Korn and Kruse, 2004), the forward price of real zero-coupon bonds (Kazziha, 1999; Mercurio, 2005), or inflation forward rates (Kenyon, 2008). Extensions to these models have also been developed so that risk factors other than diffusion, like stochastic volatility (Mercurio and Moreni, 2006 and 2009; Kruse, 2007; Kenyon, 2008) and jumps (Hinnerich, 2008), are incorporated. Other researchers adopted the square-root process of Cox *et al.* (1985) to model the spot inflation rate, in conjunction with a short-rate model for the nominal spot rate (see, e.g., Chen et al., 2006 and Falbo *et al.*, 2010). Nonetheless, over the years practitioners have been using a model of their own: the so-called market model which is based on the displaced diffusion dynamics for simple inflation forward rates and has not been documented in the publicly available literature. There are also researches that investigate various issues in inflation-rate modeling. Among others, Cairns (2000) considers inflation models where nominal interest rates are ensured positive; Chen et al. (2006) estimate inflation risk premium; and Belgrade and Benhamou (2004b) examine seasonality in inflation rates and manage to take the seasonality into account in their CPI-based model.

While important advancements were made over the years, a certain degree of disorder in the literature has also been created. There were at least three versions of "market models" (Beldgrade *et al.*, 2004; Mercurio-Moreni, 2006; the practitioners' model), and at least two different notions of "inflation forward rates," adopted respectively by models based on zero-coupon inflation-indexed swaps (ZCIIS) and models based on year-on-year inflation-indexed swaps (YYIIS). When using a ZCIIS-based model to price derivatives on YYIIS, the technique of convexity adjustment is used to calculate the YYIIS swap rates.

The state of disorder was ended in 2011 when a new paradigm in inflation derivatives modeling occurred (Wu, 2011). Under the new paradigm, the notion of inflation forward rate is defined as the fair rate for a forward contract on inflation rate, which is shown to be replicable statically and thus is unique, and the lognormal martingale dynamics for displaced inflation forward rates can be justified, thus allowing a rigorous rebuilding of the practitioners' model. Consequently, the market model for inflation rates is now uniquely defined. Moreover, Wu (2011) establishes a Heath-Jarrow-Morton (HJM) type equation for instantaneous inflation forward rates and, by also making use of the classic HJM equation for nominal forward rates, re-derives the HJM type equation for real forward rates as was established earlier by Jarrow and Yildirim (2003), yet drops their notion of "volatility of the Consumer Price Index", which is justified to be zero.

This chapter has several important implications. First, we show that the ZCIIS- and YYIIS-based market models are identical and the use of "convexity adjustment" is wrong and unnecessary. Second, we unify the closed-form pricing of inflation caplets, floorlets and swaptions with the Black formula for displaced-diffusion processes which allows us to quote these derivatives using the "implied Black's volatilities." Finally, we provide a proper platform for developing smile models.

The rest of the chapter is organized as follows. In Section 13.1, we introduce major inflation derivatives and highlight real zero-coupon bonds, which is part of our primitive state variables. In Section 13.2, we define the notion of forward inflation rates, rebuild the extended market model, and develop a Heath-Jarrow-Morton-type model in terms of continuous compounding forward nominal and inflation rates. Section 13.3 is devoted to pricing major inflation-indexed derivatives under the market model, where we produce closed-form formulae for caps, floors and swaptions. In Section 13.4, we briefly discuss the comprehensive calibration of the market model, and demonstrate some calibration results with market data. In Section 13.5, we demonstrate the construction of smile models, in particular, using the SABR methodology.

13.1   CPI Index and Inflation Derivatives Market

Inflation-rate security markets have evolved steadily over the past twenty years. Among developed countries, the outstanding notional value of inflation-linked government bonds has grown from about 50 billion dollars in 1997 to over 3.2 trillion dollars in 2016.[1] There are inflation-linked securities in most major currencies, including the Pound sterling, Canadian dollar, yen and, of course, the Euro and the U.S. dollar. The global daily turnover was about $10 billion a day on average in 2009, which is largely dominated by Euro- and dollar-denominated securities. For more information on global inflation-indexed security markets, we refer readers to the Annual Report by Barclays Capital. By comparing the size of inflation markets to the sizes of Treasury security or credit markets, one has to conclude that the interest on inflation securities has been tepid in the past. Nonetheless, since the 2007-08 financial crisis, there has been a concern for potential high inflation, which could possibly be caused by the expansionary monetary policy adopted across the globe (Jung, 2008).

The payoff functions of inflation-linked securities depend on inflation rates, which are defined as the percentage change rates of the CPI. The CPI represents an average price of a basket of services and goods, and the average price is compiled by official statistical agencies of central governments. The evolution of CPI indexes in both Europe and the United States is displayed in Figure 13.1, which shows that both indexes increase steadily over years, with however noticeable acceleration in 2022.[2]

---

[1]Source: Components of Barclays Universal Government Inflation-lined All Maturities Bond Index, 2016.

[2]Sources: U.S. Bureau of Labor Statistics (ftp://ftp.bls.gov/pub/special.requests/cpi /cpiai.txt) and European Central Bank (http://appsso.eurostat.ec.europa.eu/nui/show. do?dataset=prc_hicp_midx&lang=en).

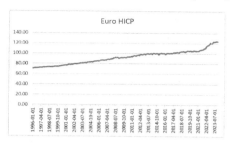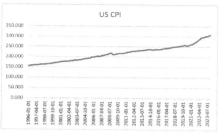

FIGURE 13.1: Consumer Price Indexes of the United States and the Euro zone.

The inflation rate of a country is defined in terms of its CPI. Denote by $I(t)$ the CPI of time t, then the inflation rate over the time period $[t, T]$ is defined as the percentage change of the index:

$$\hat{i}(t, T) = \frac{I(T)}{I(t)} - 1.$$

For purpose of comparison, we will more often use the annualized inflation rate,

$$i(t, T) = \frac{1}{T - t} \left(\frac{I(T)}{I(t)} - 1 \right).$$

Suppose the limit of the annualized inflation rate exists for $T \downarrow t$, we obtain the so-called instantaneous inflation rate, $i(t)$, which will be used largely for mathematical and financial arguments instead of modeling. An important feature that distinguishes inflation rates from interest rates is that the former can be either positive or negative, while the latter should be positive in any normal circumstances as otherwise we are in a situation of arbitrage.

Inflation-indexed bonds are mainly issued by central governments, while inflation-rate derivatives are offered and traded in the OTC markets by various financial institutions. The dollar-denominated inflation-linked securities have been predominately represented by TIPS, followed by ZCIIS and YYIIS. In recent years, caps, floors and swaptions on inflation rates have been gaining popularity. In the following subsections, we will describe these inflation-linked securities at length. Unlike the earlier practices of the markets which took YYIIS as the underlying securities of the inflation derivatives markets, we take ZCIIS as the underlying securities and on top of them construct the term structure of "inflation forward rates."

13.1.1   TIPS

TIPS are coupon bonds with fixed coupon rates but floating principals, and the latter is adjusted according to the inflation rate over the accrual period of a coupon payment. Note that typically there is a floor on the principal value of a TIPS, which is often the initial principal value. The existence of floors, as a matter of fact, turns TIPS into coupon bonds with embedded options. So strictly speaking, the pricing of TIPS should need a model.

Note that the CPI index is measured with a two-month lag. Yet this lagged index plays the role of the current index for the principal adjustments of TIPS and the payoff calculations of inflations derivatives. From a modeling point of view, lagging or not does not make a difference. With this understanding in mind, we will treat the lagged index as the current index throughout the paper.

13.1.2   ZCIIS

The ZCIIS is a swap contract between two parties with a single exchange of payments. Suppose that the contract was initiated at time t and will be expired at $T > t$, then the payment of one party equals to a notional value times the inflation rate over the contract period, i.e.

$$Not. \times \hat{i}(t, T),$$

while the counterparty makes a fixed payment in the amount

$$Not. \times \left((1 + K(t, T))^{T-t} - 1 \right),$$

where $Not.$ stands for the notional value of the contract and $K(t, T)$ is the quote for the contract. Because the value of the ZCIIS is zero at initiation, ZCIIS directly renders the price of the so-called real discount bond, which pays inflation-adjusted principal:

$$P_R(t, T) = E^Q \left[e^{- \int_t^T r_s ds} \frac{I(T)}{I(t)} \middle| \mathcal{F}_t \right] = P(t, T)(1 + K)^{T-t}. \tag{13.1}$$

Here, $P(t, T)$ is the nominal discount factor from T back to t. For real zero-coupon bonds with the same maturity date T but an earlier issuance date, say, $T_0 < t$, the price is

$$P_R(t, T_0, T) = E^Q \left[e^{- \int_t^T r_s ds} \frac{I(T)}{I(T_0)} \middle| \mathcal{F}_t \right] = \frac{I(t)}{I(T_0)} P_R(t, T). \tag{13.2}$$

We emphasize here that $P_R(t, T_0, T)$, instead of $P_R(t, T)$, is treated as the time t price of a traded security. The latter can only be considered as the initial price of a new security.

For modeling inflation-rate derivatives, we will take the term structure of real and nominal zero-coupon bonds, $P_R(t, T_0, T)$ and $P(t, T)$ for a fixed T_0 and all $T \geq t \geq T_0$, as model primitives. Note that $\{P_R(t, T), \forall T > t\}$ alone carries information on the term structure of real interest rates only. To see that, we make use of the relationship between the instantaneous inflation rate and CPI:

$$\frac{I(T)}{I(T_0)} = e^{\int_{T_0}^T i(s) ds}. \tag{13.3}$$

Plugging Equation 13.3 into Equation 13.1 and making use of the Fisher's equation (Fisher, 1930; also see Cox, Ingersoll and Ross, 1985),

$$r(t) = R(t) + i(t), \tag{13.4}$$

where $R(t)$ is the real interest rate, we obtain

$$
\begin{aligned}
P_R(t,T) &= E^Q \left[e^{-\int_t^T (r_s - i(s))ds} | \mathcal{F}_t \right] \\
&= E^Q \left[e^{-\int_t^T R_s ds} | \mathcal{F}_t \right].
\end{aligned} \tag{13.5}
$$

According to Equation 13.5, the real zero-coupon bond implies the discount factor associated to the real interest rate. This is the reason why we use the subindex "R" for the price.

Note that the real interest rates are not good candidates for the state variable, because most inflation derivatives are written on inflation rates.

13.1.3 YYIIS

YYIIS are contracts to swap an annuity against a sequence of floating payments indexed to inflation rates over future periods. The fixed-leg payments of a YYIIS are $Not.\Delta\phi_i K, i = 1, 2, \ldots, N_x$, where $\Delta\phi_i$ is the year fractions between two consecutive payments, while the floating-leg payments are of the form

$$Not. \left(\frac{I(T_j)}{I(T_{j-1})} - 1 \right),$$

and are made at time $T_j, j = 1, 2, \ldots, N_f$. Note that the payment gaps $\Delta\phi_i = \phi_i - \phi_{i-1}$ and $\Delta T_j = T_j - T_{j-1}$ can be different, yet the terms for payment swaps are the same, i.e., $\sum_{i=1}^{N_x} \Delta\phi_i = \sum_{j=1}^{N_f} \Delta T_j$. The price of the YYIIS equals to the difference in values of the fixed and floating legs. The former can be calculated by discounting, yet the latter involves the evaluation of an expectation,

$$V_{float}^{(j)}(t) = Not. E^Q \left[e^{-\int_t^{T_j} r_s ds} \left(\frac{I(T_j)}{I(T_{j-1})} - 1 \right) \middle| \mathcal{F}_t \right].$$

We will show that, contrary to current practices, the theoretical pricing of the floating leg should be model independent.

13.1.4 Inflation Caps and Floors

An inflation cap is like a YYIIS with embedded optionality: with the same payment frequency, payments are made only when a netted cash flow to the payer (of the fixed leg) is positive, corresponding to cash flows of the following form to the cap holder

$$Not.\Delta T_i \left[\frac{1}{\Delta T_i} \left(\frac{I(T_i)}{I(T_{i-1})} - 1 \right) - K \right]^+, i = 1, \ldots, N.$$

Accordingly, the cash flows of an inflation floor is

$$Not.\Delta T_i \left[K - \frac{1}{\Delta T_i} \left(\frac{I(T_i)}{I(T_{i-1})} - 1 \right) \right]^+, i = 1, \ldots, N.$$

Apparently, the prices of caplets and floorlets depend on the variance of the future inflation rates, thus making their pricing model dependent.

13.1.5   Inflation Swaptions

An inflation swaption is an option to enter into a YYIIS swap in the future. At the maturity of the option, the holder of the option should enter into the underlying YYIIS if the option ends up in the money. The underlying security of the swaption is the YYIIS. With the establishment of the theory of this chapter, the pricing of the underlying YYIIS will become model independent, which will consequently make the pricing of inflation swaptions easier.

13.2   Rebuilt Market Model and the New Paradigm

13.2.1   Inflation Discount Bonds and Inflation Forward Rates

The cash flows of several major inflation-indexed instruments, including the YYIIS and inflation caplets and floorlets, are expressed in term rates of simple inflation rates. For pricing and hedging purposes we need to define inflation forward rates as the fair rates for forward-rate agreements on inflation rates, parallel to the definition of nominal forward rates. We know that the prices of nominal zero-coupon bonds imply nominal forward rates. Not surprisingly, the prices of nominal zero-coupon bonds and real zero-coupon bonds jointly imply the inflation forward rates. To be precise, we introduce

Definition 13.2.1. *The discount bond associated to inflation rate is defined by*

$$P_I(t, T) \stackrel{\triangle}{=} \frac{P(t, T)}{P_R(t, T)}. \tag{13.6}$$

Alternatively, with $P_I(t, T)$ and $P_R(t, T)$, we effectively factorize the nominal discount factor into real and inflation discount factors,

$$P(t, T) = P_R(t, T) P_I(t, T). \tag{13.7}$$

Note that neither $P_I(t,T)$ nor $P_R(t,T)$ is a price of a tradable security,[3] yet both are observable.

We define inflation forward rates as the *return rates implied by the inflation discount bonds.*

Definition 13.2.2. *The inflation forward rate for a future period $[T_1, T_2]$ seen at time $t \le T_2$ is defined by*

$$f^{(I)}(t, T_1, T_2) \triangleq \frac{1}{(T_2 - T_1)} \left(\frac{P_I(t, T_1)}{P_I(t, T_2)} - 1 \right). \tag{13.8}$$

There is a slight problem with the above definition: the forward inflation rate is fixed at $t = T_2$, beyond the life of the T_1-maturity inflation bond, so we need to define $P_I(t, T_1)$ for $t > T_1$. In view of Equation 13.2 and Equation 13.7, we have

$$P_I(t, T_1) = \frac{I(t)}{I(T_0)} \frac{P(t, T_1)}{P_R(t, T_0, T_1)}. \tag{13.9}$$

The second ratio on the right-hand side of Equation 13.9 is the relative price between two traded bonds with an identical maturity date. It makes sense to assume that at their common maturity, T_1, the proceeds from both bonds are deposited in the saving account, so that the relative price will stay unchanged beyond T_1. In other words, the relative price beyond T_1 can be defined by constant extrapolation, thus yielding

$$P_I(t, T_1) = \frac{I(t)}{I(T_0)} \frac{I(T_0)}{I(T_1)} = \frac{I(t)}{I(T_1)}, \quad \forall t \ge T_1. \tag{13.10}$$

Given Equation 13.10, we have the value of the forward rate at its fixing date to be

$$f^{(I)}(T_2, T_1, T_2) = \frac{1}{T_2 - T_1} \left(\frac{I(T_2)}{I(T_1)} - 1 \right), \tag{13.11}$$

so the inflation forward rate converges to inflation spot rate at maturity.

Next, we will argue that $f^{(I)}(t, T_1, T_2)$ so defined is the fair rate seen at time t for a forward contract on inflation rate over $[T_1, T_2]$. We rewrite Equation 13.8 into

$$f^{(I)}(t, T_1, T_2) = \frac{1}{(T_2 - T_1)} \left(\frac{F_R(t, T_1, T_2) P(t, T_1)}{P(t, T_2)} - 1 \right), \tag{13.12}$$

where

$$F_R(t, T_1, T_2) \triangleq \frac{P_R(t, T_2)}{P_R(t, T_1)} = \frac{P_R(t, T_0, T_2)}{P_R(t, T_0, T_1)} \tag{13.13}$$

is the relative price of two tradable securities. The following result is the corner stone of our theory.

---

[3] $P_R(t,T)$ is treated as the price of a zero-coupon bond of a virtue "foreign currency" by Jarrow and Yildirim (2003).

Proposition 13.2.1. *Let $t \leq T_1 \leq T_2$. The T_1-forward price of a real bond with maturity T_2 seen at time t is $F_R(t, T_1, T_2)$.*

Proof. Do the following zero-net transactions.

1. At time $t \geq T_0$,

   (a) long the forward contract to buy $\frac{I(T_1)}{I(T_0)}$ dollars of T_2-maturity real bond at time T_1 at the unit price $F_R(t, T_1, T_2)$;

   (b) long one unit of T_1-maturity real bond at the price of $P_R(t, T_0, T_1)$;

   (c) short $\frac{P_R(t,T_0,T_1)}{P_R(t,T_0,T_2)}$ unit(s) of T_2-maturity real bond at the unit price of $P_R(t, T_0, T_2)$.

2. At time T_1, exercise the forward contract to buy the T_2-maturity real bond (that pays $I(T_2)/I(T_1)$) at the unit price $F_R(t, T_1, T_2)$, applying all proceeds from the T_1-maturity real bond.

3. At Time T_2, close out all positions.

The net profit or loss from the transactions is

$$P\&L = \left(\frac{1}{F_R(t, T_1, T_2)} - \frac{P_R(t, T_0, T_1)}{P_R(t, T_0, T_2)} \right) \frac{I(T_2)}{I(T_0)}. \qquad (13.14)$$

For the absence of arbitrage, the forward price must be set equal to Equation 13.13. □

Note that the forward contract used in the proof does not specify the number of units of the T_2-maturity real bond for purchasing, which is the only difference from a usual forward contract. Yet the seller can still perfectly hedge the forward contract.

In view of Equation 13.12, we can treat $f^{(I)}(t, T_1, T_2)$ as the T_1-forward price for the payoff of $f^{(I)}(T_2, T_1, T_2)$ at T_2, and thus have proven

Proposition 13.2.2. *The inflation forward rate $f^{(I)}(t, T_1, T_2)$ is the unique arbitrage-free rate seen at the time t for a T_1-expiry forward contract on the inflation rate over the future period $[T_1, T_2]$.*

Proposition 13.2.2 should help to end the situation of the coexistence of multiple definitions of forward inflation rates. Note that our definition Equation 13.8 coincides with one of the definitions of inflation forward rates, $Y_i(t)$, given in Mercurio and Moreni (2009).

13.2.2   Compatibility Condition

We now proceed to the construction of dynamic models for inflation forward rates of both simple and instantaneous compounding. Under the risk neutral measure Q, $P(t, T)$ and $P_R(t, T_0, T)$ are assumed to follow the lognormal processes

$$dP(t,T) = P(t,T)\left(r_t dt + \Sigma(t,T) \cdot d\mathbf{W}_t\right),$$
$$dP_R(t,T_0,T) = P_R(t,T_0,T)\left(r_t dt + \Sigma_R(t,T) \cdot d\mathbf{W}_t\right), \tag{13.15}$$

where r_t is the risk-free nominal (stochastic) interest rate, $\Sigma(t,T)$ and $\Sigma_R(t,T)$ are d-dimensional $\mathcal{F}_t$-adaptive volatility functions of $P(t,T)$ and $P_R(t,T_0,T)$, respectively,[4] and "·" means scalar product. The volatility functions are assumed sufficiently regular in t and T so that the SDE Equation 13.15 admits a unique strong solution, and their partial derivatives with respect to T exist and have finite L_2 norms with respect to t. Moreover, there is no volatility for both nominal and real zero-coupon bonds, meaning

$$\Sigma(t,T) = \Sigma_R(t,T) = 0 \quad \text{for} \quad t \geq T. \tag{13.16}$$

By making use of the dynamics of $P_R(t,T_0,T)$ and the dynamics of the CPI,

$$dI(t) = i(t)I(t)dt,$$

we can derive the dynamics of $P_R(t,T)$:

$$dP_R(t,T) = P_R(t,T)\left((r_t - i(t))dt + \Sigma_R(t,T) \cdot d\mathbf{W}_t\right). \tag{13.17}$$

Being a T_1-forward price of a tradable security, $F(t,T_1,T_2)$ should be a lognormal martingale under the T_1-forward measure whose volatility is the difference between those of $P_R(t,T_0,T_2)$ and $P_R(t,T_0,T_1)$, i.e.,

$$\frac{dF_R(t,T_1,T_2)}{F_R(t,T_1,T_2)} = (\Sigma_R(t,T_2) - \Sigma_R(t,T_1)) \cdot (d\mathbf{W}_t - \Sigma(t,T_1)dt). \tag{13.18}$$

Note that $d\mathbf{W}_t - \Sigma(t,T_1)dt$ is (the differential of) a Brownian motion under the T_1-forward measure, Q_{T_1}, defined by the Radon-Nikodym derivative

$$\left.\frac{dQ_{T_1}}{dQ}\right|_{\mathcal{F}_t} = \frac{P(t,T_1)}{B(t)P(0,T_1)},$$

where $B(t) = \exp(\int_0^t r_s ds)$ is the unit price of the money market account.

There is an important implication by Equation 13.18. By Ito's Lemma, we also have

$$\frac{dF_R(t,T_1,T_2)}{F_R(t,T_1,T_2)} = (\Sigma_R(t,T_2) - \Sigma_R(t,T_1)) \cdot (d\mathbf{W}_t - \Sigma_R(t,T_1)dt). \tag{13.19}$$

The coexistence of Equations 13.18 and 13.19 poses a constraint on the volatility functions of the real zero-coupon bonds.

---

[4]It is not hard to see that the volatility of $P_R(t,T_0,T)$ does not depend on T_0.

Proposition 13.2.3 (Compatibility condition). *For arbitrage pricing, the volatility functions of the real bonds must satisfy the following condition:*

$$(\Sigma_R(t, T_2) - \Sigma_R(t, T_1)) \cdot (\Sigma(t, T_1) - \Sigma_R(t, T_1)) = 0. \tag{13.20}$$

Its differential version is, by letting $T_2 \to T_1 = T$,

$$\dot{\Sigma}_R(t, T) \cdot \Sigma_I(t, T) = 0, \tag{13.21}$$

where $\Sigma_I(t, T) \triangleq \Sigma(t, T) - \Sigma_R(t, T)$ is the percentage volatility of $P_I(t, T)$, and the overhead dot means partial derivatives with respect to T, the maturity.

Let us try to comprehend the compatibility condition. We understand that $\Sigma_R(t, T_2) - \Sigma_R(t, T_1)$ represents the volatility of the forward real rate defined by

$$f_R(t, T_1, T_2) \triangleq \frac{1}{T_2 - T_1} \left(\frac{P_R(t, T_1)}{P_R(t, T_2)} - 1 \right).$$

Literally, Equation 13.20 means that the price of inflation discount bond with maturity T_1 must be uncorrelated with forward real rates of any future period beyond T_1. This is reasonable and is not restrictive at all.

The differential version of the compatibility condition, Equation 13.21, will be used later to derive an HJM-type model for inflation rates.

13.2.3   Rebuilding the Market Model

For generality, we let $T = T_2$ and $\Delta T = T_2 - T_1$, then we can cast Equation 13.12 into

$$f^{(I)}(t, T - \Delta T, T) + \frac{1}{\Delta T} = \frac{1}{\Delta T} \frac{F_R(t, T - \Delta T, T) P(t, T - \Delta T)}{P(t, T)}.$$

The dynamics of $f^{(I)}(t, T - \Delta T, T)$ follows readily from those of F_R and P's.

Proposition 13.2.4. *Under the risk neutral measure, the governing equation for the simple inflation forward rate is*

$$d \left(f^{(I)}(t, T - \Delta T, T) + \frac{1}{\Delta T} \right)$$
$$= \left(f^{(I)}(t, T - \Delta T, T) + \frac{1}{\Delta T} \right) \gamma^{(I)}(t, T) \cdot (d\mathbf{W}_t - \Sigma(t, T)dt), \tag{13.22}$$

where

$$\gamma^{(I)}(t, T) = \Sigma_I(t, T - \Delta T) - \Sigma_I(t, T)$$

is the percentage volatility of the displaced inflation forward rate.

In formalism, Equation 13.22 is just the practitioners' model. Yet in applications, practitioners bootstrap the inflation forward rates from YYIIS and calibrate the model to inflation caps/floors for $\gamma^{(I)}(t,T)$. Let us present the market model for inflation rates in comprehensive terms. The state variables consist of two streams of spanning SOFR forward rates and forward inflation rates, $f_j(t) \overset{\triangle}{=} f(t,T_{j-1},T_j)$ and $f_j^{(I)}(t) \overset{\triangle}{=} f^{(I)}(t,T_{j-1},T_j), j = 1,2,\ldots,N$, that follow the following dynamics:

$$\begin{cases} df_j(t) = f_j(t)\gamma_j(t) \cdot (d\mathbf{W}_t - \Sigma(t,T_j)dt) , \\ d\left(f_j^{(I)}(t) + \dfrac{1}{\Delta T_j}\right) = \left(f_j^{(I)}(t) + \dfrac{1}{\Delta T_j}\right) \gamma_j^{(I)}(t) \cdot (d\mathbf{W}_t - \Sigma(t,T_j)dt), \end{cases}$$
$$(13.23)$$

where

$$\Sigma(t,T_j) = - \sum_{k=\eta_t+1}^{j} \frac{\Delta T_{k+1} f_k(t)}{1 + \Delta T_{k+1} f_k(t)} \gamma_k(t)$$

and $\eta_t = \max\{j|T_j \le t\}$. Note that according to Equation 13.16, there must be $\gamma_j(t) = \gamma_j^{(I)}(t) = 0$ for $t \ge T_j$. The initial nominal and inflation forward rates are derived from prices of nominal and real discount bonds. We want to highlight here that $f_j^{(I)}(t)$ is also a martingale under the T_j-forward measure, its own "cash flow measure."

13.2.4 The New Paradigm

Analogously to the definition of nominal forward rates, we define the instantaneous inflation forward rates as

$$f^{(I)}(t,T) \overset{\triangle}{=} -\frac{\partial \ln P_I(t,T)}{\partial T}, \quad \forall T \ge t, \tag{13.24}$$

or

$$P_I(t,T) = e^{-\int_t^T f^{(I)}(t,s)ds}.$$

By Ito's Lemma, we have

$$\begin{aligned} - d\ln P_I(t,T) &= d\ln\left(\frac{P_R(t,T)}{P(t,T)}\right) \\ &= -\left(i(t) + \frac{1}{2}\|\Sigma_I(t,T)\|^2\right) dt - \Sigma_I(t,T) \cdot (d\mathbf{W}_t - \Sigma(t,T)dt) . \end{aligned} \tag{13.25}$$

Differentiating the above equation with respect to T and making use of the compatibility condition, Equation 13.21, we then have

$$df^{(I)}(t,T) = -\dot{\Sigma}_I \cdot (d\mathbf{W}_t - \Sigma(t,T)dt) . \tag{13.26}$$

Equation 13.26 shows that $f^{(I)}(t,T)$ is a Q_T-martingale and its dynamics are fully specified by the volatilities of the nominal and inflation forward rates.

In an HJM context, the volatilities of nominal and inflation forward rates, $\sigma(t,T) = -\dot{\Sigma}(t,T)$ and $\sigma^{(I)}(t,T) = -\dot{\Sigma}_I(t,T)$, are first prescribed, and the volatilities of the zero-coupon bonds follow from

$$\Sigma(t,T) = -\int_t^T \sigma(t,s)ds \quad \text{and} \quad \Sigma_I(t,T) = -\int_t^T \sigma^{(I)}(t,s)ds.$$

Then, the extended HJM model with nominal and inflation forward rates is

$$\begin{cases} df(t,T) = \sigma(t,T) \cdot d\mathbf{W}_t + \sigma(t,T) \cdot \left(\int_t^T \sigma(t,s)ds \right) dt, \\[4mm] df^{(I)}(t,T) = \sigma^{(I)}(t,T) \cdot d\mathbf{W}_t + \sigma^{(I)}(t,T) \cdot \left(\int_t^T \sigma(t,s)ds \right) dt, \end{cases} \quad (13.27)$$

which takes the initial term structures of nominal and inflation forward rates as inputs.

If we treat Equation 13.27 as a framework of no-arbitrage models, then the market model Equation 13.22 fits in the framework with the volatility function

$$\sigma^{(I)}(t,T) = -\dot{\Sigma}_I(t,T) = \frac{\partial}{\partial T} \left(\sum_{k=0}^{\lceil \frac{T-t}{\Delta T} \rceil} \gamma^{(I)}(t, T - k\Delta T) \right),$$

where $\lceil x \rceil$ is the integer ceiling of x.

13.2.5 Unifying the Jarrow-Yildirim Model

According to their definitions, nominal, inflation and real forward rates for continuous compounding satisfy the relationship

$$f_R(t,T) = f(t,T) - f^{(I)}(t,T).$$

Subtracting the two equations of 13.27 and applying the compatibility condition, Equation 13.21, we then arrive at

$$df_R(t,T) = \sigma_R(t,T) \cdot d\mathbf{W}_t + \sigma_R(t,T) \cdot \left(\int_t^T \sigma_R(t,s)ds \right) dt, \quad (13.28)$$

where

$$\sigma_R(t,T) = \sigma(t,T) - \sigma^{(I)}(t,T) = -\dot{\Sigma}_R(t,T).$$

In contrast, under our notations the equation established by Jarrow and Yildirim (2003) for the real forward rates is

$$df_R(t,T) = \sigma_R(t,T) \cdot d\mathbf{W}_t + \sigma_R(t,T) \cdot \left(\int_t^T \sigma_R(t,s)ds - \sigma_I(t) \right) dt, \quad (13.29)$$

where $\sigma_I(t)$ is the volatility of the CPI index. Given that $\sigma_I(t) \equiv 0$, the two equations are identical.

Even if the CPI volatility were not zero, we can still re-derive the Jarrow and Yildirim model by recognizing that the volatility of $P_R(t, T_0, T)$ satisfies $\Sigma_R(t, t) = \sigma_I(t)$ and redoing the arguments. Based on the above analysis, we claim that the market model is consistent with the framework of the "foreign currency analogy."

In a two-currency economy, the exchange rate is usually stochastic with volatility. Having no volatility, the CPI behaves like a money market account instead of an exchange rate. Hence, we emphasize here that inflation derivatives modeling is not completely analogous to cross-currency derivatives modeling.

13.3   Pricing Inflation Derivatives

We have established for the first time that simple inflation forward rates are lognormal martingales under respective forward measures. As a result, the current practices on pricing some inflation derivatives must undergo some changes.

13.3.1   YYIIS

The price of a YYIIS is the difference in value of the fixed leg and floating leg. While the fixed leg is priced as an annuity, the floating leg is priced by discounting the expectation of each piece of payment:

$$
\begin{aligned}
V_{float}^{(j)}(t) &= Not.P(t, T_j)E_t^{Q_j}\left[\left(\frac{I(T_j)}{I(T_{j-1})} - 1\right)\right] \\
&= Not.\Delta T_j P(t, T_j)E_t^{Q_j}\left[f_j^{(I)}(T_j)\right] \\
&= Not.\Delta T_j P(t, T_j)f_j^{(I)}(t),
\end{aligned}
\tag{13.30}
$$

where we have made use of the martingale property of the inflation forward rates. The value of the floating leg is just a summation, and the value of the YYIIS is the difference between the values of the fixed and floating legs.

In the marketplace, YYIIS are treated as another set of securities parallel to ZCIIS, and the "inflation forward rates" implied by YYIIS and ZCIIS can be different. In existing literature, pricing YYIIS using a ZCIIS-based model goes through the procedure of "convexity adjustment," which is shown to be flawed. Our theory, for the first time, suggests that such differences should create arbitrage opportunities.

13.3.2 Caps and Floors

In view of the displaced diffusion processes for simple forward inflation rates, we can price a caplet with $1 notional value straightforwardly as follows:

$$
\Delta T_j E_t^Q \left[e^{-\int_t^{T_j} r_s ds} (f_j^{(I)}(T_j) - K)^+ \right]
$$

$$
= \Delta T_j P(t, T_j) E_t^{Q_j} \left[\left(\left(f_j^{(I)}(T_j) + \frac{1}{\Delta T_j} \right) - \left(K + \frac{1}{\Delta T_j} \right) \right)^+ \right] \quad (13.31)
$$

$$
= \Delta T_j P(t, T_j) \{ \mu_j(t) \Phi(d_1^{(j)}(t)) - \tilde{K}_j \Phi(d_2^{(j)}(t)) \},
$$

where $\Phi(\cdot)$ is the standard normal accumulative distribution function, and

$$
\mu_j(t) = f_j^{(I)}(t) + 1/\Delta T_j, \quad \tilde{K}_j = K + 1/\Delta T_j,
$$

$$
d_1^{(j)}(t) = \frac{\ln \mu_j / \tilde{K}_j + \frac{1}{2} \sigma_j^2(t)(T_j - t)}{\sigma_j(t) \sqrt{T_j - t}}, \quad d_2^{(j)}(t) = d_1^{(j)}(t) - \sigma_j(t) \sqrt{T_j - t},
$$

with $\sigma_j(t)$ to be the mean volatility of $\ln(f_j^{(I)}(t) + \frac{1}{\Delta T_j})$:

$$
\sigma_j^2(t) = \frac{1}{T_j - t} \int_t^{T_j} \| \gamma_j^{(I)}(s) \|^2 ds. \quad (13.32)
$$

Equation 13.31 is the Black's formula for inflation caplet.

The inflation-indexed cap with maturity T_N and strike K is the sum of a series of inflation-indexed caplets with the cash flows at T_j for $j = 1, \cdots, N$. Denote by $\text{IICap}(t; N, K)$ the price of the inflation-indexed cap at time $t < T_1$. Based on Equation 13.31, we have

$$
\text{IICap}(t; N, K) = \sum_{j=1}^{N} \Delta T_j P(t, T_j) \{ \mu_j(t) \Phi(d_1^{(j)}(t)) - \tilde{K}_j \Phi(d_2^{(j)}(t)) \}. \quad (13.33)
$$

The value of inflation-indexed floor with maturity T_N and strike K follows from the call-put parity, such that

$$
\text{IIFloor}(t; N, K) = \sum_{j=1}^{N} \Delta T_j P(t, T_j) \{ \tilde{K}_j \Phi(-d_2^{(j)}(t)) - \mu_j(t) \Phi(-d_1^{(j)}(t)) \}.
$$

$$
(13.34)
$$

Equation 13.31 is like an old bottle filled with new wine: the input inflation forward rates should be implied by ZCIIS instead of YYIIS. Given inflation caps of various maturities, we can consecutively bootstrap $\sigma_j(t)$, the "implied caplet volatilities" in either a parametric or a non-parametric way. With additional information on correlations between inflation rates of various maturities, we can determine $\gamma_j^{(I)}$, the volatility of inflation rates and thus fully specify the displace-diffusion dynamics for inflation forward rates. We may also include inflation swaption prices to the input set for determining $\gamma_j^{(I)}$'s.

13.3.3 Swaptions

The discussions on the pricing of inflation swaptions have been rare (Hinnerich, 2008). An inflation swaption is an option to enter into a YYIIS at the option's maturity. Without loss of generality, we consider here an underlying swap which has the same payment frequency for both fixed and floating legs. Similar to the situation of swaps on nominal interest rates, it is straightforward to show that the market prevailing inflation swap rate (that nullifies the value of a swap) is

$$S_{m,n}(t) = \frac{\sum_{i=m+1}^{n} \Delta T_i P(t, T_i) f_i^{(I)}(t)}{\sum_{i=m+1}^{n} \Delta T_i P(t, T_i)}. \tag{13.35}$$

The above expression can be recast into

$$S_{m,n}(t) + \frac{1}{\Delta T_{m,n}} = \sum_{i=m+1}^{n} \omega_i \mu_i(t), \tag{13.36}$$

for

$$w_i(t) = \frac{\Delta T_i P(t, T_i)}{A_{m,n}(t)}, \quad A_{m,n}(t) = \sum_{i=m+1}^{n} \Delta T_i P(t, T_i),$$

and

$$\frac{1}{\Delta T_{m,n}} = \sum_{i=m+1}^{n} w_i(t) \frac{1}{\Delta T_i}.$$

We have the following results on the dynamics of the swap rate.

Proposition 13.3.1. *The displaced forward swap rate* $S_{m,n}(t) + \frac{1}{\Delta T_{m,n}}$ *is a martingale under the swap measure* $\mathbb{Q}_{m,n}$ *corresponding to the numeraire* $A_{m,n}(t)$. *Moreover,*

$$d\left(S_{m,n}(t) + \frac{1}{\Delta T_{m,n}}\right) = \left(S_{m,n}(t) + \frac{1}{\Delta T_{m,n}}\right)$$
$$\times \sum_{i=m+1}^{n} \left[\alpha_i(t)\gamma_i^{(I)}(t) + (\alpha_i(t) - w_i(t))\Sigma_i(t)\right] \cdot d\mathbf{W}_t^{(m,n)}, \tag{13.37}$$

where $d\mathbf{W}_t^{(m,n)}$ *is a* $\mathbb{Q}_{m,n}$-*Brownian motion,* $\Sigma_i(t) \triangleq \Sigma(t, T_i)$, *and*

$$\alpha_i(t) = \frac{w_i(t)\mu_i(t)}{\sum_{j=m+1}^{n} w_j(t)\mu_j(t)}.$$

Proof. Differentiate Equation 13.36, we obtain the dynamics of the displaced swap rate as

$$d\left(S_{m,n}(t) + \frac{1}{\Delta T_{m,n}}\right) = \sum_{i=m+1}^{n} \mu_i(t)dw_i(t) + w_i(t)d\mu_i(t) + dw_i(t)d\mu_i(t). \tag{13.38}$$

One can easily show that

$$dw_i(t) = w_i(t)(\Sigma_i(t) - \Sigma_{m,n}(t)) \cdot (d\mathbf{W}_t - \Sigma_{m,n}(t)dt), \tag{13.39}$$

where $\Sigma_{m,n}(t) = \sum_{i=m+1}^{n} w_i \Sigma_i(t)$. Making use of Equation 13.23 and Equation 13.39, we obtain

$$d\left(\sum_{i=m+1}^{n} w_i(t)\mu_i(t)\right)$$

$$= \sum_{i=m+1}^{n} w_i(t)\mu_i(t)\left[(\Sigma_i(t) - \Sigma_{m,n}(t)) \cdot (d\mathbf{W}_t - \Sigma_{m,n}(t)dt)\right.$$

$$\left. + \gamma_i^{(I)}(t) \cdot (d\mathbf{W}_t - \Sigma_i(t)dt) + \gamma_i^{(I)}(t) \cdot (\Sigma_i(t) - \Sigma_{m,n}(t))dt\right]$$

$$= \sum_{i=m+1}^{n} w_i(t)\mu_i(t)\left(\Sigma_i(t) - \Sigma_{m,n}(t) + \gamma_i^{(I)}(t)\right) \cdot (d\mathbf{W}_t - \Sigma_{m,n}(t)dt)$$

$$= \left(\sum_{i=m+1}^{n} w_i(t)\mu_i(t)\right)$$

$$\times \left[\sum_{i=m+1}^{n} \alpha_i(t)\left(\gamma_i^{(I)}(t) + \Sigma_i(t)\right) - \Sigma_{m,n}(t)\right] \cdot (d\mathbf{W}_t - \Sigma_{m,n}(t)dt),$$

which is Equation 13.37.

$\square$

The martingale property of the swap rate is easy to see because it is the relative value between the floating leg and the annuity, the latter is the numeraire asset of the measure $\mathbb{Q}_{m,n}$, both are tradable securities.

By appropriately freezing coefficients of Equation 13.37, the displaced forward inflation swap rate $S_{m,n}(t) + \frac{1}{\Delta T_{m,n}}$ becomes a lognormal variable, and closed-form pricing of inflation swaptions will then follow. Consider a T_m-maturity swaption on the YYIIS over the period $[T_m, T_n]$ with strike K, we can derive its value as

$$V_t = A_{m,n}(t)\left[\left(S_{m,n}(t) + \frac{1}{\Delta T_{m,n}}\right)\Phi(d_1^{(m,n)}) - \tilde{K}_{m,n}\Phi(d_2^{(m,n)})\right], \tag{13.40}$$

where

$$\tilde{K}_{m,n} = K + \frac{1}{\Delta T_{m,n}},$$

$$d_1^{(m,n)} = \frac{\ln\left(S_{m,n}(t) + 1/\Delta T_{m,n}\right)/\tilde{K}_{m,n} + \frac{1}{2}\sigma_{m,n}^2(t)(T_m - t)}{\sigma_{m,n}(t)\sqrt{T_m - t}},$$

$$d_2^{(m,n)} = d_1^{(m,n)} - \sigma_{m,n}(t)\sqrt{T_m - t},$$

$$\sigma_{m,n}(t) = \frac{1}{T_m - t}\int_t^{T_m} \left\|\sum_{i=m+1}^{n}\left[\alpha_i(t)\gamma_i^{(I)}(s) + (\alpha_i(t) - w_i(t))\Sigma(s, T_i)\right]\right\|^2 ds.$$

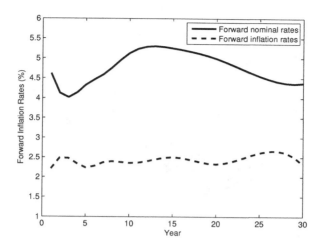

FIGURE 13.2: Term structure of the forward nominal rates and forward inflation rates.

Equation 13.40 is the Black's formula for inflation swaptions.

The swaption formula, Equation 13.40, implies a hedging strategy for the swaption. At any time t, the hedger should long $\Phi(d_1^{(m,n)})$ units of the underlying inflation swap for hedging. Proceeds from buying or selling the swap may go in or out of a money market account.

Finally, we emphasize that, with Black's formula (for displaced diffusion processes), inflation caps, floors, and swaptions can be quoted using implied volatilities, regardless the sign (i.e., either positive or negative) of input inflation forward rates or swap rates.

13.4 Model Calibration

A comprehensive calibration of the inflation-rate model, Equation 13.23, means the simultaneous determination of volatility vectors for inflation forward rates, based on market data of YYIIS, inflation caps and inflation swaptions. For non-parametric calibration, one can adopt the methodology for the calibration of the LIBOR market model developed by Wu (2003).

As a demonstration, we have calibrated the two-factor market model to the prices of Euro ZCIIS and (part of the) inflation caps as of April 7, 2008,[5] and observed good performance. Figure 13.2 shows the term structures of

---

[5]For brevity the data are not presented here, which are however available upon request.

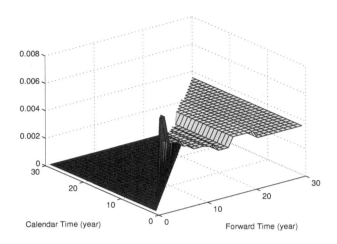

FIGURE 13.3: Calibrated local volatility surface, $\gamma_i^{(I)}(t)$.

inflation forward rates as well as nominal forward rates. Figure 13.3 shows the local volatility function obtained by calibrating the model to implied cap volatilities of various maturities but a fixed strike $K = 2\%$.

13.5 Smile Modeling

With the dynamics of displaced diffusions only, the market model cannot price volatility smiles in cap/floor markets. For that purpose, we should extend or modify the current model in ways parallel to the extensions to the SOFR market model we have made in Chapter 11. The most desirable solution at this point perhaps is to adopt the SABR (Hagen et al., 2002) dynamics for the expected displaced forward inflation rates, $\mu_j(t)$, and consider the following model:

$$\begin{cases} d\mu_j(t) = \mu_j^{\beta_j}(t)\gamma_j(t)dW_t^{(j)}, \\ d\gamma_j(t) = \nu_j\gamma_j(t)dZ_t^j, \quad \gamma_j(0) = \alpha, \end{cases} \tag{13.41}$$

where β_j and ν_j are constants, and both $W_t^{(j)}$ and $Z_t^{(j)}$ are one-dimensional (correlated) Brownian motions under the T_j-forward measure. Note that SABR model had been adopted for modeling $f_j^{(I)}(t)$, the inflation forward rates themselves (Mecurio and Mereni, 2009). Such a model, however, will become invalid once we have deflation in our economy.

Chapter 14

Market Model for Credit Derivatives

In the years before 2008, investors of credit markets had witnessed a rapid growth of liquidity in credit default swaps (CDS), options on the credit default swaps (or credit swaptions), tranches of collateralized debt obligations (CDO), as well as an exotic portfolio of credit derivatives like bespoke single-tranche CDOs and CDOs of CDOs (so-called CDO^2). As the property market heated up in many areas of the United States, speculations in credit derivatives drove the property prices to a staggering level, eventually ending in a crash that almost brought down the U.S. banking system. Following the 2008 financial crisis, the credit derivatives market shrank considerably. Still, some of the derivative instruments, including CDS and CDO, survived the crisis. In this chapter, we mostly address the pricing of CDS and credit swaptions, and use the model so developed in conjunction with copulas to price CDOs.

Various pricing theories have emerged in the pre-crisis markets. For the pricing of single-name swaptions, market practitioners and a series of researchers, including Schönbucher (2000), Arvanitis and Gregory (2001), Jamshidian (2002), and Hull and White (2003), among others, had led to the adoption of the Black formula as the market standard. For the pricing of CDO tranches, where the modeling of dependent defaults is essential, there are survival-time copula models represented by Li (2000), Gregory and Laurent (2003), Andersen, Sidenius and Basu (2003), and Giesecke (2003); as well as the structural models represented by Zhou (2001) and Hull et al. (2005). The Gaussian copula model formulated by Gregory and Laurent (2003) and later improved by Andersen et al. (2003) was well received and had once become the market standard. Despite the successes, there were major problems remaining. One fundamental problem, from a modeling point of view, was the detachment of the single-name credit derivatives market from the portfolio credit derivatives market, as there had been no unified pricing framework for both markets yet. As a result, consistent pricing and hedging across the two markets has been difficult to achieve.

There are also limitations with the popular portfolio credit derivatives models mentioned above. One of the major limitations of the survival-time copula approach for CDOs is that it does not constitute a proper dynamic model. Although the Gaussian copula model can take the default probabilities estimated from the CDS markets as inputs, it has no capacity to utilize either spread dynamics or spread correlations observed in the CDS markets for CDO pricing. The implication is that although this kind of models can be used to price CDO tranches (e.g., exponential-copula model of Giesecke (2003)) but not option on spreads of either single names or portfolios. With the Gaussian copula model, the market standard, default-time correlations are incorporated through mapping default times into normal random variables. There is, however, no economic intuition on how to input the correlations for these normal random variables. Structural models, on the other hand, do not have this problem, as under these models default is triggered when the firm value breaches a barrier (Black and Cox, 1976; Hull *et al.*, 2005). However, structural models are less convenient to use, as model users have to specify a number of model parameters, including drifts, volatility and correlations of the firm values, as well as default barriers. Given that firm values are not observable, a proper specification of the model is a daunting task by itself. Moreover, firm-value-based structural models do not naturally take the price information like credit spreads and implied spread volatilities from the single-name credit derivatives markets as inputs, hence, they do not ensure price consistency across the single-name and portfolio credit markets.

In this chapter, we deliver a unified framework of Ho and Wu (2008) for pricing the CDS options and the CDOs. Motivated by the use of Black's formula for single-name swaptions, we first establish a market model with forward hazard rates. Unlike usual structural models with firm values, the state variables of our market model are indirectly observable quantities. For the pricing of single-name credit derivatives, the new model is highly analogous to the SOFR market model. By utilizing copulas for default-time correlation, the model so developed can conveniently be applied to pricing CDO tranches, using Monte Carlo simulations. Dynamically evolving CDS rates, CDS-rate correlations, and implied volatilities of credit swaptions, which are all observable in credit markets, can all be utilized for pricing the co-dependence of defaults of credit portfolios.

This chapter is organized as follows. Sections 14.1 through 14.6 concern single-name credit derivatives. In Section 14.1, we present the definition of risky zero-coupon bonds and explain the intuition behind this definition. In Section 14.2, we define risky forward rates and forward spreads. In Section 14.3, we deliver swap-rate formulae for two types of credit swaps. Section 14.4 is devoted to the calculation of par CDS rates of both floating-rate and fixed-rate bonds. In Section 14.5, we demonstrate the estimation of implied survival rates and implied recovery rates. In Section 14.6, we present the market model for single-name swaptions. Through Sections 14.1 to 14.6, we lay down the foundation for the pricing of portfolio credit derivatives. In Section 14.7, we introduce CDOs, describe the Monte Carlo method for CDO pricing under the market model, and demonstrate the capacity of the model for accommodating spread correlations and default-time correlations. Pricing examples with standardized CDOs will be presented. Finally in Section 14.8, we provide some ending remarks.

14.1 Pricing of Risky Bonds: A New Perspective

Without loss of generality, we model the default time as the first jump time of a Cox process with an intensity (or hazard rate) process, $\lambda(t)$. For technical convenience and clarity of presentation, we assume independence between credit spreads and U.S. Treasury yields, as well as independence between hazard rate and the recovery rate of the credit (with the understanding that this may be against some empirical findings, e.g., Duffee (1998)). We also assume a single seniority across all bonds under the same credit name.

We begin with bond pricing. A defaultable coupon bond pays regular coupons until default or maturity, whichever comes first. In case of a default, the market convention is that a creditor will receive a final payment that consists of a fraction of both the principal and the last accrued interest. The schedule of the final payment varies. Without loss of generality, we assume that (1) the final payment is made at the next coupon date following the default, and (2) the last coupon accrues until the final payment date.[1] Let c be the coupon rate of a risky bond with term[2] $[T_m, T_n]$, τ be the default time and R_τ be the recovery rate at the default time. Then, the cash flow at $T_j, j = m+1, \ldots, n$ can be expressed as

$$\Delta T c \mathbf{1}_{\{\tau > T_j\}} + R_\tau (1 + \Delta T c) \mathbf{1}_{\{T_{j-1} < \tau \leq T_j\}},$$

where $\Delta T = 1/(\text{coupon frequency})$ and $\mathbf{1}_{\tau \in \Omega}$ is the indicator function that equals to 1 if $\tau \in \Omega$, or 0 otherwise. According to the arbitrage pricing theory (APT) (Harrison and Pliska, 1981), the bond is then priced as the risk-neutral expectation of discounted cash flows:

$$B^c(t) = \sum_{j=m+1}^{n} E_t^Q \left[\frac{B(t)}{B(T_j)} \left\{ \Delta T_j c \mathbf{1}_{\{\tau > T_j\}} + R_\tau (1 + \Delta T_j c) \mathbf{1}_{\{T_{j-1} < \tau \leq T_j\}} \right\} \right]$$

$$+ E_t^Q \left[\frac{B(t)}{B(T_n)} \mathbf{1}_{\{\tau > T_n\}} \right]$$

$$= \sum_{j=m+1}^{n} P_j(t) \Delta T_j c E_t^{Q_j} \left[\mathbf{1}_{\{\tau > T_j\}} + R_\tau \mathbf{1}_{\{T_{j-1} < \tau \leq T_j\}} \right]$$

$$+ \sum_{j=m+1}^{n} P_j(t) E_t^{Q_j} \left[R_\tau \mathbf{1}_{\{T_{j-1} < \tau \leq T_j\}} \right] + P_n(t) E_t^{Q_n} \left[\mathbf{1}_{\{\tau > T_n\}} \right],$$

$$(14.1)$$

where we write $P_j(t) = P(t, T_j)$ for simplicity. The expression for payment

---

[1]The protection payment made only at a coupon date is a harmless idealization. There are other default payment schedules as well. For example, the last payment may occur at $\tau + 90$ days. Most payment schedules can be accommodated by adjusting the recovery rate.

[2]The coupon dates are $\{T_j\}_{j=m+1}^n$. The bond is said to be "forward starting" if $m > 0$.

upon a default, $R_\tau(1 + \Delta T_j c)$, conforms well with industry practice: the compensation to the creditor is determined by the outstanding principal and the accrued interest of the defaulted bond, which are treated as in the same asset class, and future coupons are not taken into consideration (Schönbucher, 2004). The second line of Equation 14.1 results from the changes of measures, followed by a regrouping of the PVs of the coupons and the principal. The payout at time T_j is priced by using the T_j-forward measure.

Parallel to the U.S. Treasury market, we introduce here the "C-strip" and the "P-strip" of risky zero-coupon bonds, which are backed separately by coupons and principals. It is not hard to see that the C-strip zero-coupon bonds should be defined as

$$
\begin{aligned}
\bar{P}_j(t) &= E_t^Q \left[\frac{B(t)}{B(T_j)} \left(1_{\{\tau > T_j\}} + R_\tau 1_{\{T_{j-1} < \tau \le T_j\}} \right) \right] \\
&= P_j(t) E_t^{Q_j} \left[1_{\{\tau > T_j\}} + R_\tau 1_{\{T_{j-1} < \tau \le T_j\}} \right] \stackrel{\triangle}{=} P_j(t) D_j(t),
\end{aligned}
\tag{14.2}
$$

while the "P-strip" zero-coupon bonds are defined as in the last line of Equation 14.1. Here in Equation 14.2, a new variable $D_j(t)$ is defined. Unlike their Treasury counterparts, the risky zero-coupon bonds of the two strips have apparently different cash-flow structures. Given only the prices of risky coupon bonds, we cannot identify the PVs of zero-coupon bonds in either strip unless additional information is provided or further assumptions are made.

In principle, zero-coupon bonds of the two strips can be generated through marketing the cash flows of the coupons and the principals separately. In reality, existing risky zero-coupon bonds are in the form of commercial papers, which only have very short maturities. As we shall see, with additional information like the CDS rates, we are able to extract the prices of risky zero-coupon bonds of both strips from the prices of associated coupon bonds. The notion of C-strip zero-coupon bonds, in particular, is important for the construction of our model.

14.2   Forward Spreads

To understand the product nature of CDS, we need to clarify the notion of "risky forward rates" adopted earlier by Schönbucher (2000) and Brigo (2005). A risky forward rate should be defined as the fair rate on a defaultable loan for a future period of time, say, $(T_{j-1}, T_j]$, that is backed by the coupon flows of defaultable bonds of an entity. If a default of the bond occurs before T_{j-1}, the loan ceases to exist. If a default occurs between T_{j-1} and T_j, then a recovery value proportional to the recovery rate of the bonds applies. Assume the notional of the loan to be \$1. According to the APT, the risky forward rate, denoted as $\hat{f}_j(t)$, must nullify the PV of the cash flows of the risky loan:

$$
0 = E_t^Q \left[\frac{B(t)}{B(T_{j-1})} \mathbf{1}_{\{\tau > T_{j-1}\}} \right]
$$

$$
- E_t^Q \left[\frac{B(t)}{B(T_j)} (1 + \Delta T_j \hat{f}_j(t))(\mathbf{1}_{\{\tau > T_j\}} + R_\tau \mathbf{1}_{\{T_{j-1} < \tau \le T_j\}}) \right] \quad (14.3)
$$

$$
= P_{j-1}(t) E_t^{Q_{j-1}} \left[\mathbf{1}_{\{\tau > T_{j-1}\}} \right] - P_j(t) D_j(t)(1 + \Delta T_j \hat{f}_j(t))
$$

$$
\stackrel{\triangle}{=} P_{j-1}(t) \Lambda_{j-1}(t) - P_j(t) D_j(t)(1 + \Delta T_j \hat{f}_j(t)),
$$

where $\Lambda_{j-1}(t)$ is the $\mathbb{Q}_{j-1}$ probability of survival until T_{j-1}, and it is equal to the $\mathbb{Q}$ probability of survival until T_j given the independence between U.S. Treasury yields and the default probability of the entity. Equation 14.3 gives rise to

$$
\hat{f}_j(t) = \frac{1}{\Delta T_j} \left[\frac{P_{j-1}(t)}{P_j(t)} \frac{\Lambda_{j-1}(t)}{D_j(t)} - 1 \right]
$$

$$
= \frac{1}{\Delta T_j} \left[\frac{\bar{P}_{j-1}(t)}{\bar{P}_j(t)} \frac{\Lambda_{j-1}(t)}{D_{j-1}(t)} - 1 \right]. \quad (14.4)
$$

The two lines in the equation above lead to two alternative expressions of $\hat{f}_j(t)$. The first expression is in terms of the SOFR term rates,

$$
\hat{f}_j(t) = \frac{1}{\Delta T} \left[\frac{P_{j-1}(t)}{P_j(t)} - 1 \right] + \frac{1}{\Delta T} \frac{P_{j-1}(t)}{P_j(t)} \left(\frac{\Lambda_{j-1}(t)}{D_j(t)} - 1 \right)
$$

$$
= f_j(t) + \frac{1}{\Delta T} \frac{P_{j-1}(t)}{P_j(t)}
$$

$$
\times \left(\frac{E_t^{Q_j} \left[\mathbf{1}_{\{\tau > T_{j-1}\}} \right] - E_t^{Q_j} \left[\mathbf{1}_{\{\tau > T_j\}} + R \mathbf{1}_{\{T_{j-1} < \tau \le T_j\}} \right]}{D_j(t)} \right) \quad (14.5)
$$

$$
= f_j(t) + \frac{1}{\Delta T} \frac{P_{j-1}(t)}{P_j(t)} \left(\frac{E_t^{Q_j} \left[(1 - R)\mathbf{1}_{\{T_{j-1} < \tau \le T_j\}} \right]}{D_j(t)} \right)
$$

$$
= f_j(t) + \frac{1 + \Delta T_j f_j(t)}{\Delta T} \left(\frac{E_t^{Q_j} \left[(1 - R)\mathbf{1}_{\{T_{j-1} < \tau \le T_j\}} \right]}{D_j(t)} \right),
$$

where $f_j(t)$ is the default-free SOFR forward rate for the period (T_{j-1}, T_j) seen at time t, defined by

$$
f_j(t) = \frac{1}{\Delta T_j} \left(\frac{P_{j-1}(t)}{P_j(t)} - 1 \right), \quad t \le T_j.
$$

The second expression for the risky forward rate is

$$
\hat{f}_j(t) = \frac{1}{\Delta T_j} \left[\frac{\bar{P}_{j-1}(t)}{\bar{P}_j(t)} - 1 \right] - \frac{1}{\Delta T_j} \frac{\bar{P}_{j-1}(t)}{\bar{P}_j(t)} \left[1 - \frac{\Lambda_{j-1}(t)}{D_{j-1}(t)} \right]
$$

$$
= \bar{f}_j(t) - \frac{1 + \Delta T_j \bar{f}_j(t)}{\Delta T_j} \left(\frac{E_t^{Q_j} \left[R \mathbf{1}_{\{T_{j-1} < \tau \le T_j\}} \right]}{D_{j-1}(t)} \right), \quad (14.6)
$$

where

$$\bar{f}_j(t) = \frac{1}{\Delta T_j} \left(\frac{\bar{P}_{j-1}(t)}{\bar{P}_j(t)} - 1 \right), \quad t \leq T_j.$$

is called the "defaultable effective forward rate" (Schönbucher, 2000). Note that $\bar{f}_j(t)$ should be understood as the effective rate of return over (T_{j-1}, T_j) provided that no default occurs until T_{j-1}. It is pointed out in Brigo (2004) that $\bar{f}_j(t)$ in general does not link directly to a financial contract. Putting Equation 14.5 and Equation 14.6 together, we have the order

$$f_j(t) \leq \hat{f}_j(t) \leq \bar{f}_j(t). \tag{14.7}$$

Note that $\hat{f}_j(t)$ achieves the upper bound and the lower bound when $R_\tau = 0$ and $R_\tau = 1$, respectively. The bounds on $\hat{f}_j(t)$ should be regarded as no-arbitrage constraints.

Here, we reiterate the insight of "non-separability" pointed out by Duffie and Singleton (1999), which means that the hazard rate and the loss rate cannot be simultaneously determined from bond prices alone. In view of Equation 14.4, we can say that complete term structures of the survival probability and the recovery rate, $\Lambda_j(t)$ and $E_t^Q [R_\tau | T_{j-1} < \tau \leq T_j]$, $j = 1, 2, \ldots$, can be uniquely determined from the term structures of $\hat{f}_j(t)$ and $\bar{f}_j(t)$. These in fact, as we soon shall demonstrate, are informatively equivalent to the term structures of risky-bond yields and CDS rates.

Intuitively, a "forward spread" is defined as the difference between a risky forward rate and its corresponding risk-free forward rate:

$$S_j(t) = \hat{f}_j(t) - f_j(t), \quad j = 1, 2, \ldots.$$

From Equation 14.5, we obtain

$$S_j(t) = (1 + \Delta T_j f_j(t)) \frac{E_t^{Q_j} \left[(1 - R_\tau) \mathbf{1}_{\{T_{j-1} < \tau \leq T_j\}} \right]}{\Delta T_j D_j(t)} \tag{14.8}$$

$$\triangleq (1 + \Delta T_j f_j(t)) H_j(t).$$

According to its definition, $H_j(t)$ can be interpreted as the "expected loss per risky dollar over $(T_{j-1}, T_j]$." Equation 14.8 can be rewritten as

$$1 + \Delta T_j \hat{f}_j(t) = (1 + \Delta T_j f_j(t)) (1 + \Delta T_j H_j(t)), \quad j = 1, 2, \ldots.$$

By comparing definitions, we can say that $H_j(t)$ is the discrete-tenor version of the "mean loss rate" introduced in Duffie and Singleton (1999).

14.3 Two Kinds of Default Protection Swaps

A default protection swap consists of a fee leg (or premium leg) and a protection leg. Before the default of the reference entity, the protection buyer pays the protection seller a string of fees at regular time intervals. Upon default, the protection buyer either delivers the bond to the protection seller in exchange for par (so-called physical settlement), or receives from the protection seller a payment that is equal to the loss incurred (so-called cash settlement). In this section, we consider two kinds of swaps: swaps for fixed-rate bonds and swaps for floating-rate bonds. Without loss of generality, we assume the notional value of the bonds to be \$1, and the coupon rate to be c or SOFR term rate, $f_j(T_j)$, respectively. For the swap on the fixed-rate bond, the protection payment is $(1 - R_\tau)(1 + \Delta T c)$, while for the swap on the floating-rate bond, the protection payment is $(1 - R_\tau)(1 + \Delta T_j f_j(T_j))$. The payments simply reflect the loss to the bond holders in case of a default.[3] Note that the swaps of the second kind depend only on the default status of the reference entity.

We proceed to the determination of the fair rate for the default swap of the first kind. Let us denote a swap rate by $\bar{s}$. In the CDS markets, the contractual cash flows of the fee leg (for the protection of \$1 notional) are typically

$$\bar{s} \Delta T_j \left[\mathbf{1}_{\{\tau > T_j\}} + \frac{(\tau - T_{j-1})}{\Delta T_j} \mathbf{1}_{\{T_{j-1} < \tau \leq T_j\}} \right], \quad j = 1, 2, \ldots,$$

i.e., in case of a default occurring between T_{j-1} and T_j, the protection buyer makes the final payment that is proportional to the time elapsed between the last fee payment and the default. From a financial engineering point of view, such a fee specification is troubling because the cash flow cannot be synthesized at the initiation of the swap contract. To make the swap contract replicable, we propose a slight modification to the fee leg: we let the cash flows be generated from the payout of risky zero-coupon bonds:

$$\bar{s} \Delta T_j \left[\mathbf{1}_{\{\tau > T_j\}} + R_\tau \mathbf{1}_{\{T_{j-1} < \tau \leq T_j\}} \right].$$

This modified definition only changes the last piece of fee payment after default, yet it allows us to write the value of the fee leg as

$$
\begin{aligned}
PV_{fee} &= \bar{s} \sum_{j=m+1}^{n} \Delta T_j P_j(t) E_t^{Q_j} \left[\mathbf{1}_{\{\tau > T_j\}} + R_\tau \mathbf{1}_{\{T_{j-1} < \tau \leq T_j\}} \right] \\
&= \bar{s} \sum_{j=m+1}^{n} \Delta T_j \bar{P}_j(t).
\end{aligned}
\tag{14.9}
$$

---

[3] The quantity $(1 + \Delta T_j f_j(T_j))$ can be regarded as "a constant dollar" seen at time T_j.

Let

$$\bar{A}_{m,n}(t) := \sum_{j=m+1}^{n} \Delta T_j \bar{P}_j(t),$$

which is now a tradable annuity, analogous to the fixed leg of default-free swaps.

For a swap of the first kind, the cash flow of the protection seller can be written as

$$V_{prot} = (1 + \Delta T c) \sum_{j=m+1}^{n} (1 - R_\tau) \mathbf{1}_{\{T_{j-1} \leq \tau \leq T_j\}}.$$

The PV of the protection payment is then

$$PV_{prot} = (1 + \Delta T c) \sum_{j=m+1}^{n} P_j(t) E_t^{Q_j} \left[(1 - R_\tau) \mathbf{1}_{\{T_{j-1} \leq \tau \leq T_j\}} \right]. \qquad (14.10)$$

By equating the values of the fee leg and the protection leg and making use of Equation 14.8, we obtain

$$\bar{s}_1 = (1 + \Delta T c) \frac{\sum_{j=m+1}^{n} P_j(t) E_t^{Q_j} \left[(1 - R_\tau) \mathbf{1}_{\{T_{j-1} \leq \tau \leq T_j\}} \right]}{\bar{A}_{m,n}(t)}$$

$$= (1 + \Delta T c) \sum_{j=m+1}^{n} \bar{\alpha}_j H_j(t), \qquad (14.11)$$

where

$$\bar{\alpha}_j = \frac{\Delta T_j \bar{P}_j(t)}{\bar{A}_{m,n}(t)}.$$

Note that the case $c = 0$ corresponds to a prototypical default swap, which only depends on the default status of the reference entity and actually dominates the liquidity of single-name credit derivatives markets.

For credit default swaps of the second kind, the fair swap rate can be derived analogously as

$$\bar{s}_2 = \frac{\sum_{j=m+1}^{n} P_j(t)(1 + \Delta T_j f_j(t)) E_t^{Q_j} \left[(1 - R_\tau) \mathbf{1}_{\{T_{j-1} \leq \tau \leq T_j\}} \right]}{\bar{A}_{m,n}(t)}$$

$$= \sum_{j=m+1}^{n} \bar{\alpha}_j S_j(t). \qquad (14.12)$$

Here, we have made use of the independence between credit spreads and the U.S. Treasury yields, as well as the martingale property of the forward rate: $E_t^{Q_j}[f_j(T_j)] = f_j(t)$. Note that a one-period CDS rate reduces to the forward spread.

Compared with the existing CDS rate formula (e.g., Schönbucher, 2004; Brigo, 2005), our definition simply states that a CDS rate is equal to the weighted average of credit spreads, which does not require the recovery rate as an input, and is analogous to the swap rate formula in the SOFR markets.

14.4 Par CDS Rates

A par CDS rate is a spread that, when added to a corresponding par rate of default-free bond, yields the coupon rate of a risky par bond. We will derive the par CDS rates for both floating rate bonds and fixed-rate bonds.

Typically, a defaultable floating-rate bond (which is also called a default-able floater) pays daily compounded SOFR plus a credit spread, denoted by s_F, until a default occurs, when a holder may obtain some recovered value of both principal and coupon. Hence, the cash flow of a defaultable floater at T_j can be written as

$$CF_j = (1 + \Delta T_j(f_j(T_j) + s_F)) \left(\mathbf{1}_{\{\tau > T_j\}} + R\mathbf{1}_{\{T_{j-1} < \tau \leq T_j\}}\right)$$
$$- \mathbf{1}_{\{\tau > T_j, j < n\}}.$$

The question here is: if the floater is to be priced at par, what should the fair spread rate s_F be? To answer this question, we imagine that the holder of the floater is also "long" a protection swap of the second kind. Then, his/her cash flow at time T_j is

$$CF_j = \Delta T_j f_j(T_j)\mathbf{1}_{\{\tau > T_j\}} + \mathbf{1}_{\{T_{j-1} < \tau \leq T_j\}} + \mathbf{1}_{\{\tau > T_n, j = n\}}$$
$$+ \Delta T_j(s_F - \bar{s}_2)(\mathbf{1}_{\{\tau > T_j\}} + R\mathbf{1}_{\{T_{j-1} < \tau \leq T_j\}}). \tag{14.13}$$

The first line in Equation 14.13 gives the cash flow of a rolling-forward CD that lasts until $T_{j_\tau} \wedge T_n$, where T_{j_τ} is the first fixing date after default, and such cash flows represent those of a par bond (that matures at $T_{j_\tau} \wedge T_n$). It then becomes clear that, for the defaultable floater to be priced at par, there must be

$$s_F = \bar{s}_2,$$

i.e., the default swap rate equals nothing else but the credit spread!

Given the clear relationship between a defaultable floater and a CDS, we then come up with the following hedging strategy for swaps of the second kind: once such a swap is written, the hedger goes "long" a default-free floater and goes "short" a defaultable floater, both at par. Then, the net cash flow at any fixing date is zero.

Next, we derive the par CDS rate for a corresponding fixed-rate bond with tenor $(T_m, T_n]$. We let s_X denote the par CDS rate for the fixed-rate bond. Then, s_X can be determined by equating the PVs of fixed-rate and floating-rate par bonds:

$$(R_{m,n}(t) + s_X)\bar{A}_{m,n}(t) = \sum_{j=m+1}^{n} \Delta T_j \bar{P}_j(t) [f_j(t) + \bar{s}_2]$$
$$= \sum_{j=m+1}^{n} \Delta T_j \bar{P}_j(t)\hat{f}_j(t),$$

where the PVs of the principals have been canceled, and $R_{m,n}(t)$ represents the corresponding swap rate (i.e., par rate) in SOFR derivatives markets, defined by

$$R_{m,n}(t) = \sum_{j=m+1}^{n} \alpha_j f_j(t),$$

It follows that

$$s_X = \sum_{j=m+1}^{n} \bar{\alpha}_j \hat{f}_j(t) - R_{m,n}(t)$$

$$= \bar{s}_2 + \sum_{j=m+1}^{n} (\bar{\alpha}_j - \alpha_j) f_j(t).$$

Hence, the so-called "credit spread" is different for floating-rate bonds and fixed-rate bonds, and the par CDS rate for fixed-rate bonds is close to $\bar{s}_2$ instead of $\bar{s}_1$ (when $c = 0$).

It can be verified that the short position of the default swap on the risky coupon bond with coupon rate c can be hedged by

1. being "short" the risky coupon bond,

2. being "long" a risk-free floater at par, and

3. being "long" $(c - \bar{R}_{m,n}(t))$ units of the risky annuity, $\bar{A}_{m,n}(t)$.

Here, $\bar{R}_{m,n}(t) = R_{m,n}(t) + s_X$ is the risky par yield.

14.5   Implied Survival Curve and Recovery-Rate Curve

In reality, neither $\{\hat{f}_j(t)\}$ nor $\{\bar{f}_j(t)\}$ is directly observable. For CDS pricing and other applications, it is more convenient to make use of the term structures of forward hazard rate and forward recovery rate. We define the forward hazard rate for $(T_{j-1}, T_j]$ seen at time t by

$$\lambda_j(t) = \frac{1}{\Delta T_j} \left(\frac{\Lambda_{j-1}(t)}{\Lambda_j(t)} - 1 \right), \quad \text{for } t \le T_j, \tag{14.14}$$

and the forward recovery rate for the same period by

$$R_j(t) = E_t^Q \left[R_\tau | T_{j-1} < \tau \le T_j \right], \quad \text{for } t \le T_j, \tag{14.15}$$

TABLE 14.1: Citigroup CDS Rates (7/28/2005, Bloomberg)

Maturity	1Y	3Y	5Y	10Y
Rates	0.07%	0.13%	0.19%	0.33%

TABLE 14.2: Prices of Benchmark Citigroup Bonds (7/28/2005, Bloomberg)

Maturity	Frequency	Coupon	Price
22/2/2010	Semi-annual	4.125%	98.123
1/10/2010	Semi-annual	7.25%	114.563
7/5/2015	Semi-annual	4.875%	97.563
18/5/2010	Quarterly	US LIB+15bps	99
16/3/2012	Quarterly	US LIB+12.5bps	99.80
5/11/2014	Quarterly	US LIB+28bps	100.501

and define $\eta_t = \max\{j|T_j \le t\}$. Then the survival probability $\Lambda_j(t)$ is related to the hazard rates by

$$\Lambda_j(t) = (1 + (t - T_{\eta_t})\lambda_{\eta_t}(T_{\eta_t}))^{-1} \prod_{k=\eta_t+1}^{j} (1 + \Delta T_k \lambda_k)^{-1}. \tag{14.16}$$

The standard market practice is to back out the survival probabilities from CDS of various maturities, assuming a constant recovery rate (of 40%). Instead of doing the same thing with our model, we consider simultaneously backing out the implied hazard rates and recovery rates from CDSs as well as corporate bond prices. We choose Citigroup as the credit name for a demonstration, as there is relatively richer credit information on this company. A snapshot of market quotations is provided in Tables 14.1 and 14.2, where the currency is U.S. dollars (USD), the CDS rates are for $c = 0$, and the bond prices are "clean." To build the risk-free discount curve in USD, we have used SOFR term rates up to one year, and swap rates from 2 to 20 years. The interest rate information is provided in Table 14.3. Figure 14.1 presents the forward-rate curve constructed using the data of SOFR term rates and SOFR swap rates in Table 14.3.

We determine $\{\lambda_j, R_j\}$ through reproducing the CDS rates and the bond prices of Citigroup by the swap-rate formula Equation 14.11 and the bond formula Equation 14.1, respectively. Because the problem is under-determined, we have adopted cubic-spline and linear interpolation for the hazard rates and the recovery rates, respectively, and imposed additional smoothness regularization. In our search algorithm, we took various initial guesses with $\lambda_j = 0$ and $0.0 \le R_j \le 0.6, \forall j$. Often but not always, the search ends up in one of the two solutions, depicted in Figures 14.2 to 14.4, depending on the closeness of the initial recovery rate to either $R_0 = 0.0$ or $R_0 = 0.4$. Existence of more

TABLE 14.3: SOFR term rates and SOFR swap rates (7/28/2005, Bloomberg)

SOFR	3 mth	3.6931%
	6 mth	3.8435%
	12 mth	4.1731%
Swap	2Y	4.3200%
	3Y	4.3840%
	4Y	4.4330%
	5Y	4.4690%
	7Y	4.5365%
	10Y	4.6290%
	12Y	4.6905%
	15Y	4.7630%
	20Y	4.8320%

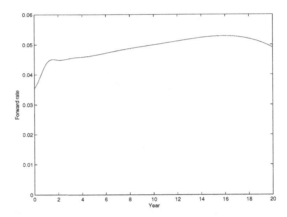

FIGURE 14.1: USD forward rates (7/28/2005).

than one solution reflects the ill-posedness nature of the calibration problem, particularly with regard to the determination of the recovery rate.

We remark here that the ill-posedness nature of the calibration problem is largely intrinsic. It is due to the insensitivity of the bond and CDS prices with respect to the change of the recovery rates,[4] particularly for a high-credit name. As a consequence of the ill-posedness, implied recovery rates and default rates may depend on initial guesses, as is shown in Figure 14.2 and 14.3. To settle down to a single solution, additional regularizations (based on financial or mathematical considerations) are needed. On the other hand, the ill-posedness may not be a great concern. As is shown in Figure 14.4, risky forward-rate curve and risky discount curve demonstrate noticeable stability.

---

[4]The insensitivity is also observed in Brigo (2005).

TABLE 14.4: GM CDS Rates (7/28/2005, Bloomberg)

Maturity	1Y	3Y	5Y	10Y
Rates	—	—	5.2117%	—

This is due to the complimentary effect between the hazard rates and the recovery rates in calibration.

By comparing the implied hazard rates with the CDS rates of Citigroup, we can say that the market quotations of CDS rates pretty much represent the risk-neutral hazard rates for zero recovery upon default. The quoted CDS spreads increase with time to maturity, and a risk-neutral hazard rate will also increase with its maturity.

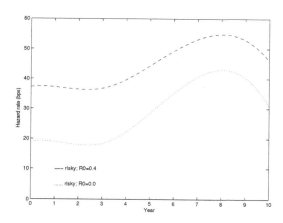

FIGURE 14.2: Implied hazard rates.

Next, we will demonstrate that when CDS rates are high, the calibration can become a well-posed problem. An example is with General Motors (GM). It had been a widely known credit event in July 2005, when the five-year CDS rate of GM shot up several hundred basis points, as shown in Table 14.4, and the yield spreads of its bonds also went up, see Table 14.5. When we calibrate the hazard rate and recovery rate of GM, we obtain almost identical results, see Figures 14.5 and 14.6, regardless of the choice of initial guesses taken for calibration. As noted earlier, CDS rates essentially reflect the hazard rates. When the hazard rates are high, the CDS rates become sensitive to the recovery rate, thus making the calibration a well-posed problem.

A remedy for the ill-posedness of the calibration can be the adoption of recovery swap quotes into the input set. A recovery swap has the following features.

- It is an agreement to swap a fixed recovery rate for a real recovery rate following a credit event.

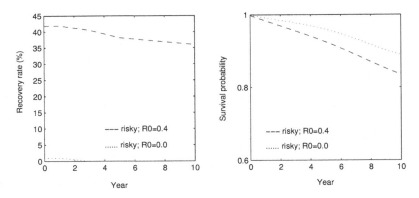

FIGURE 14.3: Implied default rates and recovery rates.

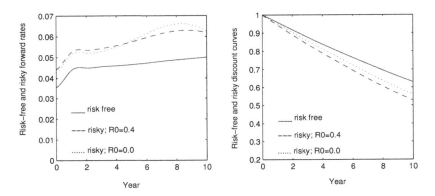

FIGURE 14.4: Forward rates and discount curves.

- It is usually traded as zero-premium credit default swaps with the reference price set at the fixed recovery rate at certain notional value.

- When no default occurs, the recovery swap expires worthlessly.

Example 14.1. *A dealer might quote a recovery swap in GM at 37/40. This means the dealer is prepared to sell a recovery swap with RS = 37% and buy at RS = 40%.*

The liquidity of the recovery swaps rose during the 2011 European debt crisis. If we could take the recovery swaps as part of the input set, all we then need to do is to back out the implied hazard rates. Unfortunately, recovery swaps usually trade only on credits that are nearing default.

TABLE 14.5: Prices of Benchmark GM Bonds (7/28/2005, Bloomberg)

Maturity	Frequency	Coupon	Price
3/15/2006	Semi-annual	7.1%	101.405
5/1/2008	Semi-annual	6.375%	98.62
4/15/2016	Semi-annual	7.7%	90.41

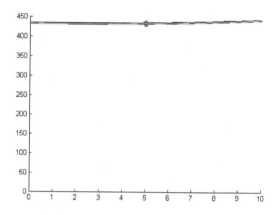

FIGURE 14.5: Implied hazard rates.

14.6 Black's Formula for Credit Default Swaptions

For either the protection buyer or the protection seller, an open position can be closed by either going "short" or going "long" of the same swap. Because the protection legs are exactly offset, the profit or loss for the pair of transactions comes from the difference of the fee legs, which is

$$(\bar{s}_{m,n}(t) - \bar{s}_{m,n}(0)) \sum_{j=m+1}^{n} \Delta T_j \bar{P}_j(t), \qquad (14.17)$$

where $\bar{s}_{m,n}(t)$ is the prevailing CDS rate, either $\bar{s}_1$ or $\bar{s}_2$, seen at time $t \leq T$.

A credit swaption or CDS option, meanwhile, is a contract that gives its holder the right, not the obligation, to enter into a forward-starting swap at time $T \leq T_m$ with a predetermined swap rate, $\bar{s}^*$. Thus the profit/loss at maturity T of the swaption is

$$(\bar{s}_{m,n}(T) - \bar{s}^*)^+ \sum_{j=m+1}^{n} \Delta T_j \bar{P}_j(T). \qquad (14.18)$$

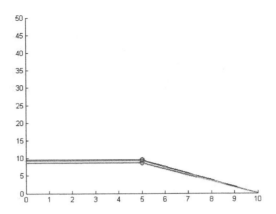

FIGURE 14.6: Implied recovery rates.

To price the CDS option (or credit swaption), we need to make a change of measures that are not equivalent but absolutely continuous.

Theorem 14.6.1 (Radon-Nikodyn). *Let* $\mathbb{P}$ *and* $\mathbb{Q}$ *be two probability measures on* $(\Omega, \mathcal{F}_{t \geq 0})$ *with* $\mathbb{Q} \ll \mathbb{P}$ *(*$\mathbb{Q}$ *absolutely continuous with respect to* $\mathbb{P}$*). Then there exists a unique* $\mathbb{P}$*-martingale* L_t *such that for all* $t \geq 0$ *and all* $\mathcal{F}_t$*-measurable and* $\mathbb{Q}$*-integrable random variable* Y*, we have*

$$E^{\mathbb{Q}}[Y] = E^{\mathbb{P}}[YL].\tag{14.19}$$

Denote

$$L_t = \left.\frac{d\mathbb{Q}}{d\mathbb{P}}\right|_{\mathcal{F}_t},\tag{14.20}$$

which satisfies

1. $L_t \geq 0$.

2. *Under* $\mathbb{Q}$*,* L_t *is almost surely bounded away from zero:* $\mathbb{Q}(\inf_{t \geq 0} L_t > 0) = 1$.

3. *If* M_t *is continuous from the right with a left limit and adaptive,* M_t *is a* $\mathbb{Q}$*-martingale iff* $L_t \times M_t$ *is a* $\mathbb{P}$*-martingale. In particular, for* $0 \leq t \leq T$:

$$L_t E^{\mathbb{Q}}[M_T|\mathcal{F}_t] = E^{\mathbb{P}}[L_T M_T|\mathcal{F}_t].$$

4. *Every non-negative, integrable* $\mathbb{P}$*-martingale* L *with* $L(0) = 1$ *defined a measure* $\mathbb{Q}$ *by Equation 14.19.*

For a proof, we refer readers to Protter (1990).

Corollary 14.6.1. *Let $A_t \geq 0$ be a dividend-free asset and $\mathbb{Q}$ a spot martingale measure (corresponding to money market account $B(t)$ as numeraire). Define a new measure corresponding to A_t by*

$$L_t = \left. \frac{d\mathbb{Q}_A}{d\mathbb{Q}} \right|_{\mathcal{F}_t} = \frac{A_t}{A_0} \bigg/ \frac{B(t)}{B(0)} 1_{A_0 > 0}.$$

Then, the A_t-relative price of any default-related asset:

$$X_t^A = \frac{X_t}{A_t} 1_{A_t > 0}$$

with the property $1_{X_t > 0} = 1_{A_t > 0}$ is a $\mathbb{Q}_A$ martingale, i.e., for all $t \leq T$,

$$X_t^A = E^{\mathbb{Q}_A} \left[X_T^A | \mathcal{F}_t \right]. \tag{14.21}$$

Proof. When $L_t = 0$, $X_t^A = A_t = A_T = X_T^A = 0$, the equality holds automatically. When $L_t > 0$, there is also $A_t > 0$. It follows that

$$
\begin{aligned}
E^{\mathbb{Q}_A} \left[X_T^A | \mathcal{F}_t \right] &= E^{\mathbb{Q}} \left[\frac{L_T}{L_t} X_T^A | \mathcal{F}_t \right] \\
&= E^{\mathbb{Q}} \left[\frac{A_T}{A_t} \frac{B(t)}{B(T)} 1_{A_t > 0} \frac{X_T}{A_T} 1_{A_T > 0} | \mathcal{F}_t \right] \\
&= \frac{1_{A_t > 0}}{A_t} E^{\mathbb{Q}} \left[\frac{B(t)}{B(T)} X_T 1_{A_T > 0} | \mathcal{F}_t \right] \\
&= \frac{1_{A_t > 0}}{A_t} E^{\mathbb{Q}} \left[\frac{B(t)}{B(T)} X_T | \mathcal{F}_t \right] \\
&= \frac{X_t}{A_t} 1_{A_t > 0} = X_t^A.
\end{aligned}
$$

$\square$

Let $\bar{A}_{m,n}(0) \geq 0$, define the *default swap measure* $\bar{\mathbb{Q}}_{m,n}$, according to

$$\left. \frac{d\bar{\mathbb{Q}}_{m,n}}{d\mathbb{Q}} \right|_{\mathcal{F}_t} = \frac{\bar{A}_{m,n}(t)}{\bar{A}_{m,n}(0)} \bigg/ \frac{B(t)}{B(0)} 1_{\bar{A}_{m,n}(0) > 0} = L_t , \tag{14.22}$$

then we have

Theorem 14.6.2 (Ho and Wu, 2008). *Under the new measure $\bar{\mathbb{Q}}_{m,n}$, there are*

1. *The $\bar{\mathbb{Q}}_{m,n}$-default probability is zero:*

$$E_t^{\bar{\mathbb{Q}}_{m,n}} \left[1_{\{\tau < T\}} \right] = 0.$$

2. *The swap rate $\bar{s}_{m,n}(t)$ of either kind of swap is a martingale:*

$$E_t^{\bar{\mathbb{Q}}_{m,n}} \left[\bar{s}_{m,n}(T) \right] = \bar{s}_{m,n}(t), \quad \forall T \in (t, T_m).$$

Proof. We proceed to prove the results one by one.

1. By definition,

$$\bar{\mathbb{Q}}_{m,n}\left[\tau \leq T\right] = E^{\bar{Q}_{m,n}}\left[1_{\tau \leq T}\right]$$

$$= E^Q\left[L_T 1_{\tau \leq T}\right]$$

$$= E^Q\left[\frac{\bar{A}_{m,n}(T)}{\bar{A}_{m,n}(0)}\frac{B(0)}{B(T)}1_{\tau > 0}1_{\tau \leq T}\right]$$

$$= \frac{1_{\tau > 0}}{\bar{A}_{m,n}(0)}E^Q\left[\frac{\bar{A}_{m,n}(T)}{B(T)}1_{\tau \leq T}\right] = 0.$$

2. The martingale property follows from taking $X_t = PV_{prot}(t)$ in the Corollary 14.6.1.

□

The first result says that the probability of default under $\mathbb{Q}_{m,n}$ is zero, this is why Schonbucher (2004) calls $\mathbb{Q}_{m,n}$ the survival measure.

In terms of the new measure, we can express the price of the default swaption at time $t \leq T$ as

$$C_t = E_t^Q\left[\frac{B(t)}{B(T)}\bar{A}_{m,n}(T)(\bar{s}_{m,n}(T) - \bar{s}^*)^+\right]$$

$$= \bar{A}_{m,n}(t)E_t^Q\left[\left(\frac{\bar{A}_{m,n}(T)}{\bar{A}_{m,n}(t)}\middle/\frac{B(T)}{B(t)}\right)1_{\bar{A}_{m,n}(t) > 0}(\bar{s}_{m,n}(T) - \bar{s}^*)^+\right] \quad (14.23)$$

$$= \bar{A}_{m,n}(t)E_t^{\bar{Q}_{m,n}}\left[(\bar{s}_{m,n}(T) - \bar{s}^*)^+\right]$$

A few comments must be noted here. $\bar{A}_{m,n}(t)$ is replicable until $t = \tau \wedge T_m$. The implication is that the price given by Equation 14.23 is an arbitrage-free price. The default swap measure defined in Equation 14.22 generalizes the definition of Schönbucher (2004), of which the annuity numeraire is non-tradable unless the recovery rate is zero. Second, in case of a default at $\tau < t$, $\bar{A}_{m,n}(t) = 0$, which implies that $\bar{\mathbb{Q}}_{m,n}$ is absolutely continuous with respect to, but not equivalent to, $\mathbb{Q}$. For swaption pricing, the lack of equivalence here is harmless. Note that Brigo (2005) provides an alternative derivation of the Black formula that preserves the measure equivalence, based on a theory of restricted filtration (Jeanblanc and Rutkowski, 2000; Bielecki and Rutkowski, 2001). According to the interpretation of Schönbucher's survival measure by Bielecki and Rutkowski (2001) (Section 15.2.2), these two approaches should be equivalent.

Based on the results of Proposition 14.6.2, we assume $\bar{s}_{m,n}(t)$ is a lognormal martingale under $\bar{\mathbb{Q}}_{m,n}$:

$$d\bar{s}_{m,n}(t) = \bar{s}_{m,n}(t)\bar{\gamma}_{m,n}\,dW_t^{(m,n)}, \quad (14.24)$$

where $W_t^{(m,n)}$ is a one-dimensional Brownian motion under $\bar{\mathbb{Q}}_{m,n}$ and $\bar{\gamma}_{m,n}$ is the swap-rate volatility. The lognormality assumption, Equation 14.24, leads readily to Black's formula for credit swaptions:

$$C = \bar{A}_{m,n}(t)\left[\bar{s}_{m,n}(t)\Phi(d_1) - \bar{s}^*\Phi(d_2)\right], \qquad (14.25)$$

with

$$d_{1,2} = \frac{\ln(\bar{s}_{m,n}(t)/\bar{s}^*) \pm \frac{1}{2}\bar{\gamma}_{m,n}^2(T-t)}{\bar{\gamma}_{m,n}\sqrt{T-t}}.$$

A hedging strategy follows from Black's formula Equation 14.25: at time $t < T$, the hedger
- maintains $\Phi(d_1)$ units of the credit default swap; and
- maintains $\bar{s}_{m,n}(t)\Phi(d_1) - \bar{s}^*\Phi(d_2)$ units of the annuity $\bar{A}_{m,n}(t)$.

14.7   Market Model with Forward Hazard Rates

Next, we show that the above swaption pricing approach can be justified under the assumptions of lognormal dynamics for either the forward hazard rates or the mean loss rates, on top of the SOFR market model:

$$\frac{df_j(t)}{f_j(t)} = \gamma_j(t) \cdot (d\mathbf{W}_t - \Sigma(t, T_j)dt), \qquad (14.26)$$

where γ_j is state independent, and

$$\Sigma(t, T_j) = -\sum_{k=\eta_t+1}^{j} \frac{\Delta T_k f_k(t)}{1 + \Delta T_k f_k(t)}\gamma_k(t).$$

We begin with

Proposition 14.7.1. *Assume that the pre-default dynamic of the discrete hazard rates is lognormal:*

$$\frac{d\lambda_j(t)}{\lambda_j(t)} = \mu_j^\lambda(t)dt + \gamma_j^\lambda(t) \cdot d\mathbf{W}_t, \qquad (14.27)$$

where $\mathbf{W}_t$ is a multi-dimensional Brownian motion under $\bar{\mathbb{Q}}_{m,n}$, then

$$\mu_j^\lambda(t) = -\gamma_j^\lambda(t) \cdot \sum_{k=\eta_t+1}^{j} \frac{\Delta T_k \lambda_k}{1 + \Delta T_k \lambda_k}\gamma_k^\lambda. \qquad (14.28)$$

Proof. According to the definition of discrete hazard rates, we have

$$\lambda_j(t) = \frac{1}{\Delta T_j} \left(\frac{\Lambda_{j-1}}{\Lambda_j} - 1 \right). \tag{14.29}$$

Assume the survival probabilities are lognormal as well:

$$d\Lambda_j(t) = \Lambda_j(t) \left\{ \Sigma_j^\Lambda \cdot d\mathbf{W}_t + \tilde{\lambda}_t dt \right\}.$$

where λ_t is the instantaneous rate for default. It then follows that

$$
\begin{aligned}
d\lambda_j(t) &= \frac{1}{\Delta T_j} d \left(\frac{\Lambda_{j-1}}{\Lambda_j} \right) \\
&= \frac{1}{\Delta T_j} \left(\frac{\Lambda_{j-1}}{\Lambda_j} \right) [\Sigma_{j-1}^\Lambda - \Sigma_j^\Lambda] \cdot [d\mathbf{W}_t - \Sigma_j^\Lambda dt],
\end{aligned} \tag{14.30}
$$

where σ_j^Λ is the percentage volatility of $\Lambda_j(t)$. Now rewrite the last equation into

$$\frac{d\lambda_j(t)}{\lambda(t)} = \frac{1 + \Delta T_j \lambda_j(t)}{\Delta T_j \lambda_j(t)} [\Sigma_{j-1}^\Lambda - \Sigma_j^\Lambda] \cdot [d\mathbf{W}_t - \Sigma_j^\Lambda dt], \tag{14.31}$$

and set

$$\gamma_j^\lambda = \frac{1 + \Delta T_j \lambda_J(t)}{\Delta T_j \lambda_j(t)} [\Sigma_{j-1}^\Lambda - \Sigma_j^\Lambda]. \tag{14.32}$$

Then there is

$$
\begin{aligned}
\Sigma_j^\Lambda &= \Sigma_{j-1}^\Lambda - \frac{\Delta T_j \lambda_j(t)}{\Delta 1 + T_j \lambda_j(t)} \gamma_j^\lambda \\
& \cdots \\
&= - \sum_{k=\eta_{t+1}}^{j} \frac{\Delta T_k \lambda_k(t)}{\Delta 1 + T_k \lambda_j(t)} \gamma_k^\lambda.
\end{aligned} \tag{14.33}
$$

Here, we take $\Sigma_{\eta_t}^\Lambda = 0$ because of no default up to t. The drift term μ_j^λ is thus

$$\mu_j^\lambda = -\gamma_j^\lambda \cdot \Sigma_j^\Lambda. \tag{14.34}$$

This completes the proof.   □

With $\{\lambda_j(t)\}$, we can construct $\{\Lambda_j(t)\}$ and $\{H_j(t)\}$. In fact, according to the definition of $H_j(t)$, in Equation 14.8, we have

$$H_j(t) = \frac{(1 - R_j)\lambda_j}{1 + R_j \Delta T_j \lambda_j}. \tag{14.35}$$

Next, we proceed to the pricing of swaptions under the extended market model. We take the first kind of swaps as an example. The swap rate is given by Equation 14.11. By Ito's Lemma, we can derive an approximate swap-rate process as follows:

$$
\begin{aligned}
d\bar{s}_{m,n}(t) &= \sum_{j=m+1}^{n} \frac{\partial \bar{s}_{m,n}(t)}{\partial H_j} H_j(t) \gamma_j^H \cdot d\mathbf{W}_t^{(m,n)} \\
&= \bar{s}_{m,n}(t) \sum_{j=m+1}^{n} \frac{\partial \bar{s}_{m,n}(t)}{\partial H_j} \frac{H_j(t)}{\bar{s}_{m,n}(t)} \gamma_j^H \cdot d\mathbf{W}_t^{(m,n)} \\
&\approx \bar{s}_{m,n}(t) \sum_{j=m+1}^{n} \frac{\partial \bar{s}_{m,n}(0)}{\partial H_j} \frac{H_j(0)}{\bar{s}_{m,n}(0)} \gamma_j^H \cdot d\mathbf{W}_t^{(m,n)} \\
&= \bar{s}_{m,n}(t) \bar{\gamma}_{m,n} \cdot d\mathbf{W}_t^{(m,n)},
\end{aligned}
\tag{14.36}
$$

where

$$
\gamma_j^H = \frac{\gamma_j^\lambda}{1 + R_j \Delta T_j \lambda_j},
$$

and

$$
\bar{\gamma}_{m,n} \triangleq \sum_{j=m+1}^{n} \bar{\omega}_j \gamma_j^H, \quad \bar{\omega}_j \triangleq \frac{\partial \bar{s}_{m,n}(0)}{\partial H_j} \frac{H_j(0)}{\bar{s}_{m,n}(0)} \approx \bar{\alpha}_j \frac{H_j(0)}{\bar{s}_{m,n}(0)},
\tag{14.37}
$$

and $\mathbf{W}_t^{(m,n)}$ is a multi-dimensional Brownian motion under $\bar{\mathbb{Q}}_{m,n}$, defined by

$$
d\mathbf{W}_t^{(m,n)} = d\mathbf{W}_t - \sum_{j=m+1}^{n} \bar{\alpha}_j \bar{\Sigma}_j(t) dt,
$$

and $\bar{\Sigma}_j$ is the volatility of $\bar{P}_j(t)$. The lognormal process for swap rates, Equation 14.36, justifies the use of Black's formula, Equation 14.25, for swaptions. Note that in the context of the SOFR market model, the approximations made in Equation 14.36 and Equation 14.37 are known to be accurate enough for application and have been justified with rigor (Brigo *et al.*, 2004). The expressions in Equation 14.36 describe the relation between forward spread volatilities and swap-rate volatilities, which can be used in practice to gauge the relative price richness/cheapness of a swaption. The model for forward hazard rates, Equation 14.27, can be calibrated to the implied volatilities of the default swaptions using the quadratic programming technology developed by Wu (2003) for market model calibrations.

Models more comprehensive than Equation 14.27 can be developed by including other risk dynamics, like jumps, stochastic volatilities, and even correlations among multiple credit names (Eberlein *et al.*, 2006). Such developments are largely parallel to existing extensions to the standard market model. Brigo (2005) and especially Schönbucher (2004) have made several extensions using swap rates as state variables.

14.8 Pricing of CDO Tranches under the Market Model

In this section, we explain how to apply the market model for pricing collateralized debt obligations (CDOs). A CDO is a way to restructure the cash flows of a portfolio of bonds (with various credit ratings). Tranches of ascending seniority are defined in terms of the percentages of the notional principal value of the portfolio and losses are always allocated to the most junior tranche that is still alive. The spread of each tranche is just the premium of protection payments which, unlike the protection payments in a single-name CDS, are made periodically until all notional value is lost. Tranches are divided by *attachment points*. Take CDX IG for example, a standardized CDO of with the attachment points 0%, 3%, 7%, 10%, 15%, 30% and, of course, 100%. The equity tranche, which is the most junior tranche with attachment points 0% and 3%, will absorb the losses to the portfolio up to the first 3% and then cease to exist. Subsequent losses will then be borne by the next tranche with attachment points of 3% and 7%, which is called the mezzanine tranche. Losses to other senior tranches are determined similarly. For each tranche, the premium of protection is calculated based on the outstanding notional principal value. To express the remaining outstanding notional principals of the tranches, we introduce the following notations:

$[P_D^i, P_U^i]$ —   the attachment points for the ith tranche, in percentage;

$D^{(k)}(T_j)$ —   the forward price of the outstanding notional value of the kth name at T_j, equal to $\mathbf{1}_{\{\tau > T_j\}} + \bar{R}\mathbf{1}_{\{T_{j-1} < \tau \leq T_j\}}$;

Then, the total outstanding notional at T_j for the portfolio in percentage is

$$D^P(T_j) = \frac{1}{K} \sum_{k=1}^{K} D^{(k)}(T_j), \tag{14.38}$$

where K be the number of credit names in the portfolio. The outstanding notional at T_j for the ith tranche can be expressed as

$$O_i(T_j) = \frac{1}{P_U^i - P_D^i} \left\{ (D^P(T_j) - 1 + P_U^i)^+ - (D^P(T_j) - 1 + P_D^i)^+ \right\}. \tag{14.39}$$

The loss to the ith tranche over $(T_{j-1}, T_j]$ is thus

$$O_i(T_{j-1}) - O_i(T_j).$$

The expression Equation 14.39 reiterates the fact that the outstanding notional for any tranche at any cash flow day can be regarded as a spread option on the total outstanding notional value of the portfolio backing the CDO.

Next, we consider the pricing of the premium rate on a tranche. Let s_i be the premium rate on the ith tranche; the value of the fee leg is then

$$PV_{fee} = s_i \sum_{j=m+1}^{n} \Delta T_j P_j(t) E_t^{Q_j} \left[O_i(T_j)\right].$$

While the value of the protection leg is

$$PV_{prot} = \sum_{j=m+1}^{n} P_j(t) E_t^{Q_j} \left[O_i(T_{j-1}) - O_i(T_j)\right].$$

The formula for the premium rate on the ith tranche is then

$$s_i = \frac{\sum_{j=m+1}^{n} P_j(t) E_t^{Q_j} \left[O_i(T_{j-1}) - O_i(T_j)\right]}{\sum_{j=m+1}^{n} \Delta T_j P_j(t) E_t^{Q_j} \left[O_i(T_j)\right]}.$$

In view of Equation 14.39, we understand that the key to CDS rate calculation lies in the valuation of a sequence of call options of the form

$$E_t^{Q_j} \left[\left(D^P(T_j) - X\right)^+\right], \quad j = m+1, \ldots, n. \tag{14.40}$$

We now consider the valuation of the above options by the Monte Carlo simulation method. In view of the definition of $D^P(T_j)$ in Equation 14.38, we need to simulate $D^{(k)}(T_j), k = 1, \ldots, K$. For simplicity, we assume a constant recovery rate, $E^Q \left[R_\tau | T_{j-1} < \tau \le T_j\right] = \bar{R} = constant$. We then can express $D^{(k)}(T_j)$ as

$$D^{(k)}(T_j) = \mathbf{1}_{\{\tau > T_j\}} \left(\mathbf{1}_{\{u > \lambda_j^{(k)}(T_j)\Delta T_j\}} + \bar{R}\mathbf{1}_{\{u \le \lambda_j^{(k)}(T_j)\Delta T_j\}}\right), \tag{14.41}$$

where u obeys the uniform distribution in $(0, 1)$, denoted by $U(0, 1; \Sigma_u)$, and Σ_u stands for the correlation matrix of u's. If taking a Gaussian copula for u's, we can proceed as follows.

1. Perform a Choleski decomposition of the input correlation matrix

$$\Sigma_g = AA^T. \tag{14.42}$$

2. Simulate $K(T_j)$ (which is the number of surviving firms at time T_j) independent standard normal random variables, $\{\tilde{\epsilon}_i\}_{i=1}^{K(T_j)}$.

3. Transform $\{\tilde{\epsilon}_i\}_{i=1}^{K(T_j)}$ to correlated normal random variables

$$\begin{pmatrix} \epsilon_1 \\ \vdots \\ \epsilon_{K(T_j)} \end{pmatrix} = A \begin{pmatrix} \tilde{\epsilon}_1 \\ \vdots \\ \tilde{\epsilon}_{K(T_j)} \end{pmatrix}.$$

4. Transform the normal random variables to uniform random variables

$$(u_1, \ldots, u_{K(T_j)}) = (\Phi(\epsilon_1), \ldots, \Phi(\epsilon_{K(T_j)})).$$

Furthermore, if we abide by the industrial convention to assume a uniform pairwise correlation such that $corr(\epsilon_k, \epsilon_j) = \rho > 0$ for any k and l, then steps 1 to 3 above are simplified into the calculations of

$$\epsilon_k = \sqrt{1-\rho}\, \tilde{\epsilon}_k + \sqrt{\rho}\tilde{\epsilon}_c, \qquad (14.43)$$

where $\tilde{\epsilon}_c$ and $\tilde{\epsilon}_k, k = 1, \ldots, K(T_j)$ are independent standard normal random variables.

The evolution of $\lambda_j^{(k)}(t)$ follows from the scheme of

$$\lambda_j^{(k)}(T_j) = \lambda^{(k)}(T_{j-1}) \exp\left((\mu_j^\lambda(T_{j-1}) - \frac{1}{2}\|\gamma_j^\lambda\|^2)\Delta T_j + \gamma_j^\lambda \Delta W^{(k)} \right). \quad (14.44)$$

Here, we have locally frozen $\mu_j^\lambda(T_{j-1})$, defined in Equation 14.28. In the evolution of $\lambda_j^{(k)}(t)$, we can incorporate the correlations of the credit spreads observed in the single-name CDS market.

We are now ready to describe the algorithm for CDO pricing. The simulation of correlated defaults is the focus of the algorithm. Note that we need to input two sets of correlations. The first set is for the correlations, Σ_H, of CDS rates, which are the state variables for the market model. This set of correlations can be observed from the market. The second set is for the correlations of default times, Σ_u, which is not quite observable and is dealt with using the technique of Gaussian copula. Let T be the maturity of the CDO, ΔT be the time interval for premium payments, $J = T/\Delta T$ be the maximal number of the premium payments, and M be the number of Monte Carlo simulation paths. We develop the following algorithm for pricing the options in Equation 14.40.

```
/* Algorithm for pricing options on D_j^P(T_j), j = 1, ..., J */

    For j = 1 : J
        V_j = 0
    end
    For m = 1 : M
        For k = 1 : K
            D^(k)(T_0) = 1
        end
        K(T_0) = K
        For j = 1 : J
            Generate {ΔW^(k)}_{k=1}^{K(T_{j-1})} ~ N(0, ΔTΣ_H)
            Generate {u_k}_{k=1}^{K(T_{j-1})} ~ U(0, 1; Σ_u)
            Put D^P(T_j) = 0
```

$l = 0$

For $k = 1 : K$ repeat

    If $D^{(k)}(T_{j-1}) = 1$, then

        $D^{(k)}(T_j) = 1$

/* Simulate default over (T_{j-1}, T_j) for the kth name */

        $l = l + 1$

        If $u_l \leq \lambda^{(k)}(T_{j-1})\Delta T_j$

            $D^{(k)}(T_j) = \bar{R}$

            $K(T_j) = K(T_j) - 1$

        end if

        $D^P(T_j) \leftarrow D^P(T_j) + D^{(k)}(T_j)$

/* Simulate the hazard rate $\lambda^{(k)}(T_j)$ according to the market
    model */

$$\lambda_j^{(k)}(T_j) = \lambda^{(k)}(T_{j-1}) \exp\left((\mu_j^\lambda(T_{j-1}) - \tfrac{1}{2}\|\gamma_j^\lambda\|^2)\Delta T_j + \gamma_j^\lambda \Delta W^{(k)}\right)$$

        end if

    end if

/* Calculate the payoff of the option */

    $V_j \leftarrow V_j + (D^P(T_j)/K - X)^+$

  end

end

/* Average payoff */

  For $j = 1 : J$

    $V_j \leftarrow V_j/M$

  end

/* The end of the algorithm */

One can see that the entire algorithm is rather easy to implement, and the computation time is about J times more than that of the Gaussian copula method of Li (2000).

Next, let us examine the ability of the model to back out correlations implied by various tranches of two standardized CDOs, namely, CDX IG and iTraxx IG. The quotes on August 24, 2004, are listed in Table 14.6.[5] The SOFR term rates and swap rates for the same day are listed in Table 14.7.[6]

Without loss of generality, we make a few reasonable simplifications in the handling of data. We assume that the curve of forward spreads is flat and equal to the index of the respective maturity, which implies that the CDS rate of any maturity is equal to the value of forward spreads. The CDS rate volatility is set at the constant level of either 50% or 100%, which represents the usual range of implied swaption volatilities (see Schöbucher (2004) and Brigo (2005)). The recovery rate is taken to be $\bar{R} = 40\%$, abiding to the industrial convention.

---

[5] The data are taken from Hull et al. (2005).

[6] We rename LIBOR to SOFR because of their identical values before their reference periods.

370   *Interest Rate Modeling: Theory and Practice*

TABLE 14.6: Quotes on 8/24/2004

	CDX IG Tranches					
	0% to 3%	3% to 7%	7% to 10%	10% to 15%	15% to 30%	Index
5-year quotes	40.02	295.71	120.50	43.00	12.43	59.73
10-year quotes	58.17	632.00	301.00	154.00	49.50	81.00

	iTraxx IG Tranches					
	0% to 3%	3% to 6%	6% to 9%	9% to 12%	12% to 22%	Index
5-year quotes	24.10	127.50	54.00	32.50	18.00	37.79
10-year quotes	43.80	350.17	167.17	97.67	54.33	51.25

TABLE 14.7: USD Yield Data (8/24/2004, Bloomberg)

SOFR	3 mth	1.760%
	6 mth	1.980%
	12 mth	2.311%
Swap	2Y	2.840%
	3Y	3.265%
	4Y	3.592%
	5Y	3.890%
	6Y	4.100%
	7Y	4.295%
	8Y	4.455%
	9Y	4.595%
	10Y	4.710%

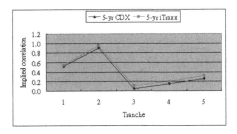
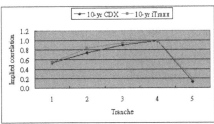

FIGURE 14.7: Implied correlations for $\|\gamma^\lambda\| = 0.5$.

Due to the lack of correlation data for CDS rates, we let the CDS rates and the Gaussian copula for default times share the same pairwise correlation. We take the number of paths to be $M = 10,000$, and the size of time stepping to be $\Delta t = 0.25$. The implied correlations of various tranches,[7] obtained through trial and error, are listed in Figures 14.7 and 14.8. The average accumulated default numbers for both CDOs, for a pairwise correlation of 20%, are shown in Figures 14.9 and 14.10.

---

[7]Note that the implied correlation for a tranche may not necessarily be unique.

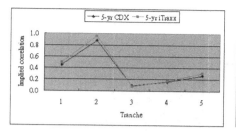 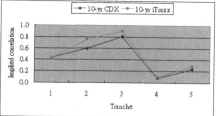

FIGURE 14.8: Implied correlations for $\|\gamma^\lambda\| = 1$.

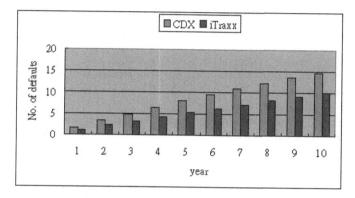

FIGURE 14.9: Accumulated no. of defaults for $\|\gamma^\lambda\| = 0.5$.

Let us comment on the results. Under the market model, the implied correlation curves do not quite look like a smile. We have let $\|\gamma^H\|$, the CDS rates volatility, vary from 50% to 100%, and have witnessed a gradual deformation of the "smile curve." Remarkably, the "smile curves" of CDX IG and iTraxx IG stay close to each other, which is interesting but an explanation is not available. The histograms of average accumulated number of defaults look reasonable. The bigger numbers of defaults for CDX IG are consistent with the bigger numbers of spreads across all tranches. The algorithm is not optimized, but it has been very robust. The pricing of an entire CDO takes about 20 seconds in a PC with an Intel Pentium 4 CPU (3.06GHz, 504MB RAM).

We tend to believe that the higher implied correlation for the mezzanine tranche is caused by the assumption of flat CDS rates across all maturities, which is against the fact that a CDS rate increases with maturity, as is seen in Table 14.1 for the CDS rates of Citigroup. If we bootstrap the forward spreads against the CDS rates, we should see that the forward spread curve has a bigger slope of increase than that of the CDS rate curve. The assumption of flat forward spread curve should have produced more defaults in the short

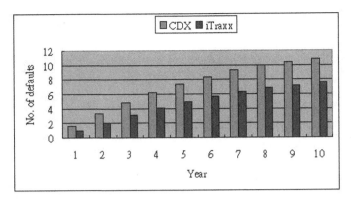

FIGURE 14.10: Accumulated no. of defaults for $\|\gamma^\lambda\| = 1$.

horizon yet fewer defaults for the long horizon, which in turn can cause higher prices for junior tranches yet lower prices for senior tranches, as reflected by the level of implied correlations.

We want to point out here that we have tried to reproduce the CDO spreads without default correlation, by letting the random variables $\{u_k\}$ be independent. The results are no good, which thus supports a widely held opinion among market participants that the spread correlations are insufficient for pricing CDOs, and thus we also need default correlations.

14.9   Notes

Our objectives of this chapter are two fold. First, we want to achieve replication pricing of CDS and CDS options. Second,to put the pricing of credit instruments, including corporate bonds, CDS, CDS options and CDOs under the same model with some properly chosen state variables. Both objectives are achieved, and the state variables are eventually chosen to be the discrete hazard rates.

In 2019, Bloomberg LLP has started to publish the recovery rates together with their corresponding CDS rates.[8] The availability of the recovery rates will make our market model with the discrete hazard rate easier to use, because extracting the implied recovery rates are often ill-posed problems.

---

[8]https://data.bloomberglp.com/professional/sites/10/Disaster-Recovery-Services-Fact-Sheet.pdf.

Chapter 15

xVA: Definition, Evaluation, and Risk Management

The 2008 financial crisis catalyzed many changes to the global derivatives market which, in particular, are seen through the proliferation of collaterals, the mandate to trade all derivatives through central counterparty clearing houses (CCP), and the emergence of various valuation adjustments, so-called xVA, in either pricing or accounting. The xVA contains a growing list of valuation adjustments (VAs): credit valuation adjustment (CVA), debit valuation adjustment (DVA), funding valuation adjustment (FVA), funding cost adjustment (FCA), collateral valuation adjustment (ColVA), margin valuation adjustment (MVA), capital valuation adjustment (KVA) etc. Yet, except for CVA and DVA,[1] the other VAs have been controversial because, in addition to a lack of unanimous definition, their inclusion in pricing or booking causes price asymmetry and asset-liability asymmetry to the trading parties. In this chapter, we will redefine the notions of CVA, DVA and FVA and derive the formulae for the rest of xVA as the expected present values of excessive cash flows due to funding spreads under the risk-neutral pricing measure. We then identify the component of the xVA to be included into the fair price defined according to IFRS 13, i.e., exit price or market price, see FASB (2011). In the end we will make our suggestions to managing the unhedgeable idiosyncratic risks behind the rest of xVA or xVA components.

The controversies started with the FVA, which is supposed to account for the funding costs for managing or manufacturing the cash flows of trades, including the funding costs for margins/collateral, capital as well as hedging. As early as 2012, a few major banks had started to adopt the notion and reported FVA as losses to their business, but such a practice has been criticized as a breach of the accounting principles that require booking with exit prices and maintaining asset-liability symmetry, see Hull and White (2012). Moreover, in a poll conducted by Risk.net in March 2015 (Sherif, 2015), about two-thirds of quants believe that banks have overstated FVA losses and the current FVA model is wrong. As a matter of fact, there is not yet a market-wide consensus on issues like how to quantify the FVA, whether FVA is part of the fair-value pricing, or whether FVA is merely an accounting entry. For the remaining xVA items, we are facing more or less the same issues.

---

[1]DVA is the CVA of the counterparty.

To understand market participants' interest in xVA, we need to get down to the FVA debate. Let us start from the first as well as a prevalent FVA model of the market, see e.g. Cameron (2013). Let B stand for a bank and x_B be its funding spread for unsecured borrowing or lending.[2] If bank B has an uncollateralized derivative trade with the time-t value $V(t)$, then to bank B the FVA of the trade is defined to be

$$\text{FVA}_B = -E_0^Q \left[\int_0^{\tau \wedge T} x_B e^{-\int_0^t r_s ds} V(t) dt \right], \tag{15.1}$$

where τ represents the time of the first bilateral default of the bank and its counterparty, T is the maturity of the derivative, r_t is the overnight risk-free rate,[3] and Q is the risk-neutral pricing measure. Note that the above formula arises from hedging an uncollateralized interest-rate swap using a fully collateralized swap, thus inducing either funding cost or funding benefit for posting or receiving cash collateral.[4] Formula 15.1 has been extended to a portfolio of trades with a single counterparty:

$$\text{FVA}_B = -E_0^Q \left[\int_0^{\tau} x_B e^{-\int_0^t r_s ds} \sum V_i(t) dt \right],$$

where the sum is over all trade with the counterparty, and $V_i(t)$ is the value of the ith trade; and to a portfolio of trades with multiple counterparties:

$$\text{FVA}_B = -E_0^Q \left[\int_0^{\tau_B} x_B e^{-\int_0^t r_s ds} \sum V_j(t) 1_{t < \tau_j} dt \right],$$

where the sum is over the netted trade values with all counterparties, and $V_j(t)$ is the netted value of all trades with counterparty j; and τ_j is the default time of counterparty j. Note that the FVA_B's defined above are not the results of arbitrage pricing, so they may better be taken as metrics for the funding costs instead of valuation adjustments for fair-value pricing. Arguably these metrics make good sense for an uncollateralized swap and a swap portfolio, but it is not necessarily so for general derivatives or portfolios of derivatives, traded either in over-the-counter (OTC) markets or through CCPs.

To tackle funding costs in generality, a number of alternative models have been developed, including Brigo et al. (2011), Crépay (2011), Burgard and Kjaer (2011), Brigo et al. (2012), Bo and Capponi (2013), Lou (2015), Bichuch et al. (2015) and Li and Wu (2015). Under these models, FVA is defined as the expected present value of funding costs for posting/receiving collaterals as well as for hedging under the risk-neutral measure, and it is often modeled and evaluated together with CVA and DVA. Burgard and Kjaer (2011), in

---

[2]The funding spread can be asymmetrical for borrowing and for lending.

[3]The proxy for the overnight risk-free rate is the Federal funds rate for USD and EONIA rate for Euro.

[4]See Funding Valuation Adjustment (FVA), Part 1: A Primer, March 20, 2014, https://www.quantifisolutions.com.

particular, pioneer the arbitrage pricing of derivatives subject to market risks, counterparty default risks and funding costs/risks, and obtain the triad of valuation adjustments (i.e., CVA, DVA and FVA) under a consistent framework. As valuation adjustments to the otherwise no-default value of a derivative, CVA and DVA are well received, yet FVA is not: it makes fair-value pricing entity dependent such that two trading parties may no longer share the same "fair price" even if they share the same "risk-neutral pricing measure."

As an accounting entry, FVA is equally problematic. First, booking an FVA-adjusted price for a trade violates the rule of IFRS 13 (IASB, 2011) that mandates the use of exit prices in fair-value accounting. Second, FVA booking for P&L accounting causes asset-liability asymmetry, unless a trading party willingly registers his counterparty's FVA loss/benefit as his FVA benefit/loss, which is not likely to happen. Nonetheless, as a measure of potential loss to their business, FVA is not ignored by major investment banks, and there have been efforts to accommodate the FVA without breaching the accounting principles. For examples, Albanese et al. (2015) and Andersen et al. (2016) advocate the reduction of FVA from a firm's equity value. A compromise, simultaneously occurring in both theory and practice, seems to arise in recent years. Hull and While (2014) have softened their stand against FVA and accepted the inclusion of the funding cost associated with the market funding liquidity risk premium into the fair value of a trade, as was first advocated by Morini and Prampolini (2010). In the marketplaces, there are increasing evidences that show banks have converged to using a uniform "market cost of funding" to price derivatives (Gregory, 2015).

Also controversial has been the DVA accounting adopted by some banks. When a bank's credit quality deteriorates, the value of its liabilities depreciates while its DVA appreciates for the same amount. As such, some banks went on to register the DVA appreciations as "profits," yet such paper profits have been disturbing and have drawn severe criticisms. As is argued in Castagna (2012), shareholders of the banks can hardly benefit from such "profits" unless the banks buy back related liabilities amid their financial distress, which is also not likely to happen. As a response to the criticisms, banks have recently excluded DVA from their Tier 1 equity capital and, according to a recent FASB revision,[5] started to register DVA under "other comprehensive incomes."

In this chapter, we present a general theory of xVA as well as derivatives pricing of Wu and Zhang (2020), which is based on the so-called bilateral replication arguments. The construction of the theory proceeds in two steps. First, we identify the risk-neutral pricing measure and then uncover the corresponding hedging or replicating strategies against the market risks and the counterparty default risks. Note that these strategies are unaffected by the funding spreads and are to be taken bilaterally by the counterparties. The cost of the replications in the absence of funding spreads is identified to be nothing else but the risk-neutral value of the derivative, and it can be decomposed into

---

[5]FASB News Release, January 5, 2016, www.fasb.org.

the no-default value and the bilateral CVAs. Second, we will figure out the additional costs caused by the funding spreads (for margins/collaterals, capitals as well as for hedging) and, following the market convention, define the FVA as the expected present values of the additional costs under the risk-neutral measure. Out of the FVA, we further identify FCA, MVA, ColVA and KVA as components. Nonetheless, it is already known that pricing in unilateral funding risks will normally cause a price spread between the counterparties (e.g. Ruiz 2013; Wu 2015; Li and Wu 2015). As a major step forward, we will show that the market funding risk premium for unsecured lending and borrowing can be bilaterally priced into derivatives trades, without causing price asymmetry. This finding justifies the emerging market practice to charge a uniform market funding liquidity risk premium in derivatives trades.

As far as the management of funding risk is concerned, we suggest the adoption of existing popular risk metrics, like VaR or CVaR. Since the realized funding cost is a path-dependent random variable, the VaR or CVaR metric can be evaluated simultaneously with derivatives fair prices under Monte Carlo simulations. Note that FVA itself is not necessarily a good measure for risk management purposes, as it is an expectation of discounted funding costs under the risk-neutral measure,[6] which can be very different from the realized funding cost. Hence, either stacking up reserve or deducting Tier 1 capital according to the FVA number is not quite justified as a proper risk management measure.

The pricing approach based on the bilateral replication has several advantages over other existing approaches. First, it is an extension of the classic unilateral replication pricing and it leads to a unique fair price that cannot be dominated by either party's trading strategy. Second, it does not take the pricing measure for granted and, instead, it treats the construction of pricing measure as a part of the pricing problem and identifies the replication strategies to be taken bilaterally by the counterparties. Third, the approach can be applied to xVA pricing under more general incomplete market models (e.g., models with stochastic interest rates, stochastic hazard rates, and jump risks), and the xVA formulations will remain quite the same. In addition, our xVA formulations can be conveniently adapted to pricing derivatives trades in CCP and in OTC markets, and the latter can be either collateralized or non-collateralized.

This chapter is organized as follows. In Section 15.1, we first describe the current conventions for posting margins, collaterals or risk capitals, then we present the bilateral replication for the fair value of a derivative in the absence of funding spread. In Section 15.2 we separate the P&L of a trade into the shortfall of payout replication and the funding cost and present the xVA formulae. In particular, we show that the market funding risk premium can be bilaterally priced into a trade, without causing price asymmetry to the

---

[6]Although it can be evaluated under the real-world measure as well.

counterparties. In Section 15.3 we present an example of xVA calculation with an interest-rate swap. In Section 15.4 we conclude.

15.1 Pricing through Bilateral Replications

15.1.1 Margin Accounts, Collaterals, and Capitals

We aim at developing a derivative pricing theory for both OTC markets and exchange markets. As of today, OTC derivative trades are largely collateralized,[7] or CSA trades[8] and collaterals are subject to periodic revisions. These are also the major features of CCP trades. As a result, we can tackle the pricing problem in the OTC market and exchange markets with the same approach.

Let us consider the pricing of a partially collateralized European option trade between two defaultable parties, B a bank and C a counterparty. If there is no default by either party until the maturity of the derivative, the party of liability will make a contractual payment to the counterparty. In case of a premature default, the party of exposure will seize the collateral and will remain entitled to a fair share of recovery values. For subsequent discussions, we introduce the following notations.

Y_T — the contractual payoff of the derivative at maturity T;

τ_i — the default time of party i, for $i = B$ and C;

$I_i(t)$ — ≥ 0, the value of the initial margin (IM) posted by party i, $i = B$ and C;

$X_i(t)$ — ≥ 0, the value of the variable margin (VM) or collaterals posted by party i, $i = B$ and C;

$K_i(t)$ — ≥ 0, the value of risk capitals allocated to party i by shareholders, $i = B$ and C;

$V(t)$ — the fair value of the derivative to B, the bank, s.t.
$V(t) > 0$ — asset,
$V(t) < 0$ — liability.

The margins, collaterals and capitals are typically in the form of cash or other Tier 1 capital assets, which cost fees to borrow and thus incurs funding costs. When one party is a clearinghouse, we may ignore its default risk and waive the requirements for margins/collaterals and capitals. Collaterals and variable margins may be lent out by the receiving party for its own funding

---

[7]ISDA Margin Survey 2014, http://www2.isda.org/functional-areas/research/surveys /margin-surveys/.

[8]A collateralized trade is subject to the Credit Support Annex (CSA), a legal document which regulates collateral for derivative transactions, and thus is called a CSA trade.

purposes, which is called rehypothecation, and thus the party who poses funds may be rewarded with higher returns. The levels of margins or the amount of collaterals are revised periodically. In fact, the determination and maintenance of the margin levels or the collateral amount are specialized topics, which are addressed in details in e.g. Gregory (2015) or Green (2016).

At the moment of the first bilateral default, $\tau = \tau_B \wedge \tau_C$, there may be a downward jump in the derivatives value. We let $V(\tau)$ denote the post-default value, such that $V(\tau) = V(\tau+)$, which is right continuous with a limit.[9] For pricing purposes, we need to specify the post-default value of the derivative, by using either the Standardized Method or the Internal Model Method (IMM). The Standardized Method consists of a series of formulae set up by regulators (see BIPRU (Financial Conduct Authority, 2014)), while the IMM gives a firm the freedom of choice for advanced models, subject to regulator's approval. In our numerical demonstration, we will use the second method, under which the default settlement value can be conveniently described. Let $M(t)$ be the mark-to-market (MtM) value of the derivative, which typically is obtained through a dealer poll mechanism, and let $c_i(t) = I_i(t) + X_i(t)$ be the total value of margins/collaterals posted by party i, then the general bilateral close-out conditions upon the first bilateral default are

$$V(\tau = \tau_B) = R_B[M(\tau_B) + c_B(\tau_B)]^- + [M(\tau_B) + c_B(\tau_B)]^+ - c_B(\tau_B),$$
$$V(\tau = \tau_C) = R_C[M(\tau_C) - c_C(\tau_C)]^+ + [M(\tau_C) - c_C(\tau_C)]^- + c_C(\tau_C),$$
$$(15.2)$$

where R_B and R_C are the recovery rates for the losses of respective defaults, the sup indices "+" and "−" mean a floor and a cap to a function at the level of zero: $f^+(\tau) = \max\{f(\tau), 0\}$ and $f^-(\tau) = \min\{f(\tau), 0\}$. For notational simplicity we write $V(\tau_i)$ for $V(\tau = \tau_i)$, $i = B$ and C.

For both mathematical or notational simplicity, we make two non-essential assumptions: (1) the risk-free rate and default intensities are deterministic functions of time, and (2) the credit default swap (CDS) rates remain constants. Also, we assume there is no default at $t = 0$, the current moment.

15.1.2   Pricing in the Absence of Funding Cost

We will model the market with probability space $(\Omega, \mathcal{G}, \mathcal{G}_t, \mathbb{P})$, where $\mathbb{P}$ is the real-world measure, $\mathcal{G}_t$ is the filtration that represents all market information up to time t, such that $\mathcal{G}_t = \mathcal{F}_t \vee \mathcal{H}_t$, where $\mathcal{F}_t$ contains all market information except the default statuses of the trading parties, while $\mathcal{H}_t$ contains only the information of the default statuses of the counterparties.

Without loss of generality, we consider the pricing of equity European options in a market where the following securities are traded: the money market account, shares or stocks, repurchasing agreements (repos), and CDS. The price dynamics of these securities are described below.

---

[9] So-called a cádlág function.

The balance of the money market account evolves according to

$$dB_t = (r_t + x_m)B_t dt, \quad \text{with} \quad B_0 = 1,$$

where x_m is the funding risk premium for the general market, so that $B_t = e^{\int_0^t (r_s + x_m)ds}$, which will be the numeraire asset for the risk-neutral measure to be defined shortly. Note that the money market account is available for firms with negligible default probability. For other firms, repo type of transactions, i.e, collaterals, can be utilized to open the account. For notational simplicity we let $\hat{r}_t = r_t + x_m$.

The share price evolves according to the usual lognormal dynamics:

$$dS_t = S_t \left[\mu_t dt + \sigma_t d\tilde{W}_t \right],$$

where $\tilde{W}_t$ is a one-dimensional Brownian motion under $\mathbb{P}$, μ_t is the expected return, and σ_t is the percentage volatility. Note that the lognormal dynamics is adopted in this chapter for the sake of simplicity, and more general dynamics can be considered with, of course, additional complexity.

The shares for hedging purposes will be "repoed in," such that the shares will be used as collaterals for borrowing. The instantaneous return from holding the repo is

$$dZ_S(t) = dS_t - (\hat{r}_t + \lambda_S - q_t)S_t dt$$
$$= S_t \left[(\mu_t - \hat{r}_t - \lambda_S + q_t)dt + \sigma_t d\tilde{W}_t \right],$$

where the second term on the RHS of the first equality is the cost of carry for the repo trade, with q_t being the dividend yield of the share and λ_S being the repo spread.[10] For a repo entered at time t, its value equals to zero: $Z_S(t) = 0$.

Let the notional value of the CDS be one dollar, then the instantaneous return of a CDS on the default of party i can be described by[11]

$$dU_i(t) = \hat{r}_t U_i(t)dt - s_i dt + L_i dJ_i, \quad i = B \text{ and } C,$$

where $s_i \geq 0$ is the annualized CDS rate, J_i is a Poisson process that jumps from 0 to 1 upon the default of party i, the $\mathbb{P}$ intensity of the jump is λ_i, and $L_i \in [0,1]$ is the corresponding loss rate, assumed to be a constant for simplicity. Note that the above CDS price dynamics holds true only if the CDS rate stays unchanged over time. Under our dynamical hedging strategy, we will always make use of the par CDS, which has zero value, when revising the hedge. For simplicity, we assume the Brownian motion and the Poisson processes are independent of one another.

---

[10]For simplicity, we have skipped discussing the details of haircuts in repo trades for a long or short stocks, which can be easily accommodated by adjusting the repo spread.

[11]We ignore the funding costs for CDS, in order to isolate the funding costs for the derivative.

According to the fundamental theory of asset pricing, there exists a measure, $\mathbb{Q}$, which is equivalent to $\mathbb{P}$ such that under $\mathbb{Q}$ the discounted prices of the repo and the CDS are martingales. For our asset price model, such a martingale measure is unique and is defined by the following Radon-Nikodym derivative:

$$\left.\frac{d\mathbb{Q}}{d\mathbb{P}}\right|_{\mathcal{F}_t} = \frac{e^{-\int_0^t \gamma_S(u)d\tilde{W}_u}}{E_0^P\left[e^{-\int_0^t \gamma_S(u)d\tilde{W}_u}\right]} \frac{e^{\gamma_B J_B(t)}}{E_0^P\left[e^{\gamma_B J_B(t)}\right]} \frac{e^{\gamma_C J_C(t)}}{E_0^P\left[e^{\gamma_C J_C(t)}\right]}$$

$$= e^{\int_0^t -\frac{1}{2}\gamma_S^2(u)du + \gamma_S(u)d\tilde{W}_u + \gamma_B J_B(t) + \tilde{\lambda}_B t(1-e^{\gamma_B}) + \gamma_C J_C(t) + \tilde{\lambda}_C t(1-e^{\gamma_C})}, \qquad (15.3)$$

with

$$\gamma_S(t) = \frac{\mu_t - \hat{r}_t - \lambda_S + q_t}{\sigma_t},$$

$$\gamma_B = \ln\frac{s_B/L_B}{\tilde{\lambda}_B} \quad \text{and} \quad \gamma_C = \ln\frac{s_C/L_C}{\tilde{\lambda}_C}.$$

Under $\mathbb{Q}$ the price dynamics of the repo and the CDS become

$$\begin{aligned} dZ_S(t) &= \sigma_t(t)S_t dW_t, \\ dU_i(t) &= \hat{r}_t U_i(t)dt + L_i\left(dJ_i - \lambda_i dt\right), \quad i = B \text{ and } C, \end{aligned} \qquad (15.4)$$

where W_t is a $\mathbb{Q}$-Brownian motion and J_i is a jump process with risk-neutral intensity $\lambda_i = s_i/L_i$. Apparently, both $Z_S(t)$ and the discounted prices of $U_i(t)$ are $\mathbb{Q}$-martingales. We call $\mathbb{Q}$ the risk-neutral measure.

According to the fundamental theorem of asset pricing (Harrison and Pliska, 1981), we have the following result on the arbitrage-free valuation of the derivative, where we have used $E_t^Q[X]$ for $E^Q[X|\mathcal{G}_t]$ for notational simplicity.

Definition 15.1.1. *The risk-neutral valuation of the derivative to the counterparties is*

$$V_f(0) = E_0^Q\left[e^{-\int_0^{\tau\wedge T} \hat{r}_s ds} V(\tau \wedge T)\right]. \qquad (15.5)$$

When Y_T is $\mathcal{F}_T$-adapted, Equation 15.5 for risk-neutral valuation contains bilateral credit valuation adjustments (BCVA) as price components. In fact, there is

$$\begin{aligned} V_f(0) &= E_0^Q[1_{\{\tau>T\}}e^{-\int_0^T \hat{r}_s ds}Y_T + 1_{\{\tau\leq T\}}e^{-\int_0^\tau \hat{r}_s ds}V_\tau] \\ &= E_0^Q[e^{-\int_0^T \hat{r}_s ds}Y_T] - E_0^Q[1_{\{\tau\leq T\}}e^{-\int_0^T \hat{r}_s ds}Y_T] \\ &\qquad\qquad + E_0^Q[1_{\{\tau\leq T\}}e^{-\int_0^\tau \hat{r}_s ds}V_\tau] \\ &= V_e(0) + E_0^Q[1_{\{\tau\leq T\}}(e^{-\int_0^\tau \hat{r}_s ds}V_\tau - e^{-\int_0^T \hat{r}_s ds}Y_T)], \end{aligned} \qquad (15.6)$$

where

$$V_e(t) = E_t^Q[e^{-\int_t^T \hat{r}_s ds}Y_T]$$

is the no-default value (NDV) of the derivative, which is defined in terms of the $\mathcal{F}_T$-adapted payoff functions. Since Y_T is $\mathcal{F}_T$-adapted and thus independent of $\mathcal{H}_T$, we have, by using the tower law,

$$
\begin{aligned}
E_0^Q[1_{\{\tau \leq T\}} e^{-\int_0^T \hat{r}_s ds} Y_T] &= E_0^Q[1_{\{\tau \leq T\}} E^Q[e^{-\int_0^T \hat{r}_s ds} Y_T | \mathcal{G}_\tau]] \\
&= E_0^Q[1_{\{\tau \leq T\}} E^Q[e^{-\int_0^T \hat{r}_s ds} Y_T | \mathcal{F}_\tau]] \qquad (15.7) \\
&= E_0^Q[1_{\{\tau \leq T\}} e^{-\int_0^\tau \hat{r}_s ds} V_e(\tau)].
\end{aligned}
$$

By substituting Equation 15.7 back to Equation 15.6 and distinguishing between $\tau = \tau_B$ and $\tau = \tau_C$, we arrive at

Corollary 15.1.1. *When Y_T is $\mathcal{F}_T$-adapted, the risk-neutral valuation of the derivative has the following decomposition:*

$$
V_f(0) = V_e(0) + CVA_B + CVA_C, \qquad (15.8)
$$

where

$$
\begin{aligned}
CVA_B &= E_0^Q[1_{\{\tau = \tau_B \leq T\}} e^{-\int_0^{\tau_B} \hat{r}_s ds} (V(\tau_B) - V_e(\tau_B))], \\
CVA_C &= E_0^Q[1_{\{\tau = \tau_C \leq T\}} e^{-\int_0^{\tau_C} \hat{r}_s ds} (V(\tau_C) - V_e(\tau_C))].
\end{aligned} \qquad (15.9)
$$

$\square$

Being $\mathcal{F}_T$-adapted means Y_T is always definable, regardless of the default statuses of the counterparties. In other words, a counterparty default will never cause the default of the underlying security. This is the case, for example, when the underlying is a stock index.

We gain two insights from formulae 15.8 and 15.9. First, the bilateral CVA is part of the risk-neutral valuation, which is symmetrical to the counterparties[12]. Second, the NDV of the derivative is the only right choice for the MtM value of the derivative when it comes to the calculation of LGD. Similar bilateral CVA formulae are also seen in other literature, e.g., Brigo et al. (2012) and Bo and Capponi (2013). The main difference between our result and theirs is that our LGD is the difference between the post-default value and the no-default value, while the LGD of the other two papers is the percentage recovery of the pre-default value.

We have the following interpretations for NDV and CVAs.

Proposition 15.1.1. *In the absence of funding cost, $V_e(0)$ is the present value (PV) of cost to replicate the payoff, and CVA_i is the $\mathbb{Q}$-expected present value of the cost to replicate the LGD by party i.*

---

[12]From the perspective of B, CVA_C is the CVA, while CVA_B is the DVA. The combined value of CVA_B and CVA_C is the bilateral CVA.

Proof. Without loss of generality, we first assume that the derivative is an asset to B and thus a liability to C. Then, C will replicate the derivative with a portfolio of repos and cash:

$$\Pi_C(t) = \delta_C Z_S(t) + \beta_C(t), \tag{15.10}$$

where $Z_S(t) = 0$,

$$\delta_C = \frac{\partial V_e(t)}{\partial S},$$

and

$$\beta_C(t) = \beta_C(0)e^{\int_0^t \hat{r}_v dv} + \int_0^t e^{\int_u^t \hat{r}_v dv} \delta_C(u) dZ_S(u). \tag{15.11}$$

To achieve perfect replication, $\beta_C(0)$ must be chosen so that

$$\beta_C(\tau \wedge T) = V_e(\tau \wedge T).$$

Taking $\mathbb{Q}$ expectation on both sides of the equation above after discounting and making use of the martingale property of $Z_S(t)$, we obtain

$$\beta_C(0) = E_0^Q \left[e^{-\int_0^{\tau \wedge T} \hat{r}_s ds} V_e(\tau \wedge T) \right] = V_e(0),$$

so that the cost to replicate the derivatives payout is just $V_e(0)$.

Party B, meanwhile, can hedge against his LGD using CDS. Note that our strategy to hedge against the counterparty credit risk (CCR) is based on the fact that, whenever liquid, CDS are the primary means for CVA desks to hedge against credit risks, see Green (2016).[13] The hedging portfolio of party B consists of CDS and cash:

$$\Pi_B(t) = \alpha_B U_C(t) + \beta_B(t),$$

where α_B is the number of CDS, defined by

$$\alpha_B(t) = -\frac{V(\tau_C = t) - V_e(t)}{L_C}, \tag{15.12}$$

and $\beta_B(t)$ is the debt in cash due to the CDS fee payment and interest accrual:

$$\beta_B(t) = \int_0^t e^{\int_s^t \hat{r}_u du} \alpha_B(s)(-s_C) ds.$$

---

[13]Names without or with illiquid CDS can be directly mapped to the liquid CDS of a name with similar profile.

The expected present value of the debt for hedging the LGD is

$$E_0^Q \left[e^{-\int_0^{\tau \wedge T} \hat{r}_s ds} \beta_B(\tau \wedge T) \right]$$

$$= E_0^Q \left[\int_0^T e^{-\int_0^t \hat{r}_s ds} \alpha_B(t)(-s_C) 1_{\{\tau > t\}} dt \right]$$

$$= E_0^Q \left[\int_0^T e^{-\int_0^t \hat{r}_s ds} (V(\tau_C = t) - V_e(t)) \lambda_C e^{-\int_0^t (\lambda_B + \lambda_C) ds} dt \right] \qquad (15.13)$$

$$= E_0^Q [1_{\{\tau = \tau_C \leq T\}} e^{-\int_0^{\tau_C} \hat{r}_s ds} (V(\tau_C) - V_e(\tau_C))],$$

which is CVA$_C$. Since the derivative is a liability to C, C does not hedge its counterparty risk, so that there is CVA$_B = 0$. It then follows that

$$E_0^Q \left[e^{-\int_0^{\tau \wedge T} \hat{r}_s ds} \beta_C(\tau \wedge T) \right] + E_0^Q \left[e^{-\int_0^{\tau \wedge T} \hat{r}_s ds} \beta_B(\tau \wedge T) \right]$$
$$= V_e(0) + \text{CVA}_C = V_f(0),$$

i.e., the fair price of the derivative is the replication cost of the payout minus the cost to the counterparty to hedge against the loss given default.

When the derivative is a liability to B, or the derivative can switch from an asset to a liability to B, C should be compensated with the present value of the cost to hedge against his LGD which, by symmetry, is CVA$_B$. Hence, in general, the fair value of a derivative to the trading parties is the cost to replicate the derivatives payout minus the bilateral costs to hedge against the loss given their counterparty's defaults. □

We make three comments here. First, while $V_e(0)$ is the actual cost to replicate the derivatives payout, CVA_i is not, and it is the expected value of the cost to replicate the LGD. Second, in the absence of funding spreads, there is always

$$E_0^Q \left[e^{-\int_0^{\tau \wedge T} \hat{r}_s ds} \beta_i(\tau \wedge T) - \beta_i(0) \right] = 0, \quad \text{for } i = B \text{ and } C, \qquad (15.14)$$

regardless of what price is taken for the trade. Yet in the presence of funding spreads, the above equality will no longer hold, as we shall witness in the following section, thus giving rise to funding costs that can affect the P&L of the trade. Third, we will confirm later that the risk-neutral value is exactly the fair price for a derivatives trade between the counterparties, with or without idiosyncratic funding costs.

15.2 The Rise of Other xVA

We now take the funding spreads for margins, collaterals and capitals into account. Once B and C enter into a trade at time $t = 0$ for a value V_0

to B, the two parties start hedging, using repos and/or CDS, posting margins/collaterals and setting aside capital according to margin, collateral and capital rules. The hedging portfolios of the two parties can be expressed as

$$\Pi_B(t) = \delta_B Z_S(t) + \alpha_B U_C(t) + \beta_B(t),$$
$$\Pi_C(t) = \delta_C Z_S(t) + \alpha_C U_B(t) + \beta_C(t),$$

(15.15)

where δ_i is the number of repo contracts held by party i, which can be

$$\delta_B = -\frac{\partial V_e(t)}{\partial S}, \quad \delta_C = \frac{\partial V_e(t)}{\partial S},$$

and α_i is the number of CDS contracts taken up by party i to hedge against LGD:

$$\alpha_B(t) = -\frac{V(\tau_C = t) - V_e(t)}{L_C}, \quad \alpha_C(t) = -\frac{V(\tau_B = t) - V_e(t)}{L_B},$$

and $\beta_i(t)$ is the total value of party i's cash in his savings account, with the initial values

$$\beta_B(0) = \Pi_B(0) = -V_0, \quad \text{and} \quad \beta_C(0) = \Pi_C(0) = V_0.$$

Here, V_0 is the initial premium payment paid by or received by B, depending on whether it is positive or negative. Note that with the above setup, the funding costs for the initial premium of the derivative is also taken into account in evaluating the derivatives values to the trading parties.

The savings account serves a number of purposes:

1. Saving or unsecured borrowing.

2. To deposit or to pay cash for the revisions of IM, VM and capital requirements.

3. To take the P&L from the revisions of hedging.

Without loss of generality, let us focus on β_B, the savings account of party B. To describe the evolution of the balance of the savings account, we need the following notations:

x_B	—	The spread for unsecured borrowing or lending over $\hat{r}_t$ for party B,
$x_B^{(I)}$	—	The funding spreads over the $\hat{r}_t$ for the initial margin of B,
$x_B^{(X)}$	—	The funding spreads over the $\hat{r}_t$ for variable margin or collateral of B,
$\gamma_B^{(K)}$	—	The funding spreads over the $\hat{r}_t$ for the capital of B (so-called net return of the capital to the shareholders),
$\phi_B(t)K_B(t)$	—	The portion of capital allocated to the savings account of B,
$\delta_B dZ_S(t)$	—	The P&L for delta hedging using repos over $(t, t+dt)$, and
$\alpha_B dU_C(t)$	—	The P&L for hedging LGD of C over $(t, t+dt)$ using CDS.

Here, $\phi_B(t) \in [0,1]$, which represents the fraction of capital borrowed from shareholders being reallocated to the savings account. Then, the change of the savings account over the time interval $(t, t+dt)$ can be described as

$$d\beta_B(t) = \Big((\hat{r}_t + x_B)[\beta_B(t) + \phi_B(t)K_B(t)] - x_B^{(I)} I_B(t) - x_B^{(X)} X_B(t)$$
$$- [\gamma_B^{(K)} + \phi_B(t)\hat{r}_t]K_B(t) \Big) dt + \delta_B dZ_S(t) + \alpha_B dU_C(t), \tag{15.16}$$

for which we need to make additional elaborations.

1. The spread for the savings account can be nonlinear and asymmetric, such that

$$x_B(\beta_B(t)) = x_B^{(l)} 1_{\{\beta_B(t) \geq 0\}} + x_B^{(b)} 1_{\{\beta_B(t) < 0\}}.$$

Here, $x_B^{(b)} \geq 0$ is the default risk premium for party B, and $x_B^{(l)} \geq 0$ is the default risk premium of the borrower. In general, there is $x_B^{(l)} \neq x_B^{(b)}$.

2. Due to a partial allocation of capital to the savings account, the interest accrual of the savings account becomes $(\hat{r}_t + x_B)[\beta_B(t) + \phi_B(t)K_B(t)]dt$, in an expense of $\hat{r}_t\phi_B(t)K_B(t)dt$ to the capital account.

3. In general, the funding spreads and the return on capital are positive, reflecting a simple reality that costs will be incurred for borrowing funds or capitals. Yet for VM or collaterals, there can be exceptions in case of rehypothecation, when the return from rehypothecation is higher than the borrowing cost, the corresponding spread turns negative, representing a funding benefit to the party who posts VM or collaterals.[14]

4. When the derivative is an asset, there will be a jump in the balance of the savings account upon counterparty default due to the CDS payment for LGD.

We now analyze the P&L of the trade to party B. The present value of the P&L is simply the PV of B's hedged portfolio's value at the termination time of the option (upon either default or maturity):

$$P\&L = e^{-\int_0^{\tau \wedge T} \hat{r}_s ds} [\beta_B(\tau \wedge T) + V(\tau \wedge T)].$$

Due to the perfect replication of the market risk and the possible LGD, there is the equality

$$V(\tau \wedge T) = V_e(\tau \wedge T) - \alpha_B L_C dJ_C(\tau \wedge T)$$
$$= V_e(0)e^{\int_0^{t \wedge T} \hat{r}_v dv} - \int_0^{t \wedge T} e^{\int_u^{t \wedge T} \hat{r}_v dv} \delta_B(u)dZ_S(u) - \alpha_B L_C dJ_C(\tau \wedge T).$$

---

[14]It should be pointed out, however, that the higher return is often due to the exposure to credit risk. Thus, for pricing purposes, one may have to adjust the spread properly to account for the credit risk.

We then rewrite the P&L into

$$P\&L = \left[(\beta_B(0) + V_e(0)) - \int_0^{t \wedge T} e^{-\int_0^u \hat{r}_s ds} \alpha_B(u) s_C du \right]$$

$$+ \left[e^{-\int_0^{t \wedge T} \hat{r}_s ds} \beta_B(\tau \wedge T) - \beta_B(0) - \int_0^{t \wedge T} e^{-\int_0^u \hat{r}_s ds} \delta_B(u) dZ_S(u) \right.$$

$$\left. - \int_0^{t \wedge T} e^{-\int_0^u \hat{r}_s ds} \alpha_B(u) dU_C(u) \right]$$

$$= I + II.$$

where I and II represent the terms enclosed by the two pairs of square brackets, respectively. We will show next that I represents the PV of the shortfall of payout replication, and II represents the PV of accumulative costs due to various funding spreads.

We first look at I. According to Equation 15.13,

$$E_0^Q[I] = \beta_B(0) + V_e(0) + CVA_C$$
$$= \beta_B(0) + V_f(0) - CVA_B.$$

If we take the trade price to be $V_0 = -\beta_B(0) = V_f(0)$, then there is

$$E_0^Q[I] = -CVA_B,$$

meaning that the debit valuation adjustment for party B is part of his P&L, and the DVA to B will result in a replication shortfall.

Next, we look at II. According to Equation 15.16, we can derive that

$$d\left(e^{-\int_0^t \hat{r}_s ds} \beta_B(t) \right) = e^{-\int_0^t \hat{r}_s ds} \left(x_B[\beta_B(t) + \phi_B(t)K_B(t)] - x_B^{(I)} I_B(t) \right.$$

$$\left. - x_B^{(X)} X_B(t) - \gamma_B^{(K)} K_B(t) \right) dt$$

$$+ \delta_B e^{-\int_0^t \hat{r}_s ds} dZ_S(t) + \alpha_B e^{-\int_0^t \hat{r}_s ds} dU_C(t).$$

Integrating the above equation from 0 to $t \wedge T$, we then have

$$e^{-\int_0^{\tau \wedge T} \hat{r}_s ds} \beta_B(\tau \wedge T) - \beta_B(0)$$

$$= \int_0^{\tau \wedge T} e^{-\int_0^t \hat{r}_s ds} \left(x_B[\beta_B(t) + \phi_B(t)K_B(t)] - x_B^{(I)} I_B(t) \right.$$

$$\left. - x_B^{(X)} X_B(t) - \gamma_B^{(K)} K_B(t) \right) dt \tag{15.17}$$

$$+ \int_0^{\tau \wedge T} \delta_B e^{-\int_0^t \hat{r}_s ds} dZ_S(t) + \int_0^{\tau \wedge T} \alpha_B e^{-\int_0^t \hat{r}_s ds} dU_C(t).$$

Substituting the right-hand side for $e^{-\int_0^{\tau \wedge T} r_s ds} \beta_B(\tau \wedge T) - \beta_B(0)$ in the expression of II, we arrive at

Proposition 15.2.1. *The realized funding cost to B is*

$$II = \int_0^{\tau \wedge T} e^{-\int_0^t \hat{r}_s ds} \left(x_B [\beta_B(t) + \phi_B(t) K_B(t)] - x_B^{(I)} I_B(t) \right.$$
$$\left. - x_B^{(X)} X_B(t) - \gamma_B^{(K)} K_B(t) \right) dt. \tag{15.18}$$

Note that II is a path-dependent random variable, except for the case of zero funding spreads for margins, collaterals and capitals, when there is $II \equiv 0$, regardless of the price V_0 taken for the trade. Following the market convention, we define FVA as the expectation of the present value of realized funding cost under the risk-neutral measure.

Definition 15.2.1. *The FVA to party B is the expected value of excess cost due to the funding spreads for various funding transactions under the risk-neutral measure:*

$$FVA_B = E_0^Q \left[\int_0^{\tau \wedge T} e^{-\int_0^t \hat{r}_s ds} \left(x_B [\beta_B(t) + \phi_B(t) K_B(t)] - x_B^{(I)} I_B(t) \right. \right.$$
$$\left. \left. - x_B^{(X)} X_B(t) - \gamma_B^{(K)} K_B(t) \right) dt \right].$$

We comment here that the FVA so defined is not so useful because neither can it be accepted for valuation adjustment for trade price, nor can it serve as a sound measure for funding risks because it cannot be replicated. Yet the funding costs and risks behind it must be managed, which will be addressed shortly.

From the FVA definition we immediately have

Definition 15.2.2. *To party B, the funding valuation adjustment is given by*

$$FVA_B = FCA_B + MVA_B + ColVA_B + KVA_B, \tag{15.19}$$

with

$$FCA_B = E_0^Q \left[\int_0^{T \wedge \tau} x_B e^{-\int_0^t r_s ds} [\beta_B(t) + \phi_B(t) K_B(t)] dt \right],$$

$$MVA_B = -E_0^Q \left[\int_0^{T \wedge \tau} x_B^{(I)} e^{-\int_0^t r_s ds} I_B(t) dt \right],$$

$$ColVA_B = -E_0^Q \left[\int_0^{T \wedge \tau} x_B^{(X)} e^{-\int_0^t r_s ds} X_B(t) dt \right], \tag{15.20}$$

$$KVA_B = -E_0^Q \left[\int_0^{T \wedge \tau} \gamma_B^{(K)} e^{-\int_0^t r_s ds} K_B(t) dt \right],$$

and $\beta_B(t)$ evolving according to Equation 15.16.

We want to emphasize here that in general FCA and FVA can depend on V_0, the trade price of the derivative, through $\beta_B(0) = -V_0$. Hence, whenever necessary, we will write $\text{FCA}(-V_0)$ or $\text{FVA}(-V_0)$ to highlight such dependence.

Different choices of V_0 will lead to some major existing results on xVA. The first choice is $V_0 = V_e(0)$. When we ignore the funding spreads for collaterals and capitals and do not hedge against counterparty default, i.e., letting $x_B^{(I)} = x_B^{(X)} = \gamma_B^{(K)} = 0$, we will then have $\beta_B(t) = -V_e(t)$ under the diffusion dynamics for the underlying security due to perfect replication, and consequently have

$$\text{FVA}_B(-V_e(0)) = \text{FCA}_B(-V_e(0)) = -E_0^Q \left[\int_0^{\tau \wedge T} x_B e^{-\int_0^t \hat{r}_s ds} V_e(t) dt \right],$$

which is identical to Equation 15.1, the FVA formula of the prevalent model.

The second choice of V_0 is the value of the derivative to a party, which is defined to be the value with which the risk-neutral expected return of the trade is the risk-free rate, $\hat{r}_t$. To party B, the value of the derivative, $V_0^{(B)}$, satisfies (Li and Wu, 2015)

$$V_0^{(B)} = V_f(0) + \text{FVA}(-V_0^{(B)}),$$

which is an implicit equation for $V_0^{(B)}$. By solving this equation we will then reproduce the xVA formulae of Green and Kenyon (2015). With additional simplifying assumptions, we can also reproduce the FCA formula of Piterbarg (2010) and the bilateral CVA and FCA formulae of Burgard and Kjaer (2011). It should be pointed out, however, the approach we have taken is much less restrictive. To derive the similar results, Burgard and Kjaer (2011) and Green and Kenyon (2015) need the "funding condition" (see Equation 6 of both papers), which is equivalent to requiring perfect unilateral replication of a derivative at all time until the first default or the maturity. Given the invalidation of "hedging own default using own bonds" (Castagna, 2012), we can see that such a "funding condition" holds only narrowly for derivatives receivables under the diffusion model for the underlying securities. In addition, our formulae have accommodated asymmetric funding spreads for unsecured borrowing and lending, which is not included in the other papers.

There is, however, normally a gap between the values to the counterparties. The value to party C, the counterparty, is implied by the equation

$$-V_0^{(C)} = -V_f(0) + \text{FVA}_C(V_0^{(C)}). \tag{15.21}$$

In Li and Wu (2015), $V_0^{(B)}$ and $V_0^{(C)}$ are called bid and ask prices, respectively. Note that the bid and ask prices so defined are subject to nonlinear pricing rules, and such nonlinearity occurs when one prices derivatives as the replication costs, as was observed in Pallavicini et al. (2011), Crépey (2011) and Bichuch *et al.* (2015). The bid and ask prices are in general different due to the idiosyncratic nature of funding costs. To see this, we let the derivative be an asset to B and assume there is no funding benefit but funding costs to both parties, then there will be $\text{FVA}_i \leq 0, i = B$ and C, thus yielding the order $V_0^{(B)} \leq V_0^{(C)}$. As has been shown by Li and Wu (2015), a trade that takes place at any price within the interval $[V_0^{(B)}, V_0^{(C)}]$ will be non-arbitrageable to both parties and, in addition, will make risk-neutral expected returns to both parties simultaneously lower than the risk-free rate.

To conclude, to derive a fair price which achieves symmetry in pricing and asset-liability symmetry in accounting, we have to ignore all idiosyncratic funding spreads, by taking $x_i^{(b)} = x_i^{(l)} = x_i^{(I)} = x_i^{(X)} = \gamma_i^{(K)} = 0, i = B$ and C and including the market funding liquidity risk premium into the risk-free rate (Gregory, 2015), such that $\hat{r}_t = r_t + x_m$, and then perform risk-neutral valuation for the defaultable claim. The result is given by Equation 15.5 or Equation 15.8 if the payoff function is $\mathcal{F}_T$-adapted.

We finish this section with several more comments. First, for P&L accounting, we suggest the booking of the idiosyncratic component(s) of the realized funding costs which, to party B, is

$$
FC_B(t) = \int_0^t e^{\int_u^t \hat{r}_s ds} \left(x_B[\beta_B(u) + \phi_B(u)K_B(u)] - x_B^{(I)} I_B(u) \right.
$$
$$
\left. - x_B^{(X)} X_B(u) - \gamma_B^{(K)} K_B(u) \right) du \tag{15.22}
$$

for a trade entered at time $t = 0$. Actually, it is straightforward to show

$$
V_t + \Pi_B(t) = FC_B(t) + CVA_B(0)e^{\int_0^t \hat{r}_s ds}, \tag{15.23}
$$

which means that the P&L of the hedged portfolios is nothing else but the realized total (idiosyncratic) funding costs plus the accrued value of the initial credit valuation adjustment.

Second, more effective methodologies should be adopted for managing the funding risk. Over the years some major banks have been setting aside reserves based on the FVA numbers (Levine, 2014). In a similar spirit, some researchers have advocated the deduction of FVA from a bank's common equity Tier 1 (CET1) capital (Albanese et al., 2015; Andersen et al., 2016). It should be pointed out that, however, as an expected value of funding cost under the risk-neutral measure, an FVA number can be far from the funding costs actually realized to a desk or a firm. For better risk management, we suggest the adoption of other established risk metrics, like the VaR or CVaR, by making use of Equation 15.16 and Equation 15.22 through Monte Carlo simulations under the real-world measure. Note that with these metrics for funding risks, we can also perform stress testing for various scenarios, including liquidity crunch.

Third, CVA is exposed to market risk also as it fluctuates with the share prices, and the CVA desks want to hedge the CVA dynamically. The hedging against share-price risk is based on the usual delta hedging, which is discussed in detail in Li and Wu (2015).

Finally, the xVA formulae 15.9 and 15.20 hold for general pricing models, including models with stochastic interest rates, stochastic hazard rates for defaults and stochastic funding spreads, and models with jumps and stochastic volatilities in their state variables.

15.3 Examples

For demonstration purposes, we consider the pricing of the xVA of a bilateral interest-rate swap, which is subject to IM, VM and capital, from the viewpoint of B. We will take the approach of IMM in the specifications of IM, VM and capital. Without loss of generality, we only consider cash as the posted assets for both IM and VM. Note that, due to the IM, there can be over-collateralization for the swap trade from time to time, which can result in very small CVA and DVA.

The initial margin requirement for potential future exposure is

$$I_B(t) \quad — \quad 99\% \text{ VaR of the P\&L of the swap for a 10-day horizon,}$$

and the initial margin account will accrue using the OIS rate.

The posting of VM is required only for the party of liability, and the amount will be revised periodically at a set of predetermined dates, $\{t_i\}$. The subsequent additions or reductions of VM are subject to (1) a threshold value, H, and (2) a minimum transfer amount, m, with $H \geq m \geq 0$. Essentially, a revision to VM is required only if (1) the liability exceeds H, and (2) the change of the liability (due to either appreciation or depreciation) exceeds m. Between two moments of margin revision, the VMs accrue interest with an interest rate no less than the OIS rate. To the counterparties, the VMs start with $X_i(t_0-) = 0, i = B$ and C, and are reset according to the following scheme: for $i = 0, 1, \ldots,$

$$Adj_B(t_i) = |V_e(t_i)| - H - X_B(t_i-),$$
$$X_B(t_i+) = \left(X_B(t_i-) - 1_{\{V_e(t_i)<0\}} \, Adj_B(t_i) \, 1_{\{|Adj_B(t_i)|\geq m\}} \right)^+,$$

and

$$Adj_C(t_i) = |V_e(t_i)| - H - X_C(t_i-),$$
$$X_C(t_i+) = \left(X_C(t_i-) + 1_{\{V_e(t_i)>0\}} \, Adj_C(t_i) \, 1_{\{|Adj_C(t_i)|\geq m\}} \right)^+,$$

where t_i- and t_i+ denote the moment immediately before and after a revision. Upon a bilateral default, the default settlement value of the swap is described in Equation 15.2, with $c_i(t) = I_i(t) + X_i(t)$ to be the total value of the collateral posted by party $i = B$ or C.

For simplicity, we consider only the capital requirement for market risk under IMM, which is

$$K_B(t) \quad — \quad 99\% \text{ VaR of the P\&L of the swap for 10-day horizon.}$$

We use no risk capital to stack up the savings account, meaning that $\phi_B = 0$ is taken.

We use the single-factor linear Gaussian model (LGM) (Hagan and Woodward, 1999), introduced in Section 6.6, for pricing purposes.

The LGM is naturally calibrated to the discount curve. The function $h(t)$ takes the form of Equation 6.83 with $\kappa = 0.03$. Before pricing options, we normally calibrate α_t to vanilla interest-rate derivatives like caps, floors or swaptions.

Our example is with an ATM receiver's swap of 20-year maturity, which receives fixed-rate interest annually and pays floating-rate interest semiannually according to 6M Euribor, out of a notional value of €10m. We make use of the actual 6M Euribor projection curve and the OIS curve of 02/05/2016, and calculate the 20-year ATM swap rate to be 1.135%. The local volatility function α_t is taken to be a piecewise linear function that is calibrated to the implied normal volatility matrix of swaptions of the same date, provided in Table 15.A in this chapter's appendix.

For calculating the xVA, we also need to specify the parameters for collateral and funding:

margin revision frequency — daily,
credit risk — $L_B = L_C = 0.6$,
$\lambda_C = 1.0\%$, $s_B = 0\% : 0.25\% : 3\%$,
collateral — $H = €100000$, $m = €0$.

The funding spreads for the lending/borrowing through savings account, for collateral and for capital are

savings account — $x_B^{(l)} = 0$, $x_B^{(b)} = \lambda_B L_B$,
IM — $x_B^{(I)} = \lambda_B L_B$,
VM — $x_B^{(X)} = 0$ (due to rehypothecation),
capital — $\gamma_B^{(K)} = 8\%$.

The xVA against the default rate of party B, the issuer of the swap, is shown in Figure 15.1. As one can see from the plot, CVA_B and CVA_C are all negligibly small due to over-collateralization, and ColVA is equal to zero due to rehypothecation (which eliminates the funding cost for VM). Compared with other xVA, KVA is significantly larger, apparently due to the 8% return rate on the capital demanded by the shareholders, which is significantly higher than other funding spreads. Finally, because a higher hazard rate shortens the expected default time of the issuer and consequently reduces the average time of accrual for the capital, the KVA in absolute terms decreases when the hazard rate increases.

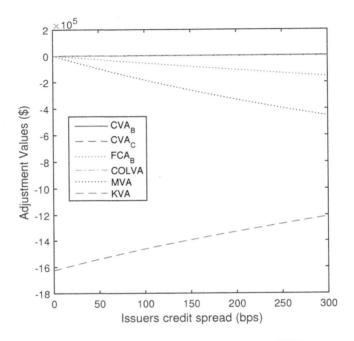

FIGURE 15.1: The xVA for the 20-year ATM swap.

15.4   Notes

As a major response to the 2007-08 financial crisis, xVA has been a focal point to regulators, practitioners, and researchers. This chapter provides a rather general yet simple framework to analyze and quantify xVA. We show that the CVA and DVA are the expected value of costs to hedge against counterparty default, and FVA is the interest accrual of funding costs due to funding spreads. We have the conclusion that the fair price of a derivative is the classical risk-neutral value in a market where the rate for unsecured lending and borrowing is the risk-free rate plus the market funding liquidity risk premium, and such a value naturally contains the bilateral CVA. There are remaining issues we did not touch upon, such as, under the (IMM) or advanced method, how to set the levels of initial margin (IM), variable margin (VM) and capital level. The level of IM, VM and capital will have impacts on the bilateral CVA, and ultimately on derivatives price. Related to IM, VM and capital, there are also the issues of collateral management, spreadsheet optimization, and management of xVA; these are the issues that are undergoing research and debate, and are beyond the scope of this book.

Appendix

TABLE 15.A: Implied normal volatilities of swaptions

Swap Maturity	1Y	2Y	3Y	4Y	5Y	7Y	10Y	15Y	20Y
Option Expiry									
1Y	0.003543	0.003889	0.004316	0.004802	0.00527	0.006123	0.006978	0.007557	0.007918
2Y	0.004732	0.00498	0.005279	0.005629	0.005957	0.006563	0.007273	0.007508	0.007752
3Y	0.00586	0.006018	0.006218	0.00642	0.006646	0.007066	0.007539	0.007663	0.007657
4Y	0.006539	0.006691	0.006784	0.006967	0.007112	0.007394	0.00773	0.007573	0.007489
5Y	0.007013	0.007117	0.007194	0.0073	0.007443	0.007648	0.00782	0.007505	0.007363
7Y	0.007519	0.007573	0.007612	0.007735	0.007807	0.007883	0.007698	0.007392	0.007117
10Y	0.007668	0.007642	0.007637	0.007882	0.007705	0.007869	0.007611	0.007116	0.006848
15Y	0.007492	0.007636	0.007657	0.007647	0.007394	0.007559	0.007199	0.006833	0.006442
20Y	0.007198	0.007364	0.007364	0.007294	0.007216	0.007141	0.006964	0.006419	0.005994

References

Albanese C, Andersen L and Iabichino S, 2015. FVA accounting, risk management and collateral trading. *Risk*, February, 64–69.

Andersen L, 2007. Efficient simulation of the Heston stochastic volatility model. Working paper, Bank of America Securities.

Andersen L and Andreasen J, 2000. Volatility skews and extensions of the LIBOR market model. *Applied Mathematical Finance* 7(1): 1–32.

Andersen L and Andreasen J, 2002. Volatile Volatilities. *Risk*, 15: 163–168.

Andersen L and Brotherton-Ratcliffe R, 2001. Extended LIBOR market models with stochastic volatility. *Journal of Computational Finance* 9(1): 1–40.

Andersen L, Duffie D and Song Y, 2016. Funding value adjustments. Working paper, http://ssrn.com/abstract=2746010.

Andersen L, Sidenius J and Basu S, 2003. All your hedges in one basket. *Risk*, November, 67–72.

Arrow K and Debreu G, 1954. Existence of an equilibrium for a economy. *Econometrica* 22: 265–290.

Arvanitis A and Gregory J, 2001. Credit: the complete guide to pricing, hedging and risk management. Risk Books.

Avellaneda M and Laurence P, 1999. *Quantitative Modeling of Derivative Securities: From Theory to Practice*. Chapman & Hall/CRC, Boca Raton, FL.

Bailey A, 2017. The future of LIBOR. https://www.fca.org.uk/news/speeches/the-future-of-libor.

Bailey A, 2018. Transforming culture in financial services. https://www.fca.org.uk/news/speeches/transforming-culture-financial-services.

Balduzzi P, Das S, Foresi S, and Sundaram R, 1996. A simple approach to three factor affine term structure models. *Journal of Fixed Income* 6: 43–53.

Barone E and Castagna A, 1997. The information content of TIPS. Internal Report. SanPaolo IMI, Turin and Banca IMI, Milan.

Baxter M and Rennie A, 1996. *Financial Calculus: An Introduction to Derivative Pricing*. Cambridge University Press, Cambridge.

Belgrade N and Benhamou E, 2004a. Reconciling Year on Year and Zero Coupon Inflation Swap: A Market Model Approach. Preprint, CDC Ixis Capital Markets. Downloadable at: http://papers.ssrn.com/sol3/papers.cfm?abstract-id=583641.

Belgrade N and Benhamou E, 2004b. Smart modeling of the inflation market: taking into account the seasonality. Preprint, CDC Ixis Capital Markets.

Belgrade N, Benhamou E and Koehler E, 2004. A Market Model for Inflation. Preprint, CDC Ixis Capital Markets. Downloadable at: http://papers.ssrn.com/sol3/papers.cfm?abstract-id=576081.

Bichuch M, Capponi A and Sturm S, 2015. Arbitrage-free pricing of xVA - part I: framework and explicit examples. Working paper, arXiv:1501.05893[q-fin.PR].

Bielecki T and Rutkowski M, 2001. *Credit Risk: Modeling, Valuation and Hedging*. Springer-Verlag.

Björk T, Di Masi G, Kabanov Y and Runggaldier W, 1997. Towards a general theory of bond markets. *Finance and Stochastics* 1: 141–174.

Björk T, Kabanov Y and Runggaldier W, 1997. Bond market structure in the presence of marked point processes. *Mathematical Finance* 7(2): 211–223.

Black F, 1976. The pricing of commodity contracts. *Journal of Financial Economics* 3: 167–179.

Black F and Cox J, 1976. Valuing corporate securities: some effects of bond indenture provision. *Journal of Finance* 31: 351–367

Black F, Derman E and Toy W, 1990. A one-factor model of interest rates and its application to Treasury bond options. *Financial Analysts Journal* 46(1): 33–39.

Black F and Karasinski P, 1991. Bond and option pricing when short rates are lognormal. *Financial Analysts Journal* 47(4): 52–59.

Black F and Scholes M, 1973. The pricing of options and corporate liabilities. *Journal of Political Economy* 81: 637–659.

Brace A, Gatarek D, and Musiela M, 1997. The market model of interest rate dynamics. *Mathematical Finance* 7: 127–154.

Brace A and Womersley R, 2000. Exact fit to the swaption volatility matrix using semi-definite programming. Working paper, National Australia Bank and University of New South Wales.

Brigo D, 2004. Candidate market models and the calibrated CIR++ stochastic intensity model for credit default swap options and callable floaters. *Proceedings of the 4th ICS Conference*, Tokyo, March 18–19.

Brigo D, 2005. Market models for CDS options and callable floaters. *Risk*, January, 89–94.

Brigo D, Capponi A, Pallavicini A and Papatheodorou D, 2011. Collateral margining in arbitrage-free counterparty valuation adjustment including rehypotecation and netting. Working paper, http://ssrn.com/abstract=1744101.

Brigo D, Capponi A and Pallavicini A (2012). Arbitrage-free bilateral counterparty risk valuation under collateralization and application to credit default swaps. *Mahtematical Finance* 24(1): 125–146.

Brigo D and Liinev J, 2003. On the distributional distance between the Libor and the Swap market models. Working paper, Banca IMI and University of Ghent.

Brigo D, Liinev J, Mercurio F and Rapisarda F, 2004. On the distributional distance between the lognormal LIBOR and Swap market models. Working paper, Banca IMI, Italy.

Brotherton-Ratcliffe R and Iben B, 1993. Yield curve applications of swap products. In: RJ Schwartz and C Smith (eds), *Advanced Strategies in Financial Risk Management*, 400–450. New York Institute of Finance, New York.

Burgard C and Kjaer M, 2011. Funding cost adjustments for derivatives. *Asia Risk*, November, 63–67.

Burne K and Eisen B, 2016. Libor Alternatives Have Their Own Issues. *Wall Street Journal*, November 7.

Cairns A, 2000. A multifactor model for the term structure and inflation for long-term risk management with an extension to the equities market. Preprint. Heriot-Watt University, Edinburgh.

Cameron M, 2013. The black art of FVA. *Risk*, April, 15–18.

Carr P and Madan D, 1999. Option valuation using the fast Fourier transform. *Journal of Computational Finance* 3: 463–520.

Carr P and Wu L, 2008. Variance risk premiums. *Review of Financial Studies*, Advance Access published April 10.

Carriere J, 1996. Valuation of early exercise price of options using simulations and nonparametric regression. *Insurance: Mathematics and Economics* 19: 19–30.

Castagna A, 2012. The impossibility of DVA replication. *Risk*, November, 72–75.

Chen N and Glasserman P, (2007). Additive and multiplicative duals for American option pricing. *Finance Stoch* 11:153–179.

Chen R, 1996. *Understanding and Managing Interest Rate Risks*. World Scientific, Singapore.

Chen R, Liu B and Cheng X, 2006. Pricing the Term Structure of Inflation Risk Premia: Theory and Evidence from TIPS. Working paper, Rutgers Business School.

Chen R and Scott L, 2001. Stochastic volatility and jumps in interest-rates: an empirical analysis. Working paper, Rutgers University and Morgan Stanley.

Chen R and Scott L, 1993. Maximum likelihood estimation for a multifactor equilibrium model of the term structure of interest rates. *Journal of Fixed Income* 3: 14–31.

Clement E, Lamberton D and Protter P, 2002. An analysis of a least squares regression algorithm for American option pricing. *Finance and Stochastics* 6: 449–471.

Cox J and Rubinstein M, 1985. *Option Theory*. Prentice-Hall, Englewood Cliffs, NJ.

Cox J, Ross S and Rubinstein M, 1979. Option pricing: A simplified approach. *Journal of Financial Economics* 7: 229–264.

Cox J, Ingersoll J and Ross S, 1985. A theory of the term structure of interest rates. *Econometrica* 53(2): 385–407.

Crépey S, 2011. A BSDE approach to counterparty risk under funding constraints. Available at grozny.maths.univ-evry.fr/pages perso/crepey.

Dai Q and Singleton K, 2000. Specification analysis of affine term structure models. *Journal of Finance* 55: 1943–1978.

Duffie D, Filipovic D and Schachermayer W, 2002. Affine processes and applications in finance. Working paper, Stanford University.

Duffie D and Kan R, 1996. A yield-factor model of interest rates. *Mathematical Finance* 6: 379–406.

Duffie D, Pan J and Singleton K, 2000. Transform analysis and asset pricing for affine jump diffusions. *Econometrica* 68: 1343–1376.

Duffie D and Singleton K, 1999. Modeling term structure of defaultable bonds. *Review of Financial Studies* 12: 687–720.

Duffie D and Stein J, 2015. Reforming LIBOR and Other Financial Market Benchmarks. *Journal of Economic Perspectives* 29(2): 191–212.

Eberlein E, Kluge W and Schönbucher P, 2006. The Lèvy LIBOR model with default risk. *Journal of Credit Risk* 2(2): 3–42.

Eberlein E and Özkan F, 2005. The Levy Libor model. *Finance Stochastic* 9: 327–348.

El-Jahel L, Lindberg H and Perraudin W, 1997, Interest Rate Distributions, Yield Curve Modelling and Monetary Policy, Dempster and Pliska eds, *Mathematics of Derivative Securities*, Cambridge University Press.

Fabozzi F, 2007. *Bond Markets, Analysis, and Strategies*, 7th edition. Prentice Hall: Upper Saddle River, NJ.

FASB, 2011. Accounting standards update. Financial Accounting Standards Board, 04.

Falbo P, Paris F and Pelizzari C, 2010. Pricing inflation-link bonds. *Quantitative Finance* 10(3): 279–293.

Feller W, 1971. *An Introduction to Probability Theory and Its Applications*, Vol. 2. Wiley, New York.

Filipovic W, 2001. A general characterization of one factor affine term structure models. *Finance and Stochastics* 5(3): 389–412.

Fisher I, 1930. *The Theory of Interest*. The Macmillan Company. ISBN13 978-0879918644.

Flesaker B and Hughston L, 1996. Positive interest, *Risk Magazine*, January.

Giancarlo C and Powell J, 2017. How to Fix Libor Pains. *The Wall Street Journal*, August 3.

Giesecke K, 2003. A simple exponential model for dependent defaults. *Journal of Fixed Income* 13(3): 74–83.

Girsanov I, 1960. On transforming a certain class of stochastic processes by absolute substitution of measures. *Theory of Probability and Applications* 5: 285–301.

Glasserman P, 2004. *Monte Carlo Methods in Financial Engineering*. Springer, Berlin.

Glasserman P and Kou S, 2003. The term structure of simple forward rates with jump risk. *Mathematical Finance* 13(3): 383–410.

Glasserman P and Merrener N, 2003. Numerical solution of jump-diffusion LIBOR market models. *Journal of Computational Finance* 7(1): 1–27.

Glasserman P and Zhao X, 2000. Arbitrage-free discretization of lognormal forward LIBOR and swap rate models. *Finance and Stochastics*, 4: 35–68.

Green A, 2016. *xVA: Credit, Funding and Capital Valuation Adjustments*. The Wiley Finance Series.

Green A and Kenyon C, 2015. MVA: initial margin valuation adjustment by replication and regression. Working paper, arXiv:1405.0508[q-fin.PR].

Gregory J, 2015. *The xVA Challenge: Counterparty Credit Risk, Funding, Collateral and Capital*, 3rd edition, Wiley.

Gregory J and Laurent JP, 2003. I will survive. *Risk*, June, 103–107.

Hagan P, Kumar D, Lesniewski A and Woodward D, 2002. Managing smile risks. *Wilmott Magazine* 1: 84–108.

Hagan P, 2018. Building Curves Using Area Preserving Quadratic Splines. *Wilmott Magazine* 95: 60-62.

Hagan P and West G, 2006. Interpolation methods for curve construction. *Applied Mathematical Finance* 13(2): 89–129.

Hagan P and Woodward D, 1999. Markov interest rate models. *Applied Mathematical Finance* 6(4): 233–260.

Harrison J and Kreps D, 1979. Martingales and arbitrage in multiperiod securities markets. *Journal of Economic Theory* 20: 381–408.

Harrison J and Pliska S, 1981. Martingales and stochastic integrals in the theory of continuous trading. *Stochastic Processes and their Applications* 11: 215–260.

Heath D, Jarrow R and Morton A, 1992. Bond pricing and the term structure of interest rates: a new methodology. *Econometrica* 60: 77–105.

Henry-Labordère P, 2005. A general asymptotic implied volatility for stochastic volatility models April, available at http://ssrn.com/abstract=698601.

Heston S, 1993. A closed-form solution for options with stochastic volatility with applications to bond and currency options. *The Review of Financial Studies* 6(2): 327–343.

Hinnerich M, 2008. Inflation indexed swaps and swaptions. *Journal of banking and Finance*, forthcoming.

Ho S and Wu L, 2008. Arbitrage pricing of credit derivatives. *Credit Risk – Models, Derivatives and Management*, Chapter 22, 427–456. Financial Mathematics Series, Vol. 6, ed. W. Niklas. Chapman & Hall / CRC Boca Raton, London, New York.

Ho T and Lee S, 1986. Term structure movements and pricing interest rate contingent claims. *Journal of Finance* 42: 1129–1142.

Hou D and Skeie, 2014. LIBOR: Origins, Economics, Crisis, Scandal, and Reform. Federal Reserve Bank of New York Staff Reports, No, 667.

Hughston L, 1998. Inflation Derivatives. Working paper. Merrill Lynch.

Hull J, 1998. *Introduction to Futures and Options Markets.* Prentice-Hall, Upper Saddle River, NJ.

Hull J and White A, 1989. Pricing interest-rate derivative securities. Working paper, University of Toronto.

Hull J and White A, 2003. The valuation of credit default swap options. Working paper, Rothman School of Management.

Hull J and White A, 2012. The FVA debate. *Risk*, August, 25(7): 83–85.

Hunter C, Jäckel P, and Joshi M, 2001. Cutting the drift. *Risk*, July.

IASB, 2011. IFRS 13 fair value measurement. *International Accounting Standard Board.*

Ikeda N and Watanabe S, 1989. *Stochastic Differential Equations and Diffusion Processes*, 2nd edition. North Holland-Kodansha, Amsterdam.

Inui K and Kijima M, 1998. A Markovian framework in multi-factor Heath–Jarrow–Morton models. *Journal of Financial and Quantitative Analysis* 33(3): 423–440.

Itô K, 1942. On stochastic processes, I. (Infinitely divisible laws of probability). *Japanese Journal of Mathematics* 18: 261–301.

Jamshidian F, 1989. An exact bond option pricing formula. *Journal of Finance* 44: 205–209.

Jamshidian F, 1997. LIBOR and swap market models and measures. *Finance and Stochastics* 1(4): 293–330.

Jamshidian F, 2002. Valuation of credit default swaps and swaptions. Preprint, NIB Capital Bank.

Jarrow R, Li H and Zhao F, 2003. Interest rate caps "smile" too! But can the LIBOR market model capture it? Working paper, Cornell University.

Jarrow R and Yildirim Y, 2003. Pricing treasury inflation protected securities and related derivatives using an HJM model. *Journal of Financial and Quantitative Analysis* 38(2): 409–430.

Jeanblanc M and Rutkowski M, 2000. Default risk and hazard process. Mathematical Finance Bachelier Congress 2000, edited by Geman, Madan, Pliska and Vorst, Springer-Verlag.

Jin Y and Glasserman P, 2001. Equilibrium positive interest rates: A unified view. *The Review of Financial Studies* 14(1): 187–214.

Johannes M, 2004. The statistical and economic role of jumps in continuous-time interest rate models. *The Journal of Finance* 59(1): 227–260.

Jung J, 2008. Real Growth. *RISK*, February.

Kac M, 1949. On Distributions of Certain Wiener Functionals. *Transactions of the American Mathematical Society* 65(1): 1–13.

Karatzas I and Shreve S, 1991. *Brownian Motion and Stochastic Calculus*, 2nd edition. Springer, Berlin.

Karlin S and Taylor H, 1981. *A Second Course in Stochastic Processes*. Academic Press, New York.

Kazziha S, 1999. Interest Rate Models, Inflation-based Derivatives, Trigger Notes And Cross-Currency Swaptions. PhD Thesis, Imperial College of Science, Technology and Medicine. London.

Keller-Ressel M, Papapantoleon A and Teichmann J, 2013. The affine libor models. *Mathematical Finance* 23(4): 627–658.

Kendall M, 1994. *Advanced Theory of Statistics*, 6th edition. London: Edward Arnold; New York: Halsted Press.

Kenyon C, 2008. Inflation is normal. *Risk*, July, 76–82.

Korn R and Korn E, 2000. *Option Pricing and Portfolio Optimization: Modern Methods of Financial Mathematics*. American Mathematical Society, Providence, RI.

Korn R and Kruse S, 2004. A simple model to value inflation-linked financial products, (in German), Blatter der DGVFM, XXVI (3): 351–367.

Kruse S, 2007. Pricing of Inflation-Indexed Options under the Assumption of a Lognormal Inflation Index as well as under stochastic volatility. Working paper, University of Applied Sciences–Bonn, Germany.

Kyprianou A, 2008. Lévy processes and continuous-state branching processes: part I. Lecture Notes, University of Bath.

Lee R, 2004. Option pricing by transform methods: Extensions, unification, and error control. *Journal of Computational Finance* 7: 51–86.

Levine M, 2014. It cost JPMorgan \$1.5 billion to value its derivatives right. Bloomberg, https://www.bloomberg.com/view/articles /2014-01-15/it-cost-jpmorgan-1-5-billion-to-value-its-derivatives-right-draft.

Lévy P, 1954. Théorie de l'addition des variables aléatoires, second edition. Gaulthier-Villars, Paris.

Lewis A, 2000. *Option Valuation under Stochastic Volatility.* Finance Press, Newport Beach.

Li D, 2000. On default correlations: a copula approach. *Journal of Fixed Income* 9: 43–54.

Li C and Wu L, 2015. CVA and FVA under Margining. *Studies in Economics and Finance* 32(3): 298–321.

Litterman R and Scheinkman J, 1991. Common factors affecting bond returns. *Journal of Fixed Income* 1(1): 54–63.

Longstaff F and Schwartz E, 2001. Valuing American options by simulation: A simple least-squares approach. *The Review of Financial Studies* 14: 113–147.

Lou W, 2015. Funding in option pricing: the Black-Scholes framework extended. *Risk*, April, 1–6.

Macaulay F, 1938. Some theoretical problems suggested by the movement of interest rates, bond yields, and stock prices in the U.S. since 1856. National Bureau of Economic Research, New York.

Manning S and Jones M, 2003. *Modeling inflation derivatives - a review.* The Royal Bank of Scotland Guide to Inflation-Linked Products.

Mercurio F, 2005. Pricing inflation-indexed derivatives. *Quantitative Finance* 5(3): 289–302.

Mercurio F, 2010a. A LIBOR market model with a stochastic basis. *Risk*, December, 84–89.

Mercurio F, 2010b. Interest rates and the credit crunch: new formulas and market models. Bloomberg Portfolio Research Paper No. 2010-01-FRONTIERS.

Mercurio F and Moreni N, 2006. Inflation with a smile. *Risk* March, Vol. 19(3): 70–75.

Mercurio F and Moreni N, 2009. Inflation modelling with SABR dynamics. *Risk* June, 106–111.

Merton R, 1973. Theory of rational option pricing. *Bell Journal of Economics and Management Science* 4(Spring): 141–183.

References

Merton R, 1976. Option pricing when underlying stock returns are discontinuous. *Journal of Financial Economics* 3: 125-144.

Mikosch T, 1998. *Elementary Stochastic Calculus.* World Scientific, Singapore.

Miltersen K, Sandmann K and Sondermann K, 1997. Closed-form solutions for term structure derivatives with lognormal interest rates. *Journal of Finance* 52: 409–430.

Morton A, 1988. A class of stochastic differential equations arising in models for the evolution of bond prices. Technical Report, Cornell University.

Newburg A, 1978. Financing in the Euromarket by U.S. Companies: A Survey of the Legal and Regulatory Framework. The Business Lawyer. American Bar Association. 33(4): 2177–82. JSTOR 40685905.

Oksendal B, 1992. *Stochastic Differential Equations: An Introduction with Applications*, 3rd edition. Springer, Berlin.

Oksendal B, 2003. *Stochastic Differential Equations, an Introduction with Applications*, 6th edition. Springer, Berlin.

Pallavicini A, Perini D and Brigo D, 2011. Funding valuation adjustment: a consistent framework including CVA, DVA, collateral, netting rules and rehypothecation. Working paper, http://ssrn.com/abstract=1969114.

Paulot L, 2015. Asymptotic Implied Volatility at the Second Order with Application to the SABR Model. In: P K Friz *et al.* (eds), *Large Deviations and Asymptotic Methods in Finance*, Springer Proceedings in Mathematics & Statistics 110.

Pedersen M, 1999. Bermudan Swaptions in the LIBOR market model. SimCorp Financial Research Working Paper.

Pietersz R, Pelsser A and Regenmortel M, 2004. Fast drift approximated pricing in the BGM model. *Journal of Computational Finance* 8(1): 93–124.

Piterbarg V, 2003. A Stochastic Volatility Forward LIBOR Model with a Term Structure of Volatility Smiles. Working paper, Barclays Capital.

Piterbarg V, 2010. Funding beyond discounting: collateral agreements and derivatives pricing. *Risk*, February, 97–102.

Press W, Teukolsky S, Vetterling W and Flannery B, 1992. *Numerical Recipes in C: The Art of Scientific Computing*, 2nd edition. Cambridge University Press, Cambridge.

Protter P. 1990. *Stochastic Integration and Differential Equations.* Springer.

Raible S, 2000. Lévy process in Finance: Theory, Numerics, and Empirical Facts, Ph.D. thesis, University of Freiburg.

Rebonato R, 1999. On the pricing implications of the joint lognormal assumption for the swaption and cap markets. *Journal of Computational Finance* 2(3): 57–76.

Ritchken P and Sankarasubramanian L, 1995. Volatility structures of forward and the dynamics of the term structure. *Mathematical Finance* 5: 55–72.

Rogers L, 1997. The potential approach to the term structure of interest rates and foreign exchange rates. *Mathematical Finance* 7: 157–176.

Rudin W, 1976. *Principles of Mathematical Analysis*, 3rd edition. The McGraw-Hill Companies Inc., New York.

Ruiz I, 2013. FVA demystified: CVA, DVA, FVA and their interaction (Part I). Working paper.

Sato K, 1999. *Levy Processes and Infinitely Divisible Distributions*. Cambridge University Press, Cambridge.

Schrager D and Pelsser A, 2006. Pricing swaptions and coupon bond options in affine term structure models. *Mathematical Finance* 16(4): 673–694.

Schönbucher P, 2000. A LIBOR market model with default risk. Working paper, Bonn University.

Schönbucher P, 2004. A measure of survival. *Risk*, August, 79–85.

Sherif N, 2015. FVA models overstate costs - Risk.net poll. *Risk.net*, May.

Siegel J. 1972. Risk, interest rates and the forward exchange. *The Quarterly Journal of Economics* 86 (2), 303–309.

Sidenius J, 2000. LIBOR market model in practice. *Journal of Computational Finance* 3(3): 5–26.

Skov J and Skovmand D, 2021. Dynamic Term Structure Models for SOFR Futures. *The Journal of Futures Markets* 41: 1520–1544.

Steel J, 2000. *Stochastic Calculus and Financial Applications*. Springer, Berlin.

Trolle A and Schwartz E, 2008. A general stochastic volatility model for the pricing of interest rate derivatives. *Review of Financial Studies* Advance Access published online on April 28.

Varadhan S, 1980. *Diffusion Processes and Partial Differential Equations*. Tata Institute Lectures, Springer, New York.

Vasicek O, 1977. An equilibrium characterization of the term structure. *Journal of Financial Economics* 5: 177–188.

van Bezooyen J, Exley C and Smith A, 1997. A market-based approach to valuing LPI liabilities. Downloadable at: http://www.gemstudy.com/DefinedBenefitPensionsDownloads.

Wu L, 2003. Fast at-the-money calibration of the LIBOR market model through Lagrange Multipliers. *Journal of Computational Finance* 6(2): 39–77.

Wu L, 2011. A New Paradigm for Inflation Derivatives Modeling. Derivative Security pricing and Modeling, eds. Batten J and Wagner N, Contemporary Studies in Economics and Financial Analysis series, Emerald.

Wu L, 2015. CVA and FVA to Derivatives Trades Collateralized by Cash. *International Journal of Theoretical and Applied Finance* 18(5), 1550035.

Wu L and Zhang D, 2020. xVA: Definition, Evaluation and Risk Management. *International Journal of Theoretical and Applied Finance* 23(1), 2050006.

Wu L and Zhang F, 2008. Fast swaption pricing under the market model with a square-root volatility process. *Quantitative Finance* 8(2): 163–180.

Wu T, 2012. Pricing and hedging the smile with SABR: Evidence from the interest rate caps market. *The Journal of Futures Markets* 32(8): 773–791.

Xia W and Wu L, 2024. ATSM for SOFR Derivatives, working paper, The Hong Kong University of Science and Technology.

Index

Printed in the United States
by Baker & Taylor Publisher Services